LAND AS A WASTE MANAGEMENT ALTERNATIVE

Proceedings of the 1976 Cornell Agricultural Waste Management Conference

RAYMOND C. LOEHR
Editor

Director, Environmental Studies Program
College of Agriculture and Life Sciences
Cornell University
Ithaca, New York

ANN ARBOR SCIENCE
PUBLISHERS INC
P.O. BOX 1425 • ANN ARBOR, MICH. 48106

Second Printing, 1977

Copyright © 1977 by Ann Arbor Science Publishers, Inc.
230 Collingwood, P.O. Box 1425, Ann Arbor, Michigan   48106

Library of Congress Catalog Card Number 76–46019
ISBN   0-250-40140-1

Manufactured in the United States of America
All Rights Reserved

PREFACE

Waste disposal approaches are essentially those of relocation of residues from the point of generation to a more suitable and nondetrimental site. The soil can serve as a receptor of organic and inorganic residues if land application methods are based on an understanding of the applicable scientific and engineering fundamentals, and if the actual methods are well designed and managed.

Of the three locations for the ultimate disposal of wastes—surface waters, atmosphere, and land—the land represents not only an appropriate location for wastes but also an opportunity to manage wastes with minimum adverse environmental impacts. The land cannot function as a neglected waste sink. It is important to avoid the environmental mistakes made in the use of air and water as waste receptors when land is considered as a waste management alternative. The soil and the wastes must be managed carefully as a total system to obtain the best use of both resources.

This volume focuses on ways to appropriately use the land as an acceptor of wastes. The question of how long and with what waste loading and crop response a soil can accept wastes before adverse conditions will occur, such as a buildup of toxic material or a breakthrough of pollutants, is addressed by a number of the authors. The application of wastewaters, sludges, and animal wastes to land is discussed in detail. The emphasis of the volume, however, is not on the type of waste that may be applied but on environmental impacts, the health aspects, and the transformations that occurred.

Numerous case studies of the land application of municipal wastewaters, industrial wastes, and sludges are included so that the actual design of systems and their effect can be perceived. Included are evaluations of systems that have been in operation for many years, including one that has been in existence for 88 years.

This volume represents a comprehensive attempt to examine critically the state of knowledge in this area. Complete and un-

iii

equivocal answers to the questions about this alternative are not yet available. The pace of activity in this field is great, and it is expected that this volume will provide·some insight on feasible directions and use for this important waste management alternative.

The success of the Conference from which these papers originated was due in equal measure to the quality of the authors and their papers, the interest and discussion of the participants, and the skill of the moderators. The interest of these individuals and the assistance of many others, such as Dr. William Jewell of Cornell, and Mr. Lynn Shuyler and Mr. Will LaVeille of the Environmental Protection Agency, who helped identify and obtain the papers that were presented, was key to the success of the Conference. The special diligence and interest of Ms. Colleen Raymond and others in the College of Agriculture and Life Sciences is gratefully acknowledged.

Raymond C. Loehr, Director
Environmental Studies Program
New York State College of Agriculture
    and Life Sciences
Conference Chairman

ACKNOWLEDGMENTS

The 1976 Waste Management Conference, "Land as a Waste Management Alternative," was sponsored by the New York State College of Agriculture and Life Sciences, a Statutory College of the State University at Cornell University, Ithaca, New York. This Conference was the eighth in the annual series of conferences sponsored by the College of Agriculture and Life Sciences on various waste management topics.

Financial support for the Conference was provided in part by the College of Agriculture and Life Sciences and in part by the Environmental Protection Agency, grant R 8044495010. This support is gratefully acknowledged.

Any opinions, findings, conclusions, or recommendations expressed in this publication are those of the authors and do not necessarily reflect the views of either the Environmental Protection Agency or the College of Agriculture and Life Sciences.

INTRODUCTION AND
OPENING REMARKS

W. K. Kennedy
Dean, New York State College of
Agriculture and Life Sciences

It is a privilege to welcome you to the Eighth Annual Waste Management Conference, which has as its focus "Land as a Waste Management Alternative." The attendees and participants in this year's conference represent a wider range of professions, public and private agencies and countries than at any of the previous seven conferences. We are especially pleased that about 20 are from other countries, and we are confident that you will offer valuable information and insights to our discussions. We also are especially pleased with the sizeable number of participants from industry and engineering consulting firms as well as representatives from federal, state and municipal agencies. As always, we welcome the excellent representation we have from other educational institutions.

In discussing land as a waste management alternative, we are seeking inputs from three different sources. One source is the research and technical people in the different disciplines of engineering, soils, and animal and plant sciences. Another important group is the practitioners from cities and countries who have the immediate problem of disposing of wastes. A third source is the representatives from the regulatory agencies who can provide valuable input to those engaged in educational and research programs and who, in turn, can gain greater perspective by interacting with those who handle and manage wastes.

The magnitude of the waste handling problem is enormous and it is increasing. It has been estimated that the 5 million dry tons of municipal wastewater sludge per year will double in the next eight to ten years. The use of the soils as a means of waste disposal has tremendous potential as well as a number of complex problems. Many wastes have valuable nutrients that can be used to good advantage in the production of crops and other vegetation. One of the heavy uses of energy in agriculture is the natural gas required to produce synthetic nitrogen fertilizer. Both agriculture and the general public will benefit from the recycling of the nitrogen in wastes through the production of crops, thus reducing the demands for synthetic nitrogen fertilizer. We must learn how to use these sludges on agricultural lands without contaminating the food supply.

The soil also is an effective means for removing heavy metal and other contaminants that can cause serious problems when emptied into streams and lakes.

Utilizing land as a waste management alternative is not without its problems. Many wastes have undesirable odors and contain large volumes of water. Soils vary tremendously in their ability to handle heavy applications of water. The amount of water that can be applied to the land depends on temperature, rainfall and other climatic variables, factors that influence the persistance of undesirable odors and the rate of application of sludges.

Some of our soils are able to handle the introduction of salt and heavy metals better than others. The effects of increased moisture level and the concentration of nutrients, salts and metals depend upon the crops being grown and the stage of their growth cycle. These problems do not deter us, but rather serve to increase the challenge on how to use the land more effectively for waste disposal and, at the same time, to utilize these wastes to enhance crop and animal production. Collectively you have the expertise and experience to turn waste into a valuable resource.

Once again, welcome to our Eighth Annual Waste Management Conference and best wishes for a stimulating and profitable conference.

CONTENTS

ix

Section V   Application of Sludges to Land

Section VI   Industrial and Agricultural Wastes

SECTION I

REGULATORY AND FUNDAMENTAL ASPECTS

A PERSPECTIVE ON LAND AS A
WASTE MANAGEMENT ALTERNATIVE

J. D. Freshman
Public Works Committee
United States Senate, Washington, D.C.

One of the largest problems with the use of land as a waste management alternative is the institutional acceptance of such systems for water pollution control. Such systems can be effective and should be utilized more broadly. With this realization, the way we have always done things, and the condition of our waters, especially in the East, it may be more appropriate to have conferences entitled "Receiving Water as a Waste Management Alternative" or "Aerate, Chlorinate and Dump" as a waste management alternative to land treatment. Because land waste management systems seem to make the most sense, they should be considered as the standard.

Although I do not pretend to be familiar with all land treatment systems, as far as I do know, every single operational municipal land treatment system is an economic and environmental success, which is substantially more than you can say for the standard treatment plant.

With a land waste management system, you do not:

1. Overload a stream by assembling all the water pollution in one place, treat it and dump it with most of the nutrients still in the effluent;

2. Lose track of the pollutants in the usual "out of sight, out of mind" tradition;

3. Put all the major communities potentially into bankruptcy in operations and maintenance expenses;

4. Waste valuable nutrients.

Despite all of these advantages, including significant environmental benefits, the potential of land waste management systems is not being utilized fully, and municipalities, industries and governmental organizations continue to debate the proper role of such systems. While there certainly are technological and potential health effect problems that need continued research and development, most problems are not technical, they are institutional—practical, if you will. More scientific studies are certainly necessary, but will never be sufficient to solve every concern about land treatment systems nor see that they are built. Many more talents are needed to solve the institutional problems.

Treat and dump is the current water pollution control approach with land being considered as a possible alternative. Yet materials which are pollutants in the water are valuable nutrients on the land. Those of us who are nonscientists are often accused of oversimplification of complex, technical issues, but I cannot overstate the impact of the simplicity of the reasons for land treatment, given our agricultural/fertilizer/nutrient cycle. Farmers spend billions of dollars in petrochemically based, energy-intensive fertilizers (nutrients) while municipalities and industries spend billions of dollars disposing of, usually at the expense of the aquatic environment, unwanted nutrients (fertilizers). The facts related to general nutrients, phosphorus and energy are available to illustrate the logic for land waste management systems.

GENERAL NUTRIENTS

Twenty-four billion gallons of treated, domestic sewage are discharged every day. The nutrient content, aggregated yearly, is substantial.

Total Nitrogen: 1,614,000,000 lb/yr
Total Phosphate (P_2O_5):   1,482,000,000 lb/yr
Total Potassium (K_2O):   942,000,000 lb/yr

According to 1975 U.S. Department of Agriculture figures, the wasted nutrients are a significant percentage of the nation's fertilizer consumption: nitrogen 9.3 percent, phosphorus 16.0 percent, and potassium 11.0 percent. The value of these nutrients, according to October 1975 prices, is substantial: $199 million dollars of nitrogen, $288 million of phosphorus, and $74 million dollars of potassium, totaling one-half billion dollars, are dumped into the water each year. These figures only include domestic waste currently treated and do not include untreated sewage, urban runoff, or any industrial sources.

PHOSPHORUS

Phosphorus is the *sine qua non* for all forms of life. It is not substitutable, it is geologically unusual and scarce, and it is difficult and inefficient to artificially reconstitute once it is dispersed. U.S. use has increased in the recent past at exponential rates. If present trends continue, exhaustion of all known U.S. reserves will occur by 2001. The Bureau of Mines has reported: "depletion of U.S. reserves of phosphate rock, particularly from Florida and Tennessee is projected to occur before or shortly after the end of this century." There are no secondary recovery sources of phosphate, yet over a billion pounds per year is wasted by dumping it into our lakes and streams.

ENERGY

In 1974 farmers spent $5.2 billion on nutrients. If the recent past trends continue, that figure will increase astronomically. To make nitrogen fertilizer, 2.5 percent of the U.S. natural gas production is used each year. From 1947 to 1970, general energy use on the farm doubled, but energy used for fertilizer quadrupled.

Isn't it time to close the loop and to emphasize resource utilization and/or recovery? Among other things, isn't using and re-using our resources what conservation, especially energy conservation, is all about?

Yet, land treatment is still considered just an alternative, rarely implemented, and institutionally in its infant stage of development. A potential land treatment system bears an enormous burden of proof before it is accepted and implemented, a burden not borne by most other waste treatment systems. When a land treatment system is proposed, it must meet a rigorous series of tests, tests that a conventional system is rarely subjected to. Many of the tests are along the lines of the papers presented at this conference—microbiological and other health hazards, transfer of water and chemicals through soils and vegetation, and removal of nutrients and trace metals.

CONGRESSIONAL VIEW

The following is a brief review of what the Congress said about land treatment and what has actually happened, although the latter is difficult to exactly determine.

Section 201 of PL 92–500, the Water Pollution Control Act Amendments of 1972 (the Act), makes the clear statement:

> (b) Waste treatment management plans and practices
> shall provide for the application of the best practicable

> waste treatment technology before any discharge into re-
> ceiving waters, including reclaiming and recycling of
> water, and confined disposal of pollutants so they will not
> migrate to cause water or other environmental pollution
> and shall provide for consideration of advanced waste
> treatment techniques.

That is the positive statement in the major section of the Act.
By implication, there are a whole series of provisions in the Act
which clearly show Congressional intent toward land treatment.
The cost-effectiveness requirement taking into account all costs,
the requirement that alternatives be fully examined, and the
movement to progressively stricter levels of pollution removal
all point in that direction.

Congress did not establish a construction grants program only
to build secondary treatment plants, although there is a secondary treatment requirement for municipal wastewaters. It established a grants program in the context of solid and strict ecological principles. Concepts such as "recycling and reclaiming"
and the "full examination of alternatives" are the major emphasis of key parts of the Act.

Thirteen of the fifteen paragraphs used by the Senate report
which served as a basis for the Act and which described the
municipal grants program concern land treatment. Only a few
of the relevant statements will be cited:

> The statement of purpose, coupled with the burden of
> proof should . . . cause a re-orientation of waste treatment
> technology toward more sustained and productive results.
> Alternative waste treatment methods, which require the
> return of pollutants to natural cycles, are only new in the
> sense that they have re-emerged for application.

And a citation from an EPA report:

> Water reaching water courses after passage through the
> filtering and decomposition afforded by soil is far purer—
> provided that soil loading rates are not exceeded—than
> any waste treatment process short of distillation.

Although the House and Senate disagree on quite a few issues,
they were united on land treatment as the following citation from
the House report shows:

> Particular attention should be given to treatment and
> disposal techniques which recycle organic matter and nu-
> trients within the ecological cycle.

These are, to me, clear statements of purpose and intent—a
series of increasingly more stringent requirements of effluent

reduction, directives to the administrator to encourage land treatment, and money to follow up this purpose and intent.

Unfortunately, it is difficult to determine how this mandate has been carried out. We do not know exactly how many land treatment projects have been initiated since passage of 92–500. A fair guess is around 32 out of 1800. We do know that EPA had no policy on land acquisition for land treatment until last fall— three years after the law was passed—even though the law states: "... and any works, including site acquisition of the land that will be an integral part of the treatment process...."

From October 1972 to January 1, 1976, with PL 92–500 authorization money, EPA has funded 32 land treatment systems that are recorded on their GICS system. This is approximately 0.99 percent (0.009) of all the money they spent on new sewage treatment facilities or upgrading old ones. It is fair to say that despite the strong language of the Act, and the clear intent of Congress, that percentage has not increased due to the 1972 Act.

Perhaps the story can best be told in the questions that cannot or have not been answered:

1. We do not know the acreage that has been purchased for land treatment.

2. We still have not resolved some fundamental policy questions such as the eligibility of leasing or pretreatment standards.

There are major hurdles to widespread implementation of land treatment systems which need to be more adequately addressed, but are not insurmountable. The first problem that has been raised against land treatment systems is a sociological one. People say that citizens will not stand for land treatment systems in their backyards. They cite this as a sociological problem, a fact of life that any advocate of such systems must concede. Actually we do not know if this problem exists for land treatment systems to any great extent or not. We do, on the other hand, know for a fact that public resistance to conventional treatment systems, even ones which promise no smell, beautiful buildings, and seclusion, can be virulent.

Montgomery County, Maryland provides an interesting example. In this affluent Washington suburb local officials have been trying to begin a sewage treatment plant for almost seven years. Opposition to specific sites from various interests in the county —sometimes environmentalists, sometimes homeowners—has stalled the project every time.

Although land as a waste treatment alternative was given cursory examination, it was rejected on institutional grounds saying that it was politically impossible. It appears that the advanced waste treatment plant is equally impossible.

Russell Train, Administrator of EPA, just announced his opposition to the latest plans, and said a plant in Prince George's County, Montgomery's less affluent cousin, should be expanded to handle Montgomery's future flows. Short of a federal bludgeon, there is no political impossibility like asking Prince George's County to accept the sewage for Montgomery County's growth.

Another of land treatment system's biggest obstacles may be that the field of water pollution control is dominated by people who view sewerage as a plumbing problem, not a resource problem. A 1974 article, quoting a far-sighted sanitary engineer, illustrates this obstacle:

> Land treatment encompasses the broad technical domains of the sanitary engineer, the hydraulic and hydrological engineers, the agronomist and soil scientists, the irrigation engineer, the virologist, metallurgist, and geologist.... Perhaps for this reason alone, the professionals have been somewhat recalcitrant, knowing that the overall design control of a (sewage irrigation) project might necessarily fall into the hands of a generalist or a water resources planner, rather than the traditional technical specialist.

As I read this conference program, it appears that you represent the disciplines that writer mentions. You need to be involved during the decision-making process—not only at the design phase but also at the planning phase where the decisions on kinds of processes are being made.

If consulting engineers, working with other engineers in the state and federal regulatory agencies make all the decisions, land waste management systems will never be utilized properly or broadly.

There is obviously a place for capital intensive, advanced waste treatment facilities, just as there is a place for current technology secondary and primary plants. But our knowledge of these technologies, our experience with these technologies, and our engineering comfort with these technologies should not and cannot mean an irrevocable commitment to these technologies.

Land treatment is an established waste treatment technique. It has a significant role to play in much of America. It can be an end process for disposition of waste, or it can be the beginning of reclamation programs. It may be a part of a productive process. Most importantly, land treatment returns nutrients to their natural cycles, and disposes of wastes in a manner in which any long-term adverse impacts can be watched, can be controlled, and can be corrected. It is this feature which places land waste management systems at the top of our alternatives list, and it is this feature which should carry this concept to widespread utilization throughout this country.

2

REGULATIONS AND GUIDELINES FOR LAND APPLICATION OF WASTES— A 50-STATE OVERVIEW

C. E. Morris
Department of Education
W. J. Jewell
Department of Agricultural Engineering
Cornell University, Ithaca, New York

INTRODUCTION

State governments frequently legislate regulations above and beyond those promulgated by the federal government. But state laws rarely become so strict that they jeopardize a federal mandate. This is, however, the situation that appears to be evolving in the area of land application of wastes.

The Federal Water Pollution Control Act Amendments of 1972 (PL 92–500) focused attention on the need to develop waste management techniques which are cost-effective and environmentally sound. Among the methods cited, the application of wastewaters and sludges to land is promoted as being a major alternative. Many states have responded to the renewed interest in land application by developing regulations or guidelines for this management practice. In some cases, these guidelines are so restrictive that they eliminate land application as a practicable treatment technology. The Environmental Protection Agency's requirement that the chosen system be cost-effective presents a particular challenge to the use of land treatment systems, especially where extensive pretreatment is required prior to land application.

This overview of state regulations and guidelines was developed as part of an educational program on land application of wastes at Cornell University. The study was designed to obtain legislation or guidelines for land application of wastes from each of the 50 states. These regulations and guidelines were evaluated to determine the extent to which this waste management practice is regulated. Major trends which the states are following in developing guidelines were identified. Finally, state regulations and guidelines were assessed in relation to PL 92–500 and its ramifications for land application of wastes.

PROCEDURE

The pollution control agency of each of the 50 states was contacted by mail with a request for a copy of current regulations and guidelines for the practice of applying wastewaters and sludges to land. A second mailing was sent to those states which had not responded after several weeks. The letter included a space where personnel could indicate if their state has no current guidelines. States which did not respond to the second mailing were contacted by telephone. This approach resulted in a response from all 50 states.

The agencies contacted were identified from the Conservation Directory (1). Several states sent information about associated waste management practices. Such information was not used in the following evaluation. Only information about application of wastewater and sludge to land has been included in this paper.

OVERVIEW OF REGULATIONS AND GUIDELINES

Of all 50 states answering this survey, about half (24 states) currently do not have regulations or guidelines for land application. Of these, five states presently are preparing guidelines. The states which do not currently have specific guidelines or regulations for land application are listed in Table 1.

Table 1. States which currently do not have regulations or guidelines for land application of wastes.

Alabama	Kentucky	North Carolina
Alaska	Louisiana	North Dakota[a]
Arizona	Maryland[a]	Oklahoma[a]
Arkansas	Massachusetts	Rhode Island
Connecticut	Mississippi	Tennessee
Hawaii	Montana	Utah
Indiana	Nevada	Washington[a]
Iowa	New Mexico[a]	Wyoming

[a] Denotes states which currently are preparing regulations or guidelines.

A glance through a selection of state guidelines for land application of wastes indicates that certain aspects of this waste management alternative have received more regulatory attention than others. It is helpful to categorize regulations and guidelines into three parts: System Design, Pre-Application Water Quality, and General Information.

Guidelines pertaining to system design include the following:

Loading rate	Cover crop
Application system	Storage
Buffer zone	Public access
Monitoring	Effluent quality

Pre-application water quality is the quality of the wastewater required before it is applied to the land. Topics regulated by some states include:

Type of pretreatment requirement	Dissolved oxygen (DO)
Five-day biochemical oxygen demand (BOD_5)	Nitrates
Solids	Organics
Toxic elements	Chlorine residual
pH	Coliform concentration

General information on regulations and guidelines include:

Type of waste applied	Factors which prevent
Philosophy and major concern of state	land application
Site characteristics	Weather restrictions

There is a wide range in the number of areas that the state regulations address. In some cases, emphasis may be placed on only one or a few areas. For example, Michigan has very specific monitoring guidelines which are relied on to protect the ground water (2). On the other hand, some states such as Florida have regulations or guidelines for nearly all the factors listed (3). It is important to note that relatively few states have legislated into law regulations for land application systems. In most cases, the states issue guidelines which must be used when designing a land application system. The design is submitted to the appropriate agency in the state for evaluation on a case-by-case basis.

Figure 1 illustrates several of the aspects of land application systems which are addressed by many of the states. In designing these systems, the engineer should be aware of requirements for pretreatment, storage capacity, type of hardware, extent of buffer zone, application rate for waste, appearance

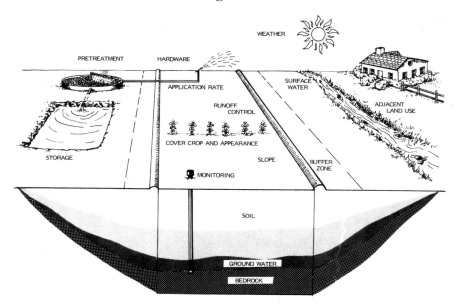

Figure 1. Several aspects of land application systems are generally regulated by the states.

of the site, runoff control, slope of the site, soil, bedrock and hydrology of the site, monitoring program, and adjacent surface water and land use. Several states prohibit waste application during certain weather events.

Discussion of Regulatory Trends

In reviewing state guidelines for land application, several trends can be observed. This is in part due to the fact that many of the states which have developed guidelines have used similar reference materials. The Ten States Standards Addendum #2 (4) is one such widely used reference. A second is the EPA publication entitled *Evaluation of Land Application Systems* (5).

One nationwide trend has special significance to the future of the land application alternative. This is the requirement which 21 out of 26 states mandate—that is, secondary treatment of waste prior to application to land. The secondary treatment standard has been set by EPA as 30 mg/l of BOD_5 and suspended solids, and no more than 200 fecal coliform organisms per 100 ml wastewater (6). Several states require 20 mg/l BOD_5 and 20 mg/l suspended solids, however. The importance of the secondary treatment requirement to the future of land application lies in the emphasis placed by EPA on treatment alternatives which are cost-effective. It is difficult to design a land application system with a secondary plant for pretreatment, while remaining within the bounds of cost-effectiveness. The states which require secondary pretreatment are listed in Table 2.

Table 2.   States that require secondary treatment prior to land application.

California[a]	Missouri	South Carolina
Colorado	New Hampshire	South Dakota
Delaware	New Jersey	Texas[b]
Florida	New York	Vermont
Georgia	Ohio	Virginia
Illinois	Oregon[b]	West Virginia
Minnesota	Pennsylvania	Wisconsin

[a] Pretreatment requirement depends on intended use of water.

[b] Requires water quality of 20 mg/1 BOD_5 and 20 mg/1 suspended solids.

The requirement for secondary treatment stems from the differing views held among waste management personnel toward land application as a waste treatment mechanism. The states which require secondary treatment tend to view land application as a means of disposing of wastewater. The states which do not require pretreatment consider the soil itself to be a treatment medium, recognizing the renovation capabilities of the soil.

In evaluating land application sites, several factors automatically eliminate a site from consideration in most states. These factors include:

1. Possibility of creating a ground water mound close to the soil surface
2. Excessive slope of land
3. Proximity of site to surface waters or drainage ditches which could be contaminated by runoff
4. Location of site on a flood plain
5. Extreme soil characteristics, such as high rate of permeability
6. Faults or fractures in the underlying bedrock
7. Creation of an unaesthetic appearance, health problem, or nuisance on the site or on adjacent property

Nearly half of the states require that land application sites incorporate methods to control surface runoff, and to prevent off-site runoff from flowing onto the site. Ditches and berms are the most commonly used method of runoff control. Additionally, erosion from these sites must be controlled.

The slope of the site is an important consideration in land application. Generally the slope for cultivated fields which are irrigated with wastewater should be no greater than 4 percent. For sodded fields, a maximum of 8 percent slope is typically suggested. For year-round application to forested slopes, some states require a slope no greater than 8 percent, while others allow this operation on slopes of up to 20 percent. Seasonal operation on terraced, forested land is allowed on slopes of up to 25 percent in New Jersey, although most other states limit this practice to slopes under 15 percent. Guidelines for site characteristics are summarized in Table 3.

Operation of land application systems is required to cease under severe weather conditions in several states. In these states, irrigation should not be carried out during or after heavy rainstorms. Irrigation must cease in winter in Missouri and Nebraska. New York, Minnesota, New Hampshire and South Dakota limit the irrigation season to 18–32 weeks per year, effectively eliminating the option of winter irrigation. States which allow winter irrigation often require that a) pipes be capable of being drained to prevent freeze-up, b) runoff from snow or ice build-up be controlled, and/or c) irrigation be stopped when the ground is frozen or snow-covered. This last requirement pertains to Maine, Virginia and West Virginia. Table 4 summarizes recommended design criteria for land application systems.

Monitoring guidelines are set by 85 percent of the states. The majority of these stipulate that at least one monitoring well be placed in each direction of major ground water flow. Several states also specify which parameters are to be analyzed, where wells are to be located, frequency of sampling, and equipment checks.

Representative Guidelines

The guidelines and regulations received during the course of the survey varied greatly. Three states are discussed here in detail. California is of interest because of its extensive use of primary wastewater for irrigation. Pennsylvania is one of the few states which views the soil as a treatment mechanism. Minnesota has particularly stringent design criteria.

California Guidelines

In California the Board of Health is responsible for regulations controlling the use of reclaimed wastewater for irrigation (7, 8). Regulations for the use of sewage sludge on agricultural

Table 3. Recommended site characteristics required by various states for land application of wastes.

| | Slope Guidelines | | | Minimum Depth to Ground Water, (ft) (Not Underdrained) |
	None Given	Dependent on Cover Crop	Maximum Allowable Slope	
California	x			
Colorado	x			
Delaware	x			2
Florida			5%	5[b]
Georgia		Yes	20%	10
Idaho	x			
Illinois		Yes[a]	14%	5-10[c]
Kansas	x			
Maine			8%	
Michigan	x			
Minnesota			2%	10[d]
Missouri			2%	
Nebraska		Yes		5
New Hampshire			10%	
New Jersey		Yes	25%	
New York		Yes	14%	
Ohio			12%	
Oregon	x			
Pennsylvania		Yes[a]	14%	10
South Carolina		Yes	20%	5
South Dakota	x			6[e]
Texas	x			
Vermont			25%	
Virginia			15%	5
West Virginia		Yes[a]	14%	
Wisconsin	x			

[a]Slope guidelines are 4% for irrigation of cultivated fields, 8% for irrigation of sodded fields, 8% year-round irrigation on forested areas, and 14% on seasonal irrigation of forested areas.

[b]Except for recharge basins where a minimum depth of 18 in. is required.

[c]Dependent on infiltration rate of soil.

[d]Where underdrains are installed, minimum depth to groundwater 4 ft.

land are in preparation. Table 5 summarizes water quality required for various uses in California.

Waste discharge requirements for wastewater or sludge applied to land are determined on a case-by-case basis. These regulations, which take into account existing pertinent legislation, are oriented toward protecting the waters of the state.

Surface irrigation of food crops may utilize wastewater that has been disinfected and oxidized. Disinfection is considered adequate if at some point in the process the median number of coli-

Table 4. Summary of several common aspects of state guidelines for land application of wastewater which have major impact on the final design.

State	Storage Requirement	Buffer Zone Requirement	Cover Crop	Hardware	Loading Rate (Max. in./ac-wk)
California	NI[a]	500 ft to water supply wells	NI	Maximum attainable separation	NI
Colorado	NI	NI	NI	NI	NI
Delaware	Reserve area required	200 ft	Crop planted before irrigation begins	Minimize aerosol formation	Not to exceed infiltration rate
Florida	7-day flow plus 3 ft freeboard	150 ft to houses 200 ft to water wells	Crop planted before irrigation begins	Minimize aerosol formation	4
Georgia	30-day design flow	At least 300 ft	Woodlands or nonfood crop	Fixed distribution system	1
Idaho	NI	NI	NI	NI	NI
Illinois	To accommodate flows in excess of irrigation	150 ft to water supplies; 200 ft to surface water	NI	Stationary systems capable of being drained	NI
Kansas	NI	NI	NI	NI	NI
Maine	NI	300 ft for spray irrigation	No application on bare soil	NI	NI
Michigan	NI	NI	NI	NI	NI
Minnesota	210 days	To extent possible	Acceptable vegetative cover	Minimize aerosol drift; automatic shutoff during rain	2 (4 during July and August)

State					
Missouri	Winter flow plus allowance for wet spring	NI	Reed canary-grass, tall fescue	Drains to prevent freeze-up	NI
Nebraska	180 days	50 ft	NI	No cross connections between potable and re-claimed water pipes	3
New Hampshire	2/3 yearly flow	Site-dependent	NI	NI	2
New Jersey	Site-dependent	200 ft	No application on bare soil	Portable systems unacceptable	2
New York	To handle maximum reasonable variation in flow	Buffer zone required	Cover crop required; harvest when necessary	Stationary systems preferred	NI
Ohio	NI	NI	No application on root crops, leafy vegetables	NI	NI
Oregon	Adequate storage	NI	No irrigation of crops for human or dairy cattle consumption	NI	NI
Pennsylvania	To handle maximum reasonable variation in flow	200 ft	Few restrictions	Fixed lines for winter irrigation; low spray trajectory	2
S. Carolina	3-day flow	100 ft	No application on bare soil or crops for human or grazing animal consumption	Drains to prevent freeze-up	2
S. Dakota	210-day flow	To extent possible	"Suitable"	Minimize wind drift and aerosol formation	2

Table 4. (continued)

State	Storage Requirement	Buffer Zone Requirement	Cover Crop	Hardware	Loading Rate (Max. in./ac-wk)
Texas	NI	500 ft to water supplies	No irrigation of food crops for raw consumption	No cross connections between potable and re-claimed water pipes	NI
Vermont	April–May flows	100 ft to surface waters	NI	Pump system must deliver daily flow within 8 hrs	2
Virginia	60 days	200 ft for forested site; 600 ft for open site	No irrigation of food crops for raw consumption	Permanent spray system	2
W. Virginia	60 days	400 ft	NI	Permanent spray system, drains to prevent freeze-up	1
Wisconsin	NI	1,000 ft from public water supplies	NI	Minimize runoff, incorporate sludge	NI

[a]NI = No information given.

Table 5.   Water quality criteria for reclaimed wastewater in California is dependent upon ultimate use of the water.

Water Quality Requirement	Ultimate Use
Equivalent to primary-treated	1. Surface irrigation of orchards, vineyards 2. Food crops which are to be processed. [a] 3. Surface or spray irrigation of fodder, fiber, and seed crops
Disinfected, oxidized	1. Surface irrigation of food crops[b] 2. Irrigation of pasture for dairy animals[c] 3. Restricted recreational impoundment (noncontact water sports)[b] 4. Landscape impoundments
Disinfected, oxidized, clarified, filtered	1. Spray irrigation of food crops[d] 2. Nonrestricted recreational impoundments (contact water sports)[d]

[a]Requires permission from Department of Health.

[b]Requires median of 2.2 coliform organisms per 100 ml.

[c]Requires median of 23 coliform organisms per 100 ml.

[d]Requires median of 2.2 coliform organisms per 100 ml; no more than 23 coliform organisms per 100 ml in any 30-day period.

form bacteria does not exceed 2.2 per 100 ml for the last 7 days for which analyses have been completed.

Reclaimed water with quality equivalent to that of primary-treated effluent may be used for surface irrigation of orchards and vineyards as long as the fruit does not come in contact with the irrigating water or the ground.

Food crops which are processed in such a way as to destroy pathogens before consumption may also be irrigated with primary sewage effluent. Before such irrigation commences, however, permission must be granted by the Department of Health. To date the Department has discouraged such uses.

In designing the system, water lines for domestic water and reclaimed water should have maximum attainable separation. Valves and outlets on the reclaimed water lines should be operable only by authorized personnel.

Other general regulations in California specify that: the discharge resulting from irrigation must be confined to the site, runoff must be contained, no unaesthetic slimes or odors are allowed, ponding should be minimized, and the breeding of insect disease vectors must be controlled. The monitoring regimen is decided by the state.

Where primary wastewater is used, the site should be fenced.

For spray sites, windblown spray should not be allowed to reach areas which are accessible to the public. Drinking fountains must be protected from windblown spray. If wastewater is used at golf courses, notice of its use shall be given on scorecards and at water hazards on the golf course. Qualified supervision is required at recreational impoundments. Wastewater is not to be spread on roads, walkways, picnic tables, or other areas not under control of the user.

When wastewater is used for irrigation, enough time should pass between the final irrigation and harvesting to allow the crops and soil to dry. Irrigated areas should be thoroughly dry before grazing animals, particularly dairy animals, are released on them. Measures to prevent direct contact between edible portions of the crop and the reclaimed water should be taken. Table 6 summarizes design criteria for California. These regulations

Table 6.   Summary of design criteria for land application of wastes in California.

Parameter	Guideline
Major concern	Protection of waters of state
Pretreatment	Depends on ultimate use
Runoff control	Discharge confined to site
Hardware	Maximum attainable separation of reclaimed and domestic water lines
Buffer zone	At least 500 ft from water wells, reservoirs
Monitoring	Dependent on water quality
Cover crop	Dependent on water quality
Public access	Dependent on water quality

appear to offer adequate protection of the environment while still providing the means of converting a waste into a valuable resource.

Pennsylvania Guidelines

Pennsylvania has produced the *Spray Irrigation Manual* (9) which describes guidelines for locating and designing spray irrigation sites. It also includes a handbook which, upon completion by the engineer, becomes part of the required design report.

All spray irrigation systems in Pennsylvania are considered to be discharges to waters of the state. Therefore, such systems will require a permit from the Department of Environmental Resources.

The soil mantle is viewed as a treatment mechanism. Only waste constituents which can be treated successfully by the soil may be applied. Even so, secondary treatment is generally required prior to land application.

Irrigation should be restricted to a once-per-week application. This regimen allows the soil to dry out and re-aerate, thereby promoting the conditions necessary to continue waste degradation. The soil itself should provide for a slow but continuous downward movement of wastewater. This slow percolation allows adequate detention time in the soil matrix needed for the waste-degrading reactions to occur.

The extent of the required buffer zone is often a factor in design of land application systems. In Pennsylvania, a 200-ft buffer zone is required. In some cases the cost of land for buffer zones may be considerable. The condition of the ground water is also important in selecting a site for spray irrigation of wastewater. Usually, the maximum ground water elevation should not exceed 10 ft, and the formation of a ground water mound should be avoided. Monitoring wells should be placed in each direction of major ground water movement away from the site. In addition, one well should monitor the background quality of ground water entering the site. Ground water monitoring points may be natural springs or wells; in either case, these monitoring points must provide adequate access and water flow for sampling at all times. Chemical analyses are specified by the Pennsylvania Department of Environmental Resources.

The cover crop chosen should provide a high rate of evapotranspiration and should prevent erosion from runoff. The crop should be harvested and removed at least once a year and more often if large amounts of nitrogen are applied.

If the irrigation facilities are not designed to handle surge flows or to operate during the winter, storage must be provided. The storage facilities should be able to handle the maximum reasonable variation in flow.

For year-round sprinkler operation, solid-set piping is recommended. Year-round irrigation may only be conducted in areas of permanent vegetation, preferably forest areas. Pipes should have drains to prevent freezing. Runoff from ice accumulation must be controlled.

Adjacent land use should be protected from spray odor and degradation of ground water. Trajectory of spray should be as low as possible to prevent aerosol spread. Table 7 summarizes design criteria for Pennsylvania.

Table 7. Summary of design criteria for land application of wastes in Pennsylvania.

Parameter	Guideline
Philosophy	Considers soil to be a treatment medium
Pretreatment	Secondary treatment
Runoff control	Control runoff from ice buildup
Hardware	Fixed irrigation lines for winter operation. Drains to prevent pipe freeze-up. Trajectory of spray as low as possible.
Buffer zone	200 ft around spray field
Monitoring	One well in each direction of major underground flow; determine quality of groundwater entering site
Cover crop	Idle fields, forests, grasslands, crop areas
Storage	To handle maximum reasonable variation in flow
Slope	0–14% depending on cover crop
Application rate	Maximum 2 in./week
Ground water depth	At least 10 ft to ground water. No formation of ground water mound

Minnesota Guidelines

Minnesota has published recommended design criteria for land application of wastewater (10). Municipal effluents must receive secondary treatement prior to disposal on land, but industrial effluents may be exempted from this requirement. The water quality prior to land application must be as follows:

BOD_5=25mg/l, SS=30 mg/l, Fecal Coliform=200 MPN/100 ml

A storage lagoon with a minimum 210-day capacity must be built in accordance with criteria for waste stabilization ponds.

Land application systems must be located 1 mile from municipal water supplies, and 0.25 mile from a private water supply, human residence, state parks, or other recreation areas, and crop irrigation areas. Buffer zones should be developed by purchase of adjacent land, or control of land use in adjacent area.

The site should be diked to a) prevent runoff from leaving the site, b) aid in recapture of effluent, and c) prevent extra surface water from entering the site.

Overall application should not exceed 52 in./ac-yr. This figure is used in calculating required acreage. Maximum application rate is set at 2 in./ac-wk except for July and August, when up to 4 in./acre-wk may be allowed. Maximum application rate is 0.5 in./hr, with a maximum total application of 0.25 in./hr. The spray season is limited to 18 weeks. No part of the field may be sprayed continuously for more than 6 days.

Spray equipment should minimize wind drift and aerosol formation. There should be provision for automatic shutoff of irrigation during precipitation.

The minimum depth to ground water should be 10 ft. However, if the site is drained this requirement is reduced to 4 ft. Monitoring wells must be placed in all directions of major ground water flow from the site. The wells should be 200 ft beyond the perimeter of the site, not more than 500 ft apart, and should extend no deeper than 5 ft below the seasonal low water table. Separate wells are to be used to sample water from deeper strata. The quality and quantity of effluent discharged to the field, and the static water level of the wells, should also be measured. Routine tests required include:

Conductivity
Chlorides
Dissolved salts
Nitrates
Nitrites
Ammonia

Methylene blue active
    substance
Fecal coliform
BOD_5
Phosphorus

Acceptable vegetative cover is required. It should be harvested and disposed of at least once a year. Additional treatment and/ or site abandonment may be required if there is a significant change in ground water quality or effluent quality from the spray field. The wastewater applied to land should conform as much as possible to drinking water standards. Table 8 summarizes design criteria for Minnesota.

DISCUSSION

The states of the nation, empowered by Congress to regulate waste management activities within their boundaries, have taken varying approaches to the land application alternative. Twenty-six of fifty states have issued regulations or guidelines for this practice. Five states are currently preparing guidelines. Most of the remaining nineteen states approve land treatment design on a case-by-case basis. Existing pollution control legislation of the state, as well as existing guidelines for land application developed by other states or agencies, are frequently used for reference.

Table 8.   Summary of design criteria for land application of wastes in Minnesota.

Parameter	Guideline
Pretreatment	BOD$_5$ 25 mg/1 Suspended solids 30 mg/1 Fecal coliform 20 MPN/100 ml
Runoff control	Dikes to prevent runoff
Hardware	Automatic shutoff capability during precipitation
Buffer zone	To extent possible through purchase or control of adjacent land
Monitoring	One well in each direction of major underground flow. Wells 200 ft beyond perimeter, 500 ft apart, no deeper than 5 ft below low water table. Baseline data before operation; test for conductivity chlorine residual, total dissolved solids, nitrate-, nitrite-, and ammonia-nitrogen, phosphorus, fecal coliform, methylene blue active substance
Cover crop	Harvest once per year
Storage	210 days
Spray season	18 weeks
Application rate	2 in./week
Ground water	At least 10 ft to ground water; at least 4 ft if underdrained

Regulations and guidelines vary according to geography, demography and the economy of the states. The states which have issued guidelines are primarily those which have larger populations and are more industrialized. The popularity of land application in the state also affects the issuance of guidelines.

Several regulations may severely affect the economic viability of the land application alternative. The cost of land is a major portion of the system cost, and requirements for storage and buffer zones may impinge on the cost-effectiveness of the system. For example, using Minnesota's requirement for a 210-day storage capacity, and assuming a one million gallon per day flow, an additional 92 acres would be required for a storage lagoon 7 ft deep. Using Pennsylvania's requirement for a 200-ft buffer zone all around the irrigation area, a spray field 1 mile square would require nearly 100 acres additionally for the buffer zone.

The requirement that wastewaters receive secondary treatment prior to application on land is perhaps the most severe restriction affecting the viability of land application as a practicable and cost-effective treatment technology. About 85 percent of the states require secondary treatment prior to land application. For a 1-MGD flow, secondary treatment costs about $50,000. It is obvious that land application of wastes cannot be cost-effective compared to secondary treatment in cases where secondary treatment is required prior to land application of wastes.

The secondary treatment requirement regulates the BOD_5, suspended solids, and fecal coliform allowed in wastewater used for irrigation. However, other waste constituents which may cause problems, particularly nitrate, phosphorus and toxic substances, are not generally regulated by the states.

Land application can be a cost-effective waste treatment alternative at appropriate sites. It should be remembered that land application systems are specific to each site, and it is often difficult to generate guidelines which are applicable to all systems. In developing guidelines, states should allow sufficient latitude in order to take full advantage of the potential of land application.

Requirements for secondary treatment should depend on the ability of the site to remove waste constituents, and the management scheme to be used at the site. Loading rates should take into account nitrogen, phosphorus and toxic substances, as well as BOD_5, suspended solids, and fecal coliform. In most instances, land application system designs are approved or denied on the basis of careful, site specific evaluations. So long as this continues to be the case, land application can remain a viable waste management alternative.

ACKNOWLEDGMENTS

The authors wish to express their appreciation to the personnel of state pollution control agencies across the country for their help and interest in this study. In addition, Mr. Eugene DeMichele of the Water Pollution Control Federation was very generous in loaning to us material which he had collected during a similar study. This study was undertaken as part of a two-year project (1974–1976), supported in part by grant T–900500 from the Academic Training Section, Environmental Protection Agency.

REFERENCES

1. National Wildlife Federation. *Conservation Directory.* Washington, D.C., 205 pp. (1975).
2. Michigan Department of Natural Resources. Groundwater monitoring guidelines for on land waste disposal facilities, Environmental Protection Branch (1973).

3. State of Florida. Guidelines for the treatment and/or disposal of waste-waters by irrigation on land, Department of Pollution Control, Division of Operations, Memo No. 149 (1973).
4. Great Lakes-Upper Mississippi River Board of State Sanitary Engineers. Addendum #2, Standards for waste treatment works, Municipal Sewage Facilities (1970).
5. Environmental Protection Agency. "Evaluation of land application systems," EPA–430/9–75–001, U.S. Government Printing Office, Washington, D.C., 182 pp. (1975).
6. "Secondary treatment information," *Federal Register.* 38(159) : 22298–22299 (1973).
7. California Administrative Code. "Statewide standards for the safe direct use of reclaimed wastewater for irrigation and recreational impoundments," Title 17 (1973).
8. California Administrative Code. "Wastewater reclamation criteria," Title 22, Division 4 (1975).
9. Pennsylvania Department of Environmental Resources. "Spray Irrigation Manual," Bureau of Water Quality Management, Publication No. 31, 49 pp. (1972).
10. Minnesota Pollution Control Agency. Recommended design criteria for disposal of effluent by land application, Division of Water Quality (1972).

ADDITIONAL REFERENCES

Bauer, D. S. Maryland Environmental Health Administration. Personal communication (1975).

Bogedain, F. O., A. Adamczyk and T. J. Tofflemire. "Land disposal of wastewater in New York State," New York State Department of Environmental Conservation, 63 pp. (1974).

Bradford, W. W. West Virginia Department of Health. Personal communication (1975).

Brown, R. E., et al. "Ohio guide for land application of sewage sludge," Cooperative Extension Service, The Ohio State University, Bulletin 598 (1975).

California State Water Resources Control Board. "The Porter-Cologne Water Quality Control Act" (1974).

Chisolm, C. Mississippi Pollution Control Commission. Personal communication (1975).

Coburn, D. L. North Carolina Department of Natural and Economic Resources. Personal communication (1975).

Coerver, J. F. Louisiana Health and Human Resources Administration. Personal communication (1975).

Coleman, M. S. Oklahoma State Department of Health. Personal communication (1975).

Crane, L. E. Iowa Department of Environmental Quality. Personal communication (1975).

DeNike, L. Washington Department of Ecology. Personal communication (1975).

Dudley, J. New Mexico Environmental Improvement Agency. Personal communication (1975).

Elwood, J. R. Massachusetts Division of Pollution Control. Personal communication (1975).

Flanders, H. P. Vermont Agency of Environmental Conservation. Personal communication (1975).

Follett, R. H. Arizona Department of Health Services. Personal communication (1975).

Glowacz, M. Guidelines for domestic sewage effluent disposal by spray irrigation, District Services Division, Department of Health and Environmental Control, 8 pp. (1973).

Hannah, H. G. Arkansas Department of Pollution Control and Ecology. Personal communication (1975).

Idaho Department of Environmental and Community Services. Water Quality Standards and Wastewater Treatment Requirements (1973).

Illinois Environmental Protection Agency. Illinois advisory committee on sludge and wastewater utilization on agriculture land (February 1975).

Illinois Environmental Protection Agency. Design criteria for municipal sludge utilization on agricultural land, draft copy, technical policy WPC-3 (1975).

Kansas State Department of Health. State Board of Health Regulations: Agricultural and Related Wastes Control, Environmental Health Services (1967).

Keeney, D. R., K. W. Lee and L. M. Walsh. "Guidelines for the application of wastewater sludge to agricultural land in Wisconsin," Department of Natural Resources, Technical Bulletin No. 88, Madison, Wisconsin, 36 pp. (1975).

Kim, S. W. Indiana State Board of Health. Personal communication (1975).

Missouri Clean Water Commission. Effluent regulation as amended (1975).

Nebraska. Revised standards and guidelines for reviewing land irrigation facilities in the State of Nebraska (1974).

Nevada Bureau of Environmental Health. Personal communication (1975).

Peterson, N. L. North Dakota Division of Water Supply Supply and Pollution Control. Personal communication (1975).

Peterson, M., *et al.* "A guide to planning and designing effluent irrigation disposal systems in Missouri," The University of Missouri Extension Division, 90 pp. (1973).

Scribner, J. W. Alaska Department of Environmental Conservation. Personal communication (1975).

South Dakota Department of Environmental Protection. Recommended design criteria for disposal of effluent by irrigation, Water Quality Program (1975).

Sparks, F. C. Colorado Water Conservation Board. Personal communication (1975).

State of California. Guidelines for the use of reclaimed water for spray irrigation of crops, Department of Health (1974).

State of Delaware. Regulations governing the control of water pollution, Department of Natural Resources and Environmental Control (1974).

State of Georgia. Criteria for wastewater disposal by spray irrigation. Department of Natural Resources, Environmental Protection Division, Water Quality Control Section (1973).

State of Illinois. Wastewater land treatment site regulation act (1973).

State of Kentucky. Permits to discharge sewage, industrial and other wastes, Department of Natural Resources and Environmental Protection, Bureau of Water Quality, Division of Water Quality, 601 KAR 5:005, 015, 025, 035, 045 (1975).

State of Maine. Maine guidelines for manure and manure sludge disposal on land, The Life Sciences and Agriculture Experiment Station and the Cooperative Extension Service, University of Maine at Orono, and the Maine Soil and Water Conservation Commission, Miscellaneous Report 142, 21 pp. (1972).

State of Maine. Maine Guidelines for septic tank sludge disposal on land, The Life Sciences and Agriculture Experiment Station and the Cooperative Extension Service, University of Maine at Orono, and the Maine Soil and Water Conservation Commission, Miscellaneous Report 155, 20 pp. (1974).

State of New Hampshire. Standards of design for sewerage and waste treatment systems, Water Supply and Pollution Control Commission (1975).

State of New Jersey. Guidelines for the design of land disposal facilities utilizing spray irrigation, Draft, Division of Water Resources, Department of Environmental Protection (1975).

State of Oregon. Subsurface and Alternative Sewage Disposal, Oregon Administrative Rules Chapter 340, Division 7 (1975).

State of Rhode Island. General Laws of 1956, Title 46, Chapter 12, Water Pollution, as amended (1971).

State of Virginia. Proposed regulations—spray irrigation, State Water Control Board (1975).

Sudweeks, C. K. Utah Bureau of Water Quality. Personal communication (1975).

Taylor, R. B. Connecticut Department of Environmental Protection. Personal communication (1975).

Tennessee Water Quality Control Board. Personal communication (1975).

Texas State Department of Health. Recommendations from the staff of the division of wastewater technology and surveillance when the domestic wastewater effluent is to be used for irrigation of areas accessible to the public (1972).

Tinsley, P. S. Maryland Department of Natural Resources. Personal communication (1975).

Warr, J. W. Alabama Water Improvement Commission. Personal communication (1975).

Williamson, A. E. Wyoming Department of Environmental Quality. Personal communication (1975).

Yuen, G. Hawaii Department of Health. Personal communication (1975).

Zollman, D. M. Montana Department of Health and Environmental Sciences. Personal communication (1975).

3

ECONOMIC AND REGULATORY ASPECTS OF LAND APPLICATION OF WASTES TO AGRICULTURAL LANDS

J. B. Johnson
U.S. Department of Agriculture, East Lansing, Michigan
L. J. Connor
Department of Agricultural Economics, Michigan State
    University, East Lansing, Michigan

INTRODUCTION

Land application of animal wastes is an age-old occurrence. However, not all wastes which are or could be applied to agricultural lands are generated on the farm. Some municipal wastewaters, including wastes from both productive and consumptive activities, are applied to agricultural lands. A more recent phenomenon is the use in agriculture of (waste heat) warm water from electric power-generating plants.

Problems have emerged in the U.S. through the waste disposition practices of firms in certain sectors of the economy and certain municipalities. Because the users of the waste assimilative capacities of the environment do not bear the costs of their actions, there is a tendency to make too great a use of these capacities (1). In these situations public intervention is needed to cause off-site costs to be reflected in the waste management decisions.

In this discussion, the focus is on the methods of economic analysis and selected research findings relative to concerns of land application of wastes. First, however, the provisions of PL 92–500 are reviewed as they relate to land application and the control of water pollution from animal waste, waste heat, and municipal wastewater.

PROVISIONS OF PL 92–500

Among the several means of public intervention to affect discharges to waters are effluent charges or taxes, subsidies or incentive payments, stream standards, and effluent standards. In recent years major emphasis has been placed on the use of standards—either stream standards, effluent standards, or some combination thereof. The use of the standards has occurred for two main reasons: 1. because some of the relevant benefits from surface water quality improvement cannot be established on economic grounds, and 2. there apears to be an extreme urgency expressed by Congress, the federal government executive branches, and state agencies that led to the use of standards because they can be expediently implemented (2). The major impetus for movement in this direction has come from PL 92–500, the Federal Water Pollution Control Act Amendments of 1972.

The stated objective of this Act is "to restore and maintain the chemical, physical, and biological integrity of the Nation's waters" (3). Among the major provisions of this Act are:

1. the discharge of pollutants to navigable waters be eliminated by 1985; and establishing a time frame for achieving these major provisions;
2. federal financial assistance be provided to construct publicly owned waste treatment works; and
3. areawide waste treatment management processes be developed to assure adequate control of sources of pollutants in each State.

Several sections of 92–500 relate specifically to the types of wastes this discussion considers.

Provisions Applicable to Animal Wastes

Section 306 of PL 92–500 identifies "feedlots" as a category of point source dischargers. Section 402 of PL 92–500 provides permit requirements for point source dischargers, and Section 301 specifies the general outline for the development of effluent limitations guidelines applicable to industrial categories of point source dischargers (3).

U.S. Environmental Protection Agency (EPA) announcements indicate that the "feedlots" subject to permit requirements and effluent limitations guidelines are concentrated livestock and poultry operations. The basic permit program for feedlots requires those with capacities of 1000 head or more and a discharge to obtain permits, those with 300- to 1000-head capacity discharging into surface waters through a man-made conveyance or with waters coming in contact with the feedlot to obtain a permit, and under limited conditions, certain feedlots of less than 300-head capacity to obtain a permit (4).

Effluent limitations guidelines for these recently issued EPA permit requirements have not been announced. However, if the guidelines parallel those EPA announced in February 1974 for operations of 1000 or more animal units, major concern will be to achieve control of storm runoff and wastewater discharges leaving animal production facilities and reaching navigable waters (5). Land application concerns under such guidelines have primarily related to the disposition of controlled runoff and wastewater that is periodically applied to agricultural lands.

Section 304 of PL 92–500 calls for the development of (a) guidelines for identifying and evaluating the nature and extent of nonpoint sources of pollutants and (b) processes, procedures and methods to control pollution resulting from nonpoint sources (3). First in a list of several nonpoint sources is "agricultural and silvicultural activities, including runoff from fields and crop and forest lands."

Land application of animal wastes is included in the initial group of potential agricultural nonpoint sources of water pollution identified by EPA (6). General sources of animal wastes applied to land are (a) wastes removed from feeding and/or holding facilities, (b) runoff and other discharges controlled from animal production facilities (per requirements of Section 306), and (c) animal excretion on pasture and rangeland (6). Methods and procedures to limit the water pollution potential of animal wastes applied to land were announced by EPA.

Provisions Applicable to Waste Heat

Section 306 of PL 92–500 identifies "steam electric power plants" as an industrial category of point source dischargers. As an industrial category of point dischargers, steam electric power plants are subject to requirements of permit regulations specified in Section 402 and the effluent limitations guidelines specified in Section 301 of PL 92–500 (3). Section 502 defines pollutant to include heat discharged into water. Thus all sources discharging heat to navigable water are defined as point sources.

Provisions Applicable to Municipal Wastewater

Publically owned waste treatment works are the topic of many sections of PL 92–500. Concern within this paper is limited to provisions that affect the application of municipal wastewaters to agricultural lands.

In Section 108, which applies to demonstration projects for

rehabilitation and environmental repair of Lake Erie, specific reference is made to the consideration of alternative systems for municipal wastewater management to include advanced waste treatment technology and land disposal systems, including aerated treatment-spray irrigation technology. Title II of PL 92–500 discusses a series of federal provisions for grants to public treatment work. Section 201(d) encourages waste treatment management that results in the construction of revenue-producing facilities including the recycling of potential sewage pollutants through the production of agricultural products, and the reclamation of wastewater. Section 201(g) (2) (A) will not allow EPA to make grants for treatment works unless the grant applicant has satisfactorily demonstrated that alternative waste management techniques have been studied and evaluated (3).

Provisions Requiring Development of Waste Management Treatment Plans

Several provisions of PL 92–500 call for the development of waste management plans on a state, river basin, or areawide basis. Section 303 delineates a planning process and other activities to be carried out at the state level. Section 209 calls for the Water Resources Council to prepare a plan for all river basins prior to 1980, with priority given to basins including areawide planning activities under Section 208 of the statute. Of current concern to many are the provisions of Section 208 (3).

Section 208 calls for implementation of areawide waste treatment management plans. Governors of the states have designated particular substate areas which have substantial water quality problems as a result of urban-industrial concentrations or other factors. In addition to the substate areas previously designated by the governors of each state, a recent court decision requires that all remaining areas within each state be considered a general Section 208 planning area (7). Areawide waste treatment management plans developed for each area will contain alternatives for waste treatment management of all wastes generated within the area.

Any plan prepared for an area designated under Section 208 will be required to include several points. Among those pertinent to the wastes considered in this discussion are:

1. The identification of treatment works necessary to meet anticipated municipal and industrial waste treatment needs of the area, including analyses of alternative systems, and the specification of any requirements for the acquistion of land for treatment purposes.
2. A process to (a) identify, if appropriate, agriculturally and sil-

viculturally related nonpoint sources of pollution, including runoff from manure disposal areas, and from land used for livestock and crop production; and (b) set forth procedures and methods, including land use requirements, to control to the extent feasible such sources.

EPA may not make any grant for construction of a publicly owned treatment work within the area covered by the plan except to the waste treatment management agency and only for works conforming to the plan.

Many of the decision units—such as livestock and poultry producers, electric power plant firms, and municipal waste treatment works—subject to control by EPA (or state-approved programs) under specific provisions of PL 92–500 may be further influenced by plans developed by areawide planning agencies. It is important that planners in these agencies be aware of prior research and methods to determine the economic effects of controlling such entities under specific provisions of PL 92–500. It is also extremely important that the planners keep in mind that the decision units mentioned above may all seek to apply wastes to land as one possible alternative to limit surface water degradation.

ECONOMIC ASPECTS OF LAND APPLICATION OF WASTES

The remainder of this chapter is a discussion of the economic methodology and selected research findings relative to concerns of land application of wastes as an alternative to limit surface water degradation.

Animal Wastes

A number of studies have been conducted on the economic impacts of runoff control (or discharge control) from animal production facilities. The general findings of these studies, most of which used partial budgeting analysis, were that the per head direct costs (the sum of fixed and variable costs) of runoff control were greater for smaller capacity production systems than for larger capacity systems; per head capital outlay requirements demonstrated a similar pattern.

Empirical estimates of the per head capital outlays and direct costs are presented for a runoff control system which would have satisfied effluent limitation guidelines similar to those announced by EPA in February 1974 (Table 1).

These estimates reflect a runoff control system consisting of diversion, settling basin, retention pond, and pump-irrigation

Table 1.   Per head capital outlays and direct cost increases for feedlot
runoff control[a]

Capacity Class:	Beef Feeding[b]	
	Capital Outlay	Direct Cost
1–99	$145	$21.17
100–199	21	3.19
200–499	12	1.84
500–999	8	1.28
1,000 and more	3	0.69
	Dairy[c]	
	Capital Outlay	Direct Cost
15	187	50.00
30	69	19.00
80	34	10.00
150	25	7.00
	Swine[d]	
	Capital Outlay	Direct Cost
1–99	61	4.62
100–199	20	1.37
200–499	11	0.74
500–999	7	0.44
1,000 and more	4	0.26

[a]Sources: (8, 9, and 10).   [c]For the northern region of the U.S.

[b]For the eastern beef-feeding states.   [d]For the 15 leading hog-producing states.

equipment components. The pump-irrigation equipment was of
sufficient capacity to allow the feedlot operator to apply the contained runoff onto land the feedlot operator owned or otherwise
controlled within a 10-day period subsequent to its occurrence.

Several assumptions underlie this method of analysis. The livestock producer is assumed to have a fixed set of production facilities of sufficient capacity to meet his desired production
levels. This method of analysis also presupposes that the number
of cattle fed annually and the feed efficiency realized are not
influenced by requiring runoff control, and that adequate field
time (field conditions suitable for waste spreading and/or irrigation) and labor are available for spreading settling basin
wastes and for the pumping and irrigating of controlled runoff
(11).

Pherson suggests that the costs of controlling lot runoff are generally underestimated when a partial budgeting approach is employed (11). A partial budgeting analysis of the economic impacts of runoff control (one process of an animal production enterprise) for an animal production enterprise ignores the alternative uses of certain resources that may have higher value uses at particular times in other enterprises on the farm. Therefore, a whole farm analysis of economic impacts of runoff control is encouraged (11). Such an approach will allow the inclusion of indirect costs of pollution control.

Pherson analyzed the economic effects of runoff control on a set of Minnesota beef feedlot situations, each with a one-time capacity of 500 head located on a 500-acre corn-soybean farm. He employed a profit-maximizing linear programming model to assess the indirect costs. Critical resources specified in the model were cropland, feeding facilities, field time, and available labor. Resource requirements for each activity on the feedlot-farm situations were budgeted and the model run to select the optimum combination of beef feeding and other farm activities within the applicable resource constraints.

Net revenue (total revenue less variable costs) of a 500-head open-lot beef feedlot prior to runoff control was estimated to be $61,737 with an associated total fixed cost of $37,022. Through the imposition of runoff control, the net revenue was reduced to $61,360, and the total fixed costs were increased to $37,802. The total measured effect of runoff control is:

$$
\begin{array}{lrr}
\text{Indirect costs:} & \$61,737 - \$61,360 = & \$ \ \ 377 \\
\text{Fixed costs:} & 37,802 - \ \ 37,022 = & \underline{\ \ \ 780} \\
& & \$1,157
\end{array}
$$

Explicit recognition that runoff control may compete with other farm activities for available field time and labor resulted in a reduction of $377 in returns to labor above that which was indicated through a partial budgeting analysis (11). For the analysis of runoff control where the fixed costs reflecting the ownership costs of the control system are a large part of the total cost of runoff control, the partial budgeting analysis of the runoff control process of the animal production enterprise does not yield results that differ greatly from the whole farm linear programming analysis.

Estimates of the economic effects of regulating the land application of livestock wastes to limit water pollution potential can vary substantially. Such variation is introduced by the modeling techniques employed (partial budgeting or whole farm analysis) and restrictions placed on the land application of animal wastes. Under the assumption that winter spreading of dairy farm wastes would be prohibited as a means of reducing the potential for field runoff in northern dairy regions, Johnson *et al.* considered the economic effects of adding winter storage to systems which made a historical practice of spreading daily (12). Considering 40- and 80-cow herd stanchion barn dairy systems, it was shown that the addition of stacker and storage facilities increased the per head capital requirements by $167 and $144, respectively. Likewise, total production costs were increased by $19 and $11 per head, suggesting economies of size in the adjustment process. Through winter storage, the total hours of labor required for waste handling were reduced. However, with labor valued at a uniform rate of $3 per hour, the labor savings are far outweighed by the additional ownership costs incurred for the stacker and storage facilities. The result is a net increase in the costs incurred for waste management.

An analysis of a similar rule was conducted for beef-feeding operations in Michigan (13). Employing a whole farm simulation model with optimizing subroutines, this analysis demonstrated diseconomies of size in adjusting the prohibition of winter spreading. Lower per head capital outlays were incurred by small-capacity producers than large-capacity producers. Production cost increases demonstrated a similar pattern. Diseconomies of size occurred because the larger operations were required to invest in a large waste-loading and spreading equipment complement to be capable of spreading wastes during the limited time periods available for such activities (14).

The following summary observations can be made relative to the economic effects of regulating the land application of animal wastes:

1. If winter application of animal wastes is restricted and winter storage is required, control of nonpoint sources of water pollution associated with animal wastes is generally more costly than the control of point sources (runoff from feedlots) of water pollution associated with animal wastes.
2. Partial budgeting and whole farm analyses yield similar economic effects for two firms of similar resource base when a restriction of controlling lot runoff is analyzed.
3. With reference to proposed regulations to limit water pollution from the land application of animal wastes, partial budgeting analyses suggest economies of size in the adjustment process. Using the whole farm approach, if the farmer has relatively small quantities of waste and a large equipment complement available for handling waste, the economic effects will approximate that arrived at through partial budgeting analysis. However, if there are relatively large quantities of waste and a limited handling capacity, the whole farm analysis will probably suggest diseconomies of size in the adjustment process.

Waste Heat

Waste heat from electric-generating facilities has been utilized or proposed for a variety of agricultural and aquacultural uses. Considerable research on the economic feasibility for utilizing waste heat for such purposes has been conducted. Two characteristics typify this research: (a) most studies have assessed the economic feasibility of single or combined use systems for utilized waste heat; and (b) most studies have used some form of benefit-cost analysis.

Johns *et al.* employed a combination of partial budgeting techniques and income statements in assessing the costs and returns of single uses of waste heat for purposes of soil heating for intensive vegetable production, irrigation of cropland, and the heating of greenhouses (15). Boersma compared combined urban-agricultural systems with conventional heat dissipation methods on a present value basis (16). These and other previous studies have all used a benefit-cost approach in analyzing economic feasibility.

Benefit-cost analysis is well suited for assessing which alternative use of waste heat is most profitable. As the choice criterion, the magnitude of the net benefits for each alternative is considered. However, most benefit-cost analyses do not indicate if waste heat is allocated optimally among competing uses, nor do they provide any information on the least-cost waste heat system (17). Optimality considerations are from the utility company's point of view.

Cost-effectiveness analysis is also used to assess investment alternatives for the dissipation of waste heat. The general question with which this method deals is the determination of a least-cost combination of resources that will achieve a desired goal; with the incorporation of present value discounting techniques, this method is especially useful in long-range planning. A recent study used such an analysis in determining the optimal combination of subsystems and their sizes to dissipate a specified amount of waste heat (17). From the different alternatives available, the combination of subsystems which would minimize the net present value of capital outlays over specified time horizons was determined. The objective function used optimized the net present value of the total integrated system, including the initial capital outlays, the annual costs and net returns, and the make-up power and other costs of the general piping and distribution system.

Results of comparing the use of waste heat in an integrated agricultural and aquacultural system with other conventional systems for utilizing or dissipating waste heat are presented

(Table 2). These results assume total ownership and management by the utility company. For the time period and discount rates shown, the integrated system of agricultural and aquacultural uses is the least-cost system of those considered. Although it is still quite expensive in current dollar terms, it is much less expensive than the alternative conventional systems considered.

The economic feasibility of utilizing waste heat from power plants in agriculture and aquaculture obviously depends on the type of system and uses made of the waste heat. For a single use or a combined use system to be economically feasible, the total benefits must exceed total costs in present value terms. Since single or combined systems are seldom capable of dissipating heat on a yearly basis, the initial monetary outlays need to be considered in any such analysis. In contrast, the integrated system must meet two objectives: (a) it must dissipate a fixed amount of waste heat, and (b) it must be the least-cost alternative as compared to conventional heat dissipation methods. A number of economic factors affect the attainment of these objectives. These are: (a) the form of economic organization and integration; (b) capital and land availability; (c) types of commodities produced and market potential; and (d) the design and spatial relationships involved in the total system.

The form of economic organization and integration will greatly affect the overall feasibility of utilizing waste heat in agriculture and aquaculture. Each combination of land ownership, the number and nature of firms involved in managing the use of waste heat, and the relationship to a public utility will have varying impacts on the level of economic and physical integration, and the nature and distribution of external effects created. Fee simple acquisition and management options for utilizing waste heat will require that large amounts of capital required be provided by the electric power-generating firm. The public

Table 2. Preliminary objective function values for alternative waste heat systems.[a]

Time Horizon and Discount Rate	Integrated System	Natural Draft	Cooling Pond	Spring Canal	Mechanical Draft
30 years at 12 percent	30,834,100	56,486,961	57,526,773	50,256,078	49,518,588

[a] Source: (17).

authority option would require the least. The degree of control exercised over the waste heat facilities by a utility company would be greatest for the fee simple alternatives and the least through the public authority option (Table 3). Site-specific characteristics, rather than size of facility, dictate the importance of these considerations.

Table 3. Site acquisition and management options for utilizing waste heat from electrical generating facilities.

Fee simple acquisition
    Purchase and manage
    Purchase and lease back
    Purchase and resale on condition

Less than fee simple acquisition
    Purchase easements

Contractual agreements and real property interest
    Waste heat water cooperatives
    Contractual arrangement

Public authority (such as TVA)

[a] Source: (17).

All waste heat systems are currently restrained by the investment and operating capital requirements. The availability of adequate quality land at reasonable prices in close proximity to the electrical-generating facility is also a constraint; in fact, in many situations the unavailability of such land may preclude the utilization of waste heat for agricultural and aquacultural purposes.

Decisions concerning the choice of agricultural and aquacultural products to be raised are influenced by the capability of the existing food marketing structure in the region to process, transport, and sell those products to local or regional markets. For fresh markets and perishable commodities, the availability of local and regional markets to assimilate the additional volume without an adverse impact on prices received at off-season marketing periods is important. Finally, the design and spatial relationships for waste heat utilization systems are exceedingly crucial. These factors determine the general piping and distribution system requirements. As the initial capital costs and annual operating costs are large for the piping and distribution system, care must be taken to minimize these monetary costs wherever possible.

Municipal Wastewater

Land application of wastewater for use in crop production is an old concept. Historically wastewater has been applied to land following primary and secondary treatment; the soil and any vegetation then absorb and filter nitrates, phosphates and other elements from wastewater. The remaining water then drains through the soil to recharge the ground water or to return via underground drains to the watercourse.

A number of municipalities in the U.S. are presently utilizing land application of municipal wastewater as a treatment method (18). One of the larger land treatment projects is in Muskegon County, Michigan, where wastewater irrigation is used to treat the industrial and municipal wastewater from an area with a population of 160,000 people on 10,000 acres of land (18).

A number of studies have been conducted on the economic aspects of land application of municipal wastewaters. However, a few studies illustrate the major economic considerations.

Carlson and Young estimated the demand for land treatment of municipal wastewater in the U.S. (19). Their analysis indicated that irrigation prices, federal subsidies, and relative labor and capital prices help explain the adoption of land treatment systems. Land treatment systems were found to be better suited for small towns. There is a higher probability of cities with smaller volumes of wastewater and less annual rainfall to accept land treatment more readily than cities with other situations. Land acquisition costs were not found to be a major consideration in accepting or rejecting land treatment.

A recent report analyzed the economic feasibility of utilizing municipal wastewater for crop production in Michigan (20). The revenues of farms receiving wastewater varied with changes regarding cropping pattern, intensity of land use, yields and prices. Net revenues from irrigation were influenced by fertilizer cost savings due to nutrients in the effluent, and by the level of any shared construction and operating costs of the irrigation system. For the various assumptions specified in the study, irrigation costs were offset with net revenue increases more frequently with corn, dry beans and alfalfa than with soybeans or wheat.

The forms of economic organization and integration chosen and the design criteria for land treatment systems are important considerations. Several options are available for a wastewater authority in acquiring land and managing treatment sites (20). (These options are identical to those specified in Table 3 for waste heat utilization.) Because of the large land requirements and associated capital costs, fee simple acquisition may be a limited option for large municipalities. Since few municipalities have much agricultural expertise, it is crucial that agricultural aspects be involved in managing a crop production system utilizing wastewater. Satisfactory agreements between farmers and wastewater authorities are imperative for an operable system.

A system for applying wastewater to land can be designed for the primary objective of wastewater treatment and renovation or for the objective of maximizing crop production or some combination of both (20). A system designed to maximize wastewater renovation requires less land for a given volume of wastewater. This reduces land costs to the wastewater authority, but is more disruptive to agriculture on the acreage involved. Alternatively, the maximization of the crop production objective requires more land for irrigating wastewater, but is more compatible with the existing agricultural structure of the region. The number of years a system functions effectively in renovating wastewater is influenced by the balance between nutrients supplied and nutrients taken up by the plants and stored in the soil.

SUMMARY

An age-old practice has been to apply animal waste to agricultural lands for plant nutrient recovery and other benefits. More recently, for reasons including the limiting of water quality degradation and energy conservation, municipal wastewaters and waste heat (warm water) from electric power-generating plants have been applied to agricultural lands. PL 92–500 has specific provisions that apply to the wastes cited as examples— animal wastes, waste heat, and municipal wastewater. The regulatory aspects of the land application of these wastes to limit water quality degradation are outlined.

The economic effects and methodology used to measure the economic effects of regulating the three example types of waste are discussed. In general, animal producers have found the costs of controlling point sources of water pollution less than those of the control of water pollution from land application of animal wastes (nonpoint pollution). Partial budgeting and whole farm analyses yield similar estimates of economic effects for point source regulations and divergent estimates for regulations affecting land application of animal wastes.

Studies of waste heat have been of three general types— single use, combined use, and integrated systems. The integrated system approach appears preferable because of two main reasons: it considers the dissipation of the total annual volume of waste heat as an explicit objective; and it is directed at solving for the least-cost system, in terms of capital outlay and annual cost expenditures, rather than seeking that alternative with the most desirable benefit-cost ratio.

Studies concerning the land application of municipal wastewater have determined the economic effects of systems designed for two different objectives. Some systems are designed to maximize the water renovation potential of the system. Others are designed to maximize the crop response from agricultural lands receiving municipal wastewater applications while providing the required levels of treatment. For a given volume of wastewater, systems designed to maximize water renovation generally require lower capital outlays by the municipality than those designed to maximize crop production; however, those designed to maximize water renovation are also more disruptive to the local agricultural patterns than those designed for crop production. As with waste heat dissipation systems, the municipal wastewater utilization systems are plagued with economic organization and integration problems. It has been found that smaller municipalities are the most likely users of land treatment systems for municipal wastewater management.

PL 92–500 also has provisions that provide for planning at the state, river basin, and areawide levels. Planning groups at these levels can recommend the use of various regulatory instruments, including land use planning. Land use planning under these provisions of PL 92–500 is viewed as a means to an end— the end being the reduction of water quality degradation. In addition to the example wastes considered—animal wastes, waste heat, and municipal wastewater—other wastes such as agricultural processing plant wastes and municipal treatment plant sludge can be affected by the land use planning regulations proposed by planning agencies.

In areas where several wastes are available for land application, planning agencies will have to consider methods for determining the cost-effective use of land—considering such issues as the selection of those wastes to be applied, the portion of the agricultural land base that will be available for each waste, and the problems of timing the application of wastes to limit the potential for surface water degradation. An analysis of the competing uses for available land will be quite site-specific. Among those variables that will have to be specified for the particular geographic area are: (a) the physical resource characteristics such as location of the available agricultural lands within the planning area or land slope; (b) the technical relations such as the ratio of productive (or consumptive) activity to the waste to land applied and the land loading rates for the wastes to be applied; (c) the relative price relationships—commodity prices for the alternatives that can be grown on land to which wastes are applied and the costs of handling the various wastes; (d) the regulatory legal restrictions, generally not predictable but subject to change, relative to the management of the wastes; and (e) the institutional arrangments related to the control and management of lands to which wastes are to be applied—fee simple versus public authority, for example.

REFERENCES

1. Haveman, Robert H. "Efficiency and equity in natural resource and environmental policy," *Am. J. Agric.* 55(5): 868–878 (1973).
2. Kneese, A. V. and B. T. Bower. *Managing Water Quality, Economics, Technology, Institutions*, The Johns Hopkins Press, Baltimore, Maryland (1968).
3. Anon. Federal Water Pollution Control Act Amendments of 1972, Public Law 92–500, 92d Congress, S. 2770 (1972).
4. Anon. "Concentrated animal feeding operations," *Federal Register* 41 (54): 11458–11461.
5. Anon. "Effluent limitations guidelines and standards, feedlots point source category," *Federal Register* 39(32): 5704–5708.
6. Anon. "Methods and practices for controlling water pollution from agricultural nonpoint sources," EPA 30/9–73–015, U.S. EPA (1973).
7. Anon. U.S. District Court, District of Columbia, Decision of July 25, 1975.
8. Johnson, J. B. and G. A. Davis. "Economic impacts of implementing EPA water pollution control rules on the United States beef feeding industry," in: *Proc. 3rd International Symposium of Livestock Wastes*, pp. 49–54 (1975).
9. Buxton, B. M. and S. J. Ziegler. "Economic impact of controlling surface water runoff from U.S. dairy farms," Agric. Econ. Rpt. No. 260, ERS, U.S. Department of Agriculture (1974).
10. Van Arsdall, R. N., R. B. Smith and T. A. Stucker. "Economic impact of controlling surface water runoff from point sources in U.S. hog production," Agric. Econ. Rpt. No. 263, ERS, U.S. Department of Agriculture.
11. Pherson, C. L. "Beef waste management economics of Minnesota farmer-feeders," in: *Proc. 1974 Cornell Agricultural Waste Management Conference*, pp. 234–249 (1974).
12. Hoglund, C. R. and B. Buxton. "An economic appraisal of alternative waste management systems designed for pollution control," *J. Dairy Science* 56: 1354–1366. (1973).
13. Anon. "Beef feedlot design and management in Michigan," Res. Rpt. No. 292, Michigan State University and ERS, U.S. Department of Agriculture (1976).
14. Forster, D. L., L. J. Connor and J. B. Johnson. "Economic impacts of selected water pollution control rules on Michigan beef feedlots of less than 1,000-head capacity," Res. Rpt. No. 270, Michgan State University with ERS, U.S. Department of Agriculture (1975).
15. Johns, R. W. *et al.* "Agricultural alternatives for utilizing off peak

electrical energy and cooling water," Washington State University, Department of Agricultural Economics (1971).

16. Boersma, L. *et al.* "A systems analysis of the economic utilization of warm water discharge from power generating stations," Oregon State University, Engineering Experiment Station Bulletin 48, 257 pp. (1974).

17. Meekhof, R. L. and L. J. Connor. "Economics of waste heat utilization from power plants in agriculture and aquaculture," ASAE Paper No. 75-3528 (1975).

18. Pounds, C. E. and R. W. Crites. *Wasterwater Treatment and Reuse by Land Application*, Vol. 2, U.S. EPA, Washington, D.C. (1973).

19. Carlson, G. A. and C. E. Young. "Factors affecting adoption of land treatment of municipal wastewater," *Water Resources Research* 11(5), (1975).

20. Christensen, L. A., L. J. Connor and L. W. Libby. "An economic analysis of the utilization of municipal wastewater for crop production," Agric. Econ. Rpt. No. 292, Michigan State University (1975).

4

LAND APPLICATION AS A BEST PRACTICABLE WASTE TREATMENT ALTERNATIVE

D. A. Haith
Departments of Agricultural Engineering and
Environmental Engineering
D. C. Chapman
Department of Agricultural Engineering
Cornell University, Ithaca

INTRODUCTION

The Rivers Pollution Prevention Act of 1876 required that treatment of wastes discharged into British waterways be "the best or only practicable and available means under the circumstances" (1). One such treatment method that enjoyed considerable popularity in 1876 in Britain, Europe and the United States was sewage farming, or the disposal of wastewaters by application to cropped land areas. Nearly 100 years later, the Federal Water Pollution Control Act Amendments of 1972 (PL 92–500) established the "best practicable waste treatment technology" (BPWTT) as a 1983 minimum requirement for publicly owned treatment works (POTW) discharging wastewaters into navigable waters. Subsequent regulations have specified land application as a suitable BPWTT method (2). It should not be concluded that the U.S. is a century behind Great Britain in its attitudes towards water pollution control, but it is apparent that the philosophy of our current wastewater management programs is not new.

While PL 92–500 did build upon a historical tradition, provisions for implementing these traditions are a departure from the past. In particular, through a series of regulations and guidelines, the U.S. Environmental Protection Agency has defined in considerable detail the procedures and technologies that must be followed in order to comply with the law. Some examples are the guidelines for selection of BPWTT alternatives (2). These guidelines and accompanying regulations for construction grants for POTW's specify that municipalities must evaluate three alternative types of wastewater management:

1. alternatives employing treatment and discharge to navigable waters
2. alternatives employing land application techniques and land utilization practices
3. alternatives employing reuse.

In their wastewater management planning, municipalities are expected to compare alternatives from each category and select a plan that is cost-effective. Although "costs" in this context refer to total resources costs (including social and environmental), applicable regulations emphasize monetary costs (3):

> The most cost-effective alternative shall be the waste treatment management system determined from the analysis to have the lowest present worth and/or equivalent annual value without overriding adverse nonmonetary costs and to realize at least identifiable minimum benefits in terms of applicable Federal, State and local standards for effluent quality, water quality, water reuse and/or land subsurface disposal.

A cost-effective analysis can be considered to have two stages (4). The first stage (preliminary planning) is an initial screening of alternatives (3). This is a "broad brush" evaluation of management possibilities to determine which alternatives have "cost-effective potential." The second stage is a more refined analysis involving detailed cost estimates of the alternatives screened in stage 1. Systematic techniques involving mathematical models can be of considerable value in the first stage (preliminary planning) of a cost-effective analysis. Models may also be a useful supplement to stage 2, but the site-specific characteristics of a waste treatment system design often preclude modeling as a detailed cost estimating tool.

Cost-effectiveness models have seen application in the preliminary planning of wastewater treatment plants (5–7). Far less common have been models for planning of land application systems (8–10). The requirements for BPWTT extend somewhat beyond the capabilities of these previous models, since comparisons are now needed among alternatives of widely varying characteristics. A suitable screening model must compare the costs and performance characteristics of water discharge, land application and reuse options, as well as combinations of the three. The objective of such models would be to identify cost-effective wastewater management combinations that meet the requirements of BPWTT.

A model that meets this objective is presented in this paper. The model is capable of evaluating wastewater management alternatives involving treatment and discharge to receiving waters and/or land application. Reuse is not considered in the model, other than the implicit reuse features of land application (irrigation, ground water recharge). Additional reuse options are best dealt with on a case-by-case basis, and could be added to the model where situations warrant.

The model was used to screen BPWTT options for a hypothetical city with a 10-MGD (million gallons per day) wastewater flow presently discharged after secondary treatment into a river. The example was chosen to correspond to the Environmental Protection Agency's estimate of the most common BPWTT problem [municipalities discharging into receiving waters in which water quality standards for dissolved oxygen will not be met by secondary treatment—approximately 50 percent of the nation's POTW's (2)].

MODEL DEVELOPMENT

The typical BPWTT planning situation is shown in Figure 1. A city must make a selection from a set of wastewater treatment options involving either discharge to surface receiving waters or land application. Effluent requirements for the discharge from water-based treatment will be specified in the city's discharge permit, and will require secondary treatment plus any additional renovation needed to achieve quality standards in the receiving waters (2). For the majority of water quality limited effluents, additional removal of BOD (biochemical oxygen demand) will be required to achieve river and stream dissolved-

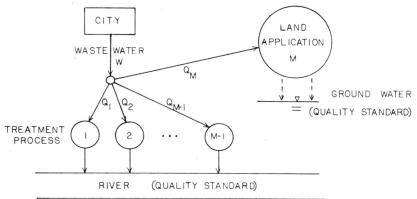

Figure 1. Best practicable waste treatment combinations.

oxygen standards, with particular emphasis on nitrogenous BOD removal. A much smaller portion of the effluents must have nutrient removal prior to discharge into lakes and estuaries (2). Water quality standards for the land application alternatives apply to the ground water into which the applied wastewater gradually percolates (assuming no discharge to surface waters via land application). "The ground water resulting from the land application of wastewater, including the affected native ground water" must generally meet the National Interim Primary Drinking Water Regulations (11).

If we define

Q_i = wastewater treated by process i (MGD)
W = wastewater flow to be treated (MGD)
M = number of possible treatment alternatives

then

$$\sum_{i=1}^{M} Q_i = W \tag{1}$$

The BPWTT planning problem is the selection of treatment options (values of Q_i) so that all wastewater is disposed of (Equation 1) and applicable surface and ground water quality standards are met at lowest monetary cost.

Treatment and Discharge to Receiving Waters

Assuming that discharge is made to a river or stream, effluent quality must be such that dissolved oxygen (DO) standards are maintained. This is typically determined using a steady-state DO model. The models are of varying complexity (12, 13), but as a minimum must include provisions for nitrogenous BOD. The simplest form is

$$C(x) = C_s \left[1 - \exp\left(\frac{-K_2 x}{u} \right) \right] + C_o \exp\left(\frac{-K_2 x}{u} \right)$$
$$- \frac{B_o K_1}{K_2 - K_1} \left[\exp\left(\frac{-K_1 x}{u} \right) - \exp\left(\frac{-K_2 x}{u} \right) \right]$$
$$- \frac{N_o K_n}{K_2 - K_n} \left[\exp\left(\frac{-K_n x}{u} \right) - \exp\left(\frac{-K_2 x}{u} \right) \right] \tag{2}$$

where x = distance downstream of waste discharge (mi)
$C(x)$ = DO concentration at x (mg/l)
C_s = saturation DO (mg/l)
C_o = initial DO after discharge (mg/l)

B_o = initial CBOD, carbonaceous (1st stage) BOD after discharge (mg/l)

N_o = initial NBOD, nitrogenous (2nd stage) BOD after discharge (mg/l)

K_n = rate of NBOD removal (day⁻¹)

u = average river velocity (mi/day)

K_2 = reaeration coefficient (day⁻¹)

K_1 = rate of CBOD removal (day⁻¹)

Additional terms can be added to Equation 2 as necessary to account for items such as benthal oxygen demands, plant and algal photosynthesis and respiration, and CBOD removal due to sedimentation (12).

Initial conditions C_o, B_o, N_o can be computed from effluent characteristics of the various treatment processes: C_i, B_i, and N_i = DO, CBOD and NBOD, respectively, of effluent from process i (mg/l). If the corresponding river parameters upstream of the discharges are C_R, B_R, and N_R, and the river flow is Q_R (MGD), then

$$C_o = \frac{C_R Q_R + \sum_{i=1}^{M-1} C_i Q_i}{Q_T} \tag{3}$$

$$B_o = \frac{B_R Q_R + \sum_{i=1}^{M-1} B_i Q_i}{Q_T} \tag{4}$$

$$N_o = \frac{N_R Q_R + \sum_{i=1}^{M-1} N_i Q_i}{Q_T} \tag{5}$$

where

$$Q_T = Q_R + \sum_{i=1}^{M-1} Q_i \tag{6}$$

If the DO standard is $\overline{C}$, then at all points downstream we must have

$$C(x) \geqslant \overline{C} \tag{7}$$

Equations similar to 1–7 have been used widely in water quality planning and constitute one component of the BPWTT model.

Land Application

The construction of a suitable model for screening land application alternatives is a challenging task. Wastewater reclamation in a soil/plant ecosystem is accomplished by a combination of physical, chemical and biological processes that are not easily described mathematically. Complete descriptions are unnecessary in a screening model, however, and a simple model that simulates the essential features of the land application system is both desirable and tractable. The key to a suitable model is the selection of these "essential" features, and the BPWTT guidelines (2) provide the following assistance.

1. Irrigation is the most common form of land application.
2. As noted earlier, a heavy emphasis is placed on protection of ground waters.
3. "The important loading rates are liquid loading in terms of inches per week, and nitrogen loading in terms of pounds per acre per year" (2).
4. "To minimize (nitrogen) buildup, the weight of total nitrogen applied in a year should not greatly exceed the weight removed by crop harvest" (2).

It seems clear that a model should apply to irrigation (rather than, say, overland flow or infiltration-percolation). It should include mass balances of water and nitrogen, and provide estimates of nitrogen losses to the ground water. Finally, it must recognize that the principal mechanism for nitrogen removal is crop growth. The above were taken as the essential features of the land application model.

The land application system has four components: pretreatment, wastewater transmission to the irrigation site, a lagoon for storage of the wastewater during time periods in which irrigation is infeasible, and the irrigation site. The relevant decision variables are

Q_M = wastewater flow to land application site (MGD)
A = irrigated land area (ac)
r = average wastewater application rate (in./wk)

The variables are not independent, since

$$\frac{Q_M}{A} = \frac{r}{258} \qquad (8)$$

If T is the number of weeks of the irrigation season and P and ET are irrigation season precipitation and evapotranspiration (in.), repsectively, the amount of water entering the ground water below the irrigation site is $7\,Q_M T/A + 0.02715\,(P - ET)$, in 10^6 gal per ac. Similarly, if n is the nitrogen concentration of the pretreated wastewater (mg/l), and NC is the nitrogen removal by the growing crop (lb/ac), an estimate of nitrogen entering the ground water from the irrigation site is $7(8.34)$

$nQ_MT/A - NC$. If we interpret the ground water standard as requiring average nitrogen concentrations in seepage water to be less than the drinking water standard of 10 mg/l, we have the constraint

$$\frac{7nQ_MT/A - NC/8.34}{7Q_MT/A + 0.02715\,(P-ET)} \leqslant 10 \tag{9}$$

which reduces to

$$\frac{Q_M}{A} \leqslant \frac{NC}{58.4\,T\,(n-10)} + \frac{0.0388\,(P-ET)}{T\,(n-10)} \tag{10}$$

We can also constrain the nitrogen application in the wastewater to be at least equal to the crop requirement, NC:

$$\frac{7\,(8.34)\,n\;\;Q_M\;\;T}{A} \geqslant NC \tag{11}$$

or

$$\frac{Q_M}{A} \geqslant \frac{NC}{58.4nT} \tag{12}$$

Equations 10 and 12 constrain the nitrogen loading rate at the land application site. The liquid loading rate will be constrained by the drainage capacity of the soil, $\bar{r}$ (in/wk):

$$r \leqslant \bar{r} \tag{13}$$

The set of Equations 8, 10, 12 and 13 thus constitute the land application component of the BPWTT screening model.

Cost Functions

Annual wastewater treatment costs ($/yr) including pretreatment for land application can be approximated (16) by functions of the form $a_i\,(Q_i)^{b_i} + c_i\,(Q_i)^{b_i}$, where a_i and b_i are constants corresponding to capital costs for treatment methods i, and c_i and d_i are analogous constants for operation and maintenance (O & M) costs.

Average annual land application costs consist of costs for transmission to the irrigation site, storage lagoons, irrigation and land purchase. The first three can be estimated from relevant cost curves (4). Annual transmission costs are $La_t\,(Q_M)^{b_t} + a_p\,(Q_M)^{b_p} + c_p\,(Q_M)^{d_p}$, where a_t and b_t are constants for pipeline costs, L is pipeline distance in miles, and a_p, b_p and c_p, d_p are appropriate constants for pumping capital and O & M costs, respectively. Storage lagoon costs are $a_s\,(Q_M)^{b_s} + c_s\,(Q_M)^{d_s}$, where again, the constants a_s, b_s and c_s, d_s refer to capital and O & M. The lagoon costs will depend on the number of weeks of storage required (4) and therefore, a_s, b_s, c_s, d_s will be functions of storage time. Irrigation costs are approximated by $K(Q_M)^e + D(Q_M)^f$, where K,e and D,f are functions of the application rate and correspond to capital and O & M.

Land purchase costs are determined by the total land required for the land application site. Assuming a 20 percent buffer area (21) that also includes required land for lagoons, land purchase cost is $1.2c_LA$ where c_L is the amortized purchase price ($/ac). A final cost element is any additional cropping costs not included in irrigation costs offset by the sale value of the crop. This cropping cost is $c_cA - pYA$, where c_c is the additional cropping cost ($/ac), p is the sale price of the crop ($/bu or $/ton) and Y is the estimated crop yield (bu/ac or ton/ac). The various costs are combined as TC, the total annual costs of the wastewater treatment system.

$$TC = \sum_{i=1}^{M}(a_iQ_i^{b_i} + c_iQ_i^{d_i}) + La_tQ_M^{b_t}$$
$$+ a_pQ_M^{b_p} + c_pQ_M^{d_p} + a_sQ_M^{b_s} + c_sQ_M^{d_s}$$
$$+ KQ_M^e + DQ_M^f + 1.2c_LA + c_cA - pYA \qquad (14)$$

Model Summary

The general form of the cost-effectiveness model is

$$MIN\ TC$$

subject to constraints noted with Equations 1-8, 10, 12 and 13 plus additional constraints, as necessary to relate a_s, b_s, c_s, d_s to weeks of storage and K,e, D and f to wastewater application rates, r. The model is highly nonlinear, and, in general, cannot be solved analytically. The independent decision variables are wastewater flows, Q_i, and irrigation area, A. A typical use of the model is to evaluate a combination of these variables (*i.e.*, a wastewater treatment plan) by computing costs (Equation 14), and determining if quality standards for surface water (Equation 7) and ground waters (Equation 10) are met.

EXAMPLE

The BPWTT planning model was applied to a hypothetical city with a 10-MGD wastewater flow, which is presently discharged after secondary treatment into a river. The river has a dissolved oxygen standard ($\bar{C}$) of 5 mg/l, and additional removal of oxygen-demanding material (CBOD and/or NBOD) will be required. Violations of the standard are a problem only during low-flow periods (July, August and September) and the city has decided on a seasonal strategy of advanced wastewater treatment. Five treatment options are considered for operation during the three month period: the existing (secondary) treatment, and the additions of filtration, nitrification, nitrification plus filtration, or land application to secondary treatment.

Water Quality Data

River parameters for the DO model are given in Table 1. Effluent quality for the four treatment options discharging to the river are listed in Table 2. Except as noted, values were

Table 1. River water quality parameters for example problem.

Parameter	Value
Q_R, river flow	30 (mgd)
u, river velocity	4.9 (mi/da)
K_1, CBOD removal rate	0.35 (da^{-1})
K_2, reaeration rate	0.50 (da^{-1})
K_n, NBOD removal rate	0.20 (da^{-1})
C_s, saturation DO	8.0 (mg/1)
C_R, river DO	8.0 (mg/1)
B_R, river CBOD	2.0 (mg/1)
N_R, river NBOD	5.0 (mg/1)

taken from the literature. Data for the land application system is in Table 3. A corn crop was chosen for irrigation. Characteristics of the land application option were chosen to be consistent with northeastern U.S. conditions. In particular, since growing season precipitation is generally sufficient for crop growth (P = ET), there are no irrigation water benefits to be obtained from the land application system. The data in Table 3 can be used to simplify the irrigation application rate constraints 10 and 12:

$$(Q_5/A) \leqslant 0.0198 \qquad (15)$$

$$(Q_5/A) \geqslant 0.00988 \qquad (16)$$

Using Equation 8, constraints 15 and 16 limit the irrigation rate

Table 2. Effluent quality for water-based treatment options.

Treatment Option	Effluent Quality (mg/l)			Reference
	CBOD, B_i	NBOD, N_i	DO, $C_i{}^a$	
Secondary	25	54	2	(15)
Secondary + filtration (microscreening)	13	50	2	(19)
Secondary + nitrification (separate single stage)	13	10	2	(20)
Secondary + nitrification + filtration	7	10	2	b

[a] All effluent DO assumed to be 2 mg/l.

[b] Assumes 50% removal of remaining CBOD.

r (in/wk) to $2.5 \leqslant r \leqslant 5.1$. The lower value supplies the minimum crop nitrogen requirements and the upper value corresponds to the limit of nitrogen in the drainage water to 10 mg/l.

Cost Data

Only added costs (above the costs of the present secondary treatment) are considered. Operation and maintenance (O&M) costs for the seasonal operation are estimated as one-third of a full year's O&M costs. All costs are updated to August, 1975.

Table 3. Water quality characteristics of land application (irrigation) option.

Parameter	Value	Explanation
ET, Irrigation season Evapotranspiration	—	Assumed that P = ET. Approximate conditions in Eastern U.S. humid areas
P, Irrigation season Precipitation	—	
T, Irrigation season	13 wk	Seasonal operation of land application
n, Nitrogen content of wastewater	20 mg/l	Ref. 15
NC, Nitrogen removal by crop growth	150 lb/ac	Rounded value for corn, Ref. 14

Cost parameters for water discharged and land treatment systems are based on literature values and are given in Tables 4 and 5, respectively. Assuming a two-mile transmission line to the land application site, total annual costs (10^3/yr) are

$$
\begin{aligned}
TC = {} & 6.9\,Q_2^{0.93} + 9.4\,Q_2^{0.55} + 22.7\,Q_3^{0.68} + 12.5\,Q_3^{0.42} \\
& + 6.9\,Q_4^{0.93} + 22.7\,Q_4^{0.68} + 9.4\,Q_4^{0.55} + 12.5\,Q_4^{0.42} \\
& + 22.8\,Q_5^{0.28} + 2.4\,Q_5^{0.78} + 0.2\,Q_5^{0.54} \\
& + [21.6 + 0.18\,(A/Q_5)]\,Q_5^{[0.74 + 0.00052(A/Q_5)]} \\
& + [8.7 + 0.077\,(A/Q_5)]\,Q_5^{[0.79 + 0.00041(A/Q_5)]} \\
& - 0.184\,A
\end{aligned}
\tag{17}
$$

Applicable constraints are noted with Equations 2-6, 15, 16 and

$$
\sum_{i=1}^{5} Q_i = 10 \tag{18}
$$

$$
C(X) \geqslant 5 \tag{19}
$$

The drainage capacity constaint, Equation 13, was not considered in the example.

Table 4.   Cost functions for water-based treatment.

Treatment Option	Flow (mgd)	Costs (10^3/yr)[a]	
		Capital	Operation and Maintenance
Secondary	Q_1	–	–
Secondary + filtration	Q_2	$6.9Q_2^{0.93}$	$9.4Q_2^{0.55}$
Secondary + nitrification	Q_3	$22.7Q_3^{0.68}$	$12.5Q_3^{0.42}$
Secondary + nitrification + filtration	Q_4	$6.9Q_4^{0.93} + 22.7Q_4^{0.68}$	$9.4Q_4^{0.55} + 12.5Q_4^{0.42}$

[a]August 1975 (ENR Construction Cost Index = 2274). Amortized at 5-5/8%, 20 yrs. Based on cost curves for 1-10 mgd flow range, Ref. 16. *O & M* costs are yearly values divided by 3. Secondary (existing) treatment costs not included.

Simulation Results

Preliminary BPWTT plans can be simulated using the screening model. A combination of treatment combinations (waste-

Table 5. Cost parameters for land application.

Cost Parameter[a]	Value	Reference	Explanation
	Transmission (Gravity Pipe)		
a_t	11.7 (10^3/mi)	(4)	Cost curves, 1–10 mgd
b_t	0.28	(4)	Cost curves, 1–10 mgd
	Storage (lagoons, no artificial lining)		
a_s	2.4 (10^3)	(4)	Cost curves, 1–10 mgd, 1 week storage
b_s	0.78	(4)	"
c_s	0.2 (10^3)	(4)	Cost curves, 1–10 mgd, 1/3 yearly O & M
d_s	0.54	(4)	Cost curves, 1–10 mgd
	Irrigation (center pivot, includes distribution, monitoring wells, roads, fencing, buildings, corn cultivation)		
K	$21.6 + 0.18 \dfrac{A}{Q_5}$ (10^3)	(4)	Cost curves, 1–10 mgd for 1, 2, 4 in./wk application rates
e	$0.74 + 0.00052 \dfrac{A}{Q_5}$	(4)	"
D	$8.7 + 0.077 \dfrac{A}{Q_5}$ (10^3)	(4)	1/3 yearly O & M
f	$0.79 + 0.00041 \dfrac{A}{Q_5}$	(4)	"
	Land Cropping Costs		
c_L	0.042 (10^3/ac-yr)	–	$500/ac purchase cost, amortized at 5-5/8%, 20 yr
c_c	0.039 (10^3/ac)	(17)	New York State cost of harvesting, storing and selling corn grain. 1973 costs increased 10% per yr
P	0.0026 (10^3/bu)	–	New York corn price, fall 1975
Y	105 (bu/ac)	(18,22)	Average from two studies: Penn. St. yields with waste-water application (108 bu/ac) and selected New York farm sites (102 bu/ac)

[a]Costs as of August 1975. Parameters chosen to give costs in 10^3/yr.

water flows, Q_i) and irrigation areas that satisfy constraints 15 and 16 are chosen; *i.e.*, $0.00988 \leqslant Q_5/A \leqslant 0.0198$. Total costs, TC, and the DO sag curve, $C(x)$, are then computed. The model is easily programmed for interactive use using the computer language APL.

Although the number of possible alternatives is infinite, simulation of a relatively small number of alternatives can give a good indication of options with "cost-effective potential." The first step, as shown in Table 6 is simulation of each of the waste-

Table 6. Simulation results—comparison of separate treatment options.

Wastewater Flows (mgd)					Irrigation Area (ac)	Irrigation Rate (in./wk)	Nitrogen Concentration in Drainage Water (mg/l)	River Dissolved Oxygen (mg/1) at x(mi) downstream x =					Annual Cost ($10³)
Q_1	Q_2	Q_3	Q_4	Q_5				5	10	15	20	25	
10	0	0	0	0	–	–	–	2.8	1.6	1.6	2.2	3.0	0
0	10	0	0	0	–	–	–	3.7	2.7	2.7	3.2	3.8	92.1
0	0	10	0	0	–	–	–	5.1	4.8	4.9	5.3	5.7	141.5
0	0	0	10	0	–	–	–	5.5	5.2	5.4	5.7	6.0	233.6
0	0	0	0	10	505	5.1	10.0	6.8	6.4	6.3	6.4	6.6	225.9

water treatment options separately. Neither secondary, secondary plus filtration nor secondary plus nitrification meets the river DO standard of 5 mg/l. Secondary plus nitrification and filtration meets the standard. Since land application ($Q_5 = 10$) results in no river discharge, it also satisfies the DO requirement. Of the five options, land application is least costly.

A second set of simulations investigates the sensitivity of the land application costs to irrigation area or irrigation rate (Table 7). It can be seen that the costs are relatively insensitive to such variations. It should be noted that the cost advantage of land application is primarily due to the nearness of the land application site (2 mi) and the fact that gravity rather than pressure transmission lines are possible. If for example, the application site was four miles from the existing sewage treatment plant, an additional cost for gravity transmission of $43,400 would be added to the land application costs given in Tables 6 and 7. Finally, given the accuracy of the cost estimates, the difference in costs between secondary treatment plus nitrification and filtration ($233,600) and the land application options ($225,900–230,800) cannot be considered significant.

A final group of simulation runs (Table 8) explores treatment combinations. Such combinations require a splitting of the waste

Table 7.   Simulation results—sensitivity of land application costs to irrigation area (rate).

Wastewater Flows (mgd)					Irrigation Area (ac)	Irrigation Rate (in./wk)	Nitrogen Concentration in Drainage Water (mg/1)	Annual Cost ($10³)
Q_1	Q_2	Q_3	Q_4	Q_5				
0	0	0	0	10	505	5.1	10.0	225.9
0	0	0	0	10	560	4.6	8.9	226.0
0	0	0	0	10	645	4.0	7.2	226.3
0	0	0	0	10	860	3.0	2.9	226.3
0	0	0	0	10	1012	2.5	0	230.8

water flow after secondary treatment into waste streams that are treated differently. As indicated in Table 8, treatment combinations can offer significant cost savings. In particular, run number 8, which sends 8 MGD of the secondary effluent to land application and discharges the remainder directly to the river, is less expensive than any other simulated wastewater treatment plan.

Additional simulations are easily run using an interactive computer model. The most promising (least expensive) of the simulated treatment plans would be carried on to the detailed planning stage of the city's wastewater management studies.

SUMMARY

The planning requirements for BPWTT (Best Practicable Waste Treatment Technology) include a preliminary screening of treatment alternatives including treatment and discharge to receiving waters, land application, and wastewater reuse. The objective of this first stage in the wastewater management planning process is the identification of treatment strategies with "cost-effective potential." The mathematical model presented in this paper can be effectively used for this screening. When programmed in an interactive mode, the model can be used by citizens groups and other nontechnical participants in the planning process, providing them with a meaningful method for evaluating alternatives. At the same time, the model provides a level of technical detail in estimates of costs and environmental impacts that is appropriate for preliminary planning.

An example problem was modeled and alternative wastewater management options were simulated. This example and the asso-

Table 8. Simulation results—treatment combinations.

Run No.	Wastewater Flows (mgd)					Irrigation Area (ac)	Irrigation Rate (in./wk)	Nitrogen Concentration in Drainage Water (mg/l)	River Dissolved Oxygen (mg/l) at x (mi) downstream x =					Annual Cost (10^3)
	Q_1	Q_2	Q_3	Q_4	Q_5				5	10	15	20	25	
1	2	2	2	2	2	129	4.0	7.2	4.7	4.0	4.1	4.5	4.0	249.1
2	0	2	3	3	2	129	4.0	7.2	5.2	4.7	4.8	5.2	5.0	287.9
3	0	0	3	3	2	258	4.0	7.2	5.8	5.5	5.5	5.8	6.1	304.8
4	0	1	3	3	3	194	4.0	7.2	5.5	5.1	5.2	5.5	5.8	300.7
5	0	1	4	1	4	258	4.0	7.2	5.6	5.1	5.2	5.5	5.8	281.3
6	1	0	4	0	5	300	4.3	8.1	5.6	5.1	5.1	5.4	5.8	231.4
7	1	0	0	0	9	581	4.0	7.2	6.3	5.8	5.7	5.9	6.2	212.9
8	2	0	0	0	8	516	4.0	7.2	5.8	5.2	5.2	5.4	5.7	198.9
9	3	0	0	0	7	452	4.0	7.2	5.4	4.7	4.6	4.9	5.3	184.1
10	0	3	0	0	7	452	4.0	7.2	5.7	5.1	5.0	5.3	5.6	220.4

ciated interactive computer program were tested on student groups at Cornell. Students with minimal computer experience and only rudimentary water quality knowledge used the model successfully to gain a greater understanding of the wastewater management planning required for BPWTT.

The increased focus in water quality planning on the comparison of alternatives requires a concomitant improvement in systematic techniques for generating and evaluating alternative treatment systems. Mathematical management models provide a means of comparing large numbers of treatment options during the preliminary planning stage. They also can be valuable teaching tools for the training of environmental planners and engineers. The model presented in this paper will not be sufficiently accurate for the *detailed* planning and design stages of wastewater management planning. Rather, combinations of process models, site-specific cost estimates and the expertise and judgment of an experienced engineer will be required.

REFERENCES

1. Rideal, S. *Sewage and Bacterial Purification of Sewage,* 3rd edition, The Sanitary Publishing Company, Ltd., London (1906), p. 19.
2. "Alternative waste management techniques for best practicable waste treatment." Report #PA–430/9–75–013, United States Environmental Protection Agency, Washington, D.C. (October, 1975).
3. United States Environmental Protection Agency. "Cost effectiveness analysis guidelines," *Fed. Reg.* 38(174), September 10, 1973.
4. Pound, C. E., R. W. Crites and D. A. Griffes. "Costs of wastewater treatment by land application," Report #EPA–430/9–75–003, United States Environmental Protection Agency, Washington, D.C. (June, 1975).
5. Lynn, W. R., J. A. Logan and A. Charnes. "Systems analysis for planning wastewater treatment plants," *J. Water Poll. Con. Fed.* 34(6): 565–581 (1962).
6. Shih, C. S., and J. A. DeFilippi. "System optimization of waste treatment process design." *Proc. Amer. Soc. Civil Engineers* 96(SA2): 409–421 (1970).
7. Smith, R. "Preliminary design of wastewater treatment systems," *Proc. Amer. Soc. Civil Engineers* 95(SA1): 117–145 (1969).
8. Haith, D. A. "Optimal control of nitrogen losses from land disposal areas," *Proc. Amer. Soc. Civil Engineers* 99(EE6): 923–937 (1973).
9. Koenig, A., and D. P. Loucks. "A management model for wastewater disposal on land," Presented at 2nd Annual National American Society of Civil Engineers Conference on Environmental Engineering Research, Development and Design, University of Florida, Gainesville (July, 1975).
10. Seitz, W. D., and E. R. Swanson. "Economic aspects of the application of municipal wastes to agricultural land," *Proceedings* of the Joint Conference on Recycling Municipal Sludges and Effluents on Land, Washington, National Association of State University and Land-Grant Colleges, pp. 175–182 (1973).

11. United States Environmental Protection Agency. "Alternative waste management techniques for best practicable treatment—supplement," *Federal Register* 41(29), February 11, 1976.
12. Thomann, R. V. *Systems Analysis and Water Quality Management.* Environmental Research and Applications, Inc., New York (1972).
13. Willis, R., D. R. Anderson and J. A. Dracup. "Steady-state water quality modeling in stream," *Proc. Amer. Soc. Civil Engineers* 101(EE2): 245–258 (1975).
14. "Evaluation of land application systems," Report #EPA–430/9–75–001, United States Environmental Protection Agency, Washington, D.C. (March, 1975).
15. Reed, S. (Ed.) "Wastewater management by disposal on the land," Special Report 171, Cold Regions Research and Engineering Laboratory, United States Army Corps of Engineers, Hanover, New Hampshire (May, 1972).
16. "Technical and economic review of advanced waste treatment processes." Report prepared for United States Army Corps of Engineers by Environmental Quality Systems, Inc., Rockville, Maryland (March, 1973).
17. Snyder, D. P. "Field crops costs and returns from farm cost accounts," Report #A.E. Research 74–14, Department of Agricultural Economics, Cornell University, Ithaca, New York (December, 1974).
18. Sopper, W. E., and L. T. Kardos. "Vegetation responses to irrigation with treated municipal wastewater." In: W. E. Sopper, and L. T. Kardos, Eds. *Recycling Treated Municipal Wastewater and Sludge Through Forest and Cropland,* Pennsylvania State University Press, University Park, Pennsylvania, Chapter 15 (1973).
19. Metcalf and Eddy, Inc. *Wastewater Engineering.* McGraw-Hill, New York (1972).
20. Smith, R. "Cost and performance estimates for wastewater treatment process trains," Unpublished memorandum, United States Environmental Protection Agency, National Environmental Research Center, Cincinnati (April 5, 1974).
21. Pound, C. E., and R. W. Crites. "Wastewater treatment and reuse by land application," Vol. II, Report #EPA–660/2–73–0066, U.S. Environmental Protection Agency, Washington, D.C. (August, 1973).
22. Herendeen, N. R. "The effect of time and rate of nitrogen application on corn in New York," Unpublished M.S. Dissertation, Cornell University, Ithaca, New York (September, 1969).

5

THE ADMISSIBLE RATE OF WASTE (RESIDUE) APPLICATION ON LAND WITH REGARD TO HIGH EFFICIENCY IN CROP PRODUCTION AND SOIL POLLUTION ABATEMENT

C. Tietjen
Institut für Pflanzenbau und Saatgutforschung
Forschungsanstalt für Landwirtschaft
Braunschweig-Voelkenrode, Germany (F.R.)

INTRODUCTION

To develop and operate waste management programs in a responsive manner, the decisions to be made will be essentially determined by four basic categories of criteria: costs, environmental factors, resource conservation, and institutional factors (1). Each category includes the following key points:

Costs	Resource Conservation
Operating and maintenance	Energy
Capital	Materials
Environmental Factors	Land
Water pollution	*Institutional Factors*
Air pollution	Political feasibility
Other health factors	Legislative constraints
Aesthetic consideration	Administrative simplicity

As the public becomes increasingly conscious of the need for resource conservation, these criteria gain in importance. To quantify the criteria is difficult as long as there is no distinctly fixed goal. At this time wastes are considered "resources out of place," and a good portion of the wastes could exercise a beneficial effect on the soil and the crop growth if applied to the land.

DISPOSAL VERSUS RECLAMATION

The decision to rank "resource conservation" as a high priority leads to the necessity to properly design some technique of land use which is different from common land disposal. To return waste into the natural cycle of transformations by land treatment does result in final waste disposal. However, at the same time, both land disposal and land treatment must be considered as separate alternatives. The goal of maximum disposal may be mutually exclusive with the goal of maximum utilization. To produce crops, land treatment of waste must accommodate the natural conditions of the site and the properties of the waste. Standards for waste utilization in crop production should provide reasonably high levels of public health protection as well as high levels of crop growth-promoting and soil-improving constituents in the waste. These levels provide the crop grower some certitude of economical success.

In one way or another most wastes return to man by means of environmental cycles. In fitting cities as well as agricultural areas into the biosphere, designs are needed to make these cycles beneficial. Along with proper engineering design and management, site selection must be recognized as an important ingredient for lasting success (2). The resource values of the wastes must be safe and provide symbiotic interfaces with areas of the landscape. Opportunities exist for using wastes to aid ecological systems, *i.e.*, crop production, rather than requiring elimination of all wastes by expensive technological processes. Biologically based recycling systems should be designed so as to optimize the production of economically and socially useful products. Further benefits of such systems may include savings in costs of soil cultivation and crop growth. Maximum crop growth is a biological proof of an ecologically balanced waste recycling system.

LAWS AND RULES

In Germany, the three important regulations are: the law of waste disposal, 1972; the law of water resources, 1957, (1976); and the "causer principle."

The "causer principle" provides the basis for regulations to define the source of pollutants and to charge the responsibility for maintaining or improving a given state and balance in the environment. This refers only to the costs for preventing the emission of pollutants. Among others, it is expected that this principle will help to change the strange situation whereby consumers pay the costs of water purification and not the causers of pollution. This is a matter of growing importance because the portion of stream and lake water which now contributes 40 percent of the water supply is increasing.

The German law of water resources is a federal skeleton law that indicates that he who changes the physical, chemical or biological quality of the water should make amends where necessary (§22). By a supplement expected in 1976, individual water management plans for surface waters are to be established and federal wastewater introduction standards are to be fixed.

The law of waste disposal defines the term waste. It also ordains that waste disposal upon cropland should not injure public welfare (§15). This is valid for wastewater, sewage sludge, feces, and compost that results from waste materials. Also included are liquid manure (urine), guelle, and stable manure if the application of these materials exceeds the "normal rate of manuring."

The details will be fixed in two orders, one for the spreading of wastewater, sewage sludge, feces and compost, and another for the by-products of livestock management. The orders are still under discussion.

Livestock Manure

The law of waste disposal distinguishes between municipal and industrial wastes and livestock wastes. No restriction on the traditional custom of utilizing livestock excrements as manure for soil conditioning and crop production is intended. The law acknowledges that the nutrients in livestock manure still contribute about 40 percent of the nutrient supply in crop production, although this relative portion is decreasing with the increasing application of commercial fertilizers (Figure 1).

By restricting the law to that part of livestock excrements which exceeds the "normal rate of manuring," a differentiation is made between livestock wastes and manure. The definition of the term "waste" given by the law includes the competence of public authorities for the waste disposal while the proper utilization of manure is a matter for experts in economical crop production.

The discrepancy in using two terms, waste and manure, for the same substrate, livestock excrements, is solved by defining the "normal rate of manuring." The traditional type of farming with a balanced ratio of crop production and livestock husbandry in the same enterprise provides the base of calculation. Knowledge and site experience of the nutrient requirements of different crops are decisive for the amount of manure application. Nitrogen and phosphorus are considered the most important con-

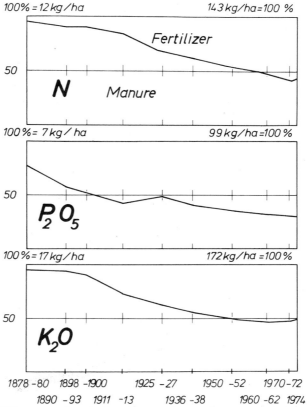

Figure 1.   Nitrogen, phosphate and potash nutrients applied in crop production from 1879 to 1974. Change of the portions applied with commercial fertilizers and with livestock manures.

stituents with concern of both crop yield and environment contamination. According to the average value of chemical analyses of manure, the different kinds of livestock are grouped into "manure livestock units." The "normal rate of manuring" is exceeded and the law becomes valid when there are more than three manure livestock units per ha (3). Table 1 defines "three manure livestock units" for different livestock.

Discussion of this approach is continuing. Soil and site characteristics, nutrient requirements of crops, different nutrient availability in liquid and solid manure, in fresh or rotted, stabilized manure, and potassium and copper surplus, are some of the arguments presented to vary the admissible ratio of manure livestock unit per ha between 1.5 and 6. To accommodate the small operations, a size limit has been suggested.

Table 1.   Ratio of manure-producing animals and land for manure is
exceeded by more than 3 "Manure Livestock Units" per ha.

3	Cattle, grown		
6	Young cattle		
72	Calves (3 months)	or	18 stable places
6	Sows with progeny		
42	Fattening pigs	or	18 stable places
300	Laying hens		
1,440	Pullets	or	600 stable places
5,400	Broiler chickens	or	900 stable places
2,475	Ducks	or	450 stable places
600	Turkeys	or	300 stable places

A survey within a region of the size and types of livestock, of the size of arable land and grassland, and of the manure capacity of the soil is needed for registration in this approach. It is recommended that a central bureau be charged with the administrative collection of manure surplus and its distribution to areas needing the surplus. A subsidy on transportation expenses increases the possibility of effective control. Since 1973 such organizations called "manure banks" have been operating successfully in the Netherlands.

An urgent question is: how extensive must be the registration? Only farms with insufficient land for manure and with more than three manure livestock units per ha are concerned. Because of inadequate statistical data, only a rough estimation is possible.

In 1971 the size of the total livestock was 12.2 million units of 500-kg live weight, corresponding to 191 million tons of excrements (feces plus urine) per year; 83 percent from cattle, 10 percent from pigs, and 4 percent from poultry. Farms with insufficient manure land had 0.1 percent of the cattle, 3 percent of the pigs, and 34 percent of the poultry. Their share of the excrements was 1.7 percent of 3 million tons. In the waste management program of the federal government (1975), this quantity of livestock excrement is considered an important amount which justifies efforts to control and to promote its utilization on cropland needing the manure.

A direct comparison with the development and situation in the United States is difficult. The aim of the basic policy is the same in both countries: health and welfare will benefit by the control and prevention of water pollution. But many differences, especially in climate and in type and size of production units, must be understood.

The establishment in the U.S. of the Environmental Protection Agency (1970) and the Federal Water Pollution Control Act (1972) are important milestones. The Act defined concentrated livestock- and poultry-growing operations (feedlots) as "point" sources of "industrial" pollution and required that discharge permits be issued for these operations. A size limit was set and only the larger feedlots are required to apply for a permit (Table 2).

Table 2. Concentrated animal feeding operations have to apply for a NPDES permit, *i.e.*, if more than the following number of animals are involved.

1,000	Feeder cattle
700	Dairy cattle
35,000	Feeder pigs
12,000	Sheep
55,000	Turkeys
180,000	Laying hens
290,000	Broiler chickens

The positive effect of this regulation can be seen in the beginning of control and in gradually excluding a large portion of the wastes which might otherwise reach the streams. The limitations of discharge are technology-based with a "no-discharge" level as a final aim wherever technically and economically feasible, predominantly as part of a crop production cycle (4).

Wastewater, Sewage Sludge, Waste Compost

In Germany the law of waste disposal (1972) distinguishes clearly between livestock wastes and municipal and industrial wastes. While livestock wastes are regarded as manure if the ratio of available land and manure livestock units is reasonable, the other group of wastes remains in the control of the law even if utilized for manuring purposes in crop production on land.

The law promotes waste disposal upon land or, more exactly, waste reuse as a resource of valuable constituents for soil improvement and crop growth. This approach is becoming increasingly necessary. However, the very great number of waste varieties requires programs of public information and education concerning both the problems and the benefits of biologically based recycling.

The order for reuse of municipal and industrial wastes in crop production on land, which is in development according to the law of waste disposal, §15, will provide for decreasing the potential of health hazards as well as the potential of environmental pollution. All wastes which do not fit the aim of soil improvement and crop nutrition because of insufficient content of valuable components are excluded from further discussion. Hazardous wastes and wastes which will be classified by the NATO group of experts under the chairmanship of the United States, subproject "Recommended Management Procedures," are not included. Thus a long list of industrial wastes which are unsuitable for reuse in crop production will be annexed to the order.

Industrial wastes, in general, are problematic wastes with regard to reuse on land for crop production because of: toxic or hazardous constituents; low content of crop nutrients and soil-improving ingredients; and unbalanced ratio of major and trace nutrient elements.

Value and suitability of industrial wastes for reuse in crop production should be determined by individual examinations. The mixing of municipal wastes, that is, household wastes and commercial wastes, together with industrial wastes does not always end up as upgrading. It requires an intensification of the control which raises the treatment costs.

A new bill of wastewater taxes is being discussed vehemently and may be passed this year (1976). It may include an improvement of the effluent quality and perhaps better control of undesired constituents in the sludge from industrial processes.

Concerning solid wastes, there is no progress to be seen to reduce the same undesired constituents.

Municipal wastes which have been declared suitable to be used in crop production contain a reasonable amount of organic matter and crop nutrients, as well as inert substances with neutral behavior. Of concern for the future, there can be ingredients present with disadvantageous or even toxic effects to the crops or the environment if applied in adequate amounts. This is a different situation for a crop grower who is familiar with the use of animal manure and its content of N, P, K, trace elements and organic matter. He would like to continue to adjust the application rate of the waste materials according to the nutrient content in order to get high yields of good quality. He will be unsuccessful if these wastes contain sufficient disadvantageous ingredients to destroy the crops or to prevent a good growth.

As a consequence, municipal wastes are suitable to be used in crop production only if the content of major nutritive elements is high, the content of trace nutritive elements is low, and the content of toxic elements is still lower.

This scheme of utilization must be controlled. The composition of wastes varies considerably. Control of the waste composition by continuous chemical analyses is not feasible. A restriction is necessary for a random sample survey, control of the waste source, and regular soil testing in order to prevent the concentration of disadvantageous elements in the soil from exceeding the tolerable level. The concentration of elements in the soil and proposed tolerable levels are noted in Table 3.

Table 3.   Tolerable amounts of some elements in soils with regard to their plant compatibility.

	Range	Total Amounts (ppm) Most Frequent	Tolerable (Proposal)
Be[a]	0.1 –10	1 –5	10
B	2 –100	5 –30	100
F	10 –500	50 –250	500
Cr	1 –100	10 –50	100
Ni	1 –100	10 –50	100
Co	1 –50	1 –10	50
Cu	2 –100	5 –20	100
Zn	10 –300	10 –50	300
As	1 –50	2 –20	50
Se	0.1 –10	1 –5	10
Mo	0.2 –10	1 –5	10
Cd[a]	0.01–1	0.1–1	5
Hg[a]	0.01–1	0.1–1	5
Pb	0.1 –10	0.1–5	100

[a]Special reservation.

A simple proposal for this level is under discussion. This proposal was developed without profound knowledge of the behavior of predominantly toxic elements and compounds in the soil and in crop nutrition. Thorough study of this behavior will require a good while. In the meantime, a number of good proposals can be used.

CROP GROWTH AND INPUT–OUTPUT RELATIONS

The waste disposal act (1972) alternative to landfill and effluent discharge into waterways is spreading upon cropland for soil improvement and crop production. No other variety of land disposal is taken into consideration. Hence we may infer that the waste applied on cropland should be adjusted to the quality and quantity requirements of the crop.

Wastewaters destined to provide nutrients and irrigation for the production of plant food and the restoration of greenery require pretreatment in order to provide reasonably high levels of public health protection as well as high levels of crop growth-promoting and soil-improving constituents. Removal of BOD, COD and solids prior to land application could be unnecessary (4). If we distinguish between water disposal and water reuse, we also acknowledge different designs of land treatment systems: the goal of maximum disposal may be mutually exclusive with the goal of maximum reutilization (5).

An example is given by Stevens (6) : at Seabrook Farms, New Jersey, a food processing plant must handle 45,000 m^3 wastewater per day on 80 ha (20.5 m per year). Because of fortuitous soil and climatic conditions this can be done by spraying the daily output directly onto 34 ha of land (132 mm per application). This procedure does not achieve the maximum irrigation potential, but it does accomplish the design goal of disposing of 45,000 m^3 without polluting the local watercourses.

On the other hand, the municipal facility at Muskegon, Michigan, is designed to handle 163,000 m^3 per day. To produce some economic benefit for the county it serves, 2,500 ha of relatively infertile land were selected for spray irrigation (2.4 m per year) to stimulate crop production and agriculture.

Another example can be given with a still further decreased load: the Sewage Utilization Association of Braunschweig, Germany, handles about 33,000 m^3 per day by spray irrigation upon about 4,000 ha of cropland, an area comparable in size to Muskegon. However, the applied wastewater of 350 mm per year is only one-seventh. It corresponds with the average water deficit in the main growing period from spring to fall at this site (7). A new thorough survey recommends this well-balanced biologically based wastewater recycling system as a model worthy of imitation (8).

Animal manure varieties, sewage sludges, and waste composts are used as sources of nutrient elements and as humus sources. Recommended rates are based on experience and research. They do not match the needs of crops, because only little information on the annual rate of biological decay of the organic materials is available for use and no prediction can be given about the availability of the organic-bound nutrients with certainty.

Recommended application rates are based mostly on the nitrogen content and on the estimated available portion. Figure 2 demonstrates the differences that can exist between the applied nitrogen and the nitrogen uptake by the crop. While about 80 percent of the nitrogen applied as NH_4NO_3 was taken up by oats

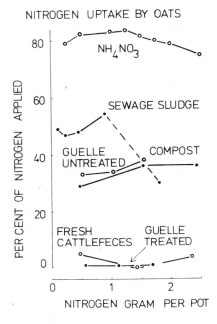

Figure 2. Nitrogen uptake by oats as percent of nitrogen applied with increasing doses of NH_4NO_3 and five organic manure varieties.

in pot experiments without loss by leaching, the same portion was 30 to 50 percent with pumpable sewage sludge, untreated cattle guelle, and compost prepared of droppings from cage hens and peat. Nitrogen in fresh cattle feces and in cattle guelle treated by aeration was nearly unavailable in the first period of growth. Perhaps, if no greater losses occur, there might be an after-effect with the next crops.

Pratt (9) developed computer programs to calculate a number of manure decay series in combination with various rates and times which demonstrate the nitrogen-accumulating effect of annual applications of manure varieties with different availability coefficients.

We may transfer this example to other nutrients, i.e., to phosphorus with a still lower availability, or even to heavy metals whether trace nutrients or predominantly toxic elements. Every accumulation of environmental concern must be controlled, but its importance can be reduced by increasing the availability or by narrowing the input-output ratio.

This can be contrary to the advice to decrease the availability, i.e., by raising the pH value, by applying lime. This procedure is effective but not useful and applicable everywhere. The best approach is to match the nutrient supply to the need of the crops

as closely as possible in order to produce maximum yields with
highest rates of uptake.

Figure 3 illustrates the yield increase as a result of available

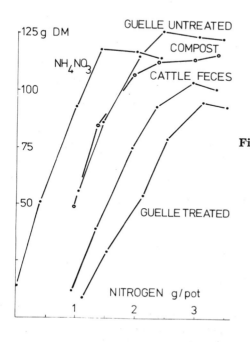

Figure 3. Yields of oats, dry
matter, in pot
experiments by
increasing amounts of
nitrogen applied with
four manure varieties
(first marks = lowest
yield) and
supplementary
NH_4NO_3 applications.

nitrogen supplements from different manures. Even the yields
resulting from fresh cattle feces and treated cattle guelle are
raised from a very low level to a more satisfactory level by additional nitrogen application. This effect on the yield of crop mass
can be transmitted to the input-output ratio of nitrogen, shown
in Figure 4.

These results with supplementary nitrogen applications were
obtained by a rather well balanced supply with all essential nutrients. The ratio of nutrients in sewage sludge is, in general,
not well balanced. There is usually a deficiency of potassium
(Table 4). This means a growth-hindering effect if the stock
of available potassium in the soil is insufficient.

Figure 5 demonstrates that increasing doses of sludge raised
the yield if supplemental potassium was applied. In the same
experiment with no change other than adding nitrogen, the importance of a balanced potassium application became evident
and resulted in much higher yields.

Another illustration of the need to balance the ratio of nu-

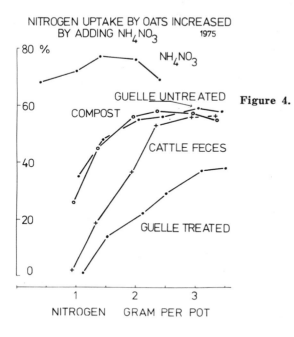

NITROGEN UPTAKE BY OATS INCREASED
BY ADDING NH_4NO_3 1975

Figure 4. Nitrogen uptake by oats as percent of nitrogen applied with four organic manure varieties (first marks = lowest values) and supplementary NH_4NO_3 applications.

Table 4. Ratio of nutrients in the applied fertilizer (stable manure + commercial fertilizer) compared to sewage sludge.

	Fertilizer	*Sewage Sludge*
N : P : K	3.3 : 1 : 3.3	3.0 : 1 : 0.3

trients is shown by an experiment with municipal compost over 15 years (Table 5). Two varieties of compost of the same origin and the same compost plant were applied: a rotted compost after a storage period in windrows over 6 to 9 months, and a fresh compost just after the fermentation period in the composting cell. A yield increase of 23 to 26 percent was brought by compost on the lowest yield level which was affected by lack of phosphorus and lack of available nitrogen. With both these nutrients added as quick-acting fertilizers, the yield level was raised by 70 percent, reducing the calculable relative amount of yield increase by compost to 4 percent.

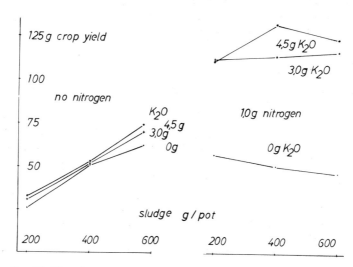

Figure 5. Yields of oats, dry matter, in pot experiments by increasing amounts of sewage sludge (200, 400 and 600 g of pumpable sewage sludge per pot with 6 kg soil). Effect of three supplementary doses of potash and one dose of nitrogen.

Table 5. Yield increase by municipal compost—average of 1959 to 1973 potatoes, rye, oats, dry matter (five compost applications total, 480 t–ha).

			No Compost (dt/ha yr)	Rotted Compost[a]	Fresh Compost[b]
				(relative values)	
No	P	fertilizer:			
No	N	fertilizer	37.9 = 100	123	126
With	N	fertilizer	59.7 = 100	106	109
With	P	fertilizer:			
No	N	fertilizer	42.5 = 100	114	118
With	N	fertilizer	61.7 = 100	104	104

[a]Rotted compost 6 to 9 months.

[b]Fresh compost 1 week.

It is evident that matching the nutrient supply with the requirements of the crop plants by balancing the ratio of nutrients is the key to maximum yields with the narrowest input-output ratio.

SUMMARY

To return wastes into the natural cycle of transformations by land application results in final disposal. However, it is important to distinguish between disposal and reuse. To produce crops, land application of wastes must accommodate the natural conditions of the site and the qualities of the wastes. Standards for waste utilization in crop production should provide reasonably high levels of public health protection as well as high levels of crop growth-promoting and soil-improving constituents.

Livestock excrements are well known for their manuring effect. Treatment of the excrements and techniques of application are decisive for they yield efficient nutrient portions. The annual application rate should not exceed the "normal rate" which is defined by the ratio of three "manure livestock units" per ha. Nitrogen and phosphorus in the excrements are the base of comparison.

The manuring and crop growth-promoting effect of municipal wastes depends on a balanced ratio of nutrients. Additional fertilizer applications may be necessary to balance the ratio of major nutrient elements and of trace elements according to the requirements of the crops in order to achieve maximum yields in crop mass with highest uptake of nutrients and a narrow input-output ratio.

Excessive amounts of trace nutrient elements and other elements including "heavy metals" are disadvantageous to crops and to the food chain. Wastes are suitable to be used in crop production by land application only if the content of major nutritive elements is high, the content of trace nutritive elements is low, and the content of predominantly toxic elements is still lower. This scheme of gradation must be controlled. The composition of wastes varies considerably. Control of waste composition by continuous chemical analyses is not feasible. Restriction is necessary for a random sample survey, regular waste source control, and regular soil testing in order to prevent the concentration of disadvantageous elements in the soil from exceeding the tolerable level.

REFERENCES

1. Colonna, R. A. "Decision-makers guide in solid waste management," U.S. EPA, Washington, D.C. (1976).

2. Hartman, W. J. "An evaluation of land treatment of municipal wastewater and physical siting of facility installations," U.S. Department of the Army, Washington, D.C. (1974).
3. Vetter, H. and A. Klasink. "Einfluss starker Wirtschaftsdüngergaben auf Boden, Wasser und Pflanzen," *Landwirtsch. Forsch.* 28(3) : 249–268 (1975).
4. Loehr, R. C. "Animal waste management with nutrient control," in: K. S. Porter, Ed., *Nitrogen and Phosphorus*, Ann Arbor Science Publishers, Ann Arbor, Michigan (1975), pp. 216–269.
5. Pound, C. E., R. W. Crites and R. E. Thomas. "Wastewater treatment and reuse by land application," U.S. EPA, Washington, D.C. (1973).
6. Stevens, R. M. "Green land—clean streams," Temple University, Philadelphia, Pennsylvania (1972).
7. Tietjen, C. "The agricultural reuse of municipal wastewater in Braunschweig, Germany," Conference on Wastewater Renovation and Reuse, Rockefeller Foundation (July 16-21, 1975).
8. Bath, S. and R. Rudolph. Gutachten zur Abwasserverregnung Braunschweig, Gesellschaft für Landeskultur Bremen (1976).
9. Pratt, P. F., F. E. Broadbent and J. P. Martin. "Using organic wastes as nitrogen fertilizers," *Calif. Agric.* 27(6) : 10–13 (1973).

Additional References

Haen, S. de. Einfluss von Klarschlamm als Düngemittel oder Bodenverbesserungsmittel auf die chemische Zusammensetzung von Gewachsen. Jah. reshauptversammlung Berlin, Verband Deutscher Landwirtschaftlicher Untersuchungs—und Forschungsanstalten, Darmstadt (1974).
Loehr, R. C. and J. D. Denit. "Effluent regulations for animal feedlots," Int. Seminar on Animal Wastes, Bratislavia, Czechoslovakia (September 28–October 5, 1975).
Walker, J. M. "Sewage sludges—management aspects for land application," *Compost Science* 16(2): 12–21 (1975).
Gesetz über die Beseitigung von Abfallen. 7. Bundesgesetzblatt I S. 873 (June 1972).
Gesetz zur Ordnung des Wasserhaushalts. 27. Bundesgesetzblatt I S. 1110 [July 1957 (1976)].
Verursacherprinzip. Umweltprogramm der Bundesregierung (1971).

6

SIGNIFICANCE OF SOIL
CHARACTERISTICS TO WASTES ON LAND

G. W. Olson
Department of Agronomy, Cornell University, Ithaca, New York

INTRODUCTION

This discussion outlines the soil characteristics that are significant to the application of wastes to land, gives criteria by which soils are rated for their limitations for accepting wastes, and introduces the concept of soil potential for matching designs of land management systems and costs with soils for waste applications in the future. Viewpoints of workers in the Cooperative Soil Survey are emphasized. The Cooperative Soil Survey is a joint effort of the U.S. Department of Agriculture, the Agricultural Experiment Stations, and other agencies in each state in the U.S. to map, describe, and classify soils, and interpret the resultant information to assist in improving all the uses of all the soils in the country. Soils are considered to be the most important resource in the country in the long run; inventory and use of that resource will determine to a large extent the survival and progress of the nation. At present, about 1,000 soil scientists are working full-time doing fieldwork in the U.S. to map more than 10,000 described soil series into more than 100,000 map units in the various landscapes. About half of the area in the more intensively used regions in the country has been mapped in detail according to modern standards, and the publication rate for soil survey reports is rapidly accelerating. This discussion is designed for users and potential users of soil maps and descriptions who might be interested in adaptation of the information to improve the application of wastes on land.

SOIL DESCRIPTIONS

Soils are discrete, describable, geographic bodies produced by interactions of climate, vegetation and fauna changing surficial geologic materials in geomorphic landscapes. Soil series are designated by names (1) which indicate specific sets of mappable characteristics and ranges of characteristics. Each soil series is unique and mappable in the field and has characteristics associated with certain geomorphic positions in local landscapes. Figure 1 is an illustration of a soil pit prepared for making a

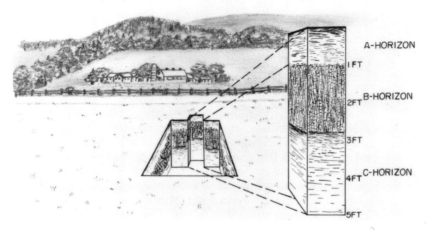

Figure 1.   Soil pit prepared for a profile description in a landscape position characteristic of the soil series. To define a series, many pits are dug and soils are described in survey areas to depths of 5 ft or more. When each soil is identified and characterized from examinations of many pits, the soils are then mapped with many auger corings and shovel-and-spade excavations.

profile description of that soil in that landscape position. In making the description, the three-dimensional (pedon) attributes of each soil are also recorded. Each soil is then classified according to its properties into a natural classification system (2). Since 1960 (3) all of the soils currently being mapped in the U.S. have been classified into the natural system. Each soil has unique properties which enable it to be identified and mapped; each soil is also unique for use and management of the areas it occupies in the landscape. The natural classification enables convenient groupings to be made to stratify the many separate soils into sets or categories that can be managed similarly for different purposes. Groupings, for example, segregate different soils into classes with similar drainage or wetness conditions, similar particle size classes, and similar mineralogy. Groupings like these have implications for development and maintenance of waste application systems and other engineering uses of soils.

Description of a soil in a field (Figure 1) involves a comparison of the properties of that soil with established descriptive standards that have been set up for describing all soils (4, 5). Both internal and external properties are described. The internal characteristics of soils include genetic horizon (layer) designations, depth and thickness of each horizon or layer, color patterns (dry, moist, wet), texture, structure, consistence (dry, moist, wet), reaction (pH), boundary distinctness and boundary surface topography, rocks and coarse fragments, concretions, organic matter, roots, aggregate (ped) coatings, soil cracking and shrink-swell behavior, temperature, water movement, cementation, compaction, leached and precipitated materials, reduced and oxidized microenvironments, iron and manganese accumulations, claypan and hardpan conditions, and other significant and special features. The external characteristics described by soil scientists include slope, landform, relief, runoff, land use, vegetation, erosion, stoniness, geology, ground water, elevation, aspect, climate and weather conditions, relation of each soil to the landscape positions of the other associated soils, and all other observable soil and land use conditions and correlations. The descriptions made in the field constitute the most valuable and important data and information about soils; at the same time that the descriptions are made, however, samples of selected soils are also collected for laboratory analyses and characterizations (6) which include analyses for particle size, bulk density, shrink-swell, water retention and release, organic matter and carbon and nitrogen, phosphorus, cations, anions, sesquioxides, acidity and alkalinity, cation exchange capacity, base saturation, clay mineralogy, and other routine and special analyses. The laboratory data are becoming increasingly important for prediction of the behavior of the soils when subjected to applications of wastes and other engineering manipulations. The laboratory data, however, are always supplementary and accessory to the data collected from the soils in place in the natural environment under field conditions.

Descriptions of soil characteristics made in the field (Figure 1) are particularly valuable because predictions about behavior under waste management systems and other land uses can be made from them. Soil properties observed, described and measured in the field are "real"; characteristics or performances predicted from those real properties are "inferred." Thus, soil characteristics like color, texture and slope are "real" properties; characteristics like wetness, high shrink-swell and erosion hazard are "inferred" properties influencing waste management techniques and other land uses. For example, if upper horizons have mottles of different colors in certain soils in humid environments, then those soils will have water tables (free water in those zones) close to the surface in the wet seasons. Clayey soils with a high content of montmorillonite will have a tendency to shrink and swell with drying and wetting as the moisture status changes; these clayey soils will also have slow permeability to wastewater effluents and other liquids. Soils on steep slopes will have greater hazards of erosion, landslides, and seepage than soils on nearly level slopes. Although water tables, soil shrinking and swelling, and soil erosion and landslides ("inferred" soil properties) may not be observed in the field at the moment they occur in the soils, the "real" soil properties influencing the "inferred" behavior can be readily described and mapped. The principle of the transfer of "real" data from observations to "inferred" performance also enables data collected at pits (Figure 1) or points of measurement to be extended to predictions of behavior to all areas where those soils are mapped—with an acceptable degree of reliability which can be expressed statistically (7). These principles of landscape description and prediction of future behavior under specified treatments are the bases for the potential usefulness of soil information for waste management systems.

Principles of soil description and mapping can best be illustrated by example. Figure 2 illustrates typical landscape of Honeoye-Lima-Kendaia-Lyons and other soils of central New York State (8). During the soil survey fieldwork in Cayuga County, many pits and auger holes were dug in these soils to describe them and delineate their boundaries. About 300 acres can be mapped in detail on an aerial photograph by a soil scientist in a day of good hard fieldwork; fewer acres can be mapped as the scale and intensity of mapping increases. Maximum use is made of aerial photographic interpretation and other remote sensing techniques, but most of the effort in mapping goes into fieldwork for which there is no substitute in the office or in the laboratory. Honeoye, Lima, Kendaia and Lyons soils all have formed in calcareous glacial till, but they are vastly different in their morphology and in their patterns of soil water states during the year. Tables 1, 2 and 3 are profile descriptions of the Honeoye, Kendaia and Lyons soils illustrating the format of the

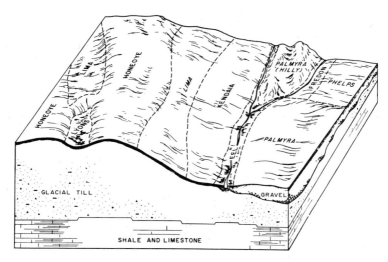

Figure 2. Soil landscape with different soils segregated into their respective landscape positions by boundaries. The lines separate the different kinds of soils that have different characteristics for use and management of the land areas for waste application. This landscape is typical of many areas mapped in Cayuga County and on the Erie-Ontario Plain in New York State [Adapted from Hutton *et al.*, (8)].

descriptions and the morphology of these contrasting soils formed in the same geologic materials. All the soils have silt loam surface horizons of aeolian origin and loam or silt loam glacial till in the deepest layers. The Honeoye, however, has silty clay loam B horizons (Table 1) with clay enrichment from pedogenic processes of weathering and leaching. The Honeoye soil does not have mottles in the profile and is classified as well drained. Although the Honeoye soil is almost always moist, it rarely has free water above 20 in. during the growing season. This moist condition, along with a high nutrient status, is nearly ideal for plant growth and is well suited for many waste application systems. The good potential of Honeoye areas for many land uses can be readily predicted by soil scientists who are familiar with the landscape and internal characteristics of the profile.

Kendaia and Lyons soils, in contrast to the Honeoye (Tables 1–3), have numerous wetness problems which would adversely affect most waste application systems. Many states have regulations about minimum depth to water table for land application of wastes, so that this soil information is particularly valuable.

Table 1. Soil profile description of typical Honeoye (Glossoboric Hapludalf; fine loamy, mixed, mesic) site.

Ap   0 to 10 in., very dark brown (10YR 3/2) silt loam with few angular coarse fragments; moderate medium granular structure; friable consistence; abundant fine
fibrous roots; pH 6.2; clear wavy boundary; 7 to 12 in. thick.

A2   10 to 15 in., brown (10YR 5/3) silt loam with pockets of dark-brown (10YR
3/2) earthworm casts and few coarse fragments of angular, subangular, and
rounded shapes; weak granular in upper part grading into moderate medium to
fine subangular blocky structure in lower part; very friable consistence; common
fine fibrous roots; pH 6.6; clear wavy boundary; 3 to 6 in. thick.

B21t   15 to 18 in., mosaic of dark-brown (10YR 4/3) silty clay loam and brown (10YR
5/3) and pale-brown (10YR 6/3) silty material with few angular coarse fragments;
clayskins; moderate medium subangular blocky structure; friable consistence;
common fine fibrous roots; pH 6.7; clear wavy boundary; 3 to 7 in. thick.

B22t   18 to 24 in., dark-brown (10YR 4/3) silty clay loam with common small pale
silty pockets and strands; clayskins on ped faces; moderate coarse subangular
blocky structure; firm consistence; few fibrous roots; pH 6.8; gradual irregular
boundary; 4 to 15 in. thick.

C   24 to 40+ in., grayish-brown (2.5Y 4/4) gravelly loam with thin brown (10YR
3/3) clayskins and sprinkling of fine silt on upper side of some plates and few
coarse fragments; weak moderate platy structure; firm consistence; pH 7.9;
strongly calcareous.

Mottles indicating perched water tables are distinct at 8 in. in the Kendaia profile, and prominent in the B2g horizon at 17 in. (Table 2). The nature of these Kendaia mottles enables soil scientists to predict that free water will rise above 20 in. in the Kendaia profile for about half of the year, and that part of that wetness period extends into the planting, growing and harvesting seasons in critical early spring and late fall periods. Based on the morphology (Table 2), Kendaia soils are classified by mappers as somewhat poorly drained and are commonly wet in the field. Although their fertility status is high, Kendaia areas have many problems for cropping and waste applications due to their wetness. With engineering structures like water diversions, terraces, and artificial drainage systems, of course, Kendaia sites can be developed or improved for waste disposal. Costs and problems of land management systems, however, are usually of much greater magnitude in the Kendaia areas than in the Honeoye ares.

Lyons soils (Table 3), in contrast to the Honeoye and Kendaia (Tables 1–2), are exceedingly wet. Lyons soils occupy the wettest parts of these landscapes (Figure 2) where water accumulates and stands for long periods. The Lyons profile

Table 2.   Soil profile description of typical Kendaia (Aeric Haplaquept; fine loamy, mixed, nonacid, mesic) site.

Ap   0 to 8 in., very dark grayish-brown (10YR 3/2) fine-textured silt loam; moderate fine and medium crumb structure; very friable consistence; many fine roots; pH 5.8; abrupt smooth boundary; 6 to 9 in. thick.

A2g   8 to 17 in., pale-brown (10YR 6/3) silt loam with light grayish-brown (10YR 6/2) ped coatings and common medium distinct yellowish-brown (10YR 5/4) and grayish-brown (2.5Y 5/2) mottles; moderate fine and medium subangular blocky structure breaking into weak medium and coarse plates; very thin discontinuous clayskins on peds; friable consistence; many fine roots; pH 5.6; gradual wavy boundary; 5 to 9 in. thick.

B2g   17 to 24 in., dark grayish-brown (10YR 4/2) silty clay loam with very dark grayish-brown (10YR 3/2) ped faces and many fine distinct to prominent yellowish-brown (10YR 5/4–5/8) and light olive-gray (5Y 6/2) mottles; thick continuous clayskins that are pale brown (10YR 6/3) in upper part and very dark grayish-brown (10YR 3/2) in lower part; strong medium and coarse angular and subangular blocky structure; firm to very firm consistence; pH 6.6; clear wavy boundary; 6 to 10 in. thick.

Cg   24 to 33+ in., dark grayish-brown (2.5Y 4/2) and grayish-brown (2.5Y 5/2) silt loam; many fine faint light olive-brown (2.5Y 5/4) mottles; gray (10YR 6/1) ped coatings; moderate coarse platy structure; firm consistence; calcareous.

(Table 3) has a thick, dark, mucky surface layer and gray reduced and mottled colors as shallow as 12 in.; the Lyons soil colors and landscape positions inform mappers that free water stands above 20 in. for most of the year in these places—and is near the surface for much of that time. As a result of the wet condition (Table 3), Lyons soils are generally not suitable for many kinds of applications of wastes. Except under special conditions, feasibility of modification of soils at a Lyons site to make it suitable to accept many kinds of wastes is most likely

Table 3.   Soil profile description of typical Lyons (Mollic Haplaquept; fine loamy, mixed, nonacid, mesic) site.

A1   0 to 12 in., very dark gray (10YR 3/1) to black (10YR 2/1) mucky silt loam; weak fine crumb structure; very friable consistence; pH 7.0; clear wavy boundary; 8 to 18 in. thick.

B2g   12 to 23 in., gray (N 5/0) fine-textured silt loam; common medium distinct grayish-brown (2.5Y 5/2) and olive-brown (2.5Y 4/4) mottles; moderate medium and coarse subangular blocky structure; friable to slightly firm consistence; pH 7.2; diffuse smooth boundary; 11 to 13 in. thick.

Cg   23 to 36+ in., olive (5Y 4/4–5/4) silt loam with gray (N 5/0) vertical streaks 18 to 24 in. apart; massive (structureless); firm consistence; calcareous.

to be negative. Lyons areas are suitable for wildlife marshes, ponds or sewage lagoons in some situations, and can be fit into a pattern of harmonious land uses if given good inventory and planning considerations.

The soil descriptions in Tables 1–3 and the landscape diagram in Figure 2 illustrate how soil conditions vary and how soil scientists delineate the external (Figure 2) and internal (Tables 1–3) characteristics into map units. The natural classification of soils, indicated in parentheses in the captions of Tables 1–3, helps to group soils into important broad categories: the first words include indications of conditions of cool (bor) humid (ud) climates, free water (aqu) wetness states, and dark friable (moll) surface horizons. The last words in parentheses in Tables 1–3 indicate the taxonomic categories of particle size (fine loamy), minerology (mixed), reaction (nonacid), and temperature (mesic).

The landscape in Figure 2 includes other soils even more contrasting than those in the Honeoye-Lima-Kendaia-Lyons drainage sequence. Palmyra soils have developed in gravelly glacial outwash; they are well drained, permeable, and generally among the best in New York State for most uses—including waste disposal by sprinkler or spreading systems. Fredon soils are also formed in gravelly glacial outwash, but are somewhat poorly

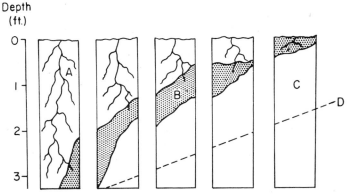

Figure 3. Soil profiles schematically arranged to show drainage classes with relative depths of rooting zone (A—Aerated soil volume), mottled zone (B—Alternating wet and dry conditions), zone with reduced conditions of oxygen deficiencies (C—Generally wet), and relative position of the fluctuating water table (D—Variable from week to week and year to year). These profiles are segregated in landscapes into areas of soil map units. Honeoye, Kendaia, and Lyons soils fit into the left, middle, and right parts of this diagram in accord with their morphological characteristics (Tables 1–3).

drained due to seepage into the low-lying landscape positions. Eel soils are typical of valley bottoms next to streams, where floodwaters create hazards for many land uses. If waste disposal systems are overloaded, then even the excellent Palmyra soils have hazards for ground water pollution. Intense competition for many different uses of Palmyra areas often limits their feasibility for use for waste disposal.

The preceding discussions of the soil profile characteristics in Tables 1–3 indicate how morphological characteristics ("real" properties) are used by soil scientists to predict behavior and performance ("inferred" properties). These field observations are critical for planning and designing waste application systems on land areas. Figure 3 summarizes the soil drainage classes or wetness states that are mapped in humid regions. To verify the predicted behavior patterns illustrated in Figure 3, numerous studies have been undertaken to precisely measure the relationship of soil morphology to inferred performance. Numerous holes were dug into Honeoye, Kendaia and Lyons sites and water level fluctuations were measured each week for several years (9). Figures 4–7 summarize some of these observa-

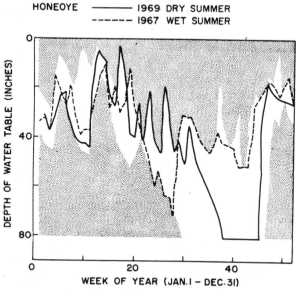

Figure 4.   Soil water fluctuations measured weekly for several years in a Honeoye site [Adapted from Fritton and Olson, (9)]. The clear area in the figure indicates the highest and lowest level recorded for each week. Data are plotted for a specific wet year and a specific dry year.

tions. Honeoye soils (Table 1; Figure 4), although classified as well drained, do have free water shallower than 20 in. in the soil profile—but water does not remain above that level for more than a few days in the wet winter and hardly ever rises to that level during the growing season. Kendaia soils (Table 2; Figure 5) have free water above 20 in. for about half of the year; these wet conditions are particularly critical for some cropping and waste application systems in the early spring and late fall. Data from two Lyons sites (Table 3; Figures 6 and 7) are illustrated;

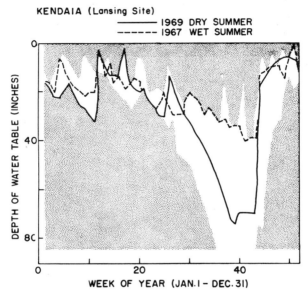

Figure 5.   Soil water fluctuations measured weekly for several years in a Kendaia site [Adapted from Fritton and Olson, (9)]. The clear area in the figure indicates the highest and lowest level recorded for each week. Data are plotted for a specific wet year and a specific dry year.

both sites had free water near the surface above 20 in. for most of the year—but the water level dropped below 40 in. at one site (Figure 7) in some years. The variation in the patterns of water fluctuations at the two Lyons sites are different (Figures 6 and 7), but not significantly different enough to justify their separation into two soil series for mapping.

Figure 8 provides an example of an actual soil map including the landscape idealized in Figure 2. This area about three miles northeast of Auburn (in the central part of New York State)

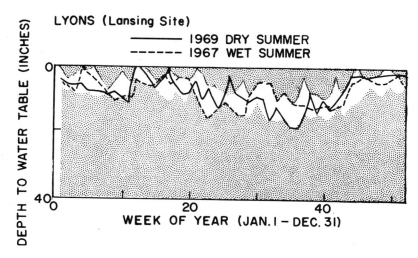

Figure 6. Soil water fluctuations measured weekly for several years in a Lyons site [Adapted from Fritton and Olson, (9)]. The clear area in the figure indicates the highest and lowest level recorded for each week. Data are plotted for a specific wet year and a specific dry year.

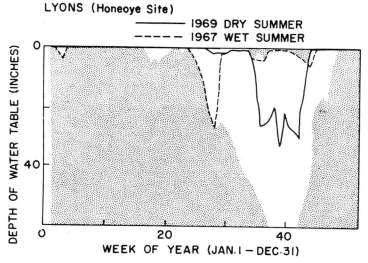

Figure 7. Soil water fluctuations measured weekly for several years in another Lyons site [Adapted from Fritton and Olson, (9)]. The clear area in the figure indicates the highest and lowest level recorded for each week. Data are plotted for a specific wet year and a specific dry year.

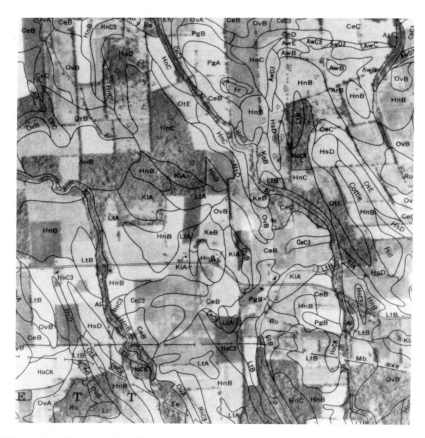

Figure 8. Portion of soil map sheet 42 from soil survey of Cayuga County [Adapted from Hutton *et al*, (8)]. This soil map area gives the actual nature of the distribution of the soils illustrated in Figure 2. Map units are labeled Hn (Honeoye), Lt (Lima), Ke (Kendaia), Kl (Kendaia and Lyons, mapped together), Pg (Palmyra), Pv (Phelps), Fr (Fredon), and Ee (Eel) on different slopes (A-E). Numerous soils other than those in Figure 2 were also mapped in this area. The part of the soil map illustrated is about a mile and a half on a side; the top of the map is north. This area is about three miles northeast of Auburn, New York.

has a complex variety of soils developed in glacial till, outwash and recent alluvium. In addition the area has some soils shallow to bedrock, and some soils developed in lacustrine (glacial lake) sediments. All of the complex properties of the soils and their interactions have many implications for applications of wastes to the land. Some of the soil areas shown in Figure 8 are only a few hundred feet across. Not only do the soils at each site affect the behavior of potential waste management systems, but soils upslope and downslope from each site also would need to be evaluated for possible runoff and attenuation capabilities. The soil map of each landscape provides a means by which waste application systems can be designed to fit the unique capabilities of the land at each treatment site.

WASTE MANAGEMENT CONSIDERATIONS

Soils are commonly grouped into land area units which can be managed similarly for a given use; these groupings are termed "technical groupings" (10) and are designed for rating areas for such waste disposal uses as septic tank seepage fields, sewage lagoons, sanitary landfill (trench and area types), and daily cover for landfills. On the basis of the soil characteristics, groupings are made of the best soils (those with slight limitations) and the worst soils (those with severe limitations) for each use; all other soils with neither slight nor severe overall limitations are assigned a rating of "moderate." These ratings are designed to indicate the magnitude of the hazards involved in using the soils, and emphasize the problems to be overcome in disposing of wastes in the different areas. The ratings also indicate relative costs and feasibility of waste application. Slight limitations indicate that the soils have properties favorable for the intended use; moderate limitations indicate moderate problems that can be overcome by special planning, design or maintenance; severe limitations can only be overcome with costly engineering designs which may make project feasibility questionable (11). Limitation ratings are based strictly on soil characteristics, but those real characteristics are related to measured and inferred performances of waste application systems in those soils.

At the present time, all of the soils in the United States are being rated for their limitations for waste disposal in septic tanks, sewage lagoons, and sanitary landfills. All of the ratings are being computerized, and the characteristics causing those limitations are also tabulated. The front page of the computer input form is reproduced in Table 4. The front page of the form gives a brief description, data, and ratings for sanitary facilities, community development, source material, and water management for each soil. The back side of the form (not illustrated here) gives ratings of the soils for recreation, farming, for-

Table 4. Front part of computer input form used to give ratings of each soil for sanitary facilities and other uses.

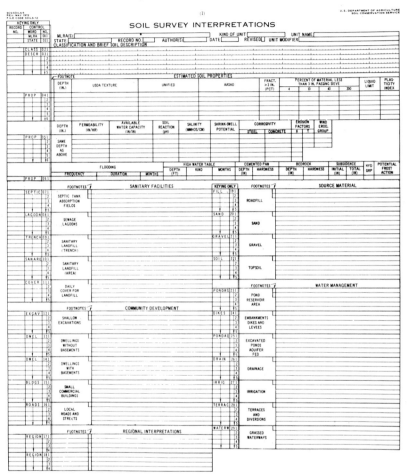

estry, wildlife and rangelands. The intent of the computer form is to give a first approximation rating of all soils for all common uses, and a listing of the major soil characteristics responsible for these ratings. As waste management is integrated into the planning process in the future, it will become more important to consider ratings also for alternative land uses. Thus, when wastes are applied to forest and farm lands, limitations and potentials of the soil must be considered for production as well as for disposal.

Tables 5–7 illustrate the criteria for ratings of limitations of soils for septic tanks, sewage lagoons, and sanitary landfills. For more than a decade, soil scientists in the U. S. have been consulting with public health officials, engineers, geologists and others to develop these guidelines to assist in using detailed soil maps for waste disposal. The single most limiting soil characteristic causes the soil to be placed in the most limiting class; thus, these categories are quite restrictive. Obviously some soil characteristics can be improved much more easily than others. These ratings are uniformly applied to all soils, however, and enable one to determine immediately by computer the best, worst and moderate soils for the various uses. From these limitations the specific hazards and the corrective measures can be determined.

Factors important in determining the criteria for the limitations of soils for seepage fields from septic tanks, listed in Table 5, include local experience and records of performance for existing seepage fields, permeability of the subsoil and substrata, depth to bedrock and impermeable layers, flooding, seasonal fluctuating and permanent water tables, stones and rocks, and slope. Failures of seepage fields are indicated by effluent rising to the soil surface, offensive odors, contamination of ground water and well waters, rank plant growth, and consequent incidence of disease and sickness in the people affected along with the environmental degradation. Seepage fields, of course, can be put on slopes steeper than 15 percent (Table 5), but costs and hazards are much greater than on gentler slopes. Seepage fields can be installed in wet clayey soils with slow permeability (Table 5) without water diversions or drainage, but under such conditions failure can be predicted in a short period of time. Ratings in Table 5 can also be useful to design of other waste application systems where liquids are injected or leached through subsoils in the treatment process.

Aerobic sewage lagoons or waste stabilization ponds (13) are shallow ponds that hold waste liquids for the time necessary for microbial decomposition. Sewage lagoons require considerations of the soils for two functions: as a vessel for the impounded area, and as soil material for the enclosing embankment. Table 6 lists the soil properties important for determining limitations for sewage lagoons, formulated after years of consultations and research (11). Enough soil material that is suitable for the structure must be available, and the completed lagoon must be capable of holding liquid wastes with minimum seepage. Soils containing moderate to high amounts of organic matter are unsuitable for the basin floor even if the floor is underlain by

Table 5.   Ratings of limitations of soils for waste disposal in septic-tank seepage fields (Adapted from Reference 11).

Item Affecting Use	Soil-Limitation Rating		
	Slight	Moderate	Severe
Permeability class[a]	Rapid,[b] moderately rapid, and upper end of moderate	Lower end of moderate	Moderately slow[c] and slow
Hydraulic-conductivity rate (Uhland core method)	>1.0 in./hr[b]	1.0–0.6 in./hr	<0.6 in./hr
Percolation rate [auger hole method; (12)]	Faster than 45 min/in.[b]	45–60 min/in.	Slower than 60 min/in.
Depth to water table[d]	>72 in.	48–72 in.	<48 in.
Flooding	None	Rare	Occasional or frequent
Slope	<8%	8–15%	>15%
Depth to hard rock,[e] bedrock, or other impervious materials	>72 in.	48–72 in.	<48 in.
Stoniness class[f]	0 and 1	2	3, 4 and 5
Rockiness class[f]	0	1	2, 3, 4 and 5

[a]Class limits are the same as those suggested by the Work Planning Conference of the National Cooperative Soil Survey (11). The suitability ratings should be related to the permeability of soil layers at and below the depth of the installed tile lines of the seepage field.

[b]Special considerations should be given to places where pollution is a hazard to water supplies.

[c]In arid or semiarid areas, soils with moderately slow permeability may have a rating of moderate.

[d]In humid areas, soil drainage classes are more useful than these depth classes because of seasonally fluctuating high water tables (9,11). Well drained soils and some moderately well drained soils generally have slight limitations, and poorly drained and very poorly drained soils have severe limitations. Somewhat poorly drained soils and some moderately well drained soils have moderate limitations, but these can be substantially improved for waste disposal by drainage and water diversion installations.

[e]These depth ratings are based on the assumption that tile in the seepage field is at a depth of about 2 ft.

[f]Class definitions are given on pages 216-223 of the Soil Survey Manual (4). Larger numbers indicate more stones and rocks.

suitable soil material. The organic matter promotes growth of aquatic plants which are detrimental to proper functioning of the lagoon. Sometimes liquids are pumped out of lagoons after

Table 6. Ratings of limitations of soils for waste disposal in sewage lagoons (Adapted from Reference 11).

Item Affecting Use	Soil-Limitation Rating		
	Slight	Moderate	Severe
Depth to permanent or fluctuating water table	>60 in.	40–60 in.[a]	<40 in.[a]
Permeability	<0.6 in./hr	0.6–2.0 in./hr	>2.0 in./hr
Depth to bedrock	>60 in.	40–60 in.	<40 in.
Slope	<2%	2–7%	>7%
Coarse fragments <10 in. diameter, % by volume	<20%	20–50%	>50%
Percent of soil surface covered by coarse fragments <10 in. diameter	<3%	3–15%	>15%
Organic matter	<2%	2–15%	>15%
Flooding[b]	None	None	Soils subject to flooding
Unified soil groups[c]	CG, SC, CL and CH	GM, ML, SM and MH	GP, GW, SW, SP, OL, OH and Pt

[a]Depth to water table can be disregarded if the floor of the lagoon is to be in nearly impermeable material at least 2 ft thick.

[b]Flooding can be disregarded if the flood water has low velocity and a depth of less than 5 ft, if it is not likely to enter or damage the lagoon embankments.

[c]The Unified soil-engineering classification system has been outlined for laymen in a publication by Olson (14) available from the Department of Agronomy (Soils) of Cornell University, Ithaca, New York, 14853; soil suitability for embankments is also outlined in the publication. The Unified soil groups are most relevant in Table 6 to suitability of soil materials for embankments retaining liquids in sewage lagoons.

treatment periods and sprayed onto adjacent soils; interpretations of the adjacent soils for spray irrigation can be made from some of the following tables. In time, solids also accumulate in the bottoms of sewage lagoons and must be cleaned out. Several of the following tables also will be helpful in determining the suitability of different soils on which the sludge might be spread.

Sanitary landfill (Table 7) in trenches consists of dug trenches in which refuse is buried and compacted at least daily, and in which the refuse is covered with a layer of soil material at least daily, and in which the refuse is covered with a layer of soil six inches thick. The material used in covering the garbage is the soil excavated in digging the trenches. A final cover of soil material at least two feet thick is placed over the landfill when

Table 7. Ratings of limitations of soils for solid-waste disposal in trench-type sanitary landfills[a] (Adapted from Reference 11).

| Item Affecting Use | Soil-Limitation Rating | | |
	Slight[b]	Moderate[b]	Severe
Depth to seasonal water table	>72 in.	>72 in.	<72 in.
Soil-drainage class	Excessively drained, somewhat excessively drained, well drained, and some[c] moderately well drained soils	Somewhat poorly drained and some[c] moderately well drained soils	Poorly drained and very poorly drained soils
Flooding	None	Rare	Occasional or frequent
Permeability[d]	Slower than 2.0 in./hr	Slower than 2.0 in./hr	Faster than 2.0 in./hr
Slope	<15%	15–25%	>25%
Dominant soil texture[e] to a depth of 60 in.	Sandy loam, loam, silt loam, and sandy clay loam	Silty clay loam,[f] clay loam, sandy clay, and loamy sand	Silty clay, clay, muck, peat, gravel and sand
Depth to hard bedrock	>72 in.	>72 in.	<72 in.
Depth to rippable bedrock	>60 in.	<60 in.	<60 in.
Stoniness class[g]	0 and 1	2	3, 4 and 5
Rockiness class[g]	0	0	1, 2, 3, 4 and 5

[a]Ratings based on soil depth of 5–6 ft commonly investigated in soil surveys. Additional geologic and engineering investigations should be made before landfills are established.

[b]If probability is high that the soil material is similar down to a depth of 10–15 ft, this could be indicated by statements like "Probably slight to a depth of 12 ft" or "Probably moderate to a depth of 12 ft" as determined by the soil survey.

[c]Soil drainage classes are related to depth of fluctuating water tables. The overlap of moderately well drained soils into two limitation classes allows some of the wetter moderately well drained soils to be given a limitation rating of moderate under certain conditions.

[d]These ratings reflect the ability of a soil to retard movements of effluents leached from landfills. In arid and semiarid areas rapid permeability may not be a severe limitation.

[e]Ratings of soil textures reflect ease of digging in soil, ease of moving soil, and trafficability of vehicles on soils in the immediate area of the trench where hard-surfaced roads are absent.

[f]Soils high in montmorillonitic clays probably need to be given a limitation rating of severe.

[g]Stoniness and rockiness class definitions are given in the Soil Survey Manual on pages 216–223 (4). Larger numbers indicate more stones and rocks.

the trenches are full. For cover material (11), soils with very friable and friable consistence are good [classes of consistence and other soil properties can be obtained from the soil profile descriptions (5)]; soils with loose and firm consistence are fair; soils with very firm and extremely firm consistence are poor. Good soil textures for cover material include sandy loam, loam, silt loam, and sandy clay loam; fair textures include silty clay, clay, muck, peat, and sand. Thicker soil profile material, of course, is more desirable if excavation and hauling is to take place. Soils with gentler slopes, without coarse fragments, and with good drainage are better as a source of cover for sanitary landfill than are steep soils with stones or than soils in wet areas. In the area-type sanitary landfills (Table 4), refuse is placed on the soil surface in successive layers and covered. The cover material generally must be imported from other soil areas. Area-type landfills can be put in places where soils are shallow to bedrock, but soil and geologic materials under these proposed sites should be investigated also so as to determine the probability that leachates from the landfill can penetrate the soil and thereby pollute water supplies. Ratings in Table 7 are applicable to any burial process where large amounts of earth are moved, and typically are adapted and refined to meet local needs and conditions (D.E. Hill, Personal communication).

In addition to the limitation ratings routinely being made for all soils in the United States (Tables 5–7), criteria are also being tested for ratings of soils for application of biodegradable solids and liquids (15). Table 8 outlines the soil properties most important for application of nontoxic solids and liquids to soils. The best soils for these applications have moderate permeability, good drainage, slow runoff, high water-holding capacity, and no flooding hazards. Soils to be avoided for waste applications are those with very rapid or very slow permeability, excessively good or very poor drainage, rapid runoff, frequent flooding, and low water-holding capacity for vegetation growth. Flood hazards are more serious for solid than liquid wastes, and a large water-holding capacity of the soil for crop growth is not as critical in arid regions where leaching losses are minimal. Ratings in Table 8 are generalized, and need to be refined to fit each local situation where specific solids or liquids are spread on soils. Special high-rate infiltration systems are best suited to highly permeable soils, and overland flow systems are adaptable to slowly permeable soils within narrow slope ranges. Success of the application of solid and liquid wastes to soil surfaces or to shallow depths is very much dependent also upon management factors. Where suitable soils are carefully prepared and where good crop

Table 8.   Ratings of limitations of soils for application of biodegradable solids and liquids (Adapted from Reference 15).

Item Affecting Use	Soil-Limitation Rating		
	Slight	Moderate	Severe
Permeability of the most restricting layer above 60 in.[a]	0.6–6.0 in./hr	6–20 and 0.2–0.6 in./hr	>20 and <0.2 in./hr
Soil drainage class[b]	Well drained and moderately well drained	Somewhat excessively drained and somewhat poorly drained	Excessively drained, poorly drained, and very poorly drained
Runoff[c]	Ponded, very slow, and slow	Medium	Rapid and very rapid
Flooding	None	None for solids; only during non-growing season allowable for liquids	Flooded during growing season (liquids) or anytime (solids)
Available water capacity from 0–60 in. or to a root-limiting layer	>8 in. (humid regions) >3 in. (arid regions)	3–8 in. (humid regions) Moderate class not used in arid regions	<3 in. (humid regions) <3 in. (arid regions)

[a]Moderate and severe limitations do not apply for soils with permeability <0.6 in./hr: (1) for solid wastes unless the waste is plowed or injected into the layers having this permeability or evapotranspiration is less than water added by precipitation and irrigation, and (2) for liquid wastes if layers having that permeability are below the rooting depth and evapotranspiration exceeds water added by precipitation and irrigation.

[b]Class definitions are given on pages 169–172 of the Soil Survey Manual (4) and illustrated in Figure 3 for humid regions.

[c]Class definitions are given on pages 166–167 of the Soil Survey Manual (4).

cover is maintained, heavy waste applications probably can be handled with resultant good crop production and low environmental hazards. Where soils with severe limitations are inadequately managed, pollution and degradation of the environment are almost certain to occur.

All of the soil criteria for limitation ratings are approximations, based on the best information that is available at the present state-of-the-art of the land application technology. Much more information is needed, especially those kinds of data which can be used to link "real" soil properties with "inferred" soil and land performances. Monitoring data are needed, for example, from waters and soils downslope from land disposal sites. Much more data are needed to link ground water and surface water pollution and geologic formations and aquifers to specific

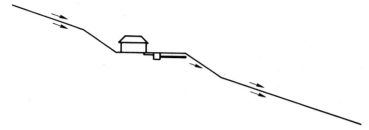

Figure 9. Schematic diagram of soil runoff and seepage affecting a septic tank seepage field site and soils downslope.

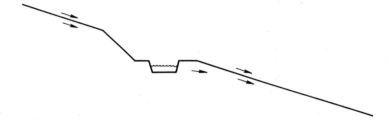

Figure 10. Schematic diagram of soil runoff and seepage affecting a sewage lagoon site and soils downslope.

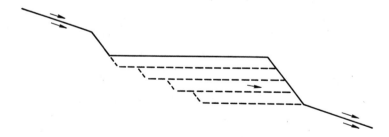

Figure 11. Schematic diagram of soil runoff and seepage affecting a sanitary landfill site and soils downslope.

soil map units. Figures 9, 10 and 11, for example, schematically illustrate how surface and seepage waters from higher-lying soils affect waste disposal sites for septic tank seepage fields, sewage lagoons, and sanitary landfills, and how runoff and seepage from the disposal sites in turn affect waters and soils at lower elevations. Almost all of these kinds of land disposal techniques are causing some off-site pollution from almost all sites in New York State and other places. Table 9 is an illustration of soil analyses of samples collected from uncontaminated and contaminated

Table 9. Analyses of surface soil samples collected from contaminated and uncontaminated points around a septic tank seepage field near Ithaca, New York, in Hudson soil material [Analyses by Cornell Soil Test Laboratory; Greweling and Peech, (16)].

Sample Number and Identification	Organic Matter (%)	pH	Exch H (me/100 g)	lb/acre Plow Layer (2,000,000 lb)										Soluble Salts $Kx10^3$
				P	K	Mg	Ca	Mn	Fe	Al	NO_3-N	NH_3-N	Zn	
1 (Contaminated, fill near sand filter outlet)	3.0	7.9	2	5	130	1,200	22,500	144	16	30	20	1	2	34
2 (Contaminated, fill over sand filter)	1.8	7.8	1	3	230	1,175	21,250	183	8	20	25	1	2	24
3 (Contaminated, fill over septic tank)	0.7	8.0	<1	2	110	1,300	28,750	165	13	45	25	<1	1	16
4 (Uncontaminated, natural soil unaffected by disposal site)	3.4	6.7	6	2	80	290	3,250	13	2	20	10	2	1	<5

soils around a septic tank seepage field near Ithaca. Samples 1, 2 and 3 collected downslope and over the filter field and septic tank have a considerable increase of soluble salts, nitrogen, and other materials above the levels of those items in the uncontaminated sample 4. Similarly, practically all sewage lagoons have seepage spots downslope where effluents or soluble materials are coming to the surface. Recent investigations of sanitary landfill sites in New York State indicate that most of these sites, also, have seepage leachate breakouts downslope which are polluting the environment; these are particularly dangerous in industrial disposal areas where heavy metals and other dangerous chemicals have been discarded. The elimination and improvement of the problems lie in a better understanding of the soil environment in which each waste application site is located, and better management of soils upslope and downslope to prevent runoff and seepage pollutions of the environment.

In the future, precision and accuracy of soil survey interpretations will be improved through refinements in the measurement and classification systems, increases in scale and intensity of mapping, and through improvements in the application of data. Concepts of soil limitations are being supplemented and replaced by concepts of soil potentials, which specify recommendations and alternatives and costs to make soils suitable as waste application sites. Table 10 is an example of such a refinement in the ratings, where five classes are used instead of three classes and where one poor soil characteristic does not necessarily force a specific soil to be placed in the most limiting category—especially if the poor soil property can be corrected relatively easily by engineering designs. Table 10 is quite specific to the environment in New York State, and is not intended for use in other areas without adaptations. Critical factors like elevation (not given on most soil maps) must be also considered because of influences of temperature and growing season on waste utilization for crop growth. Engineering designs and costs associated with the classes in Table 10 are given in Table 11; these are approximate relative costs which engineers can use also to determine alternative treatments for soils upslope and downslope from a disposal site. Thus drains, ditches, and terraces can be used upslope to cut off surface and seepage waters entering a waste disposal site (Figures 9–11); these same engineering works can be used in the disposal area itself to distribute wastes and manage the effluents; and downslope the same measures can be installed to control or prevent seepage waters from polluting the environment below the site. This kind of information, keyed to detailed soil map units and descriptions, will be increasingly

Table 10. Ratings of soil potentials for waste applications for agricultural production in New York State (Adapted from Reference 17).

Item Affecting Use	Soil Potential				
	Very Good	Good	Moderate	Poor	Very Poor
Drainage class and approximate depth in inches to permanent or fluctuating water table	Well drained >36	Moderately well drained 18–36	Somewhat poorly drained 12–18	Poorly drained 6–12	Very poorly drained <6
Total water-holding capacity [in H_2O/ rooting depth; (18)]	>6	4–6	3–4	2–3	<2
Slope (%)	<3	3–8	8–15	15–25	>25
Rooting depth (in. to root restricting horizon)	>40	30–40	20–30	10–20	<10
Trafficability [Unified soil group; (14)]	GW, GP, SW, SP, GM, GC, SM, SC, Pt (drained)	CL with PI <15	ML, CL with PI >15	OH, OL, MH	CH, Pt (undrained)
Permeability class (in./hr in least permeable horizon)	0.6–2.0	2.0–6.0	>6.0	0.6–0.06	<0.06
Erosion	None to slightly eroded	Slightly eroded	Moderately eroded	Severely eroded	Very severely eroded
Stoniness and rockiness (4)	0	1	2	3	4 and 5
pH in B horizon	>7.0	6.5–7.0	6.0–6.5	5.5–6.0	<5.5
Texture	sil	1, sicl	sl, cl	scl, c	s, ls (not irrigated)
Elevation (ft)	<400	400–1,000	1,000– 1,600	1,600– 2,000	>2,000

used as more emphasis is put on waste treatment and utilization instead of mere disposal.

Trends for future uses of soil information to improve application of wastes on land are becoming well documented in the literature. Increasingly, soils are being characterized more specifically for their capacity to assimilate and renovate specific wastes (19). Designs of waste disposal systems are being in-

Table 11. Average linear feet per acre and costs[a] for conservation and land improvement practices for waste application for agricultural production in soil potential classes of areas in New York State (Adapted from Reference 17).

Conservation Practice	Soil Potential				
	Very Good	Good	Moderate	Poor	Very Poor
Tile drains	0	100($100)	250($250)	450($450)	>600(>$600)
Diversion ditches	0	60($60)	120($120)	>120(>$120)	>120(>$120)
Terraces	0	500($500)	1,000($1,000)	Not recommended	Not recommended

[a]The average cost was calculated at $1.00 per linear foot for tile drains, diversion ditches, and terraces (terraces are not recommended for slopes greater than 12%). These estimates assume that the major limitations of poorer classes are permeability and wetness for tile drains, erosion and slope for diversion ditches, and runoff hazards for terraces.

creasingly evaluated so that "real" soil properties can be statistically related to predictable "inferred" soil performances (20, 21). Numerous workers are starting to study soils that have been subjected to waste applications for prolonged periods (22). Specific waste management designs are being developed and suggested for specific soil map units (23) and guidelines (24, 25) and regulations are being implemented in many places. With increasing levels of sophistication (26), it will be possible to develop systems of waste application to land that will maximize the recycling of the resources and protect the environment at the site as well as upslope and downslope.

SUMMARY

This discussion has concentrated on the more widespread and common waste disposal systems affected by soil properties, but other less conventional system designs, and even the designs of those systems to be developed in the future for more exotic waste can benefit from the use of soil maps, descriptions, and data when the wastes are applied to land. The potential of soils is great for renovation of waste materials; the soil profile itself is a manifestation of the recycling of natural vegetative and animal wastes through the nutrient release patterns in the soil to the current and future plant growth cycles. As resource recycling becomes more important, the soils and their properties will be increasingly considered in relation to their role in the recovery of the reusable resources from wastes applied to land.

ACKNOWLEDGMENTS

The author appreciates the many contributions of his colleagues to numerous aspects of this writing, and especially their offerings to the general philosophy and specific criteria of the soil ratings. In particular, valuable comments and reviews of this manuscript were provided by R.B. Dean and R.E. Thomas of the U.S. Environmental Protection Agency, J.E. Witty and K.K. Young of the Soil Conservation Service of the U.S. Department of Agriculture, W.J. Bauer of W.J. Bauer Consulting Engineers, D.E. Hill of the Connecticut Agricultural Experiment Station, and G.A. Garrigan, S.D. Klausner, R.C. Loehr, D.F. Smith, and R.R. Hall of Cornell University.

REFERENCES

1. Soil Survey Staff. *Soil Series of the United States, Puerto Rico, and the Virgin Islands: Their Taxonomic Classification*, Soil Conservation Service, U.S. Department of Agriculture, U.S. Government Printing Office, Washington, D.C., ∼ 400 pp. (1972).
2. Soil Survey Staff. *Soil Taxonomy: A Basic System of Soil Classification for Making and Interpreting Soil Surveys*, Soil Conservation Service, U.S. Department of Agriculture, U.S. Government Printing Office, Washington, D.C., ∼ 400 pp. (1973 preliminary abridged page proof).
3. Soil Survey Staff. *Soil Classification: A comprehensive system—The Seventh Approximation*. Soil Conservation Service, U.S. Department of Agriculture, U.S. Government Printing Office, Washington, D.C., 265 pp. (1960).
4. Soil Survey Staff. *Soil Survey Manual*, Handbook 18, U.S. Department of Agriculture, U.S. Government Printing Office, Washington, D.C. 503 pp. 1951).
5. Olson, G. W. "Criteria for making and interpreting a soil profile description: A compilation of the official USDA procedure and nomenclature for describing soils," Bulletin 212, Kansas Geological Survey, University of Kansas, Lawrence, ∼ 50 pp. (in press).
6. Soil Survey Staff. *Soil Survey Laboratory Methods and Procedures for Collecting Soil Samples*, Soil Survey Investigation Report 1, Soil Conservation Service, U.S. Department of Agriculture, U.S. Government Printing Office, Washington, D.C., 50 pp. (1967).
7. Beckett, P. H. T. and R. Webster. "Soil variability: A review," *Soils and Fertilizers* 34:1–15 (1971).
8. Hutton, F. Z. *et al.* "Soil survey of Cayuga County, New York," Soil Conservation Service, U.S. Department of Agriculture and Cornell University Agricultural Experiment Station, U.S. Government Printing Office, Washington, D.C. 208 pages of soil descriptions and interpretations and 93 soil map sheets at 1:15, 840 scale and general soil maps at 1:31, 680 scale (1971).
9. Fritton, D. D. and G. W. Olson. "Depth to the apparent water table in 17 New York soils from 1963 to 1970," New York's Food and Life Sciences Bulletin 13, Cornell University Agricultural Experiment Station, Ithaca, New York, 40 pp. (1972).
10. Orvedal, A. C. and M. J. Edwards. "General principles of technical grouping of soils," *Soil Sci. Soc. Am. Proc.* 6:386–391 (1941).

11. Soil Survey Staff. *Guide for Interpreting Engineering Uses of Soils*, Soil Conservation Service, U.S. Department of Agriculture, U.S. Government Printing Office, Washington, D.C., 87 pp. (1971).

12. Olson, G. W. "Application of soil survey to problems of health, sanitation, and engineering," Memoir 387, Cornell University Agricultural Experiment Station, Ithaca, New York, 77 pp. (1964).

13. Zall, R. R. "Lagoons: A treatment for milking center waste," Information Bulletin 52, New York State College of Agriculture and Life Sciences, Cornell University, Ithaca, New York, 12 pp. (1973).

14. Olson, G. W. "Soil survey interpretation for engineering purposes," Soils Bulletin 19, Food and Agriculture Organization of the United Nations, Rome, Italy, 24 pp. (1973).

15. Witty, J. E. and K. W. Flach. "Site selection as related to land and soil properties," presented at conference on "Soil for Management and Utilization of Organic Wastes and Wastewaters" at Muscle Shoals, Alabama 11–13 March 1975. Proceedings of symposium to be published by American Society of Agronomy, Madison, Wisconsin, 30 pp. in press).

16. Greweling, T. and M. Peech. "Chemical soil tests," Bulletin 960, New York State College of Agriculture, Cornell University, Ithaca, New York, 59 pp. (1965 revised).

17. Rogoff, M. J. "Soil ratings for farming in Central New York State," M.S. thesis, Department of Agronomy, Cornell University, Ithaca, New York, 195 pp. (1976).

18. Arnold, R. W. "Selected properties of New York soils with emphasis on rooting depth and available water-holding capacity," Agronomy Mimeo 75–10, Cornell University, Ithaca, New York, 23 pp. (1975).

19. Hill, D. E. "Wastewater renovation in Connecticut soils," *J. Env. Quality* 1:163–167 (1972).

20. Hill, D. E. and C. R. Frink. "Longevity of septic systems in Connecticut soils," Bulletin 747, Connecticut Agricultural Experiment Station, New Haven, 22 pp. (1974).

21. Huddleston, J. H. and G. W. Olson. Soil survey interpretation for subsurface sewage disposal," *Soil Sci.* 104:401–409 (1967).

22. Mielke, L. N. *et al.* "Soil profile conditions of cattle feedlots," *J. Env. Quality* 3:14–17 (1974).

23. Bouma, J. "New concepts in soil survey interpretations for on-site disposal of septic tank effluent," *Soil Sci. Soc. Am. Proc.* 38:941–946 (1974).

24. Keeney, D. R. *et al.* "Guidelines for the application of wastewater sludge to agricultural land in Wisconsin," Technical Bulletin 88, Department of Natural Resources, Madison, 36 pp. (1975).

25. Olson, G. W. "Using soils of Kansas for waste disposal," Bulletin 208, Kansas Geological Survey, University of Kansas, Lawrence, 51 pp. (1974).

26. Powers, W. L. *et al.* "Formulas for applying organic wastes to land," *J. Soil and Water Conservation* 30:286–298 (1975).

SECTION II

TRANSFORMATIONS OF NITROGEN, PHOSPHORUS AND HEAVY METALS

7

THE QUANTITATIVE DESCRIPTION OF TRANSFER OF WATER AND CHEMICALS THROUGH SOILS

M. Ungs, R. W. Cleary,* L. Boersma, and S. Yingjajaval
Department of Soil Science, Oregon State University,
    Corvallis, Oregon
*Department of Civil Engineering, Princeton University,
    Princeton, New Jersey

INTRODUCTION

An understanding of the quantitative aspects of the transfer of water, solutes and heat through soils is important for many reasons. The interest of the agricultural scientist in the development of his capability to describe transfer processes in soils was initially prompted by the need to drain soils which are too wet and irrigate soils which are too dry.

This interest has received renewed emphasis by recent concerns about the impact of agricultural practices on water quality. Agricultural development has had a great impact on the environment. Obvious examples are the removal of forests and plowing of virgin lands. The basic goal of agriculture is to manage the environment for maximum rate of production of food and fiber. Agriculture has been successful in this effort, but not without cost or sacrifice. Agricultural practices have resulted in increased erosion and have added to the natural burden of nutrient elements in streams. The use of agricultural chemicals—fertilizers, herbicides and pesticides—is one of the harmful practices cited. Some claim that the contribution of plant nutrients, particularly nitrogen, to ground water and/or surface waters is large and harmful, while others maintain that this contribution is small when measured in relation to natural processes.

While many misunderstandings about these matters can be cleared up, it is evident that agricultural scientists have not developed information or theories which allow proper quantification of the ecological impact of farming practices on the environment. If such information or predictive capabilities were more fully developed, an evaluation of the consequences and development of alternative strategies would be possible.

Further interest in prediction of transfer of water and solutes through soils was prompted by the growing desire to use land for the disposal of a wide variety of waste products ranging from household waste to hazardous chemicals. There is an urgent need to predict the long-term consequences of these practices with respect to the quality of soil and ground water.

Soil components include solid, liquid, gaseous and biological constituents. These synchronize to provide a substrate for food production, a reservoir for water, a foundation on which to live, and when used properly a filter for waste products. The soil changes constantly, with many of the changes as vital parts of the cycles on which mankind exists. A healthy soil accepts and tolerates the intruded material, but abuse of the soil system through excess loading or addition of incompatible materials may result in markedly reduced efficiency of natural soil processes.

A soil becomes polluted when its physical, chemical and biological properties are changed so that it loses part or all of its ability to serve the functions it was fit to serve in its natural state. Pollution of the soil can have profound consequences. Civilizations have declined because of it. There is thus a critical need to accurately predict consequences of all land use alternatives with respect to their potential for irreversibly polluting the soil. The basis of such predictions is the ability to describe transfer processes in the soil adequately. This report summarizes methods currently in use, or being developed for prediction of water and solute transfer in soils.

THE PROBLEM

This presentation will consider the transient, one-dimensional problem of water and solute movement in the vertical direction in the layer of soil important for plant growth. This includes problems of infiltration of rainfall or irrigation water, evaporation, redistribution of water in the soil, and uptake of water by plant roots.

The distribution of ions in soils is determined by (a) the sources and sinks, (b) the form of the exchange isotherm, (c) the exchange equilibria, (d) diffusion and dispersion, (e) convection and (f) storage and release by plant material. The part played by each of these factors is shown in Figure 1. The pores

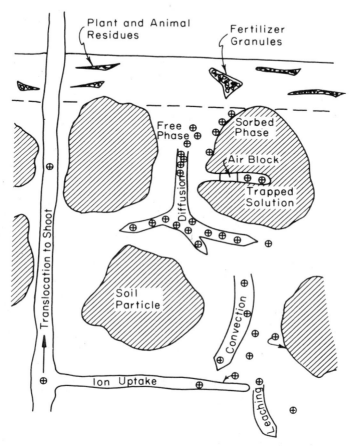

Figure 1. Schematic diagram of the distribution mechanisms for ions in soils.

are assumed to be completely filled with water with a high concentration of ions near the surface. An equilibrium exists between ions in water (free phase) and ions retained on exchange sites of soil particles (sorbed phase). As a result of concentration gradients in the pore water, diffusion occurs. Some of the ions at the moving diffusion front will be removed from the free phase by adsorption to soil particles, thus maintaining a concentration gradient. Movement of ions as a result of mass flow of water also occurs. It is accompanied by sorption at the moving front and desorption at the tail. When mass flow occurs, diffusion is usually less important. Ions which are not removed from the root zone with water passing through it are available as plant food. This sink is no longer present when vegetation is removed and the potential then exists for nutrient loss from the soil to the ground water.

The Sources of Ions

The ions present in the soil solution are obtained from many sources. The natural soil system releases ions to the soil solution as a result of the dissolving action of soil water. The rate at which ions go into solution depends on, among other variables, the pH, the temperature of the water and the rate of removal of the ions by the leaching process. In a dynamic system with plants present, ions also become available from the decaying material at the soil surface and roots below the soil surface.

The natural system can be disturbed by activities of man. Among the disturbances are application of fertilizers to vegetation, removal of trees and burning. When crops are removed much organic material is left at the soil surface. The disturbance with respect to the ion balance involves first of all the removal of the sink. Ions which were removed by plant growth are now available for leaching. The addition of an ion source at the soil surface also occurs in the form of the decaying organic material.

A different situation arises when soil is used for disposal of waste materials. These can be in the form of solid waste or liquid waste. In this case the loading of the soil with chemicals is greatly increased and the ability of the soil to absorb the materials is not clearly understood. This report presents numerical methods which may be used in the analysis of transfer problems through soils.

The Translocation Processes

The principles of diffusion and dispersion as well as exchange mechanisms between the free and sorbed phases are well understood and have been described in detail (1–10). On a field scale, the problem of ion translocation is, however, very complex and has not been treated in a quantitative manner. The greatest complications are brought about by continuous changes in soil water content. A simple series of events may be recognized. In many areas, soils are at or near freezing temperatures while a layer of snow accumulates on the surface. In early spring, snow melts and water enters the soil to gradually increase the water content. During this time, adsorbed ions go into solution, diffusion to lower soil layers occurs, and ions may become evenly distributed throughout the soil profile.

What happens as the seasons progress depends on amount of water available. In some geographic regions, there is just enough water to saturate the profile without the occurrence of deep seepage. The nutrients remain in solution where they are available for plant growth. Where more water is available it passes through the profile and carries ions with it. The sequence of events described on a seasonal scale also occurs on a shorter time base within a season as during a series of drying and wetting periods resulting from intermittent rainfall.

Statistical versus Theoretical Models

Predicting ion movement in soils is clearly a complex problem. Extensive literature exists concerning the movement of chemicals through soils (6–20). Little use has been made of this information for quantitative predictions. This lack of practical application of existing theory must be attributed to the problem of applying solutions developed for idealized steady-state systems to situations where nonisotropism is important, the boundary conditions are ill defined, and the processes are transient in nature.

Given these circumstances, it is not surprising that the advances which have been made in the development of predictive capabilities have been based on sampling techniques. In this approach, a high volume of data is collected with the hope that upon analysis, results will show sequences which can be correlated with certain physical and chemical properties of the systems (21–27).

For example, a seasonal cycle in the concentration of a certain ion in streams may be observed which is correlated with changes in physical properties of a watershed. One might then change the properties of the watershed—through vegetation removal, for example—and determine how the previously observed stream concentrations change.

Such a procedure has many drawbacks. It is costly and very slow. Several years of data collection are required to calibrate the system. Upon imposing the desired perturbations on the system many additional years of data collection are required and in the end, even upon establishment of certain correlations, it is still difficult to assign fundamental explanations to causal relationships.

The problem may also be approached through numerical analysis. In this case, a mathematical model is developed which is based on a physical description of the process to be studied. Analytical models differ from statistical models. They are based on a statement of physical laws which govern the process to be studied, and may be for steady-state or transient conditions. This report describes a procedure for the numerical analysis of the problem of water and chemical transport in soils.

ANALYSIS

Flow Equations

The general partial differential equation of soil-water transfer is given by Philip (28) as

$$\frac{\partial \theta}{\partial t} = \frac{\partial}{\partial z}(K \frac{\partial \psi}{\partial z}) - \frac{\partial K}{\partial z} \qquad (1)$$

Equation 1 is general in the sense that it applies to both homogeneous and heterogeneous soils and does not depend on any requirement that K, ψ, and θ be single-valued. In a homogeneous soil and if K and ψ are single-valued functions of θ (no hysteresis), one may introduce the water diffusivity term D, which is also a single-valued function of θ, such that

$$D = K \frac{\partial \psi}{\partial \theta} \qquad (1a)$$

Then the transient, isothermal, one-dimensional equation for vertical movement of water with a root sink term can be written as:

$$\frac{\partial \theta}{\partial t} = \frac{\partial}{\partial z} \left\{ D_z(\theta) \frac{\partial \theta}{\partial z} - K_z(\theta) \right\} + A(z,\theta,\psi_{ps}) \qquad (2)$$

where

$$D_z(\theta) = K_z(\theta) \left\{ \frac{\partial \psi}{\partial \theta} \right\} \qquad (3)$$

and θ is the volumetric water content $[L_w^3/L^3]$, ψ is the soil water potential [L], ψ_{ps} is the potential in the plant at the surface of the soil [L], $A(z,\theta,\psi_{ps})$ is the rate of water uptake by roots $[L_w^3/L^3/T]$, $D_z(\theta)$ is the vertical soil water diffusivity $[L^2/T]$, $K_z(\theta)$ is the vertical hydraulic conductivity $[L_w^3/L^2/T]$, t is time [T], and z is the vertical coordinate, positive downward [L].

There is no known analytic solution to Equation 2 because of the nonlinear dependence of D_z and K_z on the soil water content. Typically K_z and D_z will vary through five to six orders of magnitude between the wet and dry ends of the soil water content range.

The case of a zero sink term has been numerically solved by several authors (29–32).

The water content is subject to the constraints that

$$\theta_d \leqslant \theta(z,t) \leqslant \theta_s \tag{4}$$

where the subscripts d and s denote air dry and saturated conditions. For computational purposes, Feddes *et al.* (33) determined that in the absence of field data one could define θ_d as that water content which is at equilibrium with the atmosphere,

$$\theta_d = \theta(\psi_d) \tag{5}$$

such that the potential

$$\psi_d = \frac{RT}{Mg} \ln(RH) \tag{6}$$

where R is the universal gas constant [ergs/mole/°K], T is the absolute temperature [°K], M is the molecular weight of water [gr/mole], g is the acceleration due to gravity [cm/sec^2], and RH is the relative humidity of the air above the soil surface given as a ratio.

Equation 2 is usually solved as a function of water content rather than potential, to reduce the effect of hysteresis in the hydraulic conductivity function (34), and to reduce the variation of conductivity during infiltration of dry soils. Ignoring hysteresis, Equation 2 holds for homogeneous soil without ponding at the soil surface or a water table at lower depths, the so-called "semi-infinite soil profile." Under these conditions, a unique relationship exists between the soil water potential and the water content (29), namely

$$\frac{d\theta}{dt} = \frac{d\psi}{dt} \left\{ \frac{\partial\theta}{\partial\psi} \right\} \tag{7}$$

where $\{\partial\theta/\partial\psi\}$ is the "specific water capacity."

If the diffusivity and conductivity functions are known, then upon substitution of Equation 3 into Equation 7 and rearranging, one can numerically evaluate the potential in terms of known analytic functions and the water content

$$\frac{d\psi}{dt} = \frac{d\theta}{dt} \left\{ \frac{D_z(\theta)}{K_z(\theta)} \right\} \tag{8}$$

Obviously, if the potential is already given in analytic or tabular form, then Equation 8 is not needed.

To solve Equation 2, the following initial and boundary conditions are specified.

Initial Condition

$$\theta(z,0) = \theta_0(z), \quad t = 0, 0 \leqslant z \leqslant z_{max} \tag{9}$$

where z_{max} is the depth of the lower boundary and subscript 0 denotes the initial condition. Hence, the initial water content of the soil is only a function of depth and is obtained from field data. The initial soil water content $\theta_0(z)$ need not be uniform.

Boundary Conditions

a) At the lower boundary of the soil column, $z = z_{max}$, a second or Neumann-type boundary condition specifies that there is no change in the water content gradient

$$\frac{\partial \theta}{\partial z}(z_{max}, t) = 0, \quad t \geqslant o \tag{10}$$

b) At the upper boundary, $z = 0$, a general flux condition is specified

$$-D_z \left\{ \frac{\partial \theta}{\partial z} \right\} + K_z = F(0,t) \tag{11}$$

where $F(0,t)$ equals the potential surface flux $[L_w^3/L^2/T]$, after accounting for irrigation, rainfall and evaporation rates. The three most common cases considered for $F(0,t)$ are characterized by:

$$
\begin{array}{lll}
F(0,t) > 0 & \text{infiltration} & \\
F(0,t) = 0 & \text{redistribution or drainage} & \\
F(0,t) < 0 & \text{evaporation} & (12)
\end{array}
$$

It should be pointed out that potential rate of evaporation, $F(0,t)$, for a given soil depends only on atmospheric conditions and is specified as a function of time. The actual flux across the surface, $q(0,t)$, is determined by the ability of the porous medium to transmit water from below. Thus the exact boundary condition to be assigned at the soil surface can not be predicted *a priori* but is subject to the condition that

$$|q(0,t)| \leqslant |F(0,t)| \tag{13}$$

Solute Displacement Equations

The displacement of solutes by mass flow of water is described by (31, 32)

$$\frac{\partial \{\theta c\}}{\partial t} = \frac{\partial}{\partial z} \left\{ \theta D_{sz}(q_z, \theta) \frac{\partial c}{\partial z} \right\} - \frac{\partial \{q_z c\}}{\partial z} - \frac{\partial S}{\partial t} \pm G(z,t,\theta,c) \qquad (14)$$

where c is the solute concentration in solution $[M/L^3_w]$, S is the local solute concentration in the adsorbed phase $[M/L^3]$, D_{sz} is the apparent solute dispersion coefficient $[L^2/T]$, G is the solute source (+)/sink (−) term $[M/L^3/T]$, θ is the volumetric water content $[L^3_w/L^3]$, q_z is the volumetric flux $[L^3_w/L^2/T]$, t is the time [T].

The first term on the right-hand side of Equation 14 describes the flux due to the diffusion-dispersion processes. The apparent solute dispersion coefficient, D_{sz}, includes the effect of hydraulic dispersion and molecular diffusion. The second term represents the flux due to convective processes.

The equation as stated allows for diffusion to occur when the mass flow rate of water (convection) is zero. It also includes an expression for the sorption or desorption of the solute onto the soil particles and a solute source or sink term, which can describe the addition of solute to the soil in the form of fertilizers, slow decay of organic matter, addition of hazardous chemicals, or slow leaks from storage tanks for hazardous chemicals. The term can describe the uptake of solute by plant roots or the biological degradation of a chemical.

Initial Condition

$$c(z,0) = c_0(z), \quad t = 0, \quad 0 \leqslant z \leqslant z_{max} \qquad (15)$$

Upper Boundary

Several boundary conditions can be chosen. For the boundary at the soil surface, the following was used:

$$\left\{ -\theta D_{sz} \frac{\partial c}{\partial z} + q_z c \right\}\Big|_{0^+} = \left. f(t) \right|_{0^-} \qquad (16)$$

where f(t) is the solute flux at the soil surface $[M/L^2/T]$,

and for infiltration $\quad f(t) = \{ q_z(0^-,t)c(0^-,t)\}$

for redistribution $\quad f(t) = 0; \quad q_z(0^+,t) = 0$

for evaporation $\quad f(t) = 0; \quad q_z(0^+,t) < 0 \qquad (17)$

Lower Boundary

For the boundary at the lower surface

$$\frac{\partial c}{\partial z}(z_{max}, t) = 0 \tag{18}$$

Water Uptake by Roots

A model is developed to describe the extraction of water by plant roots. The leaves evaporate water in response to the evaporative demand of the atmosphere, called the potential transpiration. The water lost by the leaves is replaced by water absorbed by the roots and transferred along the plant stem. When the rate of water supply by the roots to the leaves does not equal the evaporative demand the leaves respond by decreasing the plant water potential which results in stomatal closure. The model used here is based on a "plant surface potential" which is the water potential of the plant tissue at the soil surface. Other authors (33, 35, 36) have used similar models, but several modifications are proposed here. The root model used in this report is based on the following concepts:

a) The actual transpiration demand, T_r, is some fraction of the potential transpiration demand, q_{ps}. This fraction is determined by the stomatal correction factor, c_{st}, which is a known function of the plant water potential at the soil surface. Then the actual transpiration demand is described as

$$T_r = q_{ps} \cdot c_{st}(\psi_{ps}), \quad [L_w^3/L^2/T] \tag{19}$$

The potential transpiration depends only on climatic conditions and is given as a function of time.

b) The plant surface potential, ψ_{ps}, is unknown and changes at each time step in response to the potential demand and soil supply conditions. A value of ψ_{ps} is "hunted" for until the plant root extraction over the soil column occupied by roots equals the actual transpiration demand, T_r. The plant surface potential ψ_{ps} is bounded by the condition that

$$\psi_{wilt} \leqslant \psi_{ps} \leqslant 0, \quad [L] \tag{20}$$

c) The water potential of the root, ψ_r, is equal to the sum of the plant surface potential, the distance from the soil surface, and RPF, which is the friction loss incurred by the water flowing along the root xylem to the soil surface

$$\psi_r(z, \theta, \psi_{ps}) = \psi_{ps} + z + RPF(z, \theta), \quad [L] \tag{21}$$

d) The flow of water from the soil to the root is proportional to the difference between the average soil water potential, ψ, and the potential of the root

$$(\psi_r - \psi) \tag{21a}$$

where ψ is computed from Equation 8 or tabulated for a given value of θ.

e) The hydraulic conductivity through the soil-root system, $K_{sys}(\theta)$, $[L_w^3/L^2/T]$ is proportional to the conductivities of the soil $K_z(\theta)$ and the root cortex $K_r(\theta)$ as follows

$$1/K_{sys}(\theta) = 1/K_z(\theta) + 1/K_r(\theta) \tag{22}$$

where $K_z(\theta)$ and $K_r(\theta)$ are known functions of θ.

f) Water uptake by the root at a given depth is proportional to the root absorption function, RAF. The root absorption function is determined by root density, R_d, and root growth factor, R_g, as controlled by the local soil water potential. It is also a function of root activity, R_a, which determines the actual parts of the root that actively take the water from the soil. Therefore, root absorption function

$$RAF(z,\theta,t) = R_d \cdot R_g \cdot R_a, \quad [1/L^2] \tag{22a}$$

g) The flux of water from each soil layer to the roots in that layer is given as

$$A(z,\theta,\psi_{ps}) = K_{sys}(\theta) \cdot RAF(z,\theta,t) \cdot \{\psi_r(z,\theta,\psi_{ps}) - \psi\} \tag{23}$$

or as

$$A = K_{sys} \cdot RAF \cdot (\psi_{ps}+z+RPF-\psi), \quad [L_w^3/L^3/T] \tag{24}$$

h) The actual transpiration demand, T_r, from step a) must equal the sum of the amounts of water withdrawn from all soil layer sinks

$$T_r = \sum_{i=2}^{n-1} A_i \frac{(z_{i+1} - z_{i-1})}{2}, \quad [L_w^3/L^2/T] \tag{25}$$

The sum is from $i = 2,\ldots,n - 1$ since it is assumed there is no root uptake at the top and bottom surfaces.

NUMERICAL SOLUTIONS

Water Flow

The water flow equation is solved by employing the time-centered finite difference scheme of Crank-Nicholson (37). In finite difference form, Equation 2 is evaluated about the interior *ith* space node and the k + ½ time level such that

$$\frac{\theta_i^{k+1} - \theta_i^k}{\Delta t^{k+1}} = \left[\frac{D_z(*\theta_{i+1/2}^{k+1/2})\{\theta_{i+1}^{k+1} + \theta_{i+1}^k - \theta_i^{k+1} - \theta_i^k\}}{2\Delta z_2} \right.$$

$$\left. - \frac{D_z(*\theta_{i-1/2}^{k+1/2})\{\theta_i^{k+1} + \theta_i^k - \theta_{i-1}^{k+1} - \theta_{i-1}^k\}}{2\Delta z_1} \right] \frac{1}{\Delta z_3}$$

$$- \frac{K_z(*\theta_{i+1/2}^{k+1/2}) - K_z(*\theta_{i-1/2}^{k+1/2})}{\Delta z_3} + A(z_i, *\theta_i^{k+1/2}, \psi_{ps}^{k+1/2})$$

$$i = 2, ..., n\text{-}1 \tag{26}$$

where n is the maximum number of space nodes and where the source $(+)$/sink$(-)$ term, $A(z, \theta, \psi_{ps})$, is defined by Equation 24. Subscript "i" refers to the *ith* vertical distance node, (*i.e.*, i = 1 refers to the surface, z = 0) and superscript "k" refers to the current time level t^k.

The finite difference solution of Equation 26 allows for both variable time and for variable space node locations, where the space step size is defined by

$$\Delta z_1 = z_i - z_{i-1}$$

$$\Delta z_2 = z_{i+1} - z_i$$

$$\Delta z_3 = (z_{i+1} - z_{i-1})/2 \tag{27}$$

and the time step size is defined by

$$\Delta t^k = t^k - t^{k-1}$$

$$\Delta t^{k+1} = t^{k+1} - t^k \tag{28}$$

There are a total of n space nodes but Equation 26 is evaluated only in the interior of the soil column, from i = 2, ..., n − 1 at time level k + ½. The upper boundary condition is evaluated at i = 1 + ½ at time k + 1 and the lower boundary condition is evaluated at i = n–½, and time level k + 1. Hence the n − 2 interior flow equations of Equation 26 and the upper and lower boundary expressions make a total of n equations and n unknowns.

In finite difference form of the centered Crank-Nicholson scheme the boundary and initial conditions of Equations 9, 10 and 11 reduce to:

Initial condition

$$\theta_i^{k=0} = \theta_0(z_i) \tag{29}$$

Lower boundary condition

$$\frac{\{\theta_n^{k+1} - \theta_{n-1}^{k+1}\}}{(z_n - z_{n-1})} = 0 \tag{30}$$

Upper boundary condition

$$\frac{-D_z(*\theta_{l+1/2}^{k+1})\{\theta_2^{k+1} - \theta_1^{k+1}\}}{(z_2 - z_1)} + K_z(*\theta_{l+1/2}^{k+1}) = F(0,t^{k+1}) \tag{31}$$

where $F(0,t^{k+1})$ is the potential surface flux which is known as a function of time.

The superscript "*" indicates that this variable is an estimate. The half-time step can be related to the full-time step by

$$*\theta_{i+1/2}^{k+1/2} = (*\theta_{i+1}^{k+1} + *\theta_i^{k+1} + \theta_{i+1}^k + \theta_i^k)/4$$

$$*\theta_i^{k+1/2} = (*\theta_i^{k+1} + \theta_i^k)/2$$

$$*\theta_{i-1/2}^{k+1/2} = (*\theta_i^{k+1} + *\theta_{i-1}^{k+1} + \theta_i^k + \theta_{i-1}^k)/4$$

$$*\theta_{l+1/2}^{k+1} = (*\theta_2^{k+1} + *\theta_1^{k+1})/2 \tag{32}$$

In order to evaluate K_z, D_z, A, the water content must be estimated by iteration since the diffusivity, conductivity, and root sink coefficients are nonlinear functions of water content, and since one is trying to solve for the water content values.

At the beginning of each new time level, t^{k+1}, the first estimate is made by linear extrapolation from the previous time level (38),

$$*\theta_{i+1}^{k+1} = \theta_{i+1}^k + a(\frac{\Delta t^{k+1}}{\Delta t^k})(\theta_{i+1}^k - \theta_{i+1}^{k-1})$$

$$*\theta_i^{k+1} = \theta_i^k + a(\frac{\Delta t^{k+1}}{\Delta t^k})(\theta_i^k - \theta_i^{k-1})$$

$$*\theta_{i-1}^{k+1} = \theta_{i-1}^k + a(\frac{\Delta t^{k+1}}{\Delta t^k})(\theta_{i-1}^k - \theta_{i-1}^{k-1}) \tag{33}$$

where a = 1, for linear extrapolation and a = 1.4 for accelerated linear extrapolation (29).

Using the initial estimated values of $*\theta_i^{k+1}$ from Equation 33, the coefficients in Equations 26, 30 and 31 can be evaluated and then the resulting tri-diagonal matrix can be solved by the Thomas algorithm (37) to obtain computed values of θ_i^{k+1}. Due to the nonlinear nature of the coefficients, the computed values of θ will not be the same as the estimated values. The solution can be improved by an iterative process. At each additional iteration about time level t^{k+1}, the most recent computed values of θ_i^{k+1} are used to obtain an improved estimate of $*\theta_i^{k+1}$ by setting

$$*\theta_i^{k+1} \text{ (new estimate)} = \{*\theta_i^{k+1} \text{ (old estimate)} + \theta_i^{k+1} \text{ (computed)}\} / 2$$
(34)

The iterative procedure continues until a satisfactory degree of convergence is obtained, usually within ten cycles, depending on the accuracy required. The criterion of convergence is when

$$\max \left| \frac{*\theta_i^{k+1} - \theta_i^{k+1}}{*\theta_i^{k+1}} \right| \leqslant \epsilon_w \qquad i = 1, ..., n \tag{35}$$

where ϵ_w is the maximum allowed percent difference.

The time interval is defined as the length of time required for the maximum flux of water to move in the soil from one node to the next. The defining relationship used here is (29, 31)

$$\Delta t^{k+1} \leqslant .035 \min \left| \frac{z_{i+1} - z_i}{q_{i+1/2}^{k-1/2}} \right| \qquad i = 1, ..., n-1 \tag{36}$$

or more generally

$$\Delta t^{k+1} \leqslant \min \left| \frac{\theta_{i+1/2}^{k-1/2} (z_{i+1} - z_i)}{q_{i+1/2}^{k-1/2}} \right| \qquad i = 1, ..., n-1 \tag{37}$$

where the flux, $q_{i+1/2}^{k-1/2}$, is defined at the previous time level as

$$q_{i+1/2}^{k-1/2} = K_z(\theta_{i+1/2}^{k-1/2}) - \frac{D_z(\theta_{i+1/2}^{k-1/2}) \{\theta_{i+1}^{k-1/2} - \theta_i^{k-1/2}\}}{(z_{i+1} - z_i)} \tag{38}$$

Solute Transfer

The difference equation for solute transfer evaluated for the interior points, assuming D_{sz} is only a function of θ and assuming both the adsorption/desorption term and the source/sink term are zero, then

$$\frac{\theta_i^{k+1} c_i^{k+1} - \theta_i^k c_i^k}{\Delta t^{k+1}} = \left[\theta_{i+1/2}^{k+1/2} D_{sz}(\theta_{i+1/2}^{k+1/2}) \frac{\{c_{i+1}^{k+1} + c_{i+1}^k - c_i^{k+1} - c_i^k\}}{2\Delta z_2} \right.$$

$$\left. - \theta_{i-1/2}^{k+1/2} D_{sz}(\theta_{i-1/2}^{k+1/2}) \frac{\{c_i^{k+1} + c_i^k - c_{i-1}^{k+1} - c_{i-1}^k\}}{2\Delta z_1} \right] \frac{1}{\Delta z_3}$$

$$- \frac{q_{zi+1/2}^{k+1/2} c_{i+1/2}^{k+1/2} - q_{zi-1/2}^{k+1/2} c_{i-1/2}^{k+1/2}}{\Delta z_3}$$

$$i = 2, ..., n\text{-}1 \qquad (39)$$

where

$$c_{i+1/2}^{k+1/2} = (c_{i+1}^{k+1} + c_i^{k+1} + c_{i+1}^k + c_i^k)/4$$

$$c_i^{k+1/2} = (c_i^{k+1} + c_i^k)/2$$

$$c_{i-1/2}^{k+1/2} = (c_i^{k+1} + c_{i-1}^{k+1} + c_i^k + c_{i-1}^k)/4 \qquad (40)$$

In finite difference form of the centered Crank-Nicholson scheme the initial and boundary conditions of Equations 15, 16 and 18 reduce to:

Initial condition

$$c_i^{k=0} = c_0(z_i) \qquad i = 1, ..., n \qquad (41)$$

Lower boundary condition

$$\frac{\{c_n^{k+1} - c_{n-1}^{k+1}\}}{(z_n - z_{n-1})} = 0 \qquad (42)$$

Upper boundary condition

$$-\theta_{1+1/2}^{k+1} D_{sz}(\theta_{1+1/2}^{k+1}) \frac{\{c_2^{k+1} - c_1^{k+1}\}}{(z_2 - z_1)} + (q_z c)_{1+1/2}^{k+1} = f(t^{k+1}) \qquad (43)$$

where $f(t^{k+1})$ is the solute flux at the soil surface and is defined for various surface conditions by Equation (17).

Algorithm of Iteration for Water Flow

In order to evaluate the root sink model as stated by Equation 26 at time level $k + \frac{1}{2}$, the plant surface potential, $\psi_{ps}^{k+1/2}$ and the water content, $\theta_i^{k+1/2}$ must be estimated. However, the plant surface potential and water content depend on the potential transpiration demand, the potential surface flux, the soil potential, and the root distribution. Hence one must go through an iterative algorithm to find the correct values of ψ_{ps} and θ_i for a given time step. Once these have been found, the solute concentration can be calculated by the Thomas algorithm.

The following algorithm is used:

a)   Make an initial estimate for the moisture content, $*\theta_i^{k+1}$, $i = 1, \ldots, n$ at time level $k + 1$, using Equation 33.

b)   Obtain a corresponding initial estimate of the soil potential by solving the finite difference approximation of Equation 8

$$*\psi_i^{k+1} = \psi_i^k + \{*\theta_i^{k+1} - \theta_i^k\} \frac{D_z(*\theta_i^{k+1/2})}{K_z(*\theta_i^{k+1/2})} \tag{44}$$

c)   Make an initial estimate of the plant surface potential by linear extrapolation from the previous time step

$$*\psi_{ps}^{k+1} = \psi_{ps}^k + a(\frac{\Delta t^{k+1}}{\Delta t^k})(\psi_{ps}^k - \psi_{ps}^{k-1}) \tag{45}$$

where "a" is as defined in Equation 33.

d)   Compute the actual transpiration, $*T_r^{k+1}$, based on the estimated plant surface potential and the potential transpiration as described by Equation 19.

e)   Set the actual transpiration demand equal to the sum of the root sinks according to Equation 25 and solve for the plant surface potential. This is the computed value of ψ_{ps},

$$\psi_{ps}^{k+1} = \frac{T_r(*\psi_{ps}^{k+1}) - \sum\limits_{i=2}^{n-1} K_{sys_i} \cdot RAF_i \cdot (z_i + RPF_i - *\psi_i^{k+1})(\frac{z_{i+1} - z_{i-1}}{2})}{\sum\limits_{i=2}^{n-1} K_{sys_i} \cdot RAF_i((\frac{z_{i+1} - z_{i-1}}{2})} \tag{46}$$

where K_{sys_i}, RAF_i, RPF_i are evaluated with values of $*\theta_i^{k+1}$.

f)   Compare the estimated and computed values of the plant surface potential

$$\left| \frac{*\psi_{ps}^{k+1} - \psi_{ps}^{k+1}}{*\psi_{ps}^{k+1}} \right| \leq \epsilon_{ps} \tag{47}$$

where ϵ_{ps} is the maximum allowed convergence error.

g) If a satisfactory degree of convergence is not obtained, make an improved estimate of $*\psi_{ps}^{k+1}$,

$$*\psi_{ps}^{k+1}(\text{new}) = \{*\psi_{ps}^{k+1}(\text{old}) + \psi_{ps}^{k+1}(\text{computed})\}/2 \qquad (48)$$

and return to step d. The iteration process of steps d to g continues for fixed values of $*\theta_i^{k+1}$ until the computed and estimated values of ψ_{ps} agree. If the new estimate of Equation 48 violates the feasibility conditions of Equation 20, then set ψ_{ps} to the limiting value,

$$\text{if } \psi_{ps} < \psi_{wilt}, \quad \text{set } \psi_{ps} = \psi_{wilt} \qquad (49)$$

$$\text{if } \psi_{ps} > 0, \quad \text{set } \psi_{ps} = 0 \qquad (50)$$

h) Evaluate the tri-diagonal matrix coefficients using ψ_{ps}^{k+1}, $*\theta_i^{k+1}$, and $*\psi_i^{k+1}$. Solve the tri-diagonal matrix from Equation 26, 30 and 31 by the Thomas algorithm, and compute θ_i^{k+1}.

i) Compare the estimated and computed values of θ_i^{k+1} by Equation 35. If a satisfactory degree of convergence is obtained on θ_i^{k+1}, the problem is solved for time level k + 1. Proceed to a new time level and restart on step a.

j) If a satisfactory degree of convergence is not obtained for θ_i^{k+1}, make a new estimate of $*\theta_i^{k+1}$ from Equation 34. If the new estimates violate the feasibility conditions of Equation 4, set them to the limiting value,

$$\text{if } *\theta_i^{k+1} < \theta_d, \quad \text{set } *\theta_i^{k+1} = \theta_d$$

$$\text{if } *\theta_i^{k+1} > \theta_s, \quad \text{set } *\theta_i^{k+1} = \theta_s$$

Hence the surface flux or evaporation condition of Equation 13 is always satisfied as long as there is no surface ponding or water table formation.

k) Compute the corresponding new estimates of soil water potential from Equation 44, using the θ values of step j, and return to step d.

The above algorithm consists of two iteration levels. Each time a new estimate is made of the water content at a given time level the program enters an inner iteration routine where it hunts for a value of ψ_{ps} which satisfies the transpiration mass balance condition of Equation 25, for fixed values of $*\theta_i^{k+1}$ and does not violate the wilting condition of Equation 20. Once a feasible value of ψ_{ps}^{k+1} is found, the program returns to the outer iteration routine where it makes a single improved feasible estimate of θ_i^{k+1}, for a fixed

ψ_{ps}^{k+1} and then returns back to the inner routine with the improved values of θ_i^{k+1}.

APPLICATIONS

A computer program was developed to solve the simultaneous solute and moisture flow in unsaturated soil using the finite difference solution and iteration schemes presented here. In order to test the computer programs and the numerical techniques presented in this chapter, computations were compared with data chosen from the literature. These examples are presented to show the wide range of problems that can be handled. The first example considers the simultaneous displacement of solute and water during infiltration and redistribution. The second example considers the problem of infiltration and evaporation from the soil surface. The last example reexamines the first problem, but with the addition of water uptake by roots (transpiration).

Other authors (31, 39) have presented single-valued relationships for unsaturated hydraulic conductivity, K_z, and soil water diffusivity, D_z, as functions of moisture content, θ, of a Panoche clay loam, the soil to be used in these examples.

A soil with the following properties was used:

$$K_z \text{ (cm/hr)} = 0.1958 \text{ x } 10^{-5} \text{ x } e^{35.8\theta}, \quad 0.07 \leqslant \theta \leqslant 0.38$$

$$D_z \text{ (cm}^2/\text{hr)} = e^{97.875\theta - 14.526}, \quad 0.07 \leqslant \theta \leqslant 0.15$$

$$D_z \text{ (cm}^2/\text{hr)} = 0.02625e^{25.3\theta}, \quad 0.15 < \theta \leqslant 0.36$$

$$D_z \text{ (cm}^2/\text{hr)} = e^{189.51\theta - 62.756}, \quad 0.36 < \theta \leqslant 0.37$$

$$D_z \text{ (cm}^2/\text{hr)} = 0.113 \text{ x } 10^4 \text{ x } e^{0.9\theta}, \quad 0.37 < \theta \leqslant 0.38$$

The air dry and saturated water content are given as

$$\theta_d = 0.07 \text{ cm}^3/\text{cm}^3$$

$$\theta_s = 0.38 \text{ cm}^3/\text{cm}^3$$

Infiltration and Redistribution

Using the soil properties of Panoche clay loam, the following boundary conditions (39) are used during infiltration.

Water Content

The equation used here to describe soil water movement is

$$\frac{\partial \theta}{\partial t} = \frac{\partial}{\partial z}\left\{ D_z \frac{\partial \theta}{\partial z} - K_z \right\}$$

Initial condition

$$\theta(z,0) = 0.15 + z \times 8.333 \times 10^{-4} \text{ cm}^3/\text{cm}^3, \quad 0 \leqslant z < 60 \text{ cm}$$

$$\theta(z,0) = 0.2, \quad z > 60 \text{ cm}$$

Upper boundary conditions $(z = 0)$

for infiltration $\theta(0,t) = \theta_s = 0.38 \text{ cm}^3/\text{cm}^3, \quad 0 \leqslant t < 9 \text{ hr}$

for redistribution $-D_z \frac{\partial \theta}{\partial z} + K_z = 0, \quad t > 9 \text{ hr}$

Lower boundary condition $(z = z_{max})$

$$\frac{\partial \theta}{\partial z} = 0, \quad 0 \leqslant t < \infty$$

Solute

The equation describing the solute flow is

$$\frac{\partial(\theta c)}{\partial t} = \frac{\partial}{\partial z}\left\{ \theta D_{sz} \frac{\partial c}{\partial z} \right\} - \frac{\partial(q_z c)}{\partial z}$$

Diffusion coefficient for $CaCl_2$, $D_{sz} = 4.2 \text{ cm}^2/\text{hr}$

Initial condition

$c(z, 0) = 0, t = 0$
$c(z, t) = 209 \text{ meq of } Cl^-/\text{liter}, \ 0 \ll z < 16.7 \text{ cm}$

This condition, taken from Warrick et al. (39), assumes that solute is added at time $t = 2$ hr to the layer $0 \leqslant z < 16.7$ cm in a concentration of 209 meq/liter.

Upper boundary condition $(z = 0)$
  for infiltration

$$c(0^-,t) = 209 \text{ meq/l}, \quad 2 \leqslant t < 2.8 \text{ hr}$$

$$-\theta D_{sz} \frac{\partial c}{\partial z} + q_z c = 0, \quad 2.8 \leqslant t < 9 \text{ hr}$$

  for redistribution

$$\frac{\partial c}{\partial z} = 0, \quad t \geqslant 9 \text{ hr}$$

Lower boundary condition ($z = z_{max}$)

$$\frac{\partial c}{\partial z} = 0$$

The computed soil water distributions after 3, 6 and 9 hr of infiltration are plotted in Figure 2A along with the initial distribution. The solute distribution curves corresponding to the soil water distribution (during infiltration) are given in Figure 2B. It should be noted that the chloride is released as a step function (slug) and then leached during an additional 6.2 hr with chloride-free water. Figures 3A and 3B show the water and chloride profiles after 6 hr of redistribution at time $t = 15$ hr.

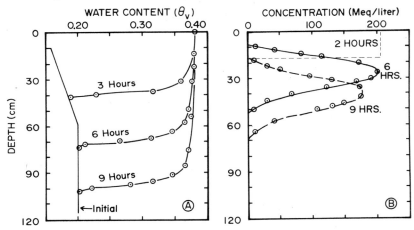

Figure 2.   A. Water content as a function of depth 3, 6, and 9 hr after starting infiltration. Soil surface remains at saturation during infiltration. B. Solute concentration in meq per liter as a function of depth. Solute initially present to a depth of 16.7 cm at $c = 209$ meq/liter. Other conditions described in text.

The computed results for water content and solute are in good agreement with those of the experimental data (39) at the end of 9 hr of infiltration.

In Figure 3A, one can see that the soil is draining faster near the top surface than at the wetting front. This indicates that drainage is more rapid in the wet zone than in the dry zone for this soil. Since hysteresis was ignored, this may or may not represent the behavior of this soil.

Evaporation

In a second example the change in water content due to evapo-

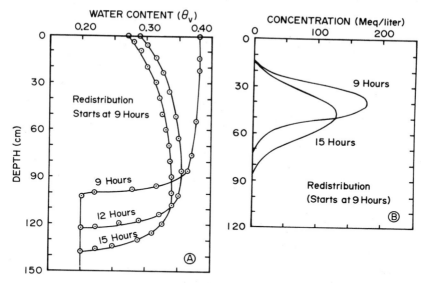

Figure 3.   A. Water content as a function of depth resulting from redistribution of the water present at 9 hr shown in Figure 2. Infiltration ceased at 9 hr. The upper boundary condition for redistri-
bution, $-D_z(\partial\theta/\partial z) + K_z = 0$ for $t > 9$ hr. B. Solute concentration
in meq per liter as a function of depth resulting from redistri-
bution of water as shown in Figure 2.

ration (Figure 4A) and the cumulative evaporation were obtained (Figure 4B). During evaporation the upper boundary
condition for a demand of 0.8 cm/hr becomes

$$-D_z \frac{\partial\theta}{\partial z} + K_z = -0.8 \text{ (cm/hr)}, \quad t > 9 \text{ hr}$$

and the upper boundary condition for the solute transport becomes

$$-\theta D_{sz} \frac{\partial c}{\partial z} + q_z c = 0, \quad t > 9 \text{ hr}$$

It should be remembered that the value of 0.8 cm/hr in the surface boundary condition represents the atmospheric demand or
the maximum rate of water loss. The actual rate of water loss
will be less than or equal to this rate.

Figure 4A shows that the surface dries out quickly and Figure
4B shows that the rate of water loss rapidly decreases below the
potential rate. Evaporation has initially little effect on the water
content near the wetting front, which is the same as shown in
Figure 3A for redistribution.

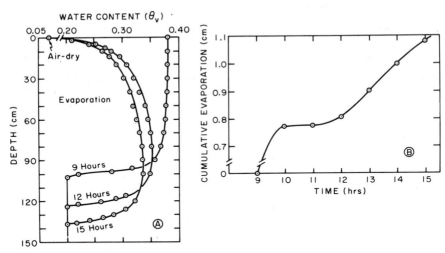

Figure 4. A. Water content as a function of depth resulting from redistribution and evaporation of the water present at 9 hr as shown in Figure 2. Evaporation started at 9 hr. B. Cumulative evaporation as a function of time.

The evaporation process started at 9 hr as shown in Figure 4B. Nearly 0.7 cm of water was evaporated during the first 45 min. Then as the soil surface became dry the evaporation rate decreased dramatically so that less than 0.05 cm of water evaporated during the next 1.75 hr (9:15 to 11:00 hr). Following that period the rate of evaporation increased and reached about 0.1 cm/hr during the 12:00 to 15:00 hr period. This increase resulted from an increased rate of water supply to the soil surface following the initial drying. A gradual decrease in the evaporation rate occurs after 15 hr as a result of continued drying at the soil surface. The responses shown are valid only for the conditions chosen for this analysis.

As the evaporative process progresses an upward movement of water is initiated and a corresponding increase in solute concentration occurs near the soil surface. For the 6 hr of evaporation this upward movement was small and is not shown.

Water Uptake by Plant Roots

In this example the problem of infiltration and redistribution is further complicated by allowing extraction of soil water by a root system to occur. Uptake of solute by the roots does not occur. The water flow equation used is

$$\frac{\partial \theta}{\partial t} = \frac{\partial}{\partial z} \left\{ D_z \frac{\partial \theta}{\partial z} - K_z \right\} + A$$

where A is the root sink term. All other initial and boundary conditions and the solute equation were stated previously. The rate of water uptake by the roots is assumed to be proportional to the water potential difference between the root xylem and the bulk soil and the hydraulic conductivity of the combined soil-root pathway, as follows

$$A = K_{sys} \cdot RAF \cdot (\psi_{ps} + z + RPF - \psi)$$

with terms defined previously. The functions K_{sys}, RAF and RPF, as well as other parameters were conceived for this calculation as reasonable approximations of known phenomena. Little experimental information is available on which to base these functions.

Potential Transpiration Rate

The actual transpiration is defined as

$$T_r = c_{st} \cdot q_{ps}$$

where c_{st} is the stomatal correction factor and q_{ps} is the atmospheric or potential transpiration demand. For this exercise the potential transpiration demand (Figure 5A) was represented by an exponential function. The potential demand is determined by climatic conditions and the daily course of the sun. The function

$$q_{ps} = -0.2e^{-0.06(h-12)^2}$$

where h is the hour of the day was chosen. It has a maximum transpiration rate of 0.2 cm/hr which occurs at noon, and a total daily transpiration of 1.2 cm/day.

Stomatal Correction Factor

The degree to which plants respond to the potential demand is in part controlled by stomatal aperture. As the ability of the soil to supply water to the plants decreases, the plants dessicate, and stomates begin to close to prevent this. Here, the stomatal closure is assumed to be two linear functions of plant water potential, ψ_{ps} (Figure 5B). These functions are

$$c_{st} = 1.0 + \psi_{ps}/80, \quad -8 \leqslant \psi_{ps} \leqslant 0 \text{ bars}$$

$$c_{st} = 1.8 + 9\psi_{ps}/80, \quad \psi_{ps} < -8 \text{ bars}$$

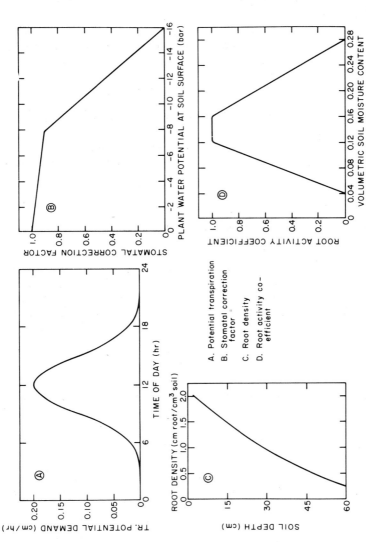

Figure 5.   Climatic and plant factors that affect the rate of water uptake by plant roots. A. Potential evaporation as determined by climatic conditions. B. Stomatal correction factor as determined by plant water stress. C. Root density distribution as a function of depth. D. Root activity coefficient as determined by soil water stress in the dry region and limited aeration in the wet region.

Hydrodynamic Resistance in Root Xylem

A function is introduced which describes the potential drop due to frictional losses in the xylem vessels. The potential decrease is a function of water content, viscosity, and length of the flow path. Little is known about friction losses in xylem vessels. We will use results obtained by Slavikova as described by Newman (40) who indicates

or

$$RPF(z) = 10 \text{ cm/cm root}$$

$$RPF(z) = 0.01 \text{ bar/cm root}$$

Hydrodynamic Resistance in Root Cortex

The main resistance to water flow is across the epidermis. This is particularly so for wet soil where the resistance of the soil itself is low. As the soil dries out, its resistance increases rapidly and becomes the major resistance.

The conductivity of the root depends on cell turgidity and age of root tissue. Values ranging from 10^{-6} to 10^{-8} cm/hr have been cited in the literature (36). In this report we will use a constant value

$$K_r = 10^{-7} \text{ cm/hr}$$

Root Absorption Function

The root absorption function is determined by root density, root growth factor, and root activity. The root density for each soil layer changes as the plant continues to grow (36, 43). The root density function used for this report (Figure 5C) is

$$R_d = 2.0 - z/30 \text{ cm root/cm}^3 \text{ soil}, \quad z \leqslant 37.5 \text{ cm}$$

$$R_d = 1.5 - z/50 \text{ cm root/cm}^3 \text{ soil}, \quad 37.5 < z \leqslant 75 \text{ cm}$$

The root density varies with time. Baldwin *et al.* (42) suggested the exponential function

$$R_d(t) = R_d(t_0)e^{kt}$$

for root growth. The time-span used for this report was so short —about 10 hr—that the growth would not be significant. The root length was therefore assumed to remain constant, *i.e.*, $R_g = 1$.

Root Activity Coefficient

This coefficient was introduced to describe the degree to which roots participate in the uptake of water and solutes. The coefficient is a function of several factors, including soil water content, age of root tissue, ion concentration, oxygen concentration, and carbon dioxide concentration.

For our calculations the activity coefficient was assumed to be a function of the soil water content only (Figure 5D). In the wet region the coefficient is low because of limited O_2 availability and in the dry region the water content limits the ability of the root cells to absorb water. The following functions were used.

$$R_a = 25\ \theta/3 \qquad\qquad 0 \leqslant \theta \leqslant 0.12$$

$$R_a = 1.0 \qquad\qquad 0.12 < \theta \leqslant 0.16$$

$$R_a = 2.0 - 25(\theta - 0.04)/3 \qquad 0.16 < \theta \leqslant 0.28$$

Wilting

Wilting is assumed to occur at

$$\psi_{wilt} = -16.0\ \text{bars}$$

RESULTS

The amount of water transpired by plants was calculated assuming the conditions and functions described above and shown in Figure 5. The transpiration process was initiated at the same time as the infiltration process. The initial water content was shown in Figure 2. Figure 6A shows a rapid decrease in the plant surface potential, ψ_{ps}. This decrease was due to increased transpiration demand as time progressed as well as decreased efficiency of water uptake by the plant roots in the rapidly wetting soil. Figure 5D shows the decrease in efficiency of water uptake in the wet region. Thus as the efficiency of water uptake decreases due to the downward movement of the wetting front, *i.e.*, the "drowning" of the upper roots, the plant surface potential must decrease to maintain a given transpiration demand. This process accounts for the rapid decrease in plant surface potential from hour 3 to 6 in Figure 6A. However, as the plant surface potential decreases, the stomates respond by closing (Figure 5B) in order to conserve water. This process reduces

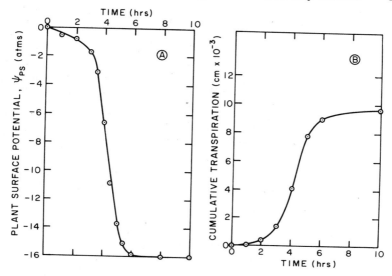

Figure 6. A. Change in plant surface potential as a function of time
B. Cumulative transpiration rate as a function of time.

the rate of potential decrease. Thus from hour 6 to 9 (Figure 6A) the change in plant surface potential is very small. Eventually the plant would reach wilting conditions and transpiration would cease.

ACKNOWLEDGMENTS

This investigation was supported by funds from the N.W. Forest and Range Experiment Station, Portland, Oregon, Grant No. 85–934, and the N.E. Forest Experiment Station, Upper Darby, Pennsylvania, Grant No. 23–586.

REFERENCES

1. Wang, J. H. and J. W. Kennedy. "Self-diffusion coefficients of sodium ion and iodide ion in aqueous sodium iodide solutions," *J. Am. Chem. Soc.* 72:2080–2083 (1950).
2. Wang, J. H. "Self-diffusion and structure of liquid water. I. Measurement of self-diffusion of liquid water with deuterium as tracer," *J. Am. Chem. Soc.* 73:510–513 (1951).
3. Wang, J. H. "Tracer-diffusion in liquids. I. Diffusion of tracer amount of sodium ion in aqueous potassium chloride solutions," *J. Am. Chem. Soc.* 74:1182–1186 (1952).
4. Wang, J. H., C. V. Robinson and I. S. Edelman. "Self-diffusion and structure of liquid water. III. Measurement of the self-diffusion of

liquid water with H^2, H^3, and O^{18} as tracers," *J. Am. Chem. Soc.* 75: 466–470 (1953).

5. Graham-Bryce, I. J. "Effect of moisture content and soil type on self-diffusion of ^{86}Rb in soils," *J. Agric. Sci.* 60:239–244 (1963).

6. Frissel, M. J. and P. Poelstra. "Chromatographic transport through soils. I. Theoretical evaluations," *Plant and Soil* 26:285–302 (1967).

7. Frissel, M. J., P. Poelstra and P. Reiniger. "Chromatographic transport through soils. III. A simulation model for the evaluation of the apparent diffusion coefficient in undisturbed soils with tritiated water," *Plant and Soil* 33:161–176 (1970).

8. Lindstrom, F. T. and L. Boersma. "Theory of chemical transport with simultaneous sorption in a water saturated porous medium," *Soil Sci.* 110:1–9 (1970).

9. Lindstrom, F. T. and L. Boersma. "A theory on the mass transport of previously distributed chemicals in a water saturated sorbing porous medium," *Soil Sci.* 111:192–199 (1971).

10. Lindstrom, F. T., L. Boersma and D. Stockard. "A theory on the mass transport of previously distributed chemicals in a water saturated sorbing porous medium: Isothermal cases," *Soil Sci.* 112:291–300 (1971).

11. Schofield, R. K. and I. J. Graham-Bryce. "Diffusion of ions in soils," *Nature* 188:1048–1049 (1960).

12. Bouldin, D. R. "Mathematical description of diffusion processes in the soil-plant system," *Soil Sci. Soc. Am. Proc.* 25:476–480 (1961).

13. Kunze, R. J. and D. Kirkham. "Deuterium and the self-diffusion coefficient of soil moisture," *Soil Sci. Soc. Am. Proc.* 25:9–12 (1961).

14. Dutt, G. R. and P. F. Low. "Diffusion of alkali chlorides in clay-water systems," *Soil Sci.* 93:233–240 (1962).

15. Olsen, S. R., W. D. Kemper and R. D. Jackson. "Phosphate diffusion to plant roots," *Soil Sci. Soc. Am. Proc.* 26:222–227 (1962).

16. Olsen, S. R., W. D. Kemper and J. C. Van Schaik. "Self-diffusion coefficients of phosphorus in soil measured by transient and steady-state methods," *Soil Sci. Soc. Am. Proc.* 29:154–158 (1965).

17. Bear, J., D. Zaslavsky and S. Irmay. "Physical principles of water percolation and seepage," UNESCO, Paris, France (1968).

18. Lai, T. M. and M. M. Mortland. "Cationic diffusion in clay minerals: I. Homogeneous and heterogeneous systems," *Soil Sci. Soc. Am. Proc.* 32:56–61 (1968).

19. Elgawhary, S. M., W. L. Lindsay and W. D. Kemper. "Effect of EDTA on the self-diffusion of zinc in aqueous solution and in soil," *Soil Sci. Soc. Am. Proc.* 34:66–70 (1970).

20. Passioura, J. B. and D. A. Rose. "Hydrodynamic dispersion in aggregated media. 2. Effects of velocity and aggregate size," *Soil Sci.* 111:345–351 (1971).

21. Gessel, S. P. and D. W. Cole. "Influence of removal of forest cover on movement of water and associated elements through soil," *J. Am. Water Works Assoc.* 57:1301–1310 (1965).

22. Johnson, N. M., G. E. Likens, F. H. Bormann and R. S. Pierce. "Rate of chemical weathering of silicate minerals in New Hampshire," *Geochim. Cosmochim. Acta* 32:531–545 (1968).

23. Bormann, F. H., G. E. Likens and J. S. Eaton. "Biotic regulation of particulate and solution losses from a forest ecosystem," *Bio Science* 19:600–610 (1969).

24. Keller, H. M. "Der Chemismus kleiner Bache in teilweise bewaldeten Einzygsgebieten in der Flyschzone lines Voralpentulas. Schweizerische Anstalt fur das Forstliche Versuchswesen," *Mitteilungen* 46:113–155 (1970).

25. Likens, G. E., F. H. Bormann, N. M. Johnson, D. W. Fisher and R. S. Pierce. "Effects of forest cutting and herbicide treatment on nutrient budgets in the Hubbard Brook Watershed ecosystem," *Ecological Monographs* 40:23–47 (1970).

26. Fredriksen, R. L. "Comparative chemical water quality—natural and disturbed streams following logging and slash burning," p. 125–137. In *Forest Land Uses and Stream Environment, Symposium Proceedings,* J. T. Krygier, Ed. Oregon State University, Corvallis (1971).

27. McColl, J. G. and D. W. Cole. "A mechanism of cation transport in a forest soil," *Northwest Science* 42:134–140 (1968).

28. Philip, J. R. "Theory of infiltration," *Advance Hydroscience* 5:215–296 (1969).

29. Hanks, R. J. and S. A. Bowers. "Numerical solution of the moisture flow equation for infiltration into layered soils," *Soil Sci. Soc. Am. Proc.* 26:530–534 (1969).

30. Rubin, J. and R. Steinhardt. "Soil water relations during rain infiltration: I. Theory," *Soil Sci. Soc. Am. Proc.* 27:246–251 (1963).

31. Dutt, G. R., M. J. Shaffer and W. J. Moore. "Computer simulation model of dynamic bio-physicochemical processes in soils," Agricultural Experiment Station, The University of Arizona, Tucson. Technical Bulletin 196 (1972).

32. Bresler, E. "Simultaneous transport of solutes and water under transient unsaturated flow conditions," *Water Resources Research* 9:975–986 (1973).

33. Feddes, R. A., E. Bresler and S. P. Neuman. "Field test of a modified numerical model for water uptake by root systems," *Water Resources Research* 10:1199–1206 (1974).

34. Hillel, D. *Soil and Water. Physical Principles and Processes.* (New York: Academic Press, 1971), 288 pp.

35. Molz, F. J. "Simulation of post-irrigation moisture movement," *J. Irrigation and Drainage Division, Proc. American Society of Civil Engineers* 98:523–532 (1972).

36. Taylor, H. M. and B. Klepper. "Water uptake by cotton root systems: An examination of assumptions in the single root model," *Soil Sci.* 120:57–67 (1975).

37. von Rosenberg, D. U. *Methods for the Numerical Solution of Partial Differential Equations* (New York: American Elsevier Publishing Co., Inc., 1969). 128 pp.

38. Neuman, S. P., R. A. Feddes and E. Bresler. "Finite element analysis of two-dimensional flow in soils considering water uptake by roots: I. Theory," *Soil Sci. Soc. Am. Proc.* 39:224–230 (1975).

39. Warrick, A. W., J. W. Biggar and D. R. Nielsen. "Simultaneous solute and water transfer for an unsaturated soil," *Water Resources Research* 7:1216–1225 (1971).

40. Newman, E. I. "Root and soil water relations, p. 363–440. In *The Plant Root and its Environment,* E. W. Carson, Ed., University Press of Virginia, Charlottesville, Virginia (1974). 691 pp.

41. Mengel, D. B. and S. A. Barber. "Development and distribution of the corn root system under field conditions," *Agron. J.* 66:341–344 (1974).

42. Baldwin, J. P., P. H. Nye and P. B. Tinker. "Uptake of solutes by multiple root systems from soil. III. A model for calculating the solute uptake by a randomly dispersed root system developing in a finite volume of soil," *Plant and Soil* 38:621–635 (1973).

43. Taylor, H. M. and B. Klepper. "Water relations of cotton. I. Root growth and water use as related to top growth and soil water content," *Agron. J.* 66:584–588 (1974).

8

NITROGEN REMOVAL FROM MUNICIPAL WASTEWATER EFFLUENT BY A CROP IRRIGATION SYSTEM

C. E. Clapp, D. R. Linden, W. E. Larson, G. C. Marten, and J. R. Nylund
Agricultural Research Service, USDA, and University of Minnesota, St Paul

That the biosystem of soil, organic crop residues, and growing crops can effectively renovate municipal wastewater effluent has been amply demonstrated (1, 2). However, information on the most efficient management of agronomic crops and soils for maximum renovation under high rates of effluent application is lacking.

In the past, agricultural science has devoted its resources to development of optimum management of crops under limited water and fertilizer regimes. However, for renovation of effluent, it is usually most economical to apply large amounts of water to limited areas of land. Application of large amounts of effluent with its associated nutrients requires development of new agronomic, irrigation, and drainage technology for most efficient renovation of effluent in harmony with agricultural production.

Treated municipal wastewater (secondary effluent) contains, on the average, about 20 mg/l phosphorus, and 10 mg/l potassium. A hectare-centimeter (ha-cm) of this effluent contains about 2 kg of nitrogen and 250 ha-cm contain about 500 kg of nitrogen. The 250 cm of treated water contains as much or more nitrogen than most crops can take up during a growing season at St. Paul, Minnesota. Crops most suitable for removing large amounts of nitrogen in the northern U.S. appear to be corn, corn with rye interseeding, and forage crops.

Under a high loading rate of effluent, however, questions arise as to the best soil and crop management. Practices needing study include selection of crop species and populations for greatest nitrogen removal, tillage and residue management for row crops, production of two crops during one season, and effectiveness of herbicides and insecticides.

The overall objective of this study was to develop agronomic practices for maximum nitrogen utilization by crops irrigated with municipal wastewater effluent. In this paper we will account for the nitrogen in the soil–water–crop system by measuring characteristic components in a field experiment.

EXPERIMENTAL

Site Selection

The experimental site selected was adjacent to the Apple Valley, Minnesota, wastewater treatment plant with effluent polishing filters (capacity 5700 m^3/day). The site was chosen because of its soil uniformity, isolation, and proximity. The Waukegan silt loam (Typic Hapludoll) soil is dark-colored, well-drained, and has a water table at about the 140- to 150-cm depth. It is formed from a silt loam material overlying neutral to calcareous outwash gravel at about 60 cm. Soil analyses before effluent application gave mean values for pH and percentages of C and N of: 5.55, 2.44, and 0.216 at 0 to 15 cm, and 5.65, 2.10, and 0.187 at 15 to 30 cm, respectively.

Drainage and Irrigation System

The 12 treatment blocks are underlain (approximately 140-cm depth) with a 10-cm-diameter perforated plastic drain at identical elevations and grades (Figure 1). Each treatment block also has a tile drain along both sides at the same elevation and grade. All tile lines are connected to a closer collector line which drains by gravity into a sump where this drainage water is pumped back to the surface and discharged into the stream. Water from the tile lines beneath treatment blocks flows free from graded closed pipe into open sampling stations which drain by gravity into the collector line. Tile drainage flow rates are measured by using a slotted tube and stilling well with water-stage recorder. The drainage system was designed to maintain a relatively uniform water table depth and to minimize crossflow between treatment blocks.

Effluent is applied to each treatment block independently

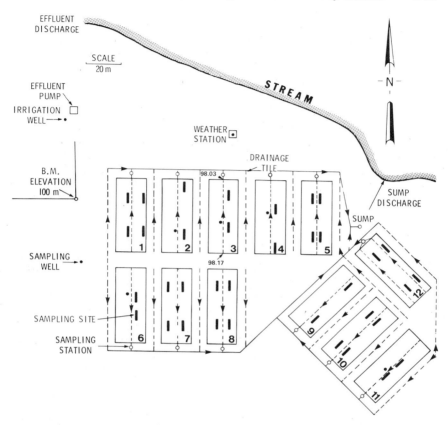

Figure 1.   General plan showing tile drainage system and water sampling for Apple Valley sewage effluent project.

through a solid set sprinkler system. Sprinklers are arranged on a 9.1-m grid for about 100 percent overlap and maximum uniformity of application. Rates and amounts of application are varied by changing nozzle sizes and sprinkling times. Irrigation amounts are measured with totalizing flow meters.

Design and Effluent Treatments

Twelve blocks of land (18.3 x 45.7 m) were arranged in a 2 by 3 factorial with two replications for comparison of crops and effluent rates (Figure 1). The six blocks planted to each crop received the following three effluent treatments during 1975:

1. *Control.* Nitrogen, phosphorus and potassium were applied as a dry fertilizer in amounts required for optimum plant growth and yields (336 kg of N, 168 kg of P, and 672 kg of K/ha). During 11 irrigations, 36 cm of irrigation water low in nutrients was applied.
2. *Low.* Corn received 109 cm of effluent (186 kg N/ha) and forage received 137 cm (234 kg N/ha).
3. *High.* Corn received 197 cm of effluent (337 kg N/ha) and forage received 240 cm (410 kg N/ha).

This was the first complete year of the wastewater effluent experiment that included both water sampling and crop harvesting. During the 1974 season, the following nitrogen applications were made: control, 336 kg/ha; low corn and low forage, 369 and 458 kg/ha; and high corn and high forage, 604 and 800 kg/ha. Crops were harvested, but water sampling was incomplete.

Cropping Treatments

The corn blocks were divided into smaller plots to compare use of a residue (corn stover) *vs.* no residue. Winter rye was used as a cover-crop for the entire corn block.

The forage blocks consisted of two replications of each treatment. The forage cropping treatments included three cutting schedules (2-time, 3-time, and 4-time) and eight forage species. Only the results of the 3-time cutting of Kentucky bluegrass and reed canarygrass are reported here.

Water Sample Collection

Soil water was sampled weekly by using porous ceramic samplers installed at soil depths of about 45 to 60 cm (at the silt loam/gravel interface) and 120 to 130 cm (15 cm above the water table) at 36 sites (Figure 1). Duplicate samplers were installed at each depth. Two sampling sites were located within each corn–residue–effluent treatment combination (four per block) and two within each reed canarygrass–effluent treatment combination (two per block).

Samples of rain water were collected in plastic measuring gauges and composited after each rainfall event for analysis.

Effluent Application and Composition

Monthly and total season applications of effluent are given for each treatment in Table 1. Irrigation began on May 16 and ended on October 29, 1975. Irrigation rates averaged 1.6 cm/hr on the forage and 1.3 cm/hr on the corn, with durations of 2 hr (low) and 4 hr (high) twice a week. Thus about 6.5 cm/week

Table 1. Effluent applications for Apple Valley sewage effluent project (1975).

Treatment	Monthly Application Amounts						Total Amount (cm)	1975 Season Totals[a]		
	May	June	July	Aug.	Sept.	Oct.		Total Nitrogen	NH$_4$+NO$_3$ Nitrogen	
			(cm)						*(kg/ha)*	
Corn										
Low	0	21.6	32.8	25.1	12.8	17.3	109	186	165	
High	0	36.1	56.4	47.5	24.4	32.5	197	337	297	
Forage										
Low	7.0	20.6	37.5	32.8	15.2	22.9	137	234	207	
High	11.4	42.4	63.1	59.9	21.3	41.5	240	410	362	

[a]Based on 1975 mean effluent nitrogen composition.

(low) and 13 cm/week (high) were applied on the forage and 5 cm/week (low) and 10 cm/week (high) were applied on the corn. The 10-cm/week application was about the maximum amount of effluent that could be applied without considerable ponding and runoff on the corn. Effluent samples were collected in plastic pans at ground level on each block and composited immediately after each irrigation for analysis. Total nitrogen applied during the season in individual application equaled that calculated from effluent composition and effluent applied. Effluent characteristics were given in Table 2; however, only the nitrogen components are of interest here.

Nitrogen Balance

Nitrogen additions as effluent, fertilizer, and in rain were calculated from concentrations and measured amounts applied. We estimated that 50 percent of the nitrogen contained in the 1974 stover residue returned to the soil surface was mineralized. Crop uptake was determined from plant tissue concentrations and yields. Drainage losses of nitrogen were obtained from concentrations measured in the soil water at the 125-cm depth (15 cm above the water table) and calculated water drainage rates. Monthly water drainage rates were obtained from the well-known water balance relationship:

Drainage = Rain + Irrigation − Evaporation − Soil Water Change

Monthly evapotranspiration for forage blocks was estimated as 80 percent of pan evaporation, but was estimated as 70 percent for the corn blocks because of early and late season differences

Table 2. Principal characteristic of effluent[a] for Apple Valley sewage effluent project (1975).

Characteristic	Mean	±	S.D.
	––––– *(mg/1)* –––––		
Total N	17.1	±	5.9
NH_4–N	13.8	±	5.6
NO_3–N	1.3	±	1.3
Total P	8.7	±	2.7
PO_4-P	7.9	±	2.4
Na	216	±	42
Ca	70.4	±	11.0
Mg	24.1	±	4.4
K	10.7	±	1.6
Cl	319	±	71
pH	8.20	±	0.10
COD	35	±	11
Suspended solids	7.4	±	18.9
Volatile solids	5.5	±	2.8
	–––– *(μmhos/cm)* ––––		
Conductivity (EC_{25})	1695	±	260

[a]Effluent from Apple Valley, Minnesota, WWTP sampled during 81 irrigations (5/16–10/29/75). Samples collected at several plot locations and composited daily.

in water use. Table 3 shows pertinent climatological data. Soil water storage did not change over monthly periods so that drainage could be calculated from the water blanace equation. Soil solution nitrogen as gain or loss was calculated from analyses of water samples collected at the 60-cm depth (assuming a constant nitrogen concentration over the 0- to 60-cm depth at 30 percent water content). Water contents were measured with a neutron probe.

The nitrogen balance for the crop irrigation system thus was estimated from the nitrogen applied (fertilizer, effluent, rain, irrigation water and residue) less the nitrogen-accounted-for (crop uptake, loss by drainage; and change in soil solution) giving a "not-accounted-for" component. This simplified nitrogen balance relationship does not consider soil organic matter transformations, denitrification, or gaseous losses, because good field estimates of these values are not available.

Table 3.   Summary of climatological observations—Apple Valley, Minnesota (1975).

	Average Temperature		Precipitation		
			Actual	Departure from Normal[a]	Pan Evaporation
	Min.	Max.			
	- - - $(^{\circ}C)$ - - -		- - - - - - - - (cm) - - - - - - -		
May	12.2	23.3	12.7	3.1	—
June	14.4	24.4	19.7	8.0	15.7
July	17.2	28.3	2.6	-7.5	23.4
August	15.0	25.6	14.1	4.8	14.8
September	8.9	18.9	4.5	-3.8	10.4
October	3.3	15.0	0.7	-4.6	8.7
Totals			54.3	0.0	73.0

[a]U.S. Weather Service normal for Farmington, Minnesota.

Analyses

Water samples were refrigerated immediately after collection. Analyses of organic components were performed in less than one week or the samples were acidified. Total N, including NO_3-N in effluent, was determined by micro-Kjeldahl (3). Total N for water, total P (after digestion), and NH_4-N, NO_3-N, PO_4-P, and Cl were determined by Technicon AutoAnalyzer.* Analyses for Na, Ca, Mg and K were performed on a Perkin-Elmer 303 atomic absorption spectrophotometer. The pH, COD, suspended solids, volatile solids, and conductivity were determined by procedures of *Standard Methods* (4).

RESULTS AND DISCUSSION

Crop Yields

Corn fodder yields in 1975 were 18.5, 12.8 and 13.6 mt/ha for the control, low and high wastewater effluent treatments; comparative grain yields were 10.8, 6.9 and 7.5 mt/ha (172, 110 and 119 bu/acre). Nitrogen deficiency on the effluent-treated blocks, especially early in the season, was one of the principal reasons for the lower fodder and grain yields.

Forage yields for Kentucky bluegrass and reed canarygrass at

---

*Company or trade names are mentioned for the reader's benefit and do not imply preference by the USDA over products not mentioned.

the 3-time cutting were 10.4 and 11.4, 7.4 and 9.9, and 9.7 and 11.6 mt/ha for the control, low and high treatments, respectively.

Soil Water

Nitrogen transformations, availability, and movement under corn and forage cropping for different effluent treatments can be followed by measuring the nitrogen concentrations in soil water. The monthly and seasonal means of total inorganic nitrogen (NO_3-N + NH_4-N) are shown in Table 4 for water collected

Table 4. Total inorganic nitrogen (NO_3-N + NH_4-N) concentrations in soil water for Apple Valley sewage effluent project.

Treatment Month	60-cm depth			125-cm depth		
	Control	Low	High	Control	Low	High
				(mg/1)[a]		
Corn						
June	107	26.5	23.8	23.4	22.6	20.5
July	141	19.6	11.4	33.0	21.2	20.4
August	131	4.0	0.4	36.2	12.0	7.4
September	128	0.8	1.0	31.0	4.7	2.4
October	119	1.6	8.0	33.0	6.4	3.9
Mean	125	10.5	8.9	31.3	13.4	10.9
Forage						
June	1.0	0.3	3.2	5.8	0.1	3.8
July	12.0	0.4	2.8	6.0	0.4	1.8
August	16.8	0.1	1.0	7.3	0.2	1.4
September	—	0.0	1.0	10.5	0.1	0.2
October	13.0	0.2	6.2	11.0	0.1	1.6
Mean	10.7	0.2	2.8	8.1	0.2	1.8

[a]Means of eight replicated samples on corn and four replicated samples on forage blocks taken by porous ceramic samplers at weekly intervals between June and October 1975.

in porous ceramic samplers at two soil depths. Of the total inorganic nitrogen concentration, nitrate-N represented 95 to 100 percent; ammonium-N values never exceeded 0.1 mg/l.

Soil water nitrogen at the 60-cm depth in the corn blocks showed an order of magnitude difference between fertilizer control and effluent treatments. Nitrogen values for the control blocks remained high throughout the season, while those of the effluent treatments decreased markedly following the period of combined higher effluent application and soil nitrification.

Increased nitrogen concentrations at the end of the season coincided with a decrease in uptake by corn before the rye cover crop was established. In the corn control blocks, nitrogen content remained essentially constant at the 125-cm depth during the season. In the corn effluent treatment blocks, nitrogen again decreased in midseason and increased at the end of the growing season.

Forage crops removed much more nitrogen than did corn, as seen by the lower monthly and seasonal mean nitrogen concentrations in soil water. Higher values on the forage control blocks at both depths reflect periodic fertilizer applications, while forage on the effluent treatment blocks removed nitrogen continuously as applied, allowing very little to move below the root zone.

Nitrogen Balance

Many attempts have been made to account for nitrogen in soil-crop systems (5). In general, these attempts have met with only limited success because quantitative data are usually not available for all the components. When a third phase, water, is added, as in a crop irrigation system, the situation becomes more complex; however, some of the components are better controlled and estimated. Our attempt to account for nitrogen in the soil–water–crop system is shown in Table 5. Nitrogen was added from fertilizer, effluent, rain or residue sources. Nitrogen was accounted for by crop uptake, drainage and change in soil solution content. Difference between applied and accounted-for components is the amount "not-accounted-for."

Nitrogen applied as fertilizer, effluent or rain can be measured accurately. The estimate of residue nitrogen requires an assumption of 50 percent mineralization when stover decomposes during the year following application. The corn blocks showed little treatment difference in nitrogen returned in residue because yields and nitrogen percentages of stover balanced each other. While other residual sources such as corn roots, rye cover-crop not removed, and forage thatch and roots are part of the total system, in our balance calculations they must remain not-accounted-for. Crop uptake nitrogen can also be measured accurately. Estimation of drainage losses, however, requires calculation of water drainage rates from water added in rainfall and irrigation less evapotranspiration (from panevaporation data). These rates are better estimates than those measured by tile flow in our system because the stream and general ground water level influence the tile flow rates. Data in the 1974 season indicated the tile flow volume from June to October in the block closest

Table 5.   Nitrogen balance for Apple Valley sewage effluent project (1975).

	Applied		Accounted-for			Not
Treatment	Fertilizer[a] or Effluent	Residue	Change in Soil Solution	Crop Uptake[b]	Drainage[c]	Accounted For
Corn	— — — — — — — — — — — — kg/ha — — — — — — — — — — — —					
Control						
Residue	341	55	-1	243	108	46
No residue	341	0	268	282	81	-290
Low						
Residue	191	60	0	152	193	-94
No residue	191	0	-29	166	137	-83
High						
Residue	342	60	13	193	233	-37
No residue	342	0	6	180	233	-77
Forage						
Control						
Kentucky bluegrass	341	—	22	288	22	9
Reed canarygrass	341	—	22	297	22	0
Low						
Kentucky bluegrass	239	—	0	185	2	52
Reed canarygrass	239	—	0	253	2	-16
High						
Kentucky bluegrass	415	—	10	271	40	94
Reed canarygrass	415	—	10	336	40	29

[a]Control blocks received 336 kg N/ha, as NH_4NO_3. All blocks received 5 kg N/ha from rain.

[b]Means of two replicated samples within one block for corn (Northrup King 476) and two replicated samples of 3-time cutting within one block for forage (Park Kentucky bluegrass and NCR-Cl reed canarygrass).

[c]Means of two replicated samples for both corn and forage.

to the stream (Block 5 of Figure 1) was 3.6 m greater than expected (from simple water balance relationships) and the block farthest from the stream (Block 6 of Figure 1) was 1.8 m less than expected. Nitrogen concentrations of water taken by porous ceramic samplers at 15 cm above the water table ($\sim$125 cm depth) are the best estimates to use with predicted drainage rates to provide values for nitrogen lost by drainage. This is true because water sampled at 125 cm comes essentially only from vertical movement through the above soil profile and has not yet been mixed and diluted with ground water coming from horizontal flow near and below the water table.

Soil solution nitrogen can be a gain or loss in the balance depending on whether nitrogen concentration increased or decreased from beginning to end of the season. Because the replicated porous ceramic samplers were not installed until early June 1975, the initial values of soil solution nitrogen came from Fall of 1974 data. Other assumptions for the soil solution nitrogen component are a constant nitrogen concentration over the 0- to 60-cm depth at 30 percent water content. The concentration assumption was valid for the effluent treatment blocks, but not for the control blocks.

For the corn blocks, the nitrogen not-accounted-for components are reasonable, but all negative except for the control-residue treatment. A negative value indicates that more nitrogen was removed or recovered from the system than was applied. Some additional source of nitrogen is implied, such as soil organic matter or previous management practices. The high value for change in soil solution nitrogen on the control-no residue plots was obtained because of a low initial nitrogen concentration coupled with a very high final value. For corn, both crop uptake and drainage components are important.

A low nitrogen uptake species, Kentucky bluegrass, and the highest uptake species, reed canarygrass, were compared by effluent treatments. Except for the low effluent-reed canarygrass treatment, the nitrogen not-accounted-for components were all positive indicating that excess nitrogen was applied. Reed canarygrass consistently removed more nitrogen than did Kentucky bluegrass in our system. Crop uptake appears to be the most important component for forages.

SUMMARY

1. Results from the first complete year of study with a crop irrigation system suggest that corn and forage crops produce high yields and remove considerable nitrogen under high wastewater effluent applications.

2. Soil water nitrogen levels were low during effluent application and plant growth periods and decreased below the rooting zone, especially on the blocks cropped to forage.

3. Perennial grass forages removed more nitrogen from effluent than did corn, during a complete growing season.

4. Nitrogen removal from effluent by the crop irrigation system used here is primarily due to crop uptake.

ACKNOWLEDGMENTS

The authors acknowledge financial support for this project from the U.S. Army Corps of Engineers and facility availability from the Metropolitan Waste Control Commission.

Contribution from the North Central Region, ARS, USDA, St. Paul, Minnesota, in cooperation with the Departments of Soil Science and Agronomy and Plant Genetics, University of Minnesota, St. Paul. Minnesota Agricultural Experiment Station Paper No. *9576*, Scientific Journal Series.

REFERENCES

1. Sopper, W. E. and L. T. Kardos, Eds. *Recycling Treated Municipal Wastewater and Sludge through Forest and Cropland*, The Pennsylvania State University Press, University Park, 479 pp. (1973).
2. Bouwer, H. and R. L. Chaney. "Land treatment of wastewater", *Advan. Agron.* 26:133–176 (1974).
3. Bremner, J. M. "Total nitrogen," in C. A. Black, *et al.* Eds., *Methods of Soil Analysis, Agronomy* 9:1149–1178 (1965).
4. Taras, M. J., A. E. Greenberg, R. D. Hoak, and M. C. Rand, Eds. *Standard Methods for the Examination of Water and Wastewater*, 13th American Public Health Association, Washington, D.C., 874 pp. (1971).
5. Allison, F. E. "The enigma of soil nitrogen balance sheets," *Advan. Agron.* 7:213–250 (1955).

9

PHOSPHATE REMOVAL BY
SANDS AND SOILS

T. J. Tofflemire
New York State Department of Environmental Conservation,
   Research Unit, and Division of Labs and Research,
M. Chen
New York State Department of Health
   Division of Labs and Research
   Albany, New York

INTRODUCTION

Cultural eutrophication of lakes is presently one of our most significant water quality problems in the northeastern United States. Phosphorus is one of the critical elements contributing to this problem. Rapid infiltration systems such as recharge basins, septic tank tile fields and intermittent sand filters when located near lakes can contribute phosphates to the lakes if the systems are not properly designed.

LITERATURE INDICATIONS

Phosphate can be removed by several mechanisms: (a) rapid removal or sorption ,(b) slow mineralization and insolubilization, (c) plant uptake, and (d) biological immobilization. For rapid infiltration systems, the first two mechanisms are most significant. The Langmuir adsorption isotherm has been used by a number of authors to describe the rapid removal mechanism (1–5). After 2.5 days (the period of an isotherm test) the additional phosphorus removal by soils slows and is termed "slow mineralization." The rate of this mechanism is not well known, but is thought to involve variscite and strengite formation in acid soils and hydroxyapatite and fluorapatite formation in alkaline soils (2, 6). Most of this slowly fixed phosphate was not readily leached by water (7, 8). Other factors, such as soil moisture tension, pH, temperature and redox potential that affect phosphate removal have also been noted (1, 9, 10).

To conduct an isotherm test, equal weights of soil are placed in containers with varying amounts of orthophosphate such as KH_2PO_4 or NaH_2PO_4 in solution in 0.01 M NaCl or 0.01 M $CaCl_2$. After periodic shaking over a period of 2 to 5 days, the samples are centrifuged or filtered and orthophosphate or radioactive phosphorus is determined. The difference between the P added and the P found in the water is the P removed in the soil. For each container solution there is an equilibrium P concentration c, and an amount of P removed per unit weight of soil, x/m (1). A typical isotherm plot using the Langmuir equation, $c/(x/m) = 1/kb + c/b$, is shown in Figure 1.

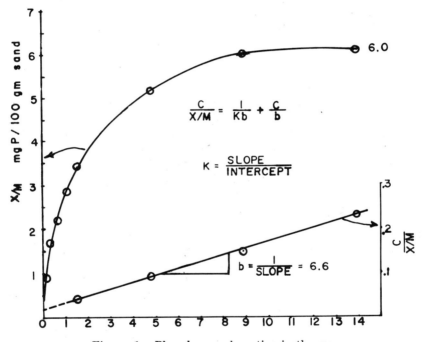

Figure 1. Phosphorus adsorption isotherms.

Plotting $c/(x/m)$ vs. c on rectangular scales generally produces a straight line. The maximum phosphate adsorption is "b max" = 1/slope, and the constant indicating bonding strength is K = slope/intercept. Some authors (11, 12) have found that at higher concentrations a second line of lesser slope may occur. As noted in Figure 1, the x/m vs. c curve approaches a horizontal asymptote near the b max value and shows whether concentrations near saturation have been approached. From these curves x/m can be found for any given P concentration.

Figure 2 indicates little change in P adsorption by soils after 5 days in isotherm tests. Enfield and Bledsoe (13) conducted isotherm tests of up to 4 months duration and gave equations for P removal vs. time. The 4-month P retention was 1.5 to 3.0 times the 5-day retention. Table 1, which summarizes literature b max values, indicates the range of 0.3 to 278 mg P/100 g soil for soils in New York State. In the preliminary evaluation of a site for phosphate retention one must consider the following factors: (a) the load of phosphate to be applied in kg P/yr;

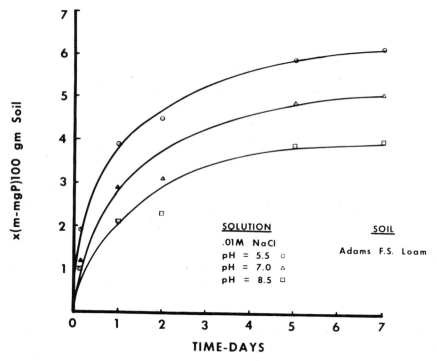

Figure 2.   Phosphorus adsorption vs. time.

(b) the unit capacity of the soil to rapidly remove phosphate in mg P/100 g soil at a given P concentration; (c) the total soil volume and weight through which the wastewater will pass; and (d) the increase in removal due to slow mineralization of the rapidly adsorped phosphate. To evaluate factor (c), data on ground water hydrology and on spreading of the wastewater

Table 1.   Summary of phosphorus adsorption values.

Location	No. Samples of Soil	Notes	Sorption Capacity or b max. mg P/100 g Soil	Reference
Michigan	29–100	Avg. for 3-ft depth	1.81–49.0	(5)
Florida	6	Avg. for 20-in. depth	nil – 28.0	(14)
New Brunswick	24	Soils from upper B horizon	227-1760	(15)
New Jersey	17	A, B & C horizons	0.165-355	(16)
Maine	3	From column tests	26-71	(17)
	2		13.3-25.9	(18)
	5		3.8-51.0	(19)
	31	Average for 31 soils	12.0	(20)
New York	240	A, B, C and deeper	0.3-278	(21)
Wisconsin	5	A, B & C horizons	2.5-20	(22)

must be gathered (1). To evaluate factor (d), equations such as those by Enfield and Bledsoe (13) which require long-term isotherm tests, can be used or a less scientific approach of multiplying the 5-day x/m by a mineralization factor can be used. Preliminary data at the Lake George, N.Y. recharge beds, which have operated for 36 years, indicated a mineralization factor of about six times the 5-day isotherm value. Column studies, to be discussed later, indicated a factor of about 2.0 to 5.0 for the ratio of the 3 to 10-mo retention/5-day retention data. Thus it is likely that the total phosphate retention by a rapid infiltration system will be at least twice the estimate based on the 5-day isotherm test. The actual percentage removal of phosphate will be very high initially and decrease with time. Equations describing the breakthrough curve or "shock layer" were given by Shah et al. (23) and Novak et al. (24).

NEW YORK SOILS DATA

Tofflemire and Chen (21) reported on 35 major soil series in New York State relative to their 5-day phosphate retention and other data. These were typical of soils in the glaciated northeastern United States. For all horizons of the 35 soils series sampled, the b max values ranged from 2.8 to 278 mg P/100 g of soil and averaged 38.2. Table 2 gives the horizon means. The amount of phosphate adsorbed at 5 mg P/l concentration is the "x/m @ 5" value. The B horizon b max and x/m @ 5 were, on the average, 2.5 times that for the C horizon. For some soils this

Table 2.   Horizon means.

	Thickness (in.)	Bulk Density (lb/ft³)	B max[a] mg P/100 g	x/m @ 5[a] mg P/100 g	CEC[a] meq/100 g
A	6.3	82.6	45.7	30.5	18.7
B	29	100.3	42.9	33.4	8.8
C	39	107.5	18.9	12.2	6.5

[a]Corrected for the % of coarse fragments (CF) greater than 2 mm; CEC = cation exchange capacity.

factor was 5, and 2 ft of B horizon were equal to 10 ft of C horizon. This is of importance because in most rapid infiltration systems the B horizon is not utilized. Modified designs to allow application of the wastewater above the B horizon are needed if phosphate removal is to be maximized.

To simplify the data for analysis, values of an A horizon, a thickness-weighted B horizon, and a thickness weighted C horizon were obtained for each soil series by calculation from the detailed horizon data. The soils were then grouped by three methods. The first grouping utilized the new national soil classification given below and had eight subgroupings:

Grouping I

Alfisols	Entisols	Inceptisols	Spodisols	Utisols
2 Aqualfs	2 Psamments	1 Aquepts	5 Fragiorthods	1 Hapludults
9 Udalfs		7 Orchepts	7 Haplorthods	

The numbers refer to the number of soils in that subgroup. Grouping System II is noted in Table 3, based on a 1/250,000 scale soil map of New York State by Arnold et al. (25). Grouping III was based on a parent materials classification adopted from a New York State soil map by Cline (26). This grouping system is provided in Table 4 and had seven subgroups. Details for Grouping III can be obtained from the senior author.

The data were then subjected to a two-way (subgroups and horizons) analysis of variance using Scheffe's approximation for unequal numbers of soils per subgroup via the NWAY 1 computer program (27). This was done for each of the three grouping systems and for each of the variables indicated in Table 5. The F values given were highly significant for most parameters with respect to horizons and with respect to subgroups. Grouping Systems II and III indicated the highest F values and significance levels.

The next step involved looking at the A, B or C horizons individually within the previous grouping systems (Table 6). Here a one-way (soil subgroups) analysis of covariance was per-

Table 3.   Second soil grouping key for weighted horizons.

Key No.	Description and Sampled Soils Included
0	Other areas not classified: 14[a] Gilpin.
1	Areas dominated by coarse textured mesic soils with sand and gravel substrata. Dominantly to excessively well drained:
	8[a] Colonie, 11 Elnora, 16 Hinkley, 36 Windsor
2	Areas dominated by medium and moderately coarse textured mesic soils with sand or gravel substrata. Dominantly well drained:
	6 Chenango, 18 Howard, 27 Palmyra, 30 Riverhead
3	Areas dominated by medium and moderately coarse and coarse textured frigid soils with sand and gravel substrata. Dominantly to excessively well drained:
	1 Adams, 9 Colton
4	Areas dominated by medium and moderately coarse textured mesic soils with compact loamy substrata. Dominantly and moderately well drained with a calcareous substrata:
	12 Erie, 17 Honeoye, 21 Lansing, 25 Ontario
5	Areas dominated by medium and moderately coarse textured mesic soils with fragipans and compact substrata. Dominantly and moderately well drained:
	13 Essex, 19 Lackawanna, 20 Langford, 28 Paxton, 33 Sodus
6	Areas dominated by medium and moderately coarse textured frigid soils with fragipans or compact substrata. Dominantly well drained soils having sand or gravel substrata:
	2 Becket No. 1, 3 Becket No. 2, 4 Bangor, 15 Hermon
7	Areas dominated by coarse to fine texture, stone-free soils with silty or clayey substrata. Dominantly to moderately well drained:
	7 Collamer, 8 Elmwood, 29 Rhinebeck, 31 Schoharie, 34 Vergennes
8	Areas dominated by moderately deep and shallow soils over bedrock. Dominantly well drained:
	5 Charlton, 22 Lordstown, 23 Mardin, 24 Morris, 26 Qguga

[a] These numbers are from an alphabetical listing of the soils from 1 to 36.

formed using the BMDO4V program (University of California Press, Berkeley, 1970) for each of the three grouping systems. These analyses apply to unit capacities, not total pound capacities, for the horizons. Either b max, x/m @ 5, or CEC was selected as the dependent variable with Bulk Density (B.D.), percent silt plus clay (S + C), and pH as the independent covariates. The F values were again generally more significant in Groupings Systems II and III. It was concluded for the B horizon that the chemical properties are significantly different for different soil

Table 4. Third soil grouping key for weighted horizons.

1. *Granite, gneiss till*		5. *Sandstone, Siltstone and Shale till*	
2,3[a]	Becket, 1,2	23	Mardin
4	Bangor	24	Morris
5	Charlton	26	Qguga
13	Hermon	19	Lackawanna
28	Paxton	20	Langford
13	Essex	22	Lordstown
2. *Sand and Gravel Outwash*		**6. *Limestone, Sandstone and Shale***	
1	Adams	*till, Calcareous*	
9	Colton	12	Erie
8	Colonie	17	Honeoye
16	Hinkley	25	Ontario
30	Riverhead	33	Sodus
36	Windsor	21	Lansing
11	Elnora		
3. *Sand and Gravel Outwash, Calcareous*		**7. *Shale and Siltstone***	
6	Chenango	14	Gilpin
27	Palmyra		
18	Howard		
4. *Silt and Clay Beds*			
31	Schoharie		
7	Collamer		
29	Rhinebeck		
34	Vergennes		
10	Elmwood		

[a] These numbers are from an alphabetical listing of the soils from 1 to 36.

subgroups. The mean values for Grouping III are given in Table 7 and the ranking of these means in Table 8. Here several important trends were noted. The best soils for removing phosphate were the acid tills. The acid tills were better than the calcareous tills, and the tills were better than the outwashes. Again within the outwashes the acid ones removed more phosphate than the calcareous ones.

The soil with the highest capacity to remove phosphate was the Becket series, which was an acid till, sampled in the New York State Adirondacks. Data for this soil series—which is classified as a Spodisol, typic fragiorthod, coarse loamy, mixed frigid, and was sampled at two different sites—are supplied in Table 9. At the other extreme a beach sand from Long Island had a b max of 0.3 mg P/100 g.

The mineralogy data for the clay, silt and sand size fraction on the C horizon were obtained. The data on the sand fraction

Table 5. F values for two-way analysis of variance.

Variable	Grouping 1			Grouping 2			Grouping 3		
	S. Groups	Horizons	Interact.	S. Groups	Horizons	Interact.	S. Groups	Horizons	Interact.
Thickness (in.)	7.0	55.2	1.92[b]	4.73	61.7	NS	3.58	40.2	NS
B.D. (lb/cu ft)	4.67	26.8	NS[e]	3.85	33.5	NS	3.72	29.5	NS
Coarse Fragments (%)	2.87[a]	NS	NS	7.09	2.57[d]	NS	25.2	5.26	NS
Silt and Clay (%)	8.42	NS	NS	46.3	3.23[c]	NS	6.15	4.34[a]	NS
pH Units	10.8	3.37[a]	NS	18.0	6.10	NS	33.7	6.17	NS
B max (mg P/100 g)	7.16	7.26	NS	14.2	15.5	NS	11.4	8.73	NS
x/m @ 5 (mg P/100 g)	5.94	4.49	NS	12.0	10.5	NS	9.06	5.73	NS
CEC (meq/100 g)	2.21[c]	6.15	NS	4.41	13.7	NS	5.68	12.1	NS
B max - (lb P)	NS	30.9	NS	6.79	70.4	1.92[b]	5.31	49.3	NS
x/m @ 5 (lb P)	2.20[c]	27.4	NS	8.28	71.7	2.25	4.43	41.2	NS
CEC (lb eq)	NS	9.9	NS	7.09	22.1	NS	8.83	23.3	2.57
Degree of Freedom	7/104	3/104	21/104	8/100	3/100	24/100	6/100	3/108	18/108

[a]Significant at 0.01.
[b]Significant at 0.025.
[c]Significant at 0.050.
[d]Significant at 0.070.
[e]Not significant; values not footnoted are significant at 0.005.

Table 6.   F values for covariance analysis with soils within subgroups as replicates with B.D., S+C and pH as covariates.

	Grouping 1			Grouping 2			Grouping 3		
	A	B	C	A	B	C	A	B	C
B max	—	3.05[a]	NS	2.81[a]	7.85	NS	4.89	4.41	NS
x/m @ 5	—	2.77[b]	NS	—	7.02	NS	—	4.10[d]	NS
CEC	—	2.88[a]	NS	NS	2.11[c]	NS	3.64	2.73[b]	1.79[e]

[a] Significant at 0.025.
[b] Significant at 0.050.
[c] Significant at 0.100.

[d] Significant at 0.006.
[e] Significant at 0.010.

Note: If not noted values are highly significant at the 0.005 level; NS = not significant.

gave the highest correlations with b max and CEC. For all the soils, percentages of quartz, potassium feldspar and plagioclase feldspar each had correlation coefficients of -0.5 to -0.7 with b max and CEC. The percentages of leachable iron and aluminum and those of silt plus clay and mixed rock fragments were positively correlated with b max and CEC. However, when these C horizon soils were broken down into subgroups according to the Grouping III parent materials classification, the absolute values of correlation coefficients decreased with the exception of percent silt plus clay which increased to the 0.7 to 0.9 range. In other words for soils of a broad diversity of parent materials, the minerals gave higher correlation with b max and CEC than for soils of a narrow diversity of parent materials.

It was noted from the detailed data on the soil horizons that, at constant percent silt plus clay, the soil b max decreased with increasing depth. This trend was further investigated by testing deep outwash and till samples (Table 10) from various locations across New York State. Thirteen depth-composited samples ranging from 25–80 ft in depth gave an average uncorrected b max of about 9.5 mg P/100 g and an x/m @ 5 value of about 4.25 mg P/100 g, which was lower than for most C horizon samples noted in Table 7.

COLUMN STUDY DATA

A long-term column study on three soils, Adams, Windsor and Honeoye, was conducted. The columns were 6 ft (1.8 m) tall and had a 4.75-in. (12-cm) i.d. The columns were dosed with 50 ml

Table 7. Grouping three means.[a]

Sub-Group	B max − mg P/100 g							
	2	1	3	4	6	7	5	Avg.
A	26.8	67.9	26.6	48.4	40.7	64.1	50.2	45.7
B	30.6	81.2	19.1	39.6	25.9	70.9	37.9	42.9
C	12.9	18.9	5.0	31.4	13.4	49.0	22.3	18.9
T	20.9	45.2	13.9	35.4	22.6	65.3	31.9	30.9
	x/m @ 5 − mg P/100 g							
A	18.5	44.9	20.3	34.1	21.9	55.2	33.1	30.5
B	23.4	68.1	14.1	27.3	15.8	59.5	28.5	33.4
C	8.6	13.7	2.8	17.7	7.8	38.4	13.9	12.2
T	15.8	35.5	10.4	22.2	13.4	54.3	22.4	22.3
	CEC − meg/100 g							
A	12.8	31.7	12.4	17.5	19.7	22.6	14.6	18.2
B	5.2	12.1	5.3	16.6	7.6	10.5	6.1	9.0
C	2.2	4.1	3.7	16.1	5.9	14.5	6.6	6.5
T	4.0	9.1	5.1	16.4	7.9	12.2	7.1	8.3
	B max − lb P/ft^2							
A	0.0098	0.0289	0.0157	0.0286	0.0202	0.0058	0.0169	0.0197
B	0.0720	0.1413	0.0616	0.0635	0.0852	0.1133	0.1008	0.0918
C	0.0435	0.0629	0.0096	0.1395	0.0431	0.0295	0.0626	0.0615
T	0.1234	0.2464	0.0887	0.2331	0.1527	0.1497	0.1861	0.1780
	x/m @ 5 − lb P/ft^2							
A	0.0076	0.0177	0.0121	0.0202	0.0105	0.0050	0.0121	0.0131
B	0.0550	0.1178	0.0474	0.0434	0.0522	0.0951	0.0713	0.0692
C	0.0257	0.0447	0.0059	0.0761	0.0240	0.0231	0.0365	0.0369
T	0.0886	0.1910	0.0669	0.1419	0.0884	0.1244	0.1253	0.1231
	CEC − lb equivalent ft^2							
A	0.0048	0.0130	0.0075	0.0106	0.0106	0.0021	0.0042	0.0080
B	0.0122	0.0210	0.0156	0.0177	0.0227	0.0168	0.0192	0.0179
C	0.0099	0.0127	0.0085	0.0773	0.0170	0.0087	0.0187	0.0228
T	0.0270	0.0512	0.0323	0.1151	0.0529	0.0280	0.0430	0.0514

[a] All values reduced by a linear correction for % CF.

of 50 mg P/l tap water solution every 30 min which gave an average application rate of 5 gpd/sq ft (1.76 l/sq md). The orthophosphate compound used was KH_2PO_4. Unsaturated flow was maintained in the columns, which are shown in Figure 3, with the black cloth removed. Black cloth was wrapped around the columns to prevent light from stimulating algae growth. The column effluent drained continuously into large sample containers. On Monday, Wednesday and Friday, the total effluent volume was measured and a portion of the 2- to 3-day composite sample analyzed, while the remainder was wasted. Table 11 gives data on the three soils, an acid one, a neutral one and a cal-

Table 8. Ranking of soil subgroups in grouping 3.

Rank for P	CEC	Group	Parent Material	Comments
1	3	1	Granite, gneiss till	Acid tills dominate in the New York Adirondacks S + C^a = 26.1%
2	4	5	Sandstone, siltstone and shale till	These soils often have high water table or rock limitations S + C = 45.4%
3	5	7	Shale and siltstone	Only one soil, "Gilpin," sampled has high rock limitation S + C = 62.9%
4	1	4	Silt and clay beds	Soils high in clay, impermeable with occasional high water tables S + C = 85.2%
5	2	6	Limestone, sandstone and shale till, calcareous	Calcareous tills, often in Western New York State S + C = 45.2%
6	7	2	Sand and gravel outwash	Outwash S + C = 12.5%
7	6	3	Sand and gravel outwash, calcareous	Calcareous outwash often in Western New York State S + C = 13.8%

^{a} Average % silt + clay for the total horizon sampled.

careous one. Pure quartz silica sand was mixed with the Honeoye soil to improve permeability and to maintain unsaturated flow. The tops of the columns had styrofoam chips to distribute the dose, while the bottoms had glass wool to prevent sand leakage.

The phosphate and pH data for the three columns are presented in Figures 4, 5 and 6. The Adams and Windsor columns first were dosed for about 10 days with tap water after which tap water with about 50 mg P/l orthophosphate was fed. The Adams soil was stored in a chemistry laboratory before placement in the column, and may have had some base poured in it which initially raised its pH. The Adams column was dosed for about 3 months, rested for about 9.8 months, and then dosed again for about 2 months. The Honeoye and Windsor columns were dosed with orthophosphate in tap water for about 8 months, rested for 4.3 months and dosed again for about 2 months. On June 20, 1975, one 3-gram dose (100 ml of 3 percent) of H_2O_2 was added to each of the columns and no change in column ef-

Table 9.   Bracket soil series data.

Horizon – Becket 1	A_p	B 21	B 22	B 23	Cx
Thickness – in.	5	3	7	10	25
pH – units	5.1	5.0	5.3	5.35	5.69
B max – mg P/100 g^a	125.0	277.2	193.4	75.5	29.6
x/m @ 5 – mg P/100 g^a	104.1	221.2	180.1	67.9	27.0
x/m @ 0.5 – mg P/100 g^a	41.6	78.5	111.1	35.8	15.2
CEC meg/100 g^a	26.5	33.7	19.8	11.8	6.5
Silt + clay % < 0.062 mm	43	43	34	23	24.5
Clay % < 0.002 mm	2	2	3	0	0.5
Coarse fragments % > 2 mm	12	12	17	32	22.5

Horizon – Becket 2	O & A	B 21	B 22	B 23	Cx
Thickness – in.	7	6	5	10	47
pH – units	4.4	4.91	4.95	5.0	5.2
B max – mg P/100 g^a	56.7	240	161.5	83.5	13.7
x/m @ 5 – mg P/100 g^a	6.3	214	135.7	88.8	10.0
x/m @ 0.5 – mg P/100 g^a	0.7	109.2	55.6	45.4	3.0
CEC meg/100 g^a	76.7	35.5	16.9	10.9	1.7
Silt + clay % < 0.062 mm	21	21	17	18	16.5
Clay % < 0.002 mm	2	2	2	1.5	0.5
Coarse fragments % > 2 mm	15	20	15	16.5	21.5

^{a} Corrected for % coarse fragments.

Table 10.   Deep soils phosphate data.

Sample	B max^a	K^a	x/m @ 5^a	x/m @ 0.5^a
DOT No. 1	16.1	0.444	11.1	2.9
2	4.8	0.348	3.05	0.71
3	4.1	0.162	1.84	0.31
4	12.5^b	0.320	7.70	1.72
5	7.3	0.343	4.62	1.07
6	5.9	0.0485	1.15	0.14
7	10.7^b	0.156	4.68	0.78
8	6.5	0.171	2.98	0.51
9	8.9	0.25	4.95	0.99
10	6.8	0.134	2.72	0.43
11	14.0	0.238	7.62	1.48
Lake George Avg.	17.5	0.0476	3.38	0.41
Paul Smith's Avg.	2.9	0.25	1.60	0.32
Deep Avg.	9.08	–	4.42	0.91

^{a} From equation 1, Langmuir equation.
^{b} For low concentration range of curve.

Figure 3

fluent phosphate or pH was noted. The columns were frequently inspected and did not exhibit clogging or slime growths. It appeared that aerobic conditions and unsaturated flow were maintained throughout the study.

All three columns showed a very rapid increase in effluent P from less than 0.01 mg/l to over 10 mg/l. This took about a month for the Adams and Honeoye soils and about 2 months for the Windsor soil. The Honeoye column was the slowest in approaching saturation after 25 mg P/l was reached. Toward the end of the study, an error in adding phosphate occurred and the

Table 11. Data on three soils—column study.

Parameter	Adams 0.11 ft	Windsor C_1	Honeoye C_1	Honeoye C_1 + 48% Sand[a]
Name	Spodosol, Typic Haplorthod, Sandy mixed mesic	Spodosol, Entic Haplorthod, Sandy mixed mesic	Alfisol, Glossobaric, Hapludalf, fine loamy, mixed mesic	
pH – 1/1-soil/H_2O	5.5	6.5	8.0	–
Dry Bulk Density lb/ft^3	100	100	110	105
% Coarse >2 mm	20	2	16	8
% Silt + Clay <0.062 mm	1	16	44	22
% Clay <0.002 mm	<0.5	<1	7	3.5
Phosphate Adsorption Values				
B max – mg P/100 g	11.6	20.7	22	10.3
x/m @ 50 mg P/1	11.0	20	18.5	8.4
k – 1/mg	0.21	0.50	0.12	–

[a] 52% Honeoye C_1 and 48% pure silica sand with an effective size = 0.35 mm, U.C. = 2.0, B max = 0.3–1.0 mg P/100 g.

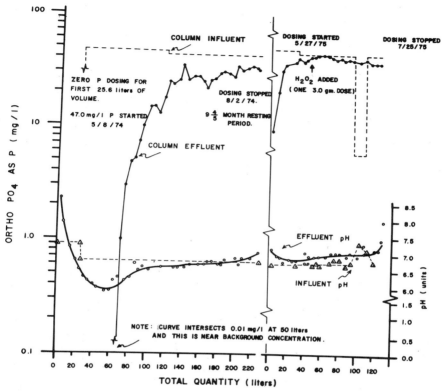

Figure 4.   Adams soil column, 0–11 ft composite.

influent contained only 6 mg P/1 for several days. During and after this period the effluent P concentration decreased only slightly from about 40 mg P/1 and leached orthophosphate from the column. However, after the 4.3-month resting period, the effluent concentrations were lower (8–25 mg P/1) and indicated that some phosphate had become nonleachable in the column. Fox and Kamprath (8) and Syers *et al.* (7) also found only partial desorption of adsorbed phosphate; the remainder was fixed. The column loadings are compared with the isotherm values in Table 12. At 10 mg P/1, the ratio of actual column P pickup to isotherm pickup was about 1/1 for the Adams and Windsor soils and 2/1 for the Honeoye. The maximum ratios varied from 2/1 to 5/1 for about 10 months dosing and 4.3 months resting. The longer resting period for the Adams soil relative to the Windsor soil may have contributed to its second phosphate pickup. Perhaps some phosphate precipitation or dif-

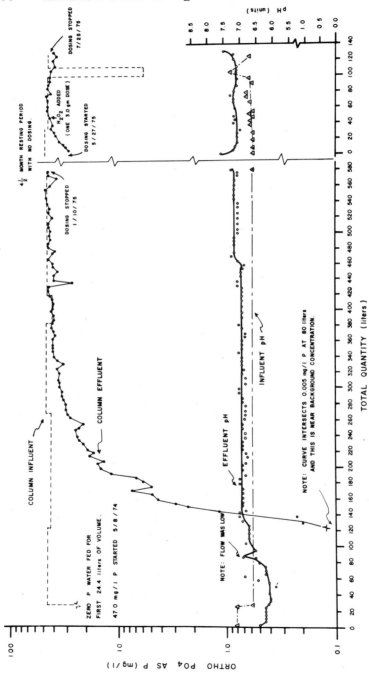

Figure 5. Windsor CI soil column.

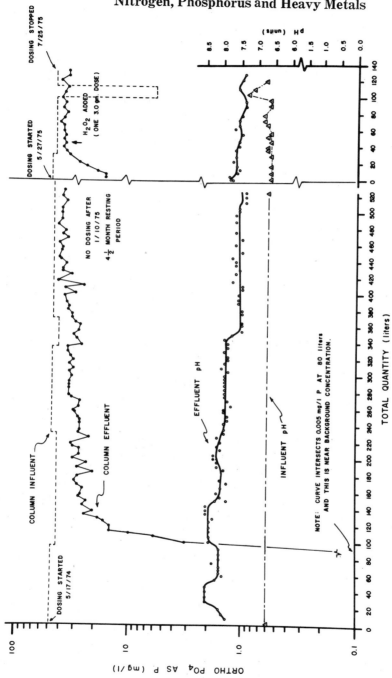

Figure 6. Honeoye soil column.

Table 12.   Column loading comparison.

Column	Adams	Windsor	Honeoye
Moisture range %	1.0–7.2	2.0–6.2	3.0–7.5
Column soil dry wt – kg	30.78	29.83	29.06
Isotherm – x/m @ 50 mg P/100 g	11.0	20.0	8.4
Isotherm P pickup – g	3.39	5.96	2.44
Actual P Pickups – g			
A. at 10 mg P/1	3.19	6.69	5.21
B. at end of first dosing period	5.47	10.75	11.03
C. max during 2nd dosing period	6.60	11.30	12.24
Ratios – Actual/Isotherm			
A. at 10 mg P/1	0.94	1.13	2.14
B. at end of 1st dosing period	1.61	1.80	4.53
C. max during 2nd dosing period	1.95	1.90	5.02

ferent reaction occurred at the high pH in the Honeoye column to boost its removals.

INTERMITTENT SAND FILTER DATA

Five intermittent slow sand filters with about a 2.0-ft (0.61-m) depth of sand and underdrains were sampled 5–7 times over a period of several months in 1975. Only three of the systems were working and were reported on in Table 13, along with the average of 18 New York sand filters sampled only once or twice in 1972. Some phosphate removal occurred in the sand filters even after long periods of operation. The total P concentrations were higher for the 18 sand filters sampled in 1972 than for those studied in 1975 because a New York law limiting phosphate in detergents was passed in 1973. These data are considered very rough and not conclusive. Brandes *et al.* (28) studied 10 pilot intermittent (1 gpd/sq ft) sand filters for 2–4 years and noted total P removal varying from 1 percent to 88 percent. P removal was highest when silt or red mud bauxite waste was mixed with the sand filter sand. The removals decreased with time however (29). The Becket B horizon soil mentioned previously, in mixture with sand filter sand would greatly improve sand filter phosphate removals although the life would still be limited. These intermittent filters are often used for treating small volumes of sewage effluents in areas where the natural soils are unsuitable. Various "mound" systems using special fill material also could be used.

Table 13.   Data on intermittent sand filters.

Filter	BC	LS	RV	Avg. of 18 N.Y.S. Plants
Sand Effective Size – mm	0.40	0.3–0.6	0.47	0.3–0.6
Years of Operation – yr	0.5	8	8	20
Application Rate – gpd/sq ft (1/sq m–day)	1.15	1.15	10	1–10
Avg. Influent Total P – mg/1	5.04[a]	3.76[a]	2.44[a]	10.1[b]
Avg. Effluent Total P – mg/1	2.24[a]	1.98[a]	2.00[a]	6.38[b]
% P Removal	56	47	18	36[b]

[a] Average of 5–7 samplings in 1975.

[b] Eighteen sand filters sampled once or twice. In some cases the influent sample was taken before an additional treatment unit such as a settling tank. The average total P for all N.Y. sewage treatment plants in 1972 was:
Influent = 12.57 mg P/1   Effluent = 9.1 mg P/1   Removal = 28%

REFERENCES

1. Tofflemire, T. J., *et al.* "Phosphate removal by sands and soils," Research Unit Technical Paper 31, N.Y. State Dept. of Environmental Conserv., 50 Wolf Rd., Albany, N.Y. 12233 (1973).
2. Reed, S., *et al.* "Wastewater management by disposal on the land," Special Report No. 171, U.S. Army Corps of Engineers, Hanover, New Hampshire (May 1972).
3. Miller, R. H. "Soil as a biological filter," paper presented at Penn State Univ. Conference on Recycling Wastewater (Aug. 21-24, 1972).
4. Rouston, R. C. and R. E. Wilding. "What happens in soil disposal of wastes?" *Chemical Engineering Progress Symposium*, Series 97, Vol. 5 (Oct. 1970).
5. Ellis, B. G. and A. E. Erickson. "Movement and transformation of various phosphorus compounds in soils," Soil Science Dept., Michigan State Univ. and Michigan Water Resource Commission (July 1967–June 1969).
6. Wright, B. and M. Peech. "Characterization of phosphate reaction products in acid soils by application of solubility criteria," *Soil Sci.* 90(1): 32–43 (1960).
7. Syers, J. K., *et al.* "Adsorption and desorption of phosphate by soils," *Soil Sci. Plant Anal.* 1: 57–62 (1970).
8. Fox, R. L. and E. J. Kamprath. "Phosphate sorption isotherms for evaluating the phosphate requirements of soils," *Soil Sci. Soc. of Amer. Proc.* 34: 902–907.
9. Monke, E. J., *et al.* "Movement of pollutant phosphorus in unsaturated soil," Technical Report No. 46, Purdue Univ., Water Resources Research Center, West Lafayette, Indiana 47907 (June 1974).
10. Barrow, N. J. and T. C. Shaw. "The slow reactions between soil and anions: 3. The effects of time and temperature on the decrease in isotopically exchangeable phosphate; and 2. The effect of time and temperature on the decrease in phosphate concentration in the soil solution," *Soil Sci.* 119: 190; 167 (1975).

11. Olsen, S. R. and F. S. Watanabe. "A method to determine a phosphorus adsorption maximum of soils as measured by the Langmuir isotherm," *Soil Sci. Soc. Amer. Proc.* 21: 144 (1957).

12. Syers, S. K., *et al.* "Phosphate sorption by soils evaluated by the Langmuir adsorption equation," *Soil Sci. Soc. of Amer. Proc.* 37: 358–363.

13. Enfield, C. G. and B. E. Bledsoe. "Kinetic model for orthophosphate reactions in mineral soils," EPA–660/2–75–022, Robert S. Kerr Environmental Research Laboratory, P.O. Box 1198, Ada, Oklahoma 74820 (1975).

14. Humphreys, F. W. and W. L. Pritchett. "Phosphorus adsorption and movement in some sandy forest soils," *Soil Sci. Soc. Amer. Proc.* 35: 495–500 (1971).

15. Saini, G. R. and A. A. MacLean. "Phosphorus retention capacities of some New Brunswick soils and their relationship with soil properties," Canada Dept. of Agriculture, Fredericton, New Brunswick, *Canadian J. Soil Science* 45: 15 (1965).

16. Toth, S. J. and F. E. Bear. "Phosphorus-adsorbing capacities of some New Jersey soils," *Soil Sci.* 64: 199 (1947).

17. Hall, M. W. "Water quality degradation of septic tank drainage," Maine Univ., Orono Water Resources Research Center (1970).

18. Willrich, T. L. and G. E. Smith. *Agricultural Practices and Water Quality,* The Iowa State Univ. Press, Ames, Iowa (1970).

19. Woodruff, J. R. and E. J. Kamprath. "Phosphorus adsorption maximum as measured by the Langmuir isotherm and its relationship to phosphorus availability," *Soil Sci. Soc. Proc.* 29: 148–150.

20. Roth, J. R. "The equilibrium between labile and crystalline phosphates in soils," Ph.D. Thesis, Cornell Univ. (1968).

21. Tofflemire, T. J. and M. Chen. "Evaluation of phosphate adsorption capacity and cation exchange capacity and related data for 35 common soil series in New York State," Technical Report No. 45, N.Y. State Dept. of Environmental Conserv., 50 Wolf Rd., Albany, N.Y. 12233 (1976).

22. Ryden, J. C., J. K. Syers, and R. F. Harris. "Nutrient enrichment of runoff waters by soils, phase I: phosphorus enrichment potential of urban soils in the City of Madison," Available from the Dept. of Soil Science, Univ. of Wisc. (1972).

23. Shah, D. B., *et al.* "A mathematical model for phosphorus movement in soils," *J. Environ. Quality,* 4(1): 87 (1975).

24. Novak, L. T., *et al.* "Phosphorus movement in soils: theoretical aspects," *J. Environ. Quality* 4(1): 93 (1975).

25. Arnold, R. W., *et al.* "Generalized soil map of New York," Scale 1:250,000, Cornell Univ., Dept. of Agronomy, Ithaca, N.Y. (1971).

26. Cline, M. G. "Soils and soil associations of New York," Extension Bulletin 930, Cornell Univ., Dept. of Agronomy, Ithaca, N.Y. (1970).

27. NWAY-1 Program 1972, Univ. of Wisconsin Academic Computing Center, Madison, Wisconsin.

28. Brandes, M., *et al.* "Experimental study of removal of pollutants from domestic sewage by underdrained soil filters," Ontario Ministry of the Environment, 135 St. Clair Ave., W., Toronto, Ontario (1974).

29. Chowdhry, N. A. "Domestic sewage treatment by underdrained filter systems," Ontario Ministry of the Environment, 135 St. Clair Ave., W., Toronto, Ontario (1974).

30. Ellis, B. G. "The soil as a chemical filter," *Michigan Agr. Exp. Station J.,* Article No. 6020, Michigan State Univ., Dept. of Soil Science. Also part of *Recycling Treated Municipal Wastewater and Sludge through Forest and Cropland,* Penn State Univ. (1973).

10

LAND APPLICATION OF WASTEWATER: FORAGE GROWTH AND UTILIZATION OF APPLIED NITROGEN, PHOSPHORUS AND POTASSIUM

A. J. Palazzo
U. S. Army Cold Regions Research and Engineering Laboratory,
Hanover, New Hampshire

INTRODUCTION

The renovation of wastewater by land application has received considerable attention in recent years. Land treatment has been gaining interest due to the provisions in Public Law 92–500 which state that "discharges of pollutants into navigable waters must be eliminated by 1985." In order to meet this requirement many of the nation's secondary sewage treatment facilities will require upgrading to either tertiary treatment or an equivalent alternative, such as land treatment. Land treatment relies on a combination of factors in the management of the system to renovate wastewater to a desired goal. Among major considerations are the application rate, plant uptake of wastewater constituents, and soil removal mechanisms. The removal capabilities of plant species are of special importance.

Vegetation is an integral part of a slow infiltration land treatment system. Sopper (1) has reported a partial list of criteria for selecting a vegatative cover to be established and maintained on a slow infiltration site. In addition to removing pollutants, plants provide other benefits to the site, such as the reduction of soil erosion and the maintenance of soil permeability.

Wastewater compositions vary among areas and most land treatment systems attempt to remove a certain percentage of the wastewater constituents, particularly nitrogen, so that a desired water quality can be obtained. Maximum application rates of wastewater to a land treatment system, without impairment to the vegetation, are desirable because less land is required to treat the wastewater. Therefore, information is needed on the capabilities of plants to renovate wastewater at various application rates. The objective of this two-year study was to characterize the effectiveness of a forage mixture in renovating wastewater when applied to land at several application rates. Nutritional changes through soil and plant analysis were determined to permit correction, where necessary, through proper management procedures.

MATERIALS AND METHODS

The research was conducted at the U.S. Army Cold Regions Research and Engineering Laboratory (CRREL), Land Treatment Research Facility, in Hanover, New Hampshire during 1974 and 1975. One component of this facility is a set of six outdoor land treatment test cells (Figure 1), which have been described in Iskandar *et al.* (2). The test cells are constructed of concrete (8.4 m x 8.4 m x 1.5 m) and are equipped with water sampling devices. The data in this report are from observation of three test cells which contain a Windsor sandy loam soil. The

Figure 1. The outdoor land treatment tests cells.

A, B and C soil horizons were separated in the field, transported to CRREL and placed in the cells. The soils were compacted to the original *in-situ* bulk density to a 1.5-m depth. Background chemical data for the upper 15 cm of the soil are shown in Table 1. Chemical analyses were run on soils that were air-dried and passed through a 2-mm sieve. Selected soil properties were determined as follows: soil pH—measurements with a glass electrode pH meter; soluble salts—conductivity on a 1:2 soil and water mixture; total phosphorus—the acid digestion procedure, extractable phosphorus—the Bray P_1 technique (3); total carbon —the wet oxidation method (4); and exchangeable cations— atomic absorption spectrometry (5).

Domestic wastewater was collected from Hanover, New Hampshire, treated and uniformly applied to the test cells by spray irrigation with a spray circle diameter of 7.6 m. The application rates were 5.0 and 15.0 cm/wk of secondary wastewater and 7.5 cm/wk of primary wastewater. As noted in Table 2, both types of wastewater were similar in the concentration of constituents which would directly affect plant growth. All wastewater was disinfected with ozone. A forage mixture that included reed canarygrass (*Phalaris arundinacea* L.), timothy (*Phleum pratense* L. var. 'Climax') and smooth bromegrass (*Bromus inermis* Leyss. var. 'Lincoln') was seeded at the rate of 12.1, 6.6, 5.5 kg/ha, respectively, on 21 May 1973.

Prior to July 1974, the wastewater applied contained an average total nitrogen and phosphorus concentration of 37.5 and 11.5 ppm, respectively. In July 1974 the wastewater was diluted in order to lower the nitrogen concentration to more closely simu-

Table 1.   Initial chemical data of the Windsor sandy loam soil (0 to 30 cm depth).

Parameter	Soil Content
Organic carbon (%)	0.99
Total phosphorus (ppm)	546.00
Exchangeable phosphorus (ppm)	67.00
pH	6.40
Soluble salts (mmho/cm)	0.30
CEC (meq/100 g)	6.76
Exchangeable cations (meq/100 g)	
Ca	2.16
Mg	0.47
K	0.10
Na	0.37

Table 2.   Average analysis of wastewater—CRREL.

Parameter	Primary Effluent	Secondary Effluent
pH	7.5	7.6
Total nitrogen (ppm)	26.4	26.9
Total phosphorus (ppm)	7.0	7.1
Potassium (ppm)	8.3	8.8
Conductivity (μmhos/cm)	394.0	402.0
Organic carbon (ppm)	56.0	47.3
Ammonium-N (ppm)	22.1	21.6
Nitrate-N (ppm)	0.6	2.4

late the outflow from a typical secondary treatment plant. An analysis of the wastewater after dilution is shown in Table 2.

The soils were sampled in spring 1974 and 1975 to a depth of 30 cm, and were analyzed for pH and exchangeable cations Ca, Mg, K and Na. The forage was cut three times per year at a height of 7.5 cm. The harvest dates included 3 June, 28 July, and 17 September in 1974 and 12 June, 23 July, and 23 September in 1975. The forage was harvested with a sickle bar mower. The vegetation within the 7.6-m spray circle or treated area was collected and weighed to obtain the total fresh weight per cell. Randomized grab samples were dried at 110°C and dry weights per plot were determined. The grab samples were analyzed for nitrogen, phosphorus and potassium. The untreated area outside the spray circle was also harvested. The dried plant tissue was ground through a 20-mesh sieve, and sample preparation for all elemental analysis was performed by dry ashing for 4 hr at 260°C. Standard analytical procedures (5) utilized were as follows: nitrogen—Kjeldahl; phosphorus—colorimetric; and potassium—atomic absorption.

RESULTS AND DISCUSSION

In September 1975 the botanical composition of the cells was observed approximately 29 months after seeding. The wastewater-treated areas contained predominantly quackgrass [Agropyron repens (L.) Beauv.]. The untreated areas contained the following species: quackgrass, smooth bromegrass, orchardgrass (Dactylis glomerata L.), timothy, and Kentucky bluegrass (Poa pratensis L.). The differences in botanical composition between the two areas can be directly related to the high amounts of water and nutrients applied to the treated areas. The presence of quackgrass in the test cells was probably due to the transportation of rhizomes of this species to the site during construction and its greater establishment rate from rhizomes as compared to the sown species from seed. Herbicides were not used to control weedy species.

Dry matter yields of the forage during 1974 and 1975 ranged from 9.63 to 13.68 metric tons/ha with the highest application rate producing the greatest yields (Figure 2). Average yields during both years showed that applications of 7.5 cm/wk produced only slightly greater yields than the forages receiving 5.0 cm/wk. Yields of forages receiving 7.5 cm/wk of wastewater were similar during both years, averaging approximately 10.65 metric tons/ha per year, while forages receiving 5 cm/wk of wastewater produced 9.63 and 11.30 metric tons/ha of dry matter in 1974 and 1975, respectively. The increase in yields in 1975 for the 5-cm/wk treatment was primarily related to the large amount of dry matter produced at the first cutting on 12 June. The data from this study were insufficient to warrant a definite conclusion and further research is being conducted to investigate the above observation.

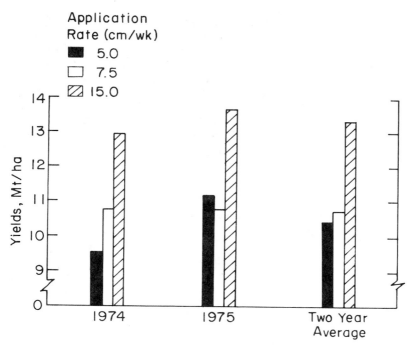

Figure 2. Yields of forages in 1974 and 1975.

Forage concentrations of nitrogen (N) and phosphorus (P) were within the recommended range for plant growth (Table 3). Nitrogen ranged from 2.91 percent to 3.50 percent and phos-

Table 3. Chemical analysis of forage tissue following treatment with three rates of wastewater.

Application Rate (cm/wk)	Nitrogen		Phosphorus		Potassium	
	74	75	74	75	74	75
				(%)		
5.0	3.20	2.91	0.35	0.27	2.67	2.10
7.5	3.50	2.93	0.32	0.29	2.37	2.22
15.0	3.47	3.39	0.35	0.32	2.72	2.47
Untreated	2.63	2.51	0.30	0.26	2.08	1.94

phorus from 0.27 percent to 0.35 percent. Potassium (K) concentrations during 1975 were marginally deficient and were lower than had occurred in 1974. The K concentrations in this study during 1975 were 2.10 percent and 2.22 percent, and 2.47 percent for the 5.0, 7.5, and 15.0 cm/wk application rates, respectively. Martin and Matocha (6) have reported that levels of K below 2.0 percent and 2.2 percent for orchardgrass and tall fescue, respectively, constitute deficient levels of these two species. Forages grown in the untreated areas contained lower concentrations of N, P and K than those treated with wastewater.

The soils in the test cells were sampled and analyzed in spring 1974 and again in spring 1975 to note changes in soil properties due to applications of wastewater. Soils sampled in spring 1975 showed reductions in soil pH and in the total amounts of ex-

Table 4. Soil analyses for the exchangeable Ca, Mg, K and Na cations at the 0–30-cm depth (average soil values for the 5.0, 7.5 and 15.0-cm/wk treatments).

Element	1974	1975	Reduction (%)
	(meq/100 g)		
Ca	2.34	0.91	61
Mg	0.47	0.10	79
K	0.10	0.06	40
Na	0.37	0.10	73
Total	3.28	1.17	64

changeable cations which included Ca, Mg, K and Na, as compared to soils sampled in spring 1974 (Table 4 and Figure 3).

After soil treatment with the various rates of wastewater, lower soil pH values were noted at the higher application rates; the wastewater had a pH of near neutrality. The pH of the soils receiving 5.0, 7.5 and 15.0 cm/wk of wastewater was 6.6, 6.1 and 5.8, respectively. The transformation of applied NH^+_4 to NO^-_3, which includes the release of H^+ in the soil, is thought to be the primary cause for the reduction in pH. In 1974, the soils contained an average of 3.28 meq/100 g of exchangeable cations and in 1975 this value was reduced 64 percent to 1.17 meq/100 g (Table 4). The reduction in the amounts of exchangeable cations can be primarily related to plant uptake and soil replacement by NH_4 which is present at a high concentration in the wastewater (Table 2).

Land applications of wastewater reduced the amount of exchangeable K in the soil at all application rates (Figure 3). The levels of K in the soil were low (0.04–0.08 meq/100 g or 15.6–31.2 ppm) in terms of supplying an adequate amount of K to plants (Doll and Lucas, 1973). The reduction in soil K is partially related to crop uptake. Crops require significant amounts of this element for growth. It is exceeded only by N in terms of crop removal of major elements. If both elements are not supplied in adequate amounts, N and K imbalances may occur within the plant resulting in reduced growth.

The wastewater utilized in this study contained approximately three times more N than K. In terms of application rates, the

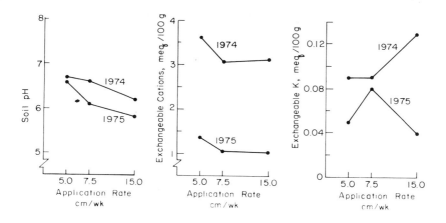

Figure 3.   Soil analyses for pH, total exchangeable cations, and exchangeable K at the 0 to 30-cm depth.

wastewater supplied approximately 114, 172 and 306 kg of K/ha per year while the forages removed an average of 247, 244 and 346 kg of K/ha per year at the 5.0, 7.5 and 15.0-cm/wk application rates, respectively. Therefore, the forages required more K than what was applied in the wastewater. The reductions in the concentrations of K in forages and soils, combined with the reduction in pH and total exchangeable cations in the soil, suggest that soil additions of dolomitic limestone to increase soil pH, Ca and Mg will have to be considered along with K fertilization in the management of the system in future growing seasons.

The contribution of the crop as a renovating mechanism in a land treatment system can be expressed by either the "total removal" or "removal efficiency" of applied wastewater constituents. Information as to the total removal and removal efficiency of applied P and, particularly, N is important when selecting a wastewater application rate to best fit the needs of the system.

Table 5 shows the data obtained on N and P removal during 1974 and 1975. Nitrogen applications ranged from 274 to 1,037 kg/ha, which goes beyond the quantities normally utilized in forage fertilization. Total removal of N and P increased during both years following increased application rates of wastewater. Forages receiving 15 cm/wk of wastewater removed 461 and 463 kg of N and 45 and 47 kg of P/ha during 1974 and 1975, respectively.

Increases in the application rate of wastewater decreased the removal efficiency of nutrients. Forage removal efficiency ranged from 43 percent to 119 percent of the N applied during the growing season (Table 5). At similar application rates, N removal

Table 5. Amounts of N and P applied to test cells and removed by forage.

Application Rate (cm/wk)	Nitrogen			Phosphorus		
	Applied (kg/ha)	Removed (kg/ha)	Removal Efficiency (%)	Applied (kg/ha)	Removed (kg/ha)	Removal Efficiency (%)
			1974			
5.0	416	308	74	116	34	28
7.5	622	369	59	174	34	20
15.0	1037	451	43	301	45	14
			1975			
5.0	274	328	119	73	31	42
7.5	422	315	75	108	31	29
15.0	819	463	57	222	44	20

efficiency was higher in 1975 than 1974. This is believed to be related to both the lower amount of N applied during the year and a greater amount of plant available N in the soil from the previous years applications. The total percentage of applied P removed by the forage was lower than that for N, which is related to the lower requirement of this element by the forage. The total percentage of P removed by the forage ranged from 14 percent to 42 percent for both 1974 and 1975. It appears from the data that other removal mechanisms within the system are required, such as soil adsorption, to remove P in the wastewater. These mechanisms are available in the soil and have been shown in other studies (2, 8).

SUMMARY

Data have been presented on the growth and chemical composition of forages when influenced by various application rates of wastewater during 1974 and 1975. The results show that the greatest average annual forage yields and N and P removal occurred at the highest application rate (15 cm/wk). However, forage removal efficiency of applied N and P was greatest at the lowest application rate of 5 cm/wk. At this rate an average of 97 percent of the applied N and 35 percent of the applied P was contained in the forage.

Analyses performed in 1974 and 1975 showed a reduction in the levels of K in the soil and forage in 1975, relative to 1974, which indicates a requirement for K fertilization for sustained productivity. The reducton in K was related to the large quantities of this element required by crops and its low concentration in the wastewater. Soil analyses also showed reductions in soil pH and total exchangeable cations to levels which could be corrected by liming.

ACKNOWLEDGMENTS

The author gratefully acknowledges the helpful suggestions and assistance by H. L. McKim, G. O. Estes, W. E. Larson, and I. K. Iskandar and the financial support provided by the Military and Civil Works Research and Investigation Projects.

REFERENCES

1. Sopper, W. E. "Crop selection and management alternatives—perennials," pp. 143–154. In: *Proc. of the Joint Conference on Recycling Municipal Sludges and Effluents on Land.* Champaign, Illinois (1973).
2. Iskandar, I. K., R. S. Sletten, D. C. Leggett and T. F. Jenkins. "Waste-

water renovation by a slow infiltration land treatment system," CRREL Report (in press).

3. Bray, R. H. "Correlation of soil tests with crop responses to added fertilizers and with fertilizer requirement," Chapter 2. In: *Diagnostic Techniques for Soils and Crops*, H. B. Kitchen (ed.), Am. Potash Institute, Washington, D.C. (1948).

4. Black, C. A. (ed.). *Methods of Soil Analysis*, American Society of Agronomy, Agronomy Series No. 9 (1965).

5. Jackson, M. L. *Soil Chemical Analysis*. Prentice Hall, Inc. Englewood Cliffs, New Jersey (1958).

6. Martin, W. E. and J. E. Matocha. "Plant analysis as an aid in the fertilization of forage crops," pp 393–426. In: *Soil Testing and Plant Analysis*. L. M. Walsh and J. D. Beaton (eds.) (1973).

7. Doll, E. C. and R. E. Lucas. "Testing soils for potassium, calcium, and magnesium," pp. 133–152. In: *Soil Testing and Plant Analysis*, L. M. Walsh and J. D. Beaton (eds.) (1973).

8. Hook, J. E., L. T. Kardos and W. E. Sopper. "Effects of land disposal of wastewaters on soil phosphorus relations," pp. 179–195. In: *Conference on Recycling Treated Municipal Wastewater through Forest and Cropland*," Environmental Protection Technology Series, EPA–600/2–74–003 (1974).

11

NITRATE RELATIONSHIPS IN THE PENN STATE "LIVING FILTER" SYSTEM

J. E. Hook

Department of Crop and Soil Sciences, Michigan State University, Lansing, Michigan

L. T. Kardos

Department of Agronomy, The Pennsylvania State University, University Park, Pennsylvania

INTRODUCTION

The fate of nitrogen has been considered critical to the treatment of municipal wastewater by land application. Miller (1) Kardos and Sopper (2), Urie (3) and others have pointed out that nitrogen removal in wastewater irrigation systems is one of the greatest limitations to land application. At the Penn State Wastewater Renovation Research sites, the fate of nitrogen has been examined under many combinations of soil-plant systems irrigated with municipal sewage effluent. The particular emphasis in the Penn State studies has been to recharge potable quality water to the ground water. This means, in terms of nitrogen, that nitrate concentration of water leaving the root zone should not exceed 10 mg N/l. Kardos *et al.* (4) reviewed the ability of several wastewater irrigation sites, treated from 1963 until 1969, to prevent excess nitrate concentration below the root zone. Hook (5) examined the ability of four soil plant systems to prevent nitrate leakage when irrigated at 5 cm/wk with sewage effluent. The purpose of this report is to tie together these and other studies of nitrogen in the Penn State wastewater application sites and to extend these studies to the present research.

The approach is to first examine the changes in nitrate occurring in the soil at a depth of 120 cm as a result of wastewater treatments. The 120-cm depth was chosen because the most intense monitoring of soil water was done at that depth—a depth generally below the biologically active zone. A comparison of the mass of nitrogen added (input) to the mass of nitrogen leaching or removed by harvest (outputs) was made to help explain the renovation at each system. Following the examination of nitrate leaching in several long-term experimental plots, the effects of alternate management of irrigation or of vegetation on each soil-plant system are presented.

LAND TREATMENT SYSTEMS

Hardwood Forests

The wastewater used at Penn State was chlorinated secondary municipal sewage effluent. The amounts of three nitrogen forms in the wastewater for the 1967 to 1974 period are shown in Table 1. The shift in predominant mineral N species from ammonium to nitrate between 1968 and 1973 was due to changes in sewage treatment plant operation. The effect of this change will be considered later.

Several hardwood forest plots in the Pennsylvania State Gamelands No. 186 have been irrigated since 1964. One site, the hardwood "new gameland" area received effluent year-round at 5 cm/wk from November 1965 until 1975. The site, located on Morrison sandy loam soil, was nearly level and had no runoff.

Results of this wastewater application on nitrate in the soil water at the 120-cm depth is shown in Figure 1. The concentration of nitrate from 1967 to 1970 exceeded 10 mg N/l year round.

Table 1. Mean annual concentration in milligrams N per liter of several nitrogen species in effluent and effluent-sludge applied to all plots.

	Year							
	1967	1968	1969	1970	1971	1972	1973	1974
Effluent								
NO_3-N	5.7	4.4	4.8	5.2	9.4	14.7	16.0	14.9
NH_4-N	15.0[a]	15.4	12.3	8.9	7.2	5.5	1.9	2.4
ORG-N	4.2[a]	5.9	6.4	6.5	3.2	2.4	3.1	5.4
Effluent-Sludge								
NO_3-N	—	—	—	—	10.7	12.3	16.7	15.8
NH_4-N	—	—	—	—	36.0	49.9	25.6	30.0
ORG-N	—	—	—	—	32.9	37.6	24.5	15.7

[a]Estimate based on 1968 concentrations.

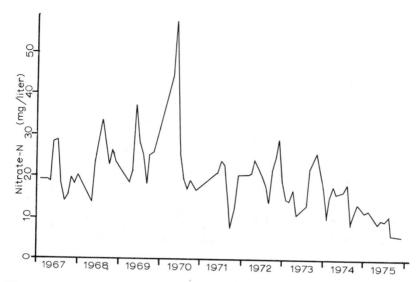

Figure 1. Mean monthly concentration of nitrate in soil water at the 120-cm depth of the hardwoods ("new gamelands") site which received 5 cm/wk of effluent year-round.

Concentrations generally were greatest in late spring and early summer. Figure 2 indicates the mass of N applied each quarter from 1967 until 1973 as well as the mass of N in the leachate. The amount of N which leached was generally equivalent to or greater than the amounts which were applied. During 3 months in 1971 and 4 months in 1972, effluent containing sludge (Table 1) was substituted for regular effluent application. Although N application during these periods was considerably greater than normal application, the amounts which leached during or subsequent to them remained about the same. For the entire period 1967 to 1973, the amount of N which leached was equivalent to 83 percent of the N applied by the wastewater.

To explain the observed ineffectiveness of this site in preventing nitrate leaching, we could at first ignore the effect of vegetation and further could assume all of the applied N is rapidly converted to nitrate. With year-round effluent additions at weekly intervals, we would expect the nitrate to rapidly move through the 120 cm of soil. The soil water would be diluted by rainfall and concentrated by evapotranspiration. Since rainfall exceeds evapotranspiration in this humid climate, the annual effect we would expect would be leaching of nitrate in amounts equivalent to nitrogen additions. The total N concentration, averaged an-

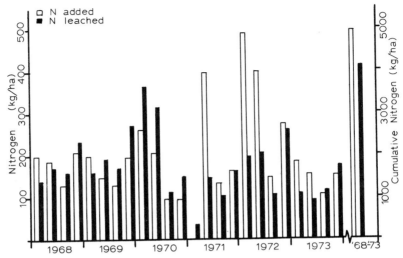

Figure 2. Quarterly and six-year totals of N applied to the hardwood site and N leached past the 120-cm depth.

nually, would be lower in the leachate than the effluent. This is essentially what we observed at the hardwoods site.

Of course, plant uptake and microbial transformations cannot be ignored, so this simple explanation is incomplete. The trees and herbaceous annuals take up N which is stored primarily in leaf tissue. Kardos *et al.* (4) reported that generally the hardwoods had increased N content in the leaves compared to control plots and also showed increased growth in response to effluent applications. The expected effect during the first year of effluent irrigation is net immobilization of N due to plant uptake. As the leaves from the first year decompose, mineralization plus second year effluent irrigation supply the needs of the plants the second year. This continues until mineralization of organic N accumulated over one or more years of treatment equals plant uptake and microbial immobilization. At this point the wastewater-soil plant system is operating at a kind of steady state. Net N accumulation by plants and microbes equals net mineralization, and N applied by wastewater equals N leaching. This steady-state system could occur in any land application system where denitrification or harvest does not occur. When dilution does not adequately lower nitrate concentration to acceptable levels, the life of a wastewater renovation system will depend on how long net accumulation exceeds net mineralization. For the hardwoods site on this sandy loam soil, this time was apparently short. A previously untreated hardwood plot adjacent to the "new gameland" site was irrigated identically with that site beginning May, 1972 (5). In the first 6 months the amount of N which leached was equivalent to 30 percent of the amount added. During the second 6 months (which included the first winter) the percentage increased to 62 percent. During the third 6 months, N leaching was slightly greater than N added. Concentration of nitrate in soil water at 120 cm during this third period averaged 20.6 mg N/l, slightly more than the total N in the effluent—19.5 mg N/l. Thus in approximately 12 to 18 months from the start of weekly irrigation the apparent steady-state condition seen in the long-term study was established.

Denitrification as suggested above is one means to prevent excess nitrate leaching even in a steady-state system. There was no evidence that denitrification had occurred. The effect of changing moisture condition on nitrate leaching was investigated by applying two applications of 2.5 cm of effluent 3 to 4 days apart rather than one 5 cm/wk application (5). The behavior of N in this application system was nearly identical to the single 5 cm/wk treatment. Perhaps higher hydraulic loads, for example 15 cm per day for 3 to 6 days followed by a rest period of a week or more, using the landscape as a rapid infiltration system similar to that of Bouwer and co-workers might result in improved N renovation but might also cause undesirable cover in the forest.

Another alternative for irrigation management on this hardwood area would be to decrease the total N load and use rainfall dilution. Beginning in 1972, one half of the 5-cm/wk treatment site received effluent at 2.5 cm/wk. During the three years which followed, nitrate concentration at the 120-cm depth decreased from an annual mean of 42.5 mg NO_3-N/l in 1970, to 21.5 in 1972, to 17.1 in 1973, 12.2 in 1974 and 10.3 in 1975.

Old Field Site

Another non-crop site at Penn State which received considerable attention was the old field–white spruce area. This site, located on Hublersburg clay loam soil, was an abandoned field which had a sparse stand of white spruce (*Picea glauca*, Moench) planted in 1955. It was treated with secondary effluent from April through November each year at a rate of 5 cm/wk from 1963 until 1971 and 7.5 cm/wk during 1972 and 1973.

The concentration of nitrate in soil water (Figure 3) at 120 cm consistently remained below 10 mg N/l during both the irrigation season and intervening winter through 1971. The mass

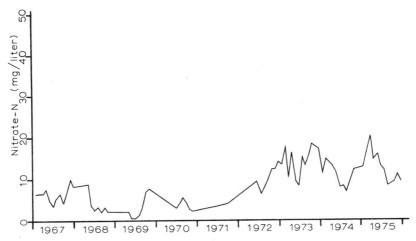

Figure 3. Mean monthly concentration of nitrate in soil water at the 120-cm depth of the old field site which received 5 to 7.5 cm/wk of effluent during April to November each year.

of N which leached annually from the site ranged from 26 to 44 percent of the mass of N applied annually in the wastewater for the period 1967 to 1971 (Figure 4). The site, when irrigated at 5 cm/wk during the warmer months with effluent containing N primarily as ammoniacal N, did an excellent job of preventing nitrate leaching and this with no harvest or residue removal.

Before this old field management can be extended to other sites, the factors which affected the N removal must be understood. In the old field site, dilution alone cannot account for the low percentage of nitrogen leaching. Net nitrogen immobilization, resulting from a build-up of soil organic matter and to some extent from increased biomass, could easily account for the N which did not appear in the leachate. From 1968 to 1971 the amount of N which did not leach was 811 kg N/ha. Richenderfer (6) reported an increase in soil N between 1963 and 1971 which would be equivalent to approximately 173 kg N/ha/yr in the upper 120 cm. Given the variability of soil organic N and the relatively small increase (11 percent) in that N which this 811 kg N/ha represents, the assumption that immobilization was the sole fate of nonleached N must be left open to question. Denitrification should be considered here because conditions were more favorable to denitrification than in the hardwoods. Soil texture was heavier, topsoil pH was higher, and herbaceous ground cover was heavy. Rapid growth of the ground cover would lower soil oxygen by root respiration and annual deposi-

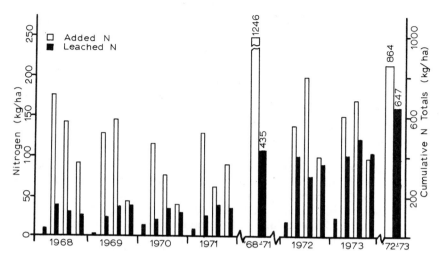

Figure 4.   Quarterly totals of N applied to the old field site and N leached past the 120-cm depth.

tion of large amounts of herbaceous residues may provide a source of energy to denitrifiers. Denitrification has not been studied directly in the field at any site at Penn State, and thus the denitrification contribution is only speculative.

Herbaceous vegetation was important in preventing nitrate leaching regardless of whether the effect was by stimulating denitrification or by immobilizing N. Kardos and Sopper (2) reported that nitrate leaching was sharply lowered following a vegetative change from red pine (*Pinus resinosa*, L.) to herbaceous weeds. The red pine plantation was located on the same soil and received the same treatment as the old field. Nitrate concentration averaged 24.2 mg N/l at the 120-cm depth in the red pine during the year preceding clearcutting. When the volunteer herbaceous vegetation became established, nitrate concentration dropped to an average of 8.3 mg N/l the first year and 3.8 mg N/l the second year following clearcutting.

Effluent irrigation at the old field at 5 cm/wk did not cause excessive nitrate leaching; however, when the irrigation rate was increased to 7.5 cm/wk in 1972 and 1973, the concentration of nitrate in soil water at 120 cm exceeded 10 mg N/l (Figure 3) and the percentage of applied N which leached increased (Figure 4).

The effect of the increased rate was somewhat confounded by a change in the relative proportion of nitrate and ammonia in the effluent (Table 1). The movement of N in soil water will depend strongly on the form of that N. The strongly adsorbed ammonium ion will move far less readily than the nitrate ion which is weakly held in some soils and not at all by others. Under equal soil water flow nitrate will move with the soil water out of the biologically active zone into the subsoil. Ammonium on the other hand would be adsorbed by soil in the biologically active zone while the water moves to the subsoil. The resulting placement of the N would affect its availability to microbial transformations and plant uptake during the week between irrigations.

Because nitrate concentrations in the soil water exceeded 10 mg N/l with the 7.5 cm/wk application, the application rate was lowered to 5.0 cm/wk in 1974 and 1975. The mean annual concentration of nitrate at 120 cm decreased to 10.9 mg NO_3–N/l in 1974 and 12.7 in 1975. During the last six months of 1975 the mean value was 9.7 mg NO_3–N/l.

Each of the above land application systems involved only irrigation rates, application periods and site effects as management

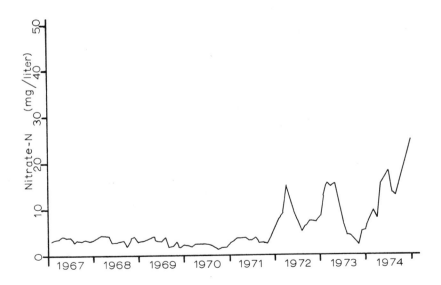

Figure 5.   Mean monthly concentration of nitrate in soil water at the 120 cm depth of the reed canarygrass site which received 5 cm/wk of effluent (1967 to 1970) or 5 cm/wk of effluent sludge (1971-1974) year-round.

alternatives. Use of harvested crops with land application of wastewater greatly expands the management alternatives which affect leachate quality. Additional benefits of usable crops and extended life of the application site make wastewater irrigation on cropland desirable even in humid climates.

Hayfield

One of the most successful wastewater renovation systems at Penn State has been the reed canarygrass (*Phalaris arundinacea*, L.) hayfield. In the initial study period (1964 to 1970), secondary municipal effluent was spray-irrigated year-round at 5 cm/wk. The grass, planted in 1964, was harvested three times each year, except 1968 (two times). The concentration of nitrate at the 120-cm depth (Figure 5) was generally below 10 mg N/l year-round. The amount of N which leached from 1967 to 1970 (Table 2) was equivalent to less than 23 percent of the amount added.

Uptake of N by the grass was primarily responsible for preventing N leaching. In the 1967 to 1970 period the amount of N removed in the harvested portion of the grass was equivalent to 56 to 71 percent of the amount added (Table 2). Yield of reed canarygrass hay as well as N content and amount of N removed by the hay are indicated in Table 3. Yields remained consistently high and N content was fairly constant throughout the study period. With the high degree of renovation provided by the crops, the effluent irrigation at 5 cm/wk could be continued indefinitely without excessive nitrate leaching.

The question of whether denitrification occurs at this site has been addressed. Soil conditions—pH, moisture, transient high water tables, low oxygen—should have been suitable; however, the mass balance suggests that if it was occurring, it was secondary to crop removal in the renovation of secondary effluent at this site.

Inasmuch as the reed canarygrass site was successful in renovating secondary effluent, the site's ability to renovate effluent containing sewage sludge was studied from 1971 to 1975. The application period, irrigation rate and the management of the grass remained the same as 1964 to 1970, but the N content of the applied wastewater increased sharply by injection of anaerobically digested liquid sludge into the secondary effluent at a dilution ratio of 1 sludge:13 effluent prior to spraying. The concentration and forms of N of the effluent-sludge mixture as irrigated is presented in Table 1.

The results of the effluent-sludge management indicated that

Table 2. Water and nitrogen relationships in the reed canarygrass (5 cm/wk) site.

		1967	1968	1969	1970	1971	1972	1973	Total
Effluent	cm	241.1	241.1	253.8	212.1	179.8	213.2	247.0	1588.1
Precipitation	cm	108.8	87.5	83.6	118.8	88.5	118.5	101.9	707.6
Leachate volume	cm	294.1	264.5	276.1	263.4	207.1	269.8	281.9	1856.8
Applied N									
Organic N	kg/ha	110.6	169.4	160.4	215.5	421.4	797.9	540.8	2416.0
Ammonium N	kg/ha	391.0	390.9	388.9	225.3	507.2	930.9	542.1	3376.3
Nitrate N	kg/ha	151.5	120.0	94.8	105.5	227.6	255.2	402.7	1357.3
Total N	kg/ha	653.1	680.3	644.1	546.3	1156.2	1984.0	1485.6	7149.6
Leached N									
Organic N	kg/ha	39.0	19.1	36.6	39.1	12.4	17.6	25.4	189.2
Ammonium N	kg/ha	10.0	12.3	22.5	13.5	13.7	11.7	7.7	91.4
Nitrate N	kg/ha	97.7	92.7	77.5	55.0	80.1	237.9	257.8	898.7
Total N	kg/ha	146.7	124.1	136.6	107.6	106.2	267.2	290.9	1179.3
Crop N	kg/ha	368.4	398.9	397.0	388.3	397.9	392.3	295.2	2638.0
Leached/Applied	%	23	18	21	20	9	14	20	16.5
Crop/Applied	%	56	60	62	71	34	20	20	37.0

Table 3. Dry matter yield, nitrogen content, and the amounts of N removed in harvests of reed canarygrass hay irrigated with sewage effluent and sludge (5 cm/wk) and in corn silage irrigated with sewage effluent (strips 2 and 4 1965-1967, all strips 1968-1974; 5 to 7.5 cm/wk plot).

		65	66	67	68	69	70	71	72	73	74
Reed Canarygrass											
Yield 1st cut	tons/ha	6.20	5.76	8.44	6.72	6.34	3.74	7.97	4.82	5.64	4.61
2nd cut	tons/ha	5.35	3.07	4.21	4.68	3.83	4.35	4.28	4.77	2.35	4.64
3rd cut	tons/ha	2.17	0.85	3.09	—	1.43	2.60	2.46	2.78	2.62	4.12
Total	tons/ha	13.73	9.68	15.74	11.40	11.60	10.69	14.71	12.37	10.61	13.37
N content 1st cut	%	2.44	3.18	2.21	4.09	3.54	4.06	2.79	2.98	2.71	3.64
2nd cut	%	3.29	3.18	2.59	2.65	3.36	3.44	1.98	3.36	2.84	3.32
3rd cut	%	3.46	2.85	2.35	—	3.06	3.34	3.69	3.18	2.88	2.75
Weighted average	%	3.00	3.15	2.34	3.50	3.42	3.63	2.70	3.17	2.78	3.25
Total N removal	kg/ha	395.4	304.9	368.4	398.9	397.0	388.3	397.9	392.3	295.0	435.2
Corn Silage											
Yield	tons/ha	9.67	12.73	10.46	2.86	14.57	12.68	7.06	6.69	8.73	—
N content	%	1.29	1.25	1.18	1.62	1.71	1.32	0.76	0.80	0.77	—
N removal	kg/ha	124.7	159.1	123.4	46.3	249.1	167.4	53.7	53.5	67.2	—

the N content and yields of the grass were approximately unchanged in 1971 to 1974 compared to previous years (Table 3). The concentration of nitrate at 120 cm (Figure 5) increased from an average of 2.9 mg N/l in the 1967–1970 period to an average of 6.9 mg N/l in the 1971–1973 period. The increase of nitrate was greatest during the winters (January to April, 1972 - 11.0 mg N/l and January to April, 1973 - 14.7 mg N/l) when it exceeded 10 mg N/l. During this period the grass is relatively inactive and soil nitrate concentration approaches the nitrate concentration of the applied effluent-sludge mixture. Although concentration of ammonium and organic N in the effluent sludge mixture ranged from 50 to 87 mg N/l, the concentration of the total N in the leachate was equivalent to only 9.2 to 19.6 percent of the amount applied during 1971 to 1973 (Table 2). Ammoniacal plus organic N in the leachate averaged only 2.4 mg N/l during the effluent-sludge period. Together leaching and crop harvest accounted for only 34 to 44 percent of the amount applied. The remainder was not accounted for, but storage of organic N of the sludge in the soil and denitrification were probable sinks. Runoff which occurs at this site has not been completely quantified. It could cause considerable changes in the estimated leachate quantity particularly during winter.

Other irrigation managements have been examined at the reed canarygrass site. During the warmer months (May to November) of 1972 and 1973, a split weekly application (2.5 cm, 3 to 4 days apart) of 5 cm of effluent-sludge mixture was made (5). This treatment resulted in essentially no difference in nitrate concentration at the 120-cm depth. Also a high treatment, 5 cm of effluent-sludge followed by 5 cm of effluent alone 3 to 4 days later each week, was examined for the same period. This treatment resulted in nitrate concentrations exceeding 10 mg N/l year-round at the 120-cm depth. This loading rate was equivalent to 2550 kg N/ha/yr, a quantity which apparently exceeded the capacity of this landscape to maintain NO_3–N below 10 mg N/l in the percolate.

Crop Field

The last of the major sites studied at Penn State is the agronomic crop field—managed from 1963 until 1967 in a 4-yr corn, oats, hay, hay strip crop rotation and from 1968 until 1975 in continuous corn. The site is located on the Hublersburg clay loam soil. It was irrigated during the growing season which was late April through October for the strip crop and mid-May to early September on the continuous corn. The strip crop rotation involved comparisons of oats, wheat, corn grain, corn silage, red clover and alfalfa under irrigations of 0, 2.5 and 5.0 cm/wk with secondary effluent. The response of the irrigated plots which received nutrients from the effluent was compared to that of the control plot which received conventional fertilizers. The application of wastewater at the rates examined resulted in economic levels of crop yield while recharging 60 to 100 percent of potable quality water to the ground water. The various crops removed from 40 to over 100 percent of the applied N. The nitrate concentration in soil water, at the 120-cm depth in 1965 to 1967, was 7.0 to 9.7 mg N/l for the 5 cm/wk and 3.8 to 8.2 for the 2.5 cm/wk treatments. The N content of all crops was within normal ranges for those crops. The N content of the hay crops was greater in effluent-treated plots than in the fertilized control plots, but the N content of the corn was at times slightly lower than in the control plots and at other times slightly higher. From 1968 until 1973, the entire strip crop area was planted to corn and harvested as silage. Effluent application rates on the high treatment were 5 cm/wk from 1968 until 1971 and 7.5 cm/wk in 1972 and 1973. The concentration of nitrate at the 120-cm depth (Figure 6) varied between approximately 5 and 15 mg N/l throughout this time. Averaged over the six years, nitrate in soil water at 120 cm was 9.2 mg N/l during winter (January to April) and 11.7 mg N/l in the summer (May to

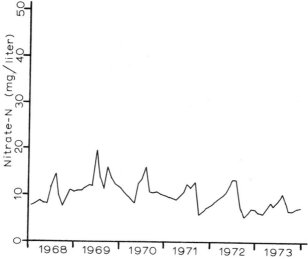

Figure 6.   Mean monthly concentration of nitrate in soil water at the 120-cm depth of the cornfield which received 5 to 7.5 cm/wk of effluent during the growing season.

September). Thus the highest nitrate concentrations occurred when the highest amounts of water were leaching during the irrigation period. Movement of this nitrate below the root zone prevented its use by plants and added nitrate to the ground water.

Harvest of corn silage removed 46 to 249 kg N/ha during the six years (Table 4). The low of 46 kg N/ha in 1968 was a result

Table 4.   Water and nitrogen relationships in the cornfield irrigated with sewage effluent (all strips; 5 to 7.5 cm/wk plots).

		1968	1969	1970	1971	1972	1973	Total
Effluent	cm	101.6	79.5	81.3	88.1	107.5	117.9	575.9
Precipitation	cm	87.5	83.6	118.8	88.5	118.5	101.9	598.8
Leachate volume	cm	125.7	102.2	133.1	115.5	164.1	152.9	793.5
Applied N								
Organic N	kg/ha	66.0	39.7	30.6	26.5	28.7	26.9	218.4
Ammonium N	kg/ha	151.7	97.1	48.2	40.0	54.0	23.3	414.3
Nitrate N	kg/ha	41.6	39.6	37.8	57.6	181.7	183.1	541.4
Total N	kg/ha	259.3	176.4	116.6	124.1	264.4	233.3	1174.1
Leached N								
Organic N	kg/ha	12.2	16.4	25.8	6.2	9.9	12.0	82.5
Ammonium N	kg/ha	5.1	6.9	8.5	3.3	2.9	2.0	28.7
Nitrate N	kg/ha	127.0	135.0	145.1	116.4	164.8	126.9	815.2
Total N	kg/ha	144.3	158.3	179.4	125.9	177.6	140.9	926.4
Crop N	kg/ha	46.3	249.1	167.4	53.7	53.5	67.2	637.2
Leached/Applied	%	56	90	154	101	67	60	79
Crop/Applied	%	18	141	132	43	20	29	54

of poor corn yields due to ineffective weed control. The amount of N removed by the crops rapidly decreased from the high in 1969. Silage yields decreased from 14.6 metric tons/ha dry matter in 1969 to 6.7 metric tons/ha in 1972 (Table 3). Nitrogen content of the plants decreased similarly from 1.71 percent in 1969 to 0.80 percent in 1972.

Together crop harvest and leaching losses removed 370 kg N/ha more than was applied from 1968 to 1973. Mineralization of soil organic N at a rate of less than 1 percent per year could supply this excess N. When the soil N was measured in fall 1968, the upper 120 cm contained 8700 kg N/ha, of which 5400 kg N/ha was in the upper 30 cm. Mineralization was apparently not rapid enough to supply enough N to produce higher yields.

Two related problems were thus encountered in this effluent-irrigated cornfield. First, the soil-plant system did not prevent nitrate leaching, and, secondly, the corn did not efficiently use added N for sustained yields. Although adequate amounts of N were applied yearly, the applied N was distributed uniformly throughout the growing season. Uptake demands, however, are not uniformly distributed. Sayer (7) showed that corn, which was planted May 20 and which yielded 13.8 metric tons/ha dry matter, had an average uptake rate of 3.1 kg N/ha/wk for the first 6 weeks following planting. Each weekly application of 5 cm of effluent containing 20 mg N/l would apply 10 kg N/ha/wk —considerably in excess of crop needs. Since there is no storage mechanism for nitrate under conventional clean-tilled corn, part of each weekly effluent N addition will be leached out by subsequent irrigations and by rainfall. Later in the growing season, corn needs sharply increase. Sayer (7) found uptake rates of 16 kg N/ha/wk during July and August. During this period the weekly application of 10 kg N/ha would be inadequate. This is aggravated by leaching since each weekly irrigation will displace at least part of the water and nitrate already in the profile. An increase in irrigation rate to meet weekly corn needs could increase leaching losses as well.

In addition to the inadequacy of this dilute N source, as applied to this plot, for meeting peak uptake demands, other factors may have affected N leaching and crop yields. Weed competition was high in some parts of the field where quackgrass (*Agropyron repens*) and fall panicum (*Panicum dichotomiflorum*) were persistent. Although pre-emergence herbicides were used, follow-up cultivation or spraying was not used. The weed competition was increased by availability of water and nutrients to the weeds while corn was establishing itself.

Stand density of the corn has not been adequate for maximum silage yields and the 38-inch row spacing has also favored weed competition.

Alternative management schemes are needed to lower nitrate concentration in the leachate and to increase yields of silage in the continuous corn. One of these schemes was begun in 1974. Several grasses and legumes were established to provide ground cover for erosion control and to take up N when corn uptake was minimal or absent. Corn was planted in 1975 with a no-tillage seeder. Grass and legume competition was controlled with herbicides. The early results suggest that the ground cover was effective in lowering nitrate leaching; however, competition of some of the grasses with the corn was still a problem and more critical herbicide management needs to be worked out.

Another aspect of the management scheme goes back to the

waste treatment prior to irrigation. Such treatment should focus on maintaining as large a ratio of NH_4–N to NO_3–N (preferably a 2 NH_4 to 1 NO_3–N ratio) as can be obtained in an aerobic treatment system. The higher NH_4 level will permit more winter storage of nitrogen in the soil on cation exchange positions and later release as nitrification proceeds during the peak crop demand period in the summer. Lagoon treatment systems usually provide low-nitrate effluents but activated sludge systems can be managed to give a low-nitrate effluent.

SUMMARY AND CONCLUSIONS

Nitrogen cycling in a land treatment system for municipal sewage effluent is site-specific and is conditioned by soil, vegetative cover, vegetation management, total nitrogen load and relative quantities of the various forms of nitrogen.

Under mixed hardwood forested conditions, a deep, well-drained sandy loam soil which received 5 cm of sewage effluent weekly attained a steady-state leakage of nitrate-nitrogen into the ground water at a concentration equivalent to the total N concentration in the applied secondary sewage effluent after two years of treatment. Decreasing the effluent and nitrogen load by 50 percent to 2.5 cm weekly for two years resulted in a 43 percent decrease in the mean annual concentration in the leachate from 21.2 to 12.2 mg NO_3–N/l. On this site it is doubtful that any nitrogen was being removed by denitrification.

On a deep, well drained clay loam soil with herbaceous vegetation and a scattering of white pine, 56 to 74 percent of the applied N was not leached into the ground water. The non-leached N was either immobilized in the soil organic matter or above-ground biomass or was denitrified. Mean annual NO_3–N concentration in the recharge remained below 10 mg/l when 5 cm of effluent was applied weekly but exceeded 10 mg/l when 7.5 cm was applied weekly.

The most effective nitrogen removal occurred on a deep, well drained clay loam soil which was irrigated weekly, year-round, with 5 cm of effluent. A vegetative cover of reed canarygrass was harvested and removed in three cuttings per season. Only when the nitrogen loading rate was increased to 2550 kg N/ha/yr by injection of sludge into the effluent pipeline did the concentration of NO_3–N in the recharge exceed 10 mg/l. Crop removal accounted for 56 to 71 percent of the N added. Immobilization as soil organic matter and removal by denitrification probably account for the remainder. When a similar soil site was irrigated only during the growing season with 5 cm of effluent at weekly intervals, a four-year rotation of corn, oats and two years of hay kept NO_3–N concentration in the recharge below 10 mg/l. When the same area was cropped continuously to corn harvested as silage, the mean annual NO_3–N concentration in the recharge hovered around 10 mg/l. Nitrogen removed in the crop and lost by leakage from the soil profile exceeded the amounts applied in the effluent. During the last three years the effluent has changed from an NH_4–N:NO_3^- ratio of 2:1 to one with ratios ranging from 1:2 to 1:5. As a result more NO_3–N has been lost from the soil profile during the irrigation period and nitrogen supply to the corn crop during peak uptake period has been inadequate.

Cropland areas from which harvests are made provide better options and opportunities for removing nitrogen than forest areas. More research is needed to define the crop management options on well drained sandy soils and to determine the suitability of moderately deep and/or moderately well drained soils, particularly fragipan soils on sloping areas, for wastewater treatment.

ACKNOWLEDGMENTS

This chapter is a contribution from the Pennsylvania State University of Agriculture Experiment Station and Institute for Research on Land and Water Resources, the Pennsylvania State University, University Park, Pennsylvania 16802, and was authorized for publication as Journal Series No. 5067 in the Pennsylvania Agricultural Experiment Station.

REFERENCES

1. Miller, R. H. "Soil microbial aspects of recycling sewage sludges and waste effluents," In: *Proc. Joint Conference on Recycling Municipal Sludges and Effluents on Land,*" Champaign, Illinois, pp. 79–90 (1973).
2. Kardos, L. T. and W. E. Sopper. "Renovation of municipal wastewater through land disposal by spray irrigation," In: *Recycling Treated Municipal Wastewater and Sludge through Forest and Cropland,* W. E. Sopper and L. T. Kardos, Eds., The Pennsylvania State University Press, University Park, Pennsylvania, pp. 148–163 (1973).
3. Urie, D. H. "Phosphorus and nitrate levels in ground water as related to irrigation of jack pine with sewage effluent," In: *Recycling Treated Municipal Wastewater and Sludge through Forest and Cropland,* W. E. Sopper and L. T. Kardos, Eds, The Pennsylvania State University Press. University Park, Pennsylvania, pp. 176–183 (1973).
4. Kardos, L. T., W. E. Sopper, E. A. Myers, R. R. Parizek and J. B. Nesbitt. "Renovation of secondary effluent for reuse as a water resource," Env. Protection Tech. Series, EPA–660/2–74–016, pp. 495 (1974).

5. Hook, J. E. "Distribution and movement of nitrogen in sites used for application of municipal sewage effluent and sludges," Ph.D. Thesis. The Pennsylvania State University (1975).
6. Richenderfer, J. L. "Effect of land disposal of treated municipal wastewater on the chemical properties of forested soils," M.S. Thesis, The Pennsylvania State University (1974).
7. Sayer, J. D. "Mineral nutrition of corn," In: *Corn and Corn Improvement*, G. R. Sprague, Ed., Academic Press, New York (1955).

12

CADMIUM TRANSFER FROM SEWAGE SLUDGE-AMENDED SOIL TO CORN GRAIN TO PHEASANT TISSUE

S. W. Melsted, T. D. Hinesly, J. J. Tyler and E. L. Ziegler
Department of Agronomy, University of Illinois,
Urbana-Champaign, Illinois

INTRODUCTION

The transfer of cadmium from sludge-amended soil to the plant and grain and on to the animal is of concern to those involved with the application of municipal wastes on land. Maximum loading rates for sewage sludges on land will be influenced by the degree to which crops will absorb hazardous elements from sludge-amended soils and the degree to which animals will absorb these same elements from the grain and forage. Similarly, monitoring disposal sites assumes the ability to predict metal movement through the food chain. In soil fertility work it has been shown that levels of available nutrients in the soil can be related to plant composition and yields in an exponential way, or as is more common through a Mitscherlich-type logarithmic expression. Many biologic reactions are logarithmic in character. This chapter presents an analysis of previously published data and attempts to show that the transfer of cadmium from the soil to corn grain and from corn grain to pheasant tissue may be expressed, and predicted, through logarithmic relationships.

BACKGROUND

The early work of Mitscherlich (1) and a few years later by Spillman and Lang (2) showed that levels of immobile nutrients in the soil could be related to crop yields through logarithmic and exponential expressions. These concepts were refined by Wilcox (3), Bray (4), Paauw (5), and many others. Later Balba and others (6–8) applied the Mitscherlich equation to crop composition with considerable success. The success of attempts to relate soil nutrient levels to crop composition and yield depends largely on the accuracy with which the quantity of the available nutrient in the soil can be determined. In this study, sludge cadmium surface applied during the previous year and plowed under in the spring appears to be reflected by the Cd content of the plant.

Numerous studies (9–14) have been conducted showing that inorganic cadmium salts when applied to soil are absorbed by plants. Several of these studies were conducted under solution culture systems in greenhouses so that the results are not directly transferable to field conditions. Generally, the Mitscherlich-Bray concepts were not applied to the data.

Animal feeding trials with cadmium have usually involved single large-dose ingestion of an inorganic salt to determine toxicity levels. As far as the authors are aware, pheasants have not been used in animal feed trials to determine cadmium absorption by various tissues. Similarly, the authors are not aware of any animal feeding trials where the daily cadmium intake was varied by feeding corn grain containing different levels of the metal. Differential cadmium levels in the ration were obtained by growing corn on highly sludge-amended soil to produce corn grain of varying cadmium composition. The corn grain was then fed to pheasants and the accumulation of cadmium in the duodenum, liver, kidney, muscle and other tissues determined. It is possible, therefore, to follow the transfer of cadmium from soil to grain to animal tissue.

METHODS AND MATERIALS

Corn was grown on a series of sludge-amended soil plots that had received sludge at 0, ¼, ½ and maximum rates. The maximum rate represented the maximum yearly amount of sludge that could be applied consistent with climatic and crop management limitations. The plots were started in 1968 and after six years the maximum accumulative sludge application rate had reached 368.6 mt/ha. In the fall of 1973 the corn grain was harvested from plots used for the sludge rate study, dried to 8 to 9 percent moisture, and used for the pheasant (*Phasianus colchicus*) feeding study. Grain samples from each of the rate plots were analyzed for cadmium and other elements.

The pheasant feeding trial involved 35 randomly selected fe-

male birds from a pen of several hundred maintained by the
Illinois Department of Conservation Game Farm, Yorkville,
Illinois. The birds were divided into eight cages of four birds.
Two cages of four birds each were randomly designated to receive corn grain harvested from plots that had received one of
the various sludge loading rates. Three birds were kept in reserve. At the start of the experiment, December 3, 1973, the
birds were all fed commercially produced corn while becoming
acclimated to their cages. Starting December 19, 1973, corn
grain from the various sludge rate treatments was introduced
and records of the amount ingested recorded. Crushed limestone
and granite were supplied as free-choice grit materials. The
birds were weighed initially and at approximately 4-week intervals during the experiment. The birds were then destroyed, dis-
sected, and the duodenum, liver, kidney, muscle, bone and other
tissue samples collected, weighed and analyzed.

RESULTS AND DISCUSSION

The full experimental data were reported by Hinesly *et al.*
(15, 16). The data used in this analysis are given in Table 1.

Table 1.   Cadmium content in corn and pheasant tissues (15, 16).

Plant and Bird Tissues (dry weight)	Sludge Treatment Rates (Cd, ppm)				
	0	*1/4-max*	*1/2-max*	*Max*	*LSD*
Corn tissue analysis					
Leaf – tassle stage	0.8	3.5	12.0	22.1	5.7
Grain	0.06	0.17	0.33	0.59	0.10
Pheasant tissue analysis					
Duodenum (0.652 g)	0.08	0.26	0.63	0.80	0.42
Liver (5.364 g)	0.66	1.24	1.72	1.99	0.84
Kidney (0.813 g)	2.84	5.57	8.66	9.41	3.29

Only the duodenum, kidney and liver tissue accumulated cadmium from the feed ration. The data are statistically significant
with sludge rate for both plant and bird tissue composition.

The relationship between annual application rates of cadmium
as a constituent of sludge and cadmium contents of corn tissues
was calculated using the Mitscherlich-Bray expression in the
logarithmic form, namely $\log (A-y) = \log A - cx$, where A is
the maximum cadmium absorbed by the plant tissue, y is the plant
composition at any given annual Cd loading rate, x, and c is a

constant. Annual application rates of Cd were amounts incorporated as a constituent of sludge by spring plowing, even though sludge was actually applied by ridge and furrow irrigation during the previous growing season. The constant, A, was calculated using the method of Balba (8) and refined by successive approximations. The value of A for corn leaf cadmium content is approximately 100 and the c is 0.0099, giving the expression $\log (100-y) = \log 100 - 0.0099x$. Similarly, the grain A value for cadmium is about 3 and c is 0.0179, giving the expression $\log (3-y) = \log 3 - 0.0179x$. Using these two expressions it is possible to select a range of values for x, the annual cadmium loading rate, and approximate the composition of the corn leaf and grain that might be expected from growing corn on soils with the various selected annual cadmium loading rates. Calculated and observed cadmium composition levels in corn leaf and grain tissue for the various cadmium application rates used in this study are given in Table 2. This table shows, for example, that if in monitor-

Table 2. Relationship of corn leaf and grain composition to previous year's surface-applied sludge Cd plowed down the spring before the crop was grown.

	Sludge-Borne Cd Applied	Leaf Composition		Grain Composition	
		Found	Calculated	Found	Calculated
		(Cd, ppm)		(Cd, ppm)	
			1971		
1970	0	0.5		0.14	
	4.25	4.9	9.3	0.70	0.48
	8.5	11.8	17.6	0.65	0.89
	17.0	25.4	32.1	0.92	1.51
			1972		
1971	0	0.7		0.14	
	7.25	8.6	15.2	0.45	0.76
	14.5	18.5	28.1	0.83	1.35
	29.0	21.9	48.4	1.10	2.10
			1973		
1972	0	0.8		0.08	
	1.0	3.5	2.3	0.15	0.12
	2.0	12.0	4.5	0.35	0.24
	4.0	22.1	8.7	0.61	0.46
			1974		
1973	0	0.2		0.09	
	1.75	1.4	3.9	0.18	0.21
	3.5	3.2	7.7	0.40	0.40
	7.0	10.9	14.7	0.81	0.75

ing a site where sludge was annually applied at rates to supply 4 kg Cd/ha, one would expect that if corn were grown on the site, the leaf at tassel stage would contain approximately 8.7 ppm Cd and the grain at harvest would contain 0.46 ppm. If one assumes a corn grain yield of 7500 kg/ha, the amount of cadmium removed by harvesting grain would be approximately 0.1 percent of the applied sludge-borne cadmium. The results generated by using the Mitscherlich-Bray expression (Table 2) show that predicted values agree well with observed values as long as annual cadmium loading rates are relatively low. The lack of predictability at higher loading rates conforms with what might be expected in view of the relationship between sludge-borne cadmium loading rates and contents in corn grain as derived from the partial regression equation (Table 3). The results in Table 3 were calculated by using coefficients which were previously determined and reported (15). Results generated from the partial regression equation indicate that when annual sludge-borne cadmium loading rates exceed 27 kg/ha, very little increase above 1.25 ppm in grain cadmium contents can be expected. The highest cadmium content of grain ever observed during a 7-year period was 1.10 ppm following a growing season in which 29 kg Cd/ha was applied as a constituent of digested sludge. Before leaving the discussion of cadmium accumulations in plant tissues, it must be pointed out that levels observed and predicted here are for one variety of corn which was used throughout the study. Levels of cadmium in tissues may be considerably different where other corn varieties are used.

Table 3. Relationship between annual sludge-borne Cd loading rates and Cd contents of corn grain (dry weight) as calculated by the partial regression equation ($y - 0.187 + 0.0744x - 0.0013x^2$, where y = ppm of Cd in grain and x = kg Cd/ha annually applied as a constituent of sludge).

Annual Cd Loading (kg/ha)	Cd Contents of Grain (ppm)
0	0.19
3	0.40
6	0.59
9	0.75
12	0.89
15	1.01
18	1.11
21	1.18
24	1.23
27	1.25

The data for cadmium depositions in bird tissues as a function of levels in diets suffer two limitations. The cadmium in the pheasant tissue at the start of the experiment is not known, and the amount of cadmium ingested from the grit was not determined. However, assuming that all of the tissue increase came from the cadmium in the grain and then calculating the intake required to give the levels in tissues of birds on the zero sludge treatment grain, the probable cadmium absorbed from the grit can be calculated by extrapolation. These recalculated values are given in Table 4. The limestone grit contained 0.77 ppm Cd.

Table 4.   Estimated amounts of Cd ingested daily by pheasants fed corn grain from sludge-amended soil. (The estimations were calculated from data presented in Table 1.)

Calculation	Sludge Treatment Rates (μg/day)			
	0	1/4-max	1/2-max	Max
Daily Cd intake from corn grain (average grain intake 38.5 g)	2.31	6.55	12.71	22.72
Daily Cd intake in excess of check	0	4.24	10.40	20.41
Calculated Cd ingested from limestone	1.18	1.18	1.18	1.18
Estimated Cd ingested daily	3.49	7.73	13.89	23.90

Again, assuming that the relationship between diet cadmium and tissue composition is logarithmic, one may use the expression $y = c \log x$, where y is the cadmium in the tissue for an x level of cadmium in the ration and c is a constant. These calculations give average c values of 6.59 for the kidney, 1.42 for the liver, and 0.459 for the duodenum. The calculated theoretical absorption lines and the actual experimental points are given in Figure 1. These calculations strongly suggest that the relationship between cadmium in animal tissue and organic cadmium in the diet are exponential in character. Extrapolating this data, it is possible to predict pheasant tissue cadmium levels from various diet levels as shown in Table 5. Again it should be noted that these estimates involve only a 100-day feeding period.

Estimates of the percentage of cadmium in the grain that was absorbed by the various pheasant tissues may be determined by subtracting the control composition from that found at the end of the experiment, or 100 days. The calculations are shown in Table 6. Note that total cadmium absorbed by the various tissues continued to increase as the total cadmium in the corn ration

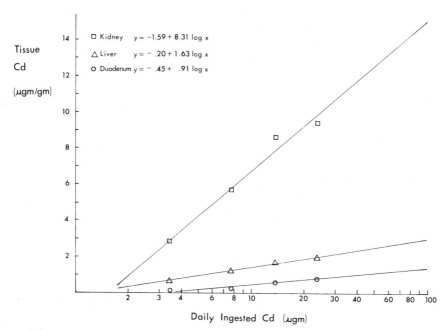

Figure 1. The calculated theoretical absorption lines, and the actual experimental points, relating daily Cd dietary levels to absorption by pheasant tissue.

increased. At the ¼ sludge treatment level about 1.29 percent of the grain cadmium was absorbed, at the ½ rate about 1.04 percent was absorbed, and at the maximum sludge rate 0.63 per-

Table 5. Predicted Cd contents in pheasant tissue after 100 days of feeding at various Cd ingestion rates.

Cd Contents of Ration (µg/day)	Tissue (µg Cd/g)		
	Kidney	Liver	Duodenum
2.0	1.89	0.43	0.14
3.0	3.14	0.68	0.22
5.0	4.16	0.99	0.32
8.0	5.95	1.28	0.42
10.0	6.59	1.42	0.46
15.0	7.75	1.67	0.54
20.0	8.57	1.84	0.60
50.0	11.20	2.41	0.78
100.0	13.18	2.83	0.92

Table 6.   Calculation of total Cd deposited in pheasant organ or tissue in 100 days under varying rates of Cd ingestion from corn grain, and percent absorbed from ingested amounts.

Calculation	Sludge Treatment Rates			
	0	1/4-max	1/2-max	Max
			(Cd, µg)	
Average total Cd ingested per bird from grain and limestone in 100 days	349	773	1389	2390
Average total Cd ingested from sludge-treated grain in 100 days		424	1040	2041
Total Cd deposited in total organ or tissue during 100 days				
Cd in Duodenum	0.05	0.17	0.41	0.51
Cd in Liver	3.54	6.65	9.22	10.67
Cd in Kidney	2.31	4.53	7.04	7.65
Total for all organs or tissue	5.90	11.35	16.67	18.84
Cd from sludge-treated grain in 100 days		5.45	10.77	12.95
Average percent of ingested grain-Cd deposited in organs and tissue of pheasants during 100 days			%	
Cd in Duodenum		0.028	0.035	0.63
Cd in Liver		0.733	0.546	0.349
Cd in Kidney		0.524	0.455	0.262
Total for all organs or tissue		1.29	1.04	0.63

cent absorbtion occurred. These data suggest that the 100-day feeding time was too short to establish a true long-term relationship between cadmium in the pheasant tissues and cadmium in the diet. The estimates made here, therefore, are probably somewhat lower than would have been obtained if the feeding had been initiated shortly after the birds were hatched. The data do show that organic dietary cadmium levels can be related to accumulations of the metal in pheasant kidney, liver, and duodenum tissue through exponential expressions.

Data presented here suggest that extremely high annual soil loading rates of sludge-borne cadmium are required to produce corn grain having very high levels of cadmium. Therefore, experimental plots which are established for the purpose of producing materials for feeding trials must be heavily treated to produce forage and grain having widely differing cadmium contents. Results of this study show that cadmium in corn grain had an availability to pheasants of only about 1 percent. One may speculate that the availability of cadmium in corn leaf tissue, or in other cereal grasses and grains, would not be much different. Whether these values obtained for pheasants will apply to poultry or other domestic animals remains to be tested, but they do provide comparative information for results obtained from other soil-plant-grain-animal foodchain studies. Hopefully the attempt to extrapolate the data will stimulate others to design future feeding studies with a view toward developing the means to predict Cd deposition in animal tissues from a wide range of cadmium levels in rations composed of natural biological materials.

Both plants and animals appear to limit the cycling of cadmium in human food sources to relatively low concentrations. At cadmium levels which healthy plants are capable of depositing in grain there is no evidence that cadmium levels would be increased in animal muscle tissue. The availability of the cadmium deposited in internal animal organs for absorption by the next successive animal in the food chain has not been investigated. Further studies are needed to assess the health significance of low level increases of cadmium in human food sources. Also, it may well be that levels of cadmium in food sources are increased during processing operations to such an extent that increases due to biological accumulations are an insignificant proportion of total levels.

ACKNOWLEDGMENTS

This contribution is from the Department of Agronomy, University of Illinois, Urbana-Champaign 61801 and is published with permission of the Agricultural Experiment Station Director. The authors gratefully acknowledge financial support provided by the Metropolitan Sanitary District of Greater Chicago and the U.S. Environmental Agency for the work reported here.

REFERENCES

1. Mitscherlich, E. A. *Bodenkunde fur Land und Forstwitre*, Paul Parey, Verlag, Berlin (1913).
2. Spillman, W. J. and E. Lang. *The Law of Diminishing Returns*, World Book Company, Chicago, Illinois (1924).
3. Wilcox, O. W. *ABC of Agrobiology*, W. W. Norton and Company, Inc., New York (1937).

4. Bray, R. H. "Correlation of soil tests with crop response to added fertilizers, Chapter 2—Diagnostic Techniques for Soils and Crops," The American Potash Institute, Washington, D.C. (1948).

5. Paauw, F. van der. "Critical remarks concerning the validity of the Mitscherlich effect law," *Plant Science* 4:97–106 (1952).

6. Balba, M. A. and R. H. Bray. "The application of the Mitscherlich equation for the calculation of plant composition due to fertilizer increments," *Soil Sci. Soc. Am. Proc.* 20:515–518 (1956).

7. Balba, M. A. and L. E. Haley. "Comparison of results obtained by the Balba-Bray equation and radioactive techniques for the determination of nutrient uptake by plants from different nutrient forms," *Soil Sci.* 82:365–368 (1956).

8. Balba, M. A. "Calculation of the uptake of different nutrient forms," *Alexandriz Agric. Res.* 6:81–92 (1958).

9. Page, A. L., F. T. Bingham and C. Nelson. "Cadmium absorption and growth of various plant species as influenced by solution cadmium concentrations," *J. Environ. Qual.* 1:288–291 (1972).

10. Haghiri, F. "Cadmium uptake by plants," *J. Environ. Qual.* 2:93–96 (1973).

11. Lagerwerff, J. V. "Uptake of cadmium, lead and zinc by radish from soil and air," *Soil Sci.* 111:129–133 (1971).

12. Chaney, R. L. "Crop and food chain effects of toxic elements in sludges and effluents," *Proc. Joint Conference on Recycling Municipal Sludges and Effluents on Land.* NASULGC, Washington, D.C. (1973).

13. John, M. K., H. H. Chuah and C. J. Van Laerhoven. "Cadmium contamination of soil and its uptake by oats," *Environ. Sci. Technol.* 6:555–557 (1972).

14. Jones, R. L., T. D. Hinesly, E. L. Ziegler and J. J. Tyler. "Cadmium and zinc contents of corn leaf and grain produced by sludge-amended soil," *J. Environ. Qual.* 4:509–514 (1975).

15. Hinesly, T. D., R. L. Jones, E. L. Ziegler and J. J. Tyler. "Effects of annual and accumulative applications of sewage sludge on the assimilation of zinc and cadmium by corn (Zea mays L.)," accepted for *Environ. Sci. Technol.* (1976).

16. Hinesly, T. D., E. L. Ziegler and J. J. Tyler. "Select chemical elements in tissues of pheasants fed corn grain from sewage-amended soil," accepted for *Agro-Ecosystems* (1976).

13

TRACE ELEMENTS IN SLUDGE ON LAND: EFFECT ON PLANTS, SOILS, AND GROUND WATER

M. B. Kirkham
Department of Plant and Soil Sciences, University of
Massachusetts, Amherst, Massachusetts

INTRODUCTION

This chapter will discuss research since 1973 on trace elements in sewage sludge applied to agricultural land. A review by Page (1) of literature on the topic printed between 1939 and 1973 was published in January 1974. Page's definition of "trace elements" will be used—elements which occur in natural systems in small amounts and when present in sufficient concentrations are toxic to living organisms. General references concerning sludge on soil have been published since 1973 (2–10), but will not be mentioned because they do not deal specifically with trace elements. This review will not discuss trace elements in effluent or composted municipal refuse (sometimes mixed with sewage sludge) or acid mine spoil amended with sewage sludge.

PLANTS

Tolerance

In July 1973, a conference entitled "Recycling of Municipal Sludge and Effluents on Land" was held in Champaign, Illinois. Research needs associated with toxic chemicals in sludge spread on soil were enumerated by a workshop chaired by Pratt (11). The first "immediate need" listed was to determine the tolerance of plants, particularly the edible parts, to trace elements. The workshop report said that the crops to be studied should include the food-, feed- and fiber-producing plants as well as forest crops, ornamentals, and brushland species. Table 1 lists concen-

Table 1. Trace element concentration of sludge-treated plants.

Plant Name Common and Latin	Sludge Source	Rate[a]	Type[b]	Soil pH	B	Ba	Cd	Co	Cr	Cu	Fe	Hg	Mn	Mo	Ni	Pb	Se	V	Zn	Ref
Corn																				
Zea mays																				
Grain	Stillwater, MN	450	S	5.3	3		0.05			1.9	28		9			<0.2			65	12
Grain	Dayton, OH	980		6.2			0.9		0.04	12	122	0.74[d]	5		1.0	<5.0			79	13
Grain	Chicago,IL	57					0.07			2.5			5.0			0.3			27	14
Grain	Oakland, CA	52	SiC	8.3			0.01			3.1					0.8				28	15
Grain	Tuscambia, AL	200	SiL	5.6			1.0								1.6	0.8			61	16
Grain	Residential	1%	SL	7.5			0.05												32	17
Grain	Chicago,IL	296	SiL				1.15												47	18
Grain	Chicago,IL	70	SiL	4.9	6.6		1.03		0.38	5.6	106	3.6[d]	18.3		3.1	0.03			152	19
Leaf	Stillwater, MN	450	S	5.3	9		1.32			8.5	225		77			4.0			293	12
Leaf	Dayton, OH	980	SiL	6.2			13.9			40	696		99		2.0	11			196	13
Leaf	Oakland,CA	52	SiC	8.3			0.5			1.3					0.6				39	15
Leaf	Tuscambia, AL	200	SiL	5.6			7.0								1.6	11.3			241	16
Leaf	Residential	1%	SL	7.5			3.9												38	17
Leaf	Chicago,IL	296	SiL				22.25												381	18
Leaf	Chicago,IL	70	SiL	4.9	43.6		11.6		4.5	8.7	112	37.9[d]	116		4.3	6.3			212	19
Leaf	4 WI cities	63-502	SL	6.8			5.9		4.2	13.0			215		5.1				216	20
Leaf		224	SL	5.5					1.4	10.0									508	21
Leaf	Washington, D.C.	230					2.9												194	9

Part	Location	No.	Soil	pH	C1	C2	C3	C4	C5	C6	C7	C8	C9	C10	C11	Ref.
Leaf	Martinsburg, WV	68	CiL	6.0					4	115	63				88	[22]
Leaf			CL	5.9–6.6			0.03		4.5			8.0			21	[23]
Husk	Dayton, OH	980	SiL	6.2			5.0		22	359	28	2	<5		202	[13]
Stem	Dayton, OH	980	SiL	6.2			5.0		1	169	25	1	<5		320	[13]
Root	Dayton, OH	980	SiL	6.2			48.7		617	11,400	467	46.7	542		1350	[13]

Barley
Hordeum vulgare

Part	Location	No.	Soil	pH	C1	C2	C3	C4	C5	C6	C7	C8	C9	C10	C11	Ref.
Grain	Lebanon, OH	22–45	L	7.8	110		<0.1			54	<0.05	<0.2	<0.2		31	[24]
Grain	Cincinnati, OH	183	G	7.2											67	[25]
Grain	Toronto, Canada							0.12								[26]
Leaf	Lebanon, OH	22–45	L	7.8	45		<0.1			137	47	<0.2	<0.2		21	[24]
Leaf	Cincinnati, OH	183		7.2			0.7								106	[25]
Leaf	Toronto, Canada							0.18								[26]
Leaf			S		120	12.0	2.4	4.0	<1.0	10.0	2.9	138	5.4	<1	65	[27]

[a] m.ton/ha or % dry wt.

[b] C=Clay; L=Loam; S=Sand; Si=Silt; G=Greenhouse medium (1:1:1, soil, sand, peat)

[c] Maximum and above-normal concentrations (1, 93) underlined. (Maximum, ppm: B, 75; Ba, 100; Cd, 0.20; Co, 0.30; Cr, 0.5; Cu, 15; Fe, 300; Hg, 0.01; Mn, 100; Mo, 1.0; Ni, 1.0; Pb, 5.0; Se, 2.0; V, 1.0; Zn, 150.)

[d] ppb

Table 1.   (continued)

Plant Name Common and Latin	Sludge Source	Rate[a]	Type[b]	Soil pH	B	Ba	Cd	Co	Cr	Cu	Fe	Hg	Mn	Mo	Ni	Pb	Se	V	Zn	Ref
Quack grass *Tritiom repens*																				
Seed	Toronto, Canada	variable	(C to S)	7.2-8.2			0.5		<1.0	9,0	50	190.	13		<1.0	5			60	28
Seed	Toronto, Canada		SL	7.5								0.13								26
Leaf	Toronto, Canada	variable	(C to S)	7.2-8.2			0.5		<1.0	14	100	200.	22		<1.0	3			45	28
Leaf	Toronto, Canada		SL	7.5								0.16								26
Burdock *Arctium minus*																				
Top	Burlington, Canada		C	7.5-8.5			1		3	18	400	0.10	125		2	8			70	28
Perennial ryegrass *Lolium perenne*																				
Leaf	England	12%	CL	4.6	36					30					350				420	29
Fodder rape *Brassica napus*																				
Leaf	Uppsala Sweden	49					0.6	1.9	4.1	8.3		0.05	41	1.7	9.2	7.7	0.06		114	1
Leaf	Uppsala Sweden	175	LS	4.8 7.2			0.72 0.38													30 30

Sample	Location	Rate	Soil	pH								
Oat												
Avena sativa												
Leaf	England	12% 176	CL	4.6 acid			2.0		400	227	502/1185	29/1
Leaf	Torrington, CT				30							
Leaf	Wolver-hampton, England	150		5.3 6.8			10.5/54	19/14		210/70		1/1
Wheat												
Priticum aestivum												
Grain	Denver, CO	100	LS	7.5	1.5	0.19	6.0	46.5	83.6	0.08	54	31
Grain	Domestic	1%	SL	6.0		0.1					45	17
Grain	Vasteras, Sweden	25	CL			0.12						32
Leaf	Domestic	1%	SL	7.5		0.1	11		43		44	17
Leaf	Newmarket, Canada		CL					1300		6	57	26
Leaf	Uppsala, Sweden	175	LS	5.8		0.12						33
Reed canarygrass												
Phalaris arundinacea												
Leaf			CL	5.9-6.6		0.16	16.9			12.0	64	23
Fescue												
Festuca arundinacea												
Leaf	Atlanta, GA	16.8	CL	6.2		36	251	168	337	550	2388	34
Festuca elatior												
Leaf	Domestic	1%	SiL	7.5		2					62	35
Unknown species												
Leaf		360		5.1-6.2 6.0-7.5							112/198	9/9

Table 1. (continued)

Plant Name Common and Latin	Sludge Source	Rate[a]	Soil Type[b]	pH	B	Ba	Cd	Co	Cr	Cu	Fe	Hg	Mn	Mo	Ni	Pb	Se	V	Zn	Ref
Rye *Seccale cereale*																				
Leaf	4 WI cities	63-502	SL	6.8			2.8		4.2	16.8			252		10				212	20
Leaf		242	SCL	No lime 8.8						20			227						775	1
				Lime 6.5						16			161						579	1
Rice *Oryza sativa*																				
Grain	Domestic	1%	SiL	7.5			0.1												39	35
Leaf	Domestic	1%	SiL	7.5			0.1												15	35
Bermudagrass *Cynodor dactylon*																				
Leaf	Domestic	1%	SiL	7.5															78	35
Sudangrass *Sorghum haleoense* Pers. var. *Sudanense Hitche*																				
Leaf	Domestic	1%	SiL	7.5			1.0												42	35
White clover *Trifolium repens*																				
Leaf	Domestic	1%	SiL	7.5															49	35

Concentration[c] (μg/g)

Plant / Part	Location	%	Soil	pH							Ref.
Alfalfa											
Medicago sativa											
Leaf	Domestic	1%	SiL	7.5						41	35
Sorghum											
Sorghum vulgare											
Leaf	Chicago, IL	1%	S		0.4	1100			37.5	430	36
Leaf	Chicago, IL	1%	SI		27	14			0.7	73	36
Leaf	Chicago, IL	1%	SiCL		14	30			3	140	36
Root	Chicago, IL	1%	S		5.8	26			7	58	36
Root	Chicago, IL	1%	SL		8.5	17			6.6	60	36
Root	Chicago, IL	1%	SiCL		57				82		36
Soybean											
Glycine max											
Seed	Domestic	1%	SL	7.5	0.6					61	17
Leaf	Domestic	1%	SL	7.5	0.4					57	17
Leaf	Chicago, IL	87	SiL		18.5						
Tulip											
Tulipa Gesnerjana											
Flower	Amherst, MA	35	G	5.0		8.6	120			15	37
Leaf	Amherst, MA	35	G	5.0		9.6	170			69	37
Bean											
Phaseolus vulgaris											
Fruit	Toronto, Canada	>30%	SL	7.2–7.8	0.1	0.3	240	6.0	20	4 / 2 / 30	28
Fruit	Domestic	1%	SL	7.8	0.05	5					
Fruit	Tuscumbia, AL	200	SiL	7.5	0.2			9.6	15.4	23 / 101	17 / 16
Fruit	Beaumont Leys, England	High		5.6	0.03	0.5		2.8	0.2	7	38,39

Table 1. (continued)

Plant Name Common and Latin	Sludge Source	Rate[a]	Type[b]	Soil pH	B	Ba	Cd	Co	Cr	Cu	Fe	Hg	Mn	Mo	Ni	Pb	Se	V	Zn	Ref
Bean (continued)																				
Leaf	Toronto, Canada	>30%	SL	7.2-7.8			0.1		1	10	450	0.3	330		2	2			30	28
Leaf	Domestic	1%	SL	7.5			0.6												47	17
Leaf	Tuscumbia, AL	200	SiL	5.6			1.3								7.1	5.8			211	16
Leaf	S. California		S	6.8	394	17	<0.5	7.3	<1.0	32				3.2	293	27		<1.0	69	27
Beet																				
Beta vulgaris																				
Edible root	Chicago, IL	1250					2.4		0.8	18			9.9	0.3	13				250	1
Edible root		65							0.8	11.7					2	0.3			53	40
Leaf	Chicago, IL	1250							1.0	10	44.2		95	0.7	16.5				>505	1
Leaf	Torrington, CT				42					>95									230	1
Cabbage																				
Brassica oleracea var. *capitata*																				
Leaf	Beaumont High Leys, England						0.05		0.03	0.6					3.4	0.14			2	38,39
Leaf	Chicago, IL	65					0.69		1.01	3.49	56.4		13.7		2.3	0.23			29	40
Carrot																				
Daucus carota																				
Edible root	Chicago, IL	65					4.0		0.7	7.1	26.1		2.9		1.1	0.25			26.1	40
Edible root	Beaumont High Leys, England						0.05		0.05	0.7					1.9	0.19			13	38,39
Edible root	Toronto, Canada	>30%	SL	7.2-7.8			0.4		0.5	5	25	0.3	15		2	2			15	28

Plant part	Location		Soil	pH												Ref.	
Carrot (continued)																	
Edible root	Stillwater, MN	450	S	5.3	30	1.15		1.5		22		18			0.9	103	12
Edible root	Toronto, Canada	1250, >30%	SL	7.2-7.8		0.1	0.06, 1	4.6, 15	0.3	120		45	0.12, 1		4	42, 30	1, 28
Leaf		1250					0.88, 3.0	9.9					0.84			99	1
Cauliflower *Brassica oleracea var. botrytis*																	
Edible part	Beaumont High Leys, England					0.05	0.06	1.9					1.5		0.16	11	38, 39
Leek *Allium porrum*																	
Edible part		1250					0.54	16.0					1.10, 7.0			135	1
Lettuce *Lactuca sativa*																	
Leaf	Beaumont High Leys, England					0.20	0.20	0.7							0.30	16	38, 39
Leaf	Stillwater, MN	450	S	5.3	28	2.67		11.9		135		79	2.7		0.8	225	39, 12
Leaf	Toronto, Canada	>30%	SL	7.2-7.8, 6.8		2	1	7	0.3	110		35	3		10	40	28
Leaf	England								0.41						3.7		10
Pea *Pisum sativa*																	
Fruit	Stillwater, MN	450	S	5.3	14	0.04		7		116		13			0.1	130	12
Pod	Stillwater, MN	450	S	5.3	26	0.06		11.1		80		29			0.6	134	12
Vine	Stillwater, MN	450	S	5.3	33	0.20		18.1		60		84			1.2	327	12

Table 1. (continued)

Plant Name Common and Latin	Source	Rate[a]	Type[b]	pH	B	Ba	Cd	Co	Cr	Cu	Fe	Hg	Mn	Mo	Ni	Pb	Se	V	Zn	Ref
												Concentration[c] ($\mu g/g$)								
Pepper *Capsicum frutescens*																				
Edible part	Chicago, IL	65					1.97		1.62	11.4	51.8		6.5		3.2	0.34			31	40
Potato *Solanum tuberosum*																				
Tuber	Stillwater, MN	450	S	5.3	6		0.23			19.0	17		7			0.6			53	12
Tuber	England			6.8								0.11				5.2				10
Tuber	Beaumont Leys, England	High							0.11	1.6					1.3	0.2			21	38,39
Tuber		1250							0.03	9.5				0.3	0.6				27	1
Leaf		1250							3.0	8.2				1.0	5.2				120	1
Radish *Raphanus sativus*																				
Edible root	Stillwater, MN	450	S	5.3	22		0.31			<0.3	24		6			0.7			98	12
Edible root	Beaumont Leys, England	High					0.10		0.30	0.7					2.9	0.3			49	38,39
Savoy *Brassica* sp.																				
Edible part	Beaumont	High							0.04	0.6					0.9	0.13			5	38,39
Spinach *Spinacia oleracea*																				
Leaf	Chicago, IL	110		Neutral	39		8.27		4.89	11.5	230		31.3		7.09	1.74			104	40
Leaf	Torrington, CT	176								65			150						815	1

Plant part	Location		Soil	pH										Ref.
Sprouts														
Edible part	Beaumont High Leys, England	High				0.30		1.4			3.5	0.40		38,39
Swiss chard *Beta vulgaris* ver. *Cicla*														
Edible part		360		5.1-6.2									1690	9
Edible part		360		6.0-7.5									627	9
Tomato *Lycopersicon esculentum*														
Fruit	Beaumont High Leys, England	High				0.03	0.08	0.2			1.1	0.3	2	38,39
Fruit	Stillwater, MN	450	S	5.3	14	0.33	<0.3	29		10		<0.4	31	12
Fruit	Chicago, IL	65				3.00	14.3	74.2		6.81	1.45	0.31	41	40
Fruit, red	Toronto, Canada	>30%	SL	7.2-7.8		0.6	5	45	12.2	10	0.6	1	35	28
Fruit, red	Newmarket, Canada		SL				12	55	12.2	14		1.2	30	26
Fruit, green	Toronto, Canada	>30%	SL	7.2-7.8		0.6	10	45	6.0	10	0.5	1	30	28
Fruit, green	Toronto, Canada		SL	7.5					6.0					26
Leaf	Burlington, Canada	C		7.5-8.5		2	16	90	0.1	300	7	4	200	28
Leaf	Toronto, Canada	>30%	SL	7.2-7.8		3	25	210	0.4	100	3	3	100	28
Leaf	S. Calif.		S	6.8	85	4.6	<1.0				209	11.0	90	27
Root	Toronto, Canada		SL	7.5	41	5.2	7.3		0.35		3.5	<1		26

Table 1. (continued)

Plant Name Common and Latin	Sludge Source	Rate[a]	Soil Type[b]	pH	Concentration[c] (µg/g) B	Ba	Cd	Co	Cr	Cu	Fe	Hg	Mn	Mo	Ni	Pb	Se	V	Zn	Ref
Turnip *Brassica rapa* Edible part	Torrington, CT	176		acid	50					>80			120						880	1

[a] m.ton/ha or % dry wt.

[b] C=Clay; L=Loam; S=Sand; Si=Silt; G=Greenhouse medium (1:1:1, soil, sand, peat)

[c] Maximum and above-normal concentrations (1, 93) underlined. (Maximum, ppm: B, 75; Ba, 100; Cd, 0.20; Co, 0.30; Cr, 0.5; Cu, 15; Fe, 300; Hg, 0.01; Mn, 100; Mo, 1.0; Ni, 1.0; Pb, 5.0; Se, 2.0; V, 1.0; Zn, 150.)

[d] ppb

trations of trace elements in plants treated with sewage sludge. Most of the data in the table come from references published since the 1973 conference. The table shows sludge source and application rate and soil type and pH, when given in the original publication. If sludge were applied at more than one rate in an experiment, the highest rate used appears in Table 1. Only plants treated with sludge that had not had trace elements added ("unspiked" sludge) are included in Table 1. Values that are underlined are upper limit or above-normal concentrations, based on the more conservative maximum value given by Allaway (41) or Melsted (42).

Table 1 shows that most experiments have been done with corn and a few with other grain crops and vegetables. Limited data are available for forest, ornamental, and brushland species. Therefore, one research need is to determine tolerance of crops which are not eaten. This is important to know in case nonedible plants, to protect the food chain, have to be grown on sludge-treated land with high concentrations of trace elements.

Table 1 also shows that plant parts most frequently analyzed are the leaves and fruit. Analyses of concentrations of elements in the roots are needed to determine amounts translocated from the soil to the root and from the root to the top of the plant.

Table 1 shows that every crop grown with sludge had excessive concentrations of one or more trace element except wheat grain, rice grain and leaf, tulip flower and leaf, pea seed and pod, and savoy. Wheat, rice and tulip probably did not accumulate high concentrations of trace elements because low sludge application rates were used. Pea fruit and savoy were treated with large amounts of sludge (>400 m.ton/ha) and appear to be nonaccumulators of trace elements.

Excessive concentrations often are correlated with high sludge application rates (Table 1). For example, except for the Stillwater, Minnesota corn, cadmium concentrations in corn grain were above-normal when 70 m.ton/ha or more sludge were used. High concentrations often are correlated with sludge composition. For example, Boswell (34) used a metal-laden sludge from Atlanta, Georgia. Consequently, the fescue he grew contained high concentrations of trace elements.

Table 2 shows the number of analyses with maximum or greater than normal concentrations divided by the total number of analyses in Table 1. There are not enough analyses in Table 1 to know the danger of cobalt, barium, molybdenum, selenium and vanadium. Table 2 shows that cadmium is an element of concern and this is well-known. Numerous investigators are studying cadmium. Nickel appears to be a problem because many

Table 2.   Percentage of analyses in Table 1 with concentrations of trace elements at maximum or above-normal concentrations (1, 93).

Element	No. of Analyses with Maximum or Greater than Normal Concentration (Total No. of Analyses) (%)
Hg	$\frac{23}{25} = 92$
Ni	$\frac{50}{60} = 83$
Cd	$\frac{60}{88} = 68$
Cr	$\frac{27}{47} = 57$
Pb	$\frac{22}{69} = 32$
Cu	$\frac{27}{86} = 31$
Zn	$\frac{33}{113} = 29$
Mn	$\frac{14}{52} = 27$
B	$\frac{4}{25} = 16$
Fe	$\frac{6}{43} = 14$

papers report high concentrations of nickel in plants grown on sludge-treated soil.

According to Table 2, mercury appears to be a hazardous element. Van Loon (26, 28), in Canada, did most of the mercury analyses reported in Table 1. He used a sludge high in mercury (25-ppm) from North Toronto and a sludge low in mercury (1 ppm) from Burlington. With both sludges, crops accumulated above normal concentrations of mercury. Because special procedures are required to measure mercury, concentrations in sludge, soil and plants often are not determined. Van Loon's data show that mercury should be measured.

Results of Yugoslav studies (43) showed that mercury levels of most plants were low (1 to 50 ppb) compared to the mercury content of soils in which they grew (100 to 600 ppm). Carrots, grasses, and some weeds had more mercury (800 ppb fresh weight) than other plants. Broadleaf plants exposed to high levels of aerial mercury contained more mercury than narrow-leaved plants. Certain species of mushroom contained more mercury than vascular plants and more than the soil in which they grew. [No data exist for trace element concentrations in mushrooms grown on sludge-treated soil. Mushrooms are a crop that should be studied because municipal garbage mixed with animal manure is being used on mushroom farms (44). Manure can contain significant concentrations of trace elements (45).] The Yugoslav workers found that some of the mercury in mushrooms was in the form of methyl mercury, which was not found in vascular plants. Aquatic plants accumulate high concentrations (243 ppm) of mercury if the mercury is present as methyl mercury (46). Jackson (47) found that mercury appearing in tops of barley and soybean plants whose roots were in mercury solutions came from mercury vapor released from roots during the experiment rather than by translocation to the tops. If mercury volatilizes from sludge-treated soils, plants could absorb it.

It has been known for a long time that plants vary widely in their tolerance to trace elements. In 1942 Sempio (48) found that one variety of wheat (frassineto) absorbed less cadmium and was more tolerant to cadmium than another (virgilio). Recently, Thomas (49, 50) showed that mockernut hickory trees accumulated rare earth elements like cerium in greater concentrations than other species (red maple, tulip trees, white oak, black gum) on the same site. Black gum trees accumulated greater concentrations of cobalt than mockernut hickory, red maple, tulip tree and white oak. Shacklette (51) reported that in coniferous trees, limber pine leaves and stems contained an average of 3.4 and 5.5 ppm cadmium, respectively, whereas lodgepole pine contained 10.5 and 16.6 ppm cadmium in leaves and stems. In deciduous trees, mountain maple leaves and stems had 1.6 and 4.9 ppm cadmium and willow leaves and stems, 15.7 and 37.6 ppm cadmium. Shagbark hickory contained five times as much cadmium as white oak grown on the same site.

In sludge studies, Dowdy and Larson (12) said that potatoes and carrots were good crops to grow on sludge-treated soil because they did *not* accumulate trace elements. However, Boswell (34) reported that sludge-treated fescue yielded better than controls despite the high concentrations of trace elements in the sludge and the plants. Fescue appeared to be a good crop to grow with sludge because the plant tolerated high concentrations of trace elements in its tissues. Smith and Bradshaw (52) showed that wild populations of *Festuca* (fescue) grew well near abandoned metal mines in England and did not accumulate lead and zinc, while a commercial variety of *Festuca* grew poorly and accumulated from two to twenty times as much lead and zinc as the wild *Festuca* plants. Length of roots of the wild plants was about two times longer than that of the commercial ones. Tolerance, therefore, appears to involve an ability to *exclude* the elements. This observation agrees with results from studies of plants grown in saline environments. Salt-tolerant plants exclude elements (Na, Cl) at the root (53). However, there are varying degrees of exclusion. Boswell's *Festuca arundinacea* took up large concentrations of trace elements compared to other plant genera. But compared to other species of *Festuca* (e.g., Smith and Bradshaw's wild populations of *Festuca rubra*), it took up relatively small amounts. Therefore, *Festuca* appears to be able to accumulate, with minimal metabolic derangement, concentrations of trace elements that other plants cannot. The following sections briefly consider what effects trace elements have on plant metabolism if they are taken up and how plants can accumulate or exclude the elements.

Physiological Studies

Transpiration

Trace elements close stomata and inhibit transpiration (54, 55). Bazzaz and co-workers (54, 55) suggested that the reason thallium was especially effective in reducing transpiration was because it is similar to potassium. Potassium is necessary for the normal functioning of stomata. It is taken up by guard cells when stomata open and released into epidermal cells when stomata close. Schnabl and Ziegler (56) showed that Al^{3+} ions prevented wilting of leaves of cut flowers because the ions closed stomata. Aluminum ions accumulated in the guard cells in place of potassium ions and inhibited starch mobilization. Wilting and loss of turgor pressure often are symptoms of trace element toxicity (57). In chrysanthemum leaves, concentrations in nutrient solution of cadmium as low as 0.10 ppm reduced transpiration rate, increased stomatal resistance, and reduced turgor pressure within 24 hours after application of the cadmium (58).

Photosynthesis

Trace elements inhibit photosynthesis (54, 55). The inhibition appears to be due in part to stomatal closure (54, 55). In addition, trace elements damage chloroplasts and cause chlorophyll degradation (59–61). The site for cadmium action appears to be system II photoreaction rather than in the electron transport chain of photosystem II near the oxygen-evolving site (60).

A common symptom of trace element toxicity is chlorosis (61–64). [There are exceptions. For example, cobalt results in dark blue-green color and nickel causes white and green banding in oat leaves (63).] The chlorosis is similar to iron deficiency. However, Root *et al.* (61) showed that iron concentration *increased* as cadmium concentration increased in corn grown in nutrient culture. Cadmium induced a zinc deficiency, which in turn caused an increase in iron concentration.

Respiration

Many divalent cations (Ba, Cd, Co, Mn, Ni, Pb, Sr) affect the rate of respiration (65, 66). The rate can be stimulated or reduced depending on the concentration and the element. At high concentrations, respiration is inhibited. Trace elements inhibit carbon dioxide evolution in lower organisms like bacteria as well as higher plants (67). The cations have an effect on membrane properties in mitochondria, the organelles in which respiration occurs in cells. The specific site of cadmium activity appears to be associated with sulfhydryl-containing enzymes in the mitochondria. The mechanism of the effect on membranes is not known, but it is important to study to understand ion selectivity and transport.

Nitrogen Metabolism

Nitrogenase activity and nitrogen fixation are inhibited in legumes by trace elements (68). Huang *et al.* (68) showed that soybean nodule ammonia, protein, and carbohydrate contents were reduced in the presence of cadmium and lead.

Hormones

No one has studied changes in hormonal levels in plants after sludge application. However, they probably occur. Effects caused by auxins are due, in part, to their chelating properties (69). Therefore, trace elements may disturb the natural chelating process and counteract auxin activity. A modifying effect of copper upon the toxicity of 2,4-D (a weed killer and an auxin) has been reported (69). Copper was more efficient than iron. Zinc is necessary for auxin synthesis (70). If levels of cadmium, which can be a zinc "antagonist," are high in sludge applied to soil, so that zinc activity is impaired, auxin levels might drop.

Abscisic acid and ethylene are produced under stress conditions (drought, flooding, heat, cold). Hence, in the presence of toxic levels of trace elements, abscisic acid and ethylene concentrations might increase in plants.

Cytokinins are known to control nutrient movement between plant parts. Müller and Leopold (71) showed that if cytokinins are applied locally to leaves, elements (P, Na) moved to the cytokinin centers. Perhaps cytokinins also would mobilize sludge-borne trace elements taken up by plants. Cytokinin applications to one leaf on a plant might be a way to concentrate the elements in one spot. Cytokinins are involved in the movement of potassium into guard cells. Nosseir (57) reported that exposure of roots of broad and kidney bean seedlings to cadmium solutions stimulated roots to release potassium and absorb calcium. Because cadmium affected potassium absorption, cytokinin concentrations probably were changed by the cadmium treatment.

Growth

The reduction in photosynthesis and other effects of trace elements on plants, including interference with mitochondrial respiration, lead to reduced growth and possible death of plants contaminated with trace elements. Some investigators have found a direct inverse relation between the rate of growth and trace element concentration (61, 72). For example, Root *et al.* showed that corn dry weight decreased as cadmium concentrations in nutrient solution increased from 1 to 40 mg/l. However, when *very* low concentrations of trace elements such as aluminum, selenium and chromium are used, stimulations in growth have been reported (70, 73). Stenlid (69) reported that elongation of roots of young wheat seedlings grown in nutrient solution was stimulated by heavy metals. The increase was 41 percent in 10^{-6} M $CuCl_2$, 28 percent in 10^{-7} M $AgNO_3$, and 10 percent in 3×10^{-7} M $HgCl_2$. Stimulations in growth of organs other than roots have been noted (69). Germination of pine seeds was stimulated by treatment for short periods with $AgNO_3$ and $HgCl_2$. Duckweed, which flowers only when days are short, flowered independently of photoperiods when small amounts of $CuSO_4$, $AgNO_3$, or $HgCl_2$ were added to the growth medium. Rooting of cuttings of *Myrica* (sweet gale family) was promoted by $AgNO_3$. Stimulation of growth of timothy occurred when 0.1 to 10 ppm mercury was added to soil. In the biochemical literature, there are many reports of stimulatory effects upon enzymes caused by low concentrations of heavy metals. Stenlid (69) suggested that the explanation of growth stimulation was allosteric effects (change in spatial arrangement) upon enzymes or the removal, inactivation or blocking of an inhibitor. The inhibitor might be an auxin, and sulfhydryl groups might play a role in the inhibition. At higher concentrations of trace elements, however, enzyme activity is inhibited (67) and in most cases after trace element addition, inhibition of growth is reported.

It is possible that with further refinements of technique and in purity of salts, more elements may be shown to be essential for growth—at least for certain plants (70, 74). Almost all naturally occurring elements have been detected in plants. Gauch (70) said that even nonessential elements might be involved in metabolism. He reported that certain enzymes are activated not only by an essential element but also by an apparently nonessential element such as cobalt or nickel. Stimulatory effects of a nonessential element reported from time to time, therefore, might be based on the beneficial role it plays as an enzyme activator when the normal essential element is deficient in the plant.

The stimulation in growth caused by a trace element might be due to effects the element has, not only on enzymes, but also on turgor pressure. In the chrysanthemum experiment (58), 0.01 ppm cadmium increased turgor pressure. Increases in turgor appear to be directly related to increased growth (75). The chrysanthemums were mature before cadmium treatments started, so no differences in yield were seen. However, with very low concentrations of cadmium (0.01 ppm), it is expected that growth would be stimulated, based on the turgor pressure results. The cadmium could have altered cytokinin concentrations in the chrysanthemums, which in turn caused the change in turgor pressure (76).

Plant Age

Age has an effect on trace element concentrations in plants. For example, work at the U.S. Plant, Soil and Nutrition Laboratory in Ithaca, New York, showed that zinc in mature legume seeds (pea and soybean) had a high level of availability for animals, although not as high as that of immature seeds. Mature pea and soybean seeds contained six times more phytic acid (a metal-binding substance) than was found in the immature seeds. However, zinc in mature legumes appeared bound to sulfhydryl compounds instead of phytic acid. Thus, the zinc apparently was not precipitated by phytic acid and was available for absorption by animals (77).

Van Loon (28) showed that spring shoots of quack grass had higher trace element contents (Cd, Cr, Cu, Fe, Hg, Mn, Ni, Pb, Zn) than mature plant leaves. Boswell's data (34) showed a tendency for trace element concentration (Cd, Cr, Cu, Mn, Pb, Zn) in fescue leaves to decrease during the growing season.

Ion Uptake and Translocation

Little is known about the mechanisms involved in trace element uptake and translocation. Tiffin (78) recently has presented a review on the topic.

Uptake. Cutler and Rains (79), using barley, determined that accumulation of cadmium involved three mechanisms. In the first one, exchange adsorption, the reversibly bound fraction of cadmium associated with roots was exchanged readily when tissue was exposed to desorption solutions. This mechanism accounted for the largest amount of cadmium taken up by roots in short-term experiments. The exchange sites were filled nonselectively with the transition-type metals (Cd, Cu, Hg, and Zn).

The second uptake mechanism, which involved a nonmetabolic binding of cadmium, appeared to be the primary way air-dried and fresh roots accumulated cadmium. This mechanism was thought to irreversibly sequester cadmium by fixing it to binding sites on the cell wall or on macromolecules in the cell. The binding apparently prevented entry of cadmium into the cell cytoplasm and reduced its toxic effects. Sequestering appeared to account for low transport of cadmium from barley roots to shoots.

Other investigators have noted that tolerance in plants involves binding of toxic metals at cell walls of roots and leaves, away from sensitive sites within the cell (67, 78, 80). Nonprotein sulfhydryl groups seem to be binding sites at cell walls (78). Mitra *et al.* (81) noted that a strain of *Escherichia coli* tolerant to cadmium developed large intracellular vacuoles and exhibited an abnormally long lag phase before cell division occurred. When proliferation began, the morphology of these cells became normal. In tolerant (accommodated) cells, 56 percent of the cadmium was associated with the cell wall, 13 percent in the membranes and 31 percent in the cytoplasm. In unaccommodated cells, the figures were 2 percent, 75 percent and 23 percent, respectively. Zwarun (67) suggested that a cell wall acted as a passive cation exchanger similar to an inert cation exchanger like a clay particle.

Cutler and Rains (79) noted that, because cadmium was transported to shoots of intact plants, cadmium had to follow a symplastic pathway (*i.e,* cross cell membranes). Diffusion, the third mechanism of accumulation, accounted for the movement of cadmium into barley root cells. The mechanism was consistent with the linear temperature response of sorption and the failure of sorption to respond to metabolic inhibitors. Active metabolic uptake was not consistent with these observations. Diffusion, coupled with sequestration, thus accounted for the accumulation and distribution of cadmium. Goren and Wanner (82) observed that uptake of lead and copper was nonmetabolic, too.

Translocation. Tanton and Crowdy (83) measured the distribution of lead chelate, a systemic fungicide, in the transpiration stream of plants. The compound was transported passively from roots to shoots. It moved in the cell walls immediately adjacent to the main veins of the leaf mesophyll cells and accumulated at evaporating surfaces, like walls of stomatal guard cells. Over short periods of treatment, the chelate was distributed evenly through the interveinal areas. During long treatments (several days), it accumulated at the edges of net-veined leaves and at the tips of parallel-veined leaves, where it scorched the tissue. They suggested that rain consequently could cause recycling of the compound from leaves to soil. Only 3 percent of the lead chelate solution was transported to the shoots. However, once inside the endodermal barrier in the root, the solute traveled in the transpiration stream to all tissues that lose water. Plant organs that had a low transpiration rate (fleshy fruits, seeds) did not accumulate the lead chelate. Flow was restricted into cereal seeds because of the xylem discontinuity between the rachella, which attaches the seed to the stem, and the pericarp, which encases the seed. In contrast, Shacklette (51) reported that fruit of apple trees sprayed with solutions of cadmium chloride accumulated cadmium as the fruit matured. Therefore, cadmium can concentrate in tissues with low transpiration rates.

Tanton and Crowdy (83) reported that distribution of lead chelate in a plant was the same whether the compound were applied to the roots or to the leaves. Buchauer (84) postulated that metal pollutants in air near smelters are taken up through stomata. However, Haghiri (62) found that soybean roots exposed to cadmium took up cadmium more efficiently than leaves treated with cadmium.

Hevesy (85) observed that roots of *Vicia faba* (broad bean) bound all lead nitrate when present in low concentration (10^{-6} N $PbNO_3$). In concentrated solutions (10^{-1} N), lead was carried up by the transpiration current.

Gordee *et al.* (86) showed that most cadmium remained in lower parts of peppermint plants treated with cadmium. The cadmium was largely in the vascular tissue of the plants. Some translocation of cadmium into new growth was observed. The low transport of trace elements from young to mature tissue may account for the decrease in concentrations of trace elements in fescue leaves sampled during the growing season (34).

Shewry and Peterson (87) followed the uptake of $^{51}CrO_4{}^{2-}$ by intact barley seedlings. Uptake increased with increased chromate concentration, but unless the concentration was high (100 μM), less than 1 percent of the isotope absorbed was transported to the shoots. Most of the isotope accumulated by the roots was present in the vacuoles. Of the total radioactivity, 86.6 percent was associated with vacuoles and 17.4 percent was bound to cell walls. They suggested that restricted chromate transport was due to localization in the root vacuoles or to lack of a specific mechanism for transport.

Monovalent cations appear to be translocated differently than cations with more than one charge. Handley and Babcock (88) observed that monovalent ions which are metabolically sequestered (Cs, K) are much less subject to loss from the tissue than those taken up passively. Cesium was metabolically (actively) accumulated and highly concentrated in veins of bean, tomato and corn. Cesium was not readily transferred from phloem (living cells) to xylem (dead cells) as was strontium, a divalent cation. They concluded that the effect of xylem flow (transpiration) upon phloem transport of ions actively accumulated was less than those passively accumulated.

In summary of the plant physiological studies, it appears that the root (cell walls and vacuoles) is an effective "binder" of trace elements. The toxic elements in sludge are accumulated passively (metabolic energy not required) in the root. Small amounts pass the endodermal barrier and are translocated by the transpiration stream to the shoot. Once they reach the leaves, they accumulate in guard cells and other evaporating surfaces. In leaf cells they damage chloroplasts and mitochondria by inactivating enzymes. They inhibit nitrogen fixation and probably alter hormonal concentrations. Because photosynthesis, respiration and other metabolic processes are upset, growth is reduced. If the trace elements are present in minute concentration (e.g., 0.01 ppm), growth may be stimulated because they may activate enzymes. Growth also may be increased by the trace amounts of metals because the ions increase turgor pressure and decrease stomatal resistance, so that more carbon dioxide can be taken up.

SOILS

The second "Immediate Need" listed by the Workshop on Toxic Chemicals (11) was to study the effects of soil factors on toxicities of chemicals. The factors known to influence the availability of a trace element are soil type, trace element concentration, pH, cation exchange capacity, organic matter, other ions present in the soil, temperature, soil water content and depth of sludge placement. Page (1) and Leeper (89) review literature on trace element availability in sludge-amended soil. This section briefly mentions a few recent papers on the topic.

Soil Type

Plants grown on peat soils and other organic substrates often are deficient in certain trace elements because trace elements exhibit a high affinity for organic material (90). Peat soils have potentially great value at sludge disposal sites. Peat has been used to renovate wastewater and remove trace elements (91–93). Chrysanthemum plants grown in peat or a standard greenhouse medium (1:1:1 soil, sand, peat) took up smaller amounts of trace elements (Cu, Fe, Zn) than plants grown in media lacking peat (M. B. Kirkham, unpublished data). Peats could be used both in the greenhouse and in the field to tie up trace elements in sludge just as they are being used to filter out trace elements in wastewater. Another material high in organic matter, bark, already has been used as a filter medium for paper mill sludge (94).

Trace Element Concentration

Many recent papers, too numerous to cite, report trace element concentrations of polluted and nonpolluted soils. Data are lacking for trace element concentrations in soils in the northeast (95). Only one paper has been published concerning the suitability of northeastern sludges for land disposal (96). Sludge disposal on soil in eastern states is not as common as it is in midwestern and western states. For example, in Massachusetts, no sewage sludge is being disposed of on land. In New England, only six communities have been located that spread sludge on land. To help alleviate this problem, a new regional research project in the northeast supported by Regional Research Fund Hatch Act, has been established. The project (NE–96) is entitled "Soil Properties Affecting Sorption of Heavy Metals from Wastes."

pH

Recent papers confirm earlier observations that pH is one of the most important factors controlling trace element availability (20, 29, 97–106). However, in a nutrient culture study, Iwai *et al.* (72) found that the amount of cadmium absorbed by corn was not affected by pH. Also, Korte *et al.* (107), studying normal soils with no sludge, did not find a correlation between soil pH and the amount of trace element removed when the soils were leached with a dilute solution of $AlCl_3$ and $FeCl_2$. The texture of a given soil was more important in determining the amount of metal leached from the soil. In general, they obtained a positive, but nonsignificant, correlation with the clay percentage and a negative correlation with the sand percentage. [In contrast to earlier publications, many recent publications on trace elements and sludge use statistical procedures to analyze results (18, 20, 108). A good knowledge of statistics is required to fully understand the papers.]

Cation Exchange Capacity

Korte *et al.* (107) found cation exchange capacity unimportant in predicting the quantity of trace element leached from soil. However, most recent papers report that cation exchange capacity controls trace element availability (104, 105).

Organic Matter

There is great debate concerning the importance of sludge organic matter in controlling the availability to plants of trace elements, particularly after sludge additions have stopped. However, it is now established that trace elements in sludge are taken up less readily than are trace elements added to sludge in the form of an inorganic salt or trace elements added directly to soil in the same concentration as present in sludge (21, 109, 110). [Organic matter makes metals less available to animals, as well as plants. Tiffin (78) reported that lead levels in organs of the meadow vole were two to nine times higher when lead nitrate was added to alfalfa diets than when the metal was fed in an organic form that had accumulated in alfalfa stems.]

Haghiri (108) found that, except for the cation exchange capacity effect of organic matter, organic matter did not influence the concentration of cadmium in oat shoots. His results indicated that the retaining power of organic matter for cadmium was predominantly through its cation exchange property rather than chelating ability. Haghiri hypothesized that when additions of organic matter were ceased at a sludge disposal site, the organic matter would be decomposed by microorganisms, and the concentration of cadmium in soil solution would increase and eventually exceed levels toxic to plant growth. Chaney (111) and Leeper (89) expressed similar views. However, this hypothesis has yet to be proved. Some sludge organic matter is broken down rapidly after sludge spreading. Miller (112) reported that 20 percent of the carbon in sludge added to soil evolved as carbon dioxide during a six-month incubation period. The rate of carbon evolution was dependent on temperature. More carbon was lost at higher temperatures. Analyses of abandoned, long-term sewage and sludge disposal sites (13, 38, 39, 113–115) show that high concentrations of trace elements do occur in soil. Above-normal concentrations of copper and cadmium in corn leaves (13), zinc in lettuce and radish (38), and zinc, chromium and nickel in forage (114) have been reported at these sites. But in no case have concentrations of elements been toxic to livestock or humans eating the food.

Zimdahl and Foster (67) felt that organic matter additions did not offer promise as a method to reduce lead availability to plants. They said that massive amounts of organic matter would be required to achieve small increases in soil organic content and exert an effect on lead accumulation by plants. The pH was more important in controlling lead uptake than organic matter.

More research is needed to determine the effect of organic matter, especially in disused sludge disposal sites, on trace element availability.

Other Ions

Phosphate

Recent results confirm that phosphate is effective in reducing trace element availability (103). Exceptions are a study by Iwai et al. (72) which found that the addition of phosphorus had no effect on cadmium uptake by corn in nutrient culture, and a report by Sandstead et al. (116) which said that phosphorus did not affect cadmium absorption in soybeans grown under field conditions in Illinois. [Studies have shown that commercial fertilizers, especially superphosphate, are contaminated with trace elements. Recent studies (117–119) have quantified the amount of contamination on different fertilizers. John et al. (120) pointed out that trace element contamination along railroad tracks can occur if trains carry commercial fertilizers.]

More and more sludges are being produced that are chemically treated. The chemicals (lime, ferric chloride, aluminum sulfate called alum, polyelectrolytes) are added to dewater the sludges or to remove phosphates from wastewater. Short-term greenhouse experiments with sludges from treatment works that remove phosphate show that phosphorus remains available for plant growth (121). However, dewatered sludges treated with lime do appear to reduce phosphate and trace element availability (37). Trace element availability at sludge disposal sites could be reduced by using limed sludges instead of spreading lime, which commonly is done.

More experiments should be carried out with chemical sludges because tremendous quantities of these sludges are being introduced into the environment. In particular, sludges treated with polyelectrolytes should be studied. Biesinger *et al.* (122) showed that aquatic organisms absorb large quantities of these polyelectrolytes and they may be dangerous if incorporated into food chains. Oliver and Cosgrove (123) said that commercial grade ferric chloride and alum added to wastewater and sludge could become significant sources of trace metals. New treatments like chlorine oxidation increase trace element availability in sludge (124). All these types of chemically treated sludges need to be studied to determine their effects on plants and soils.

Cadmium/Zinc Ratio

There are synergistic and antagonistic interactions between ionic metal species in soils that affect the absorption of elements by plant roots. One of the better known antagonistic reactions is between cadmium and zinc. Recent publications confirm that *high* zinc levels reduce cadmium uptake (12, 20, 108). Exceptions are experiments by Iwai et al. (72) and Allaway, as cited by Shacklette (51), who found that additions of zinc to nutrient solutions did not depress cadmium uptake by plants. Haghiri (108) found that additions of small amounts of zinc (5 to 50 ppm) *increased* cadmium concentrations in soybean plants. A depression in cadmium concentration occurred at the 100-ppm zinc level. He said that soil applications of zinc to reduce cadmium uptake did not appear to be practical because the suppression of cadmium occurred only when large amounts of zinc were added. At these high levels of zinc, yield was drastically reduced.

Cunningham *et al.* (20) and Dowdy and Larson (12) found that cadmium/zinc ratios of sludge generally were greater than those of plants. However, on non-sludge-treated soil, Van Hook (125) showed that cadmium/zinc ratios increased from 0.7 percent in soils to 1.8 percent in earthworms. These data indicated a biological accumulation of cadmium and zinc in earthworms relative to soil concentrations. Cadmium was accumulated more readily than zinc. Cunningham et al. (20) stated that the ratio of 1 percent cadmium/zinc in sludge proposed by Chaney of the U.S. Department of Agriculture in Beltsville, Maryland, was useful because at this ratio, zinc toxicity would occur in plants before any appreciable cadmium accumulation occurred. However, the Metropolitan Sanitary District of Greater Chicago (126) said that a cadmium/zinc ratio of 1 percent was unrealistic because metal abatement programs of cities cannot lower the cadmium content or decrease the cadmium/zinc ratio to levels proposed by Chaney. These levels are being adopted in federal guidelines for sludge disposal on land (126). If the 1 percent ratio is accepted, more than 90 percent of the sludges produced in the U.S. could not be spread on land (126).

Manganese

Three papers report that manganese concentrations decrease in plants treated with sludge (12, 40) or sewage (114). The decrease could have been due to antagonistic effects between manganese and other trace elements in the sludge. A manganese-iron antagonism has been known for a long time (74). Most papers, however, show increased concentrations of manganese in plants with sludge application (1).

Temperature

Trace element concentrations in plants increase with increased temperature in the root zone (79, 108).

Soil Water Content

Recent studies confirm that trace elements are more available under anaerobic (flooded) conditions than under aerobic conditions (25, 101, 104, 105, 127–129). However, concentrations of trace elements in solution are not altered significantly by moderate and transient changes in soil water content (129). Under dry conditions, plants are able to absorb trace elements if roots have access to subsoil water (129).

Sludge Placement

Subsurface sludge injection at depths from 30 to 45 cm is becoming a popular way to dispose of sludge, because it is more aesthetically acceptable to the public. No information has been published concerning the rate of uptake of sludge placed on the surface of the soil, or spread and then incorporated into the plow layer (20-cm depth). A few studies done with inorganic fertilizers (130, 131) show that ion uptake is dependent upon placement of the nutrient. Similar observations should be made with sludge to determine the most effective placement for nutrient uptake or for trace element exclusion. Anaerobic conditions could occur if sludge were injected below the surface, and trace elements may become less available for uptake.

Soil Versus Ocean Disposal

Ocean disposal long has been a popular method of disposal along the eastern and western seaboards because it is the cheapest method of disposal. In 1884 Barnes (a Massachusetts medical doctor), as quoted by Tarr (132), lamented:

> All the human and animal manure which the world loses, restored to the land instead of being thrown into the water would suffice to nourish the world. These heaps of garbage at the corners of the stone blocks, these tumbrils of mire jolting the streets at night, these horrid scavengers' carts, these fetid streams of subterranean slime which the pavement hides, what is all this? It is the flourishing meadow, the green grass, the thyme and sage; it is game, it is cattle, hay, corn, bread upon the table, warm blood in the veins.

In 1912 Sedgwick (133) condemned the practice of ocean disposal of sewage. He pointed out that few, if any, sewage farms existed in the east as they do in the interior part of the U.S. because sewage was put into the ocean and wasted. He said:

> If the sewage of Boston could be carried to Cape Cod, or that of New York to the sands of Long Island, Philadelphia's to the Pine Barrens of New Jersey, or that of Baltimore to the sometimes poor and thirsty soil of the Eastern Shore of Maryland, then might these comparatively desert places be made to blossom like the rose. . . . It may be that if we do not use our sewage upon the land we shall by and by be driven to seek our food more and more within the sea. The Japanese and the Chinese eat not only fish but seaweeds, and it would be strange indeed if Americans likewise should give up the land vegetables of today for the sea weeds of tomorrow.

In the United States today, much research is done to develop food from seaweed and other marine organisms.

People dispute the dangers of ocean disposal of sludge. Researchers in New York and Philadelphia report that ocean disposal is environmentally acceptable (134, 135). Papers by other investigators, too many to cite, show that aquatic organisms (algae and other plankton, fish, mussels, oysters, clams) accumulate metals in extremely high concentrations. Shellfish tend to concentrate cadmium and a biomultiplication factor of 14,400 has been reported for oysters (136). If sludge is dumped in the ocean, shellfish and other organisms probably will accumulate the metals. Therefore, as far as trace elements in sludge are concerned, land application appears to be safer than ocean disposal because the soil ties up large quantities of the trace elements.

The dangers of ocean disposal of sludge have been outlined by Ember (137).

In summary of the soil studies, pH appears to be the most important soil factor controlling trace element availability. The importance of organic matter in the sludge added to soil is debatable. Many investigators feel that after sludge additions stop, trace elements will become more available with time. Data suggest that trace element concentrations in plants from abandoned, long-term sludge disposal sites are no higher than they are in plants grown on sites still receiving sludge applications. Land disposal of sludge appears safer than ocean disposal because the soil ties up trace elements. Consequently, the elements usually are less available to land plants than they are to aquatic organisms.

GROUND WATER

A long-term need listed by the Workshop on Toxic Chemicals (11) was to study leakage of trace elements from the soil-root zone because they can be hazardous in minute amounts. Page and Chang (138) recently have reviewed the literature dealing with mobility of trace elements in sludge-treated soil and the potential for ground water pollution. They concluded that, except for boron, movement of trace elements in soils with sludge is restricted. Trace elements stay at the depth of tillage. Even though practically all trace elements are fixed in surface layers of soil, rather small increases in trace element solubility with subsequent movement to water tables could result in deterioration of ground water supplies. The increasing use of subsurface sludge injection, rather than surface spreading, could result in more leaching of trace elements to ground water.

Boswell (34) noted increased levels of zinc below 30 cm in soil with sludge. MacLean et al. (102) found that mercury leached to at least 120 cm in golf courses treated with mercurial compounds to control disease of turfgrass. Lund et al. (139) measured significant downward movement to 150 cm of trace elements (e.g., chromium) beneath disposal lagoons at two sewage treatment plants in California. Bouwer (140) reported that in the Imperial Valley of California the drain-depth ranges from about 150 cm to 300 cm and the water table depth midway between drains is about 120 cm. Therefore, if trace elements can move as far as 150 cm, then they could pollute water tables. In the Netherlands, water tables often are 50 cm or less (140). Upward flow from a water table is an important source of water for transpiring plants (141). If ground water tables are high, plants on sludge-treated land could take up the trace elements.

Ground water pollution by trace elements has occurred due to man-made causes. Mine drainage water has been shown to contain high amounts of trace elements. Leachate from landfill sites (dumps) pollute ground water (142–144), particularly in humid areas where the water table is near the surface and precipitation is abundant (145). The Environmental Protection Agency recently reported that leachate contamination in ground water caused by a dump was found two miles away from the dump and selenium in the water exceeded drinking water standards (143). Lieber and Welsch (146) reported ground waters near an electroplating industry in New York contained 3.2 μg cadmium/ml. They predicted that many years would have to elapse before rainfall and ground water dilution would eliminate cadmium contamination from ground water. In 1972 arsenic poisoning occurred in a Minnesota town which obtained its water from underground sources (147). [Half of the population in the United States obtains its water from ground water sources (147).] The arsenic came from the well water. An arsenic-based pesticide had been applied massively in 1934 to deal with a grasshopper infestation. The arsenic took 38 years to percolate down to the underlying ground water.

Ground water pollution by trace elements has occurred naturally. Shacklette (51) reported that peat bogs in Orleans County, New York were enriched with zinc, lead and cadmium by the entrance of ground water from dolomite beds that contained these elements. Thirty years after these bogs were drained, high cadmium values were reported in plants grown on the peat. MacKenzie and Viets (148) review literature giving trace element concentrations in drainage waters.

Little information exists on concentrations of trace elements in drainage water from sludge-treated sites. Analyses of ground water leachate from corn plots at Hanover Park, Illinois, after six years of spreading sludge from Chicago (up to 203 m.ton/ha dry solids) indicated that the ground water quality improved. There was a decrease in concentration of potassium, sodium, calcium, magnesium, zinc, copper, manganese, iron, sulfate, total and ammonium nitrogen, and alkalinity (126). Van Loon (28) analyzed well water on heavily sludged (rates not given) land near North Toronto, Canada. However, he could not interpret results because no wells containing uncontaminated water had been established. In general, North Toronto well results showed large variations in iron and zinc values and little variation in copper and nickel. Pike et al. (38) reported that on a long-term sludge disposal site in England nickel appeared to be more mobile than arsenic, cadmium, copper, chromium, lead and zinc.

El-Bassam *et al.* (119) found that in areas irrigated with sewage water for 80 years neither chromium nor mercury contaminated the ground water. Johnson *et al.* (114) said that on the farm in Melbourne, Australia, irrigated for 73 years with sewage, "concentration levels of several heavy metals in the farm drainage water are about the same as expected in secondary-treated municipal wastewater. Except for Cu, all of the elements included in the analyses of farm drainage water occur in higher concentration levels than would be expected in seawater."

In summary, ground water analyses of land receiving high concentrations of trace elements are variable. The results indicate that chromium, mercury, nickel and zinc can move below the 30-cm depth. To protect against trace element contamination of ground water beneath soils with sludge, more information is needed on the extent and conditions under which trace elements migrate in soil.

REFERENCES

1. Page, A. L. "Fate and effects of trace elements in sewage sludge when applied to agricultural lands. A literature review study," EPA–670/8–74–005, U.S. EPA, Cincinnati, Ohio, 98 pp. (1974).
2. Carroll, T. E., D. L. Maase, J. M. Genco and C. N. Ifaedi. "Review of land spreading of liquid municipal sewage sludge," EPA–670/2–75–049. U.S. EPA, Cincinnati, Ohio, 96 pp. (1975).
3. Council for Agricultural Science and Technology. "Utilization of animal manures and sewage sludges in food and fiber production," Rep. No. 41. Iowa State University, Ames, Iowa, 22 pp. (1975).
4. Grooms, G. C., Ed. "Ohio guide for land application of sewage sludge," Res. Bull. 1079, Ohio Agr. Res. and Develop. Ctr., Wooster, Ohio, 12 pp. (1975).
5. Hecht, N. L., D. S. Duvall and A. S. Rashidi. "Characterization and utilization of municipal and utility sludges and ashes," Vol. II. Municipal Sludges, EPA–670/2–75–033b, U.S. EPA, Cincinnati, Ohio, 231 pp. (1975).
6. Hinesly, T. D., O. C. Braids, R. I. Dick, R. L. Jones and J. E. Melina. "Agricultural benefits and environmental changes resulting from the use of digested sludge on field crops," EPA/530SW–301.1. U.S. EPA, Cincinnati, Ohio, 375 pp. (1974). Available from Nat. Tech. Information Service, Springfield, Virginia.
7. Keeney, D. R., K. W. Lee and L. M. Walsh. "Guidelines for the application of wastewater sludge to agricultural land in Wisconsin," Tech. Bull. 88, Dept. Natural Resources, Madison, Wisconsin, 36 pp. (1975).
8. Lue-Hing, C., B. T. Lynam, R. R. Rimkus and J. R. Peterson. "Digested sludge utilization in agriculture and as a soil amendment," Rep. No. 75–24, The Metropolitan Sanitary District of Greater Chicago, Dept. of Research and Development, Chicago, Illinois, 87 pp. 1975).
9. Walker, J. M. "Sewage sludge—management aspects for land application," *Compost Sci.* 16(2):12–21 (1975).
10. Webber, J. "Sludge handling and disposal practices in England," presented at Sludge Handling and Disposal Seminar, Sept. 18–19, 1974,

Toronto, Ontario, Canada. Sponsored under Canada-Ontario Agreement on Great Lakes Water Quality. Environ. Canada and Ontario Min. of the Environ. (1974).

11. Pratt, P. (Chairman). "Plant response-toxic chemicals," Workshop Group 8, pp. 228–230. In *Proc. Joint Conf. Recycling Municipal Sludges and Effluents on Land,* Champaign, Illinois, July 9–13, 1973, Nat. Assoc. State Univ. and Land-Grant Coll., Washington, D.C., 244 pp. (1973).

12. Dowdy, R. H. and W. B. Larson. "The availability of sludge-borne metals to various vegetable crops," *J. Environ. Quality* 4:278–282 (1975).

13. Kirkham, M. B. "Trace elements in corn grown on long-term sludge disposal site," *Environ. Sci. Tech.* 9:765–768 (1975).

14. Garcia, W. J., C. W. Blessin, G. E. Inglett and R. O. Carlson. "Physical-chemical characteristics and heavy metal content of corn grown on sludge-treated strip-mine soil," *Agr. Food Chem.* 22:810–815 (1974).

15. Hyde, H. C. "Utilization of wastewater sludge for agricultural soil enrichment," *J. Water Pollution Control Fed.* 48:77–90 (1976).

16. Giordano, P. M., J. J. Mortvedt and D. A. Mays. "Effect of municipal wastes on crop yields and uptake of heavy metals," *J. Environ. Quality.* 4:394–399 (1975).

17. Bingham, F. T., A. L. Page, R. J. Mahler and T. J. Ganje. "Growth and cadmium accumulation of plants grown on a soil treated with a cadmium-enriched sewage sludge," *J. Environ. Quality* 4:207–211 (1975).

18. Jones, R. L., T. D. Hinesly, E. L. Ziegler and J. J. Tyler. "Cadmium and zinc contents of corn leaf and grain produced by sludge-amended soil," *J. Environ. Quality* 4:509–514 (1975).

19. Hinesly, T. D., R. L. Jones and E. L. Ziegler. "Effects on corn by applications of heated anaerobically digested sludge," *Compost Sci.* 13(4):26–30 (1972).

20. Cunningham, J. D., D. R. Keeney and J. A. Ryan. "Yield and metal composition of corn and rye grown on sewage sludge-amended soil," *J. Environ. Quality* 4:448–454 (1975).

21. Mortvedt, J. J. and P. M. Giordano. "Response of corn to zinc and chromium in municipal wastes applied to soil," *J. Environ. Quality* 4:170–174 (1975).

22. Singh, R. N., R. F. Keefer and D. J. Horvath. "Can soils be used for sewage sludge disposal?" *Compost Sci.* 16(2):22–25 (1975).

23. Sidle, R. C., J. R. Hook and L. T. Kardos. "Heavy metals application and plant uptake in a land disposal system for waste water," *J. Environ. Quality* 5:97–102 (1976).

24. Kirkham, M. B. and G. K. Dotsor. "Growth of barley irrigated with wastewater sludge containing phosphate precipitants," pp. 97–106. In *Proc. Nat. Conf. on Municipal Sludge Management,* Information Transfer, Inc., Washington, D.C. (1974).

25. Kirkham, M. B. "Uptake of cadmium and zinc from sludge by barley grown under four different sludge irrigation regimes," *J. Environ. Quality* 4:423–426 (1975).

26. Van Loon, J. C. "Heavy metals in agricultural lands receiving chemical sewage sludges," Vol. I, Res. Rep. 9, Environ. Canada and Min. of the Environ., Toronto, Ontario, Canada, 37 pp. (1973).

27. Bradford, G. R., A. L. Page, L. J. Lund and W. Olmstead. "Trace element concentrations of sewage treatment plant effluents and sludges; their interactions with soils and uptake by plants," *J. Environ. Quality* 4:123–127 (1975).

28. Van Loon, J. C. "Heavy metals in agricultural lands receiving chemical sewage sludges," Vol. II. Res. Rep. 25, Environ. Canada and Min. of the Environ., Toronto, Ontario, Canada, 41 pp. (1975).

29. Bolton, J. "Liming effects on the toxicity to perennial ryegrass of a sewage sludge contaminated with zinc, nickel, copper and chromium," *Environ. Pollution* 9:295–304 (1975).

30. Andersson, A. and K. O. Nilsson. "Influence of lime and soil pH on Cd availability to plants," *Ambio* 3(5):198–200 (1974).

31. Sabey, B. R. and W. E. Hart. "Land application of sewage sludge: I. Effect on growth and chemical composition of plants," *J. Environ. Quality* 4:252–256 (1975).

32. Stenstrom, T. and H. Lonsjo. "Cadmium availability to wheat: a study with radioactive tracers under field conditions," *Ambio* 3(2):87–90 (1974).

33. Linnman, L., A. Andersson, K. O. Nilsson, B. Lind, T. Kjellstrom and L. Friberg. "Cadmium uptake by wheat from sewage sludge used as a plant nutrient source," *Arch. Environ. Health* 27:45–47 (1973).

34. Boswell, F. C. "Municipal sewage sludge and selected element applications to soil: effect on soil and fescue," *J. Environ. Quality* 4:267–273 (1975).

35. Bingham, F. T., A. L. Page, R. J. Mahler and T. J. Ganje. "Yield and cadmium accumulation of forage species in relation to cadmium content of sludge-amended soil," *J. Environ. Quality* 5:57–60 (1976).

36. Lu, P. Y., R. L. Metcalf, R. Furman, R. Vogel and J. Massett. "Model ecosystem studies of lead and cadmium and of urban sewage sludge containing these elements," *J. Environ. Quality* 4:505–509 (1975).

37. Kirkham, M. B. "Growth of tulips treated with sludge containing dewatering chemicals," *Environ. Pollution* 10:submitted (1976).

38. Pike, E. R., L. S. Graham and N. W. Fogden. "An appraisal of toxic metal residue in the soils of a disused sewage farm. I. Zinc, copper and nickel," *J. Assoc. Public Anal.* 13:19–33 (1975).

39. Pike, E. R., L. S. Graham and N. W. Fogden. "An appraisal of toxic metal residues in the soils of a disused sewage farm. II. Lead, cadmium and arsenic, with notes on chromium," *J. Assoc. Public Anal.* 13:48–63 (1975).

40. Lue-Hing, C., J. R. Peterson and S. J. Sedita. "West-Southwest Nu-Barth vegetable garden report," The Metropolitan Sanitary District of Greater Chicago, Dept. of Res. and Development, Chicago, Illinois, 12 pp. (1975).

41. Allaway, W. H. "Agronomic controls over the environmental cycling of trace elements," *Advance. Agron.* 20:235–274 (1968).

42. Melsted, S. W. "Soil-plant relationships, some practical considerations in waste management," pp. 121–128. In *Proc. Joint Conf. on Recycling Municipal Sludges and Effluents on Land,* Champaign, Illinois, July 9–13, 1973, Nat. Assoc. State Univ. and Land-Grant Coll., Washington, D.C. (1973).

43. United States Department of Agriculture. "Tracing mercury in a biosphere," *Agr. Res.* 24(6)12–13 (1975).

44. Franz, M. "Municipal garbage into mushroom soil," *Compost Sci.* 13(6):6–9 (1972).

45. Pfeifer, R. P. "Record yields and your operation," *Crops and Soils Mag.* 28(7):5–7 (1976).

46. Dolar, S. G., D. R. Keeney and G. Chester. "Mercury methylation by *Myriophyllum crioeum.* L.," *Environ. Letters* 1:191–198 (1971).

47. Jackson, P. C. "Uptake and release of mercury vapor by plants," *Plant Physiol.* 51(suppl.):22 (1973).

48. Sempio, C. "Mechanism of the resistance against *Oidium* produced by cadmium on wheat plants," *Ann. Facolta Agrar.*, Regia Univ. Perugia I, 4 pp.; *Chem Abstr.* 43:1463 (1942).

49. Thomas, W. A. "Accumulation of rare earths and circulation of cerium by mockernut hickory trees," *Can. J. Bot.* 53:1159–1165 (1975).

50. Thomas, W. A. "Cobalt accumulation and circulation by blackgum trees," *Forest Sci.* 22:222–226 (1975).

51. Shacklette, H. T. "Cadmium in plants," *Geol. Surv. Bull.* 1314-G, U.S. Govt. Printing Office, Washington, D.C., 28 pp. (1972).

52. Smith, R. A. H. and A. D. Bradshaw. "Stabilization of toxic mine wastes by the use of tolerant plant populations," *Inst. Mining Met. Trans.*, Sect A, 81:A230–A237 (1972).

53. Bernstein, L. "Effects of salinity on mineral composition and growth of plants," *Plant Anal. and Fertilizer Problems* 4:25–45 (1964).

54. Bazzaz, F. A., R. W. Carlson and G. L. Rolfe. "The Effect of heavy metals on plants: I. Inhibition of gas exchange in sunflower by Pb, Cd, Ni and Tl," *Environ. Pollution* 7:241–246 (1974).

55. Bazzaz, F. A., G. L. Rolfe and R. W. Carlson. "Effect of Cd on photosynthesis and transpiration of excised leaves of corn and sunflower," *Physiol. Plant.* 32:373–376 (1974).

56. Schnabl, H. and H. Ziegler. "Uber die Wirkung von Aluminiumionen auf die Stomatabewegung von *Vicia faba*-Eridermon," *Z. Pflanzenphysiologie* 74:394–403 (1975).

57. Nosseir, M. A. "Rhythmicity in the absorption and release of ions by broad and kidney beans," *Advanc. Frontiers Plant Sci.* 25:121–128 (1970).

58. Kirkham, M. B. "Internal water relations of cadmium-treated plants," Mass. Agr. Exp. Sta., NE-48 Contributor Project Progress Rep. Dynamics and energetics of the soil-plant-atmosphere continuum, 7 pp., available from Dep. Plant and Soil Sci., Univ. Mass., Amherst (1975).

59. Bazzaz, M. B. and Govindjec. "Effects of cadmium nitrate on spectral characteristics and light reactions of chloroplasts," *Environ. Letters* 6:1–12 (1974).

60. Li, E. H. and C. D. Miles. "Effects of cadmium on photoreaction II of chloroplasts," *Plant Sci Letters* 5:33–40 (1975).

61. Root, R. A., R. J. Miller, and D. E. Koeppe. "Uptake of cadmium—its toxicity, and effect on the iron ratio in hydroponically grown corn," *J. Environ. Quality* 4:473–476 (1975).

62. Haghiri, F. "Cadmium uptake by plants," *J. Environ. Quality* 2:93–96 (1973).

63. Hewitt, E. J. "Metal interrelationships in plant nutrition. I. Effects of some metal toxicities on sugar beet, tomato, oat, potato, and marrowstem kale grown in sand culture," *J. Exp. Bot.* 4:59–64 (1953).

64. Imai, I. and S. N. Siegel. "A specific response to toxic cadmium levels in red kidney bean embryos," *Physiol. Plant.* 29:118–120 (1973).

65. Bittell, J. E., D. E. Koeppe and R. J. Miller. "Sorption of heavy metal cations by corn mitochondria and the effects on electron and energy transfer reactions," *Physiol. Plant.* 30:226–230 (1974).

66. Miller, R. J., J. E. Bittell and D. E. Koeppe. "The effect of cadmium on electron and energy transfer reactions in corn mitochondria," *Physiol. Plant.* 28:166–171 (1973).

67. Zwarun, A. A. "Tolerance of *Escherichia coli* to cadmium," *J. Environ. Quality* 2:353–355 (1973).

68. Huang, C.-Y., F. A. Bazzaz and W. M. Vanderhoef. "The inhibition of soybean metabolism by cadmium and lead," *Plant. Physiol.* 54:122–124 (1974)

69. Stenlid, G. "Stimulatory effects of some heavy metals and sulfhydryl reagents upon root elongation of wheat seedlings," *Swedish J. Agr. Res.* 5:137–144 (1975).

70. Gauch, H. C. *Inorganic Plant Nutrition*, Dowden, Hutchinson and Ross, Lnc., Stroudsburg, Pennsylvania, 488 pp. (1972).

71. Muller, K. and A. C. Leopold. "The mechanism of kinetin-induced transport in corn leaves," *Planta* 68:186–205 (1966).

72. Iwai, I., T. Hara and Y. Sonoda. "Factors affecting cadmium uptake by the corn plant," *Soil Sci. Plant Nutrition* 21:37–46 (1975).

73. Huffman, E. W. D., Jr. and W. H. Allaway. "Growth of plants in solution culture containing low levels of chromium," *Plant Physiol.* 52: 72–75 (1973).

74. Gerloff, G. C. "Comparative mineral nutrition of plants," *Annu. Rev. Plant Physiol.* 14:107–124 (1963).

75. Kirkham, M. B., W. R. Gardner and G. C. Gerloff. "Regulation of cell division and cell enlargement by turgor pressure," *Plant Physiol.* 49: 961–962 (1972).

76. Kirkham, M. B., W. R. Gardner and G. C. Gerloff. "Internal water status of kinetin-treated, salt-stressed plants," *Plant Physiol.* 53:241–243 (1974).

77. United States Department of Agriculture. "New light on zinc," *Agr. Res.* 24(7):3–4 (1976).

78. Tiffin, L. O. "The form and distribution of metals in plants: an overview," paper presented at the Fifteenth Annual Hanford Life Sciences Symposium, Biological Implications of Metals in the Environment, Sept. 29–Oct. 1, 1975, Richland, Washington (1975).

79. Cutler, J. M. and D. W. Rains. "Characterization of cadmium uptake by plant tissue," *Plant Physiol.* 54:67–71 (1974).

80. Malone, C., D. E. Koeppe and R. J. Miller. "Localization of lead accumulated by corn plants," *Plant Physiol.* 53:388–394 (1974).

81. Mitra, R. S., R. H. Gray, B. Chin and I. A. Barrstein. "Molecular mechanisms of accommodation in *Escherichia coli* to toxic levels of Cd^{2+}," *J. Bacteriology* 121:1180–1188 (1975).

82. Goren, A. Von and H. Wanner. "Die Absorption von Blei und Kupfer durch Wurzeln von *Hordeum vulgare*," *Ber. Schwig. Bot. Ga. Z.* 80: 334–340 (1971).

83. Tanton, T. W. and S. H. Growdy. "The distribution of lead chelate in the transpiration stream of higher plants," *Pesticide Sci.* 2:211–213 (1971).

84. Buchauer, M. J. "Contamination of soil and vegetation near a zinc smelter by zinc, cadmium, copper and lead," *Environ. Sci. Tech.* 7: 131–135 (1973).

85. Hevesy, G. "The absorption and translocation of lead by plants," *Biochem. J.* 17:439–445 (1923).

86. Gordee, R. S., C. L. Porter and R. G. Langston. "Uptake and distribution studies of cadmium 115-m in peppermint," *Proc. Amer. Soc. Hort. Sci.* 75:525–528 (1960).

87. Shewry, P. R. and P. J. Peterson. "The uptake and transport of chromium by barley seedlings (*Hordeum vulgare* L.)," *J. Exp. Bot.* 25:785–797 (1974).

88. Handley, R. and K. L. Babcock. "Translocation of [85]Gr, ^{137}Cs and [106]Ru in crop plants," *Radiat. Bot.* 12:113–119 (1972).

89. Leeper, G. W. "Reactions of heavy metals with soils with special regard to their application in sewage wastes," prepared under contract no. DAGW75-173-C-0026 for Dept. of the Army, Corps of Engineers, 30 pp. (1972).

90. Sámsoni, Z., S. Szalay, N. Szilágyi, and A. Toth. "Investigation of trace-element uptake by plants on peat soils at Keszthely, IV. Agronkémia és Talajtan," 24:61–70 (1975).

91. Anon. "Peat moss bed filtration removes heavy metals from wastewaters," *Environ. Sci. Tech.* 8:598 (1974).

92. Osborne, J. M. "Tertiary treatment of campground wastes using a native Minnesota peat," *J. Soil and Water Conserv.* 30:235–236 (1975).

93. United States Department of Agriculture. "Byproducts scavenge metal pollutants," *Agr. Res.* 24(1):6–7 (1975).

94. Lightsey, G. R. "Bark as trickling-filter dewatering medium for pulp and paper mill sludge," Environ. Resources Center Pub. 0975, Georgia Institute of Technology, Atlanta, Georgia, 21 pp. (1975).

95. Reiners, W. A., R. H. Marks and P. M. Vitousek. "Heavy metals in subalpine and alpine soils of New Hampshire," *Oikos* 26:264–275 (1975).

96. Harter, R. D. "A survey of New Hampshire sewage sludges as related to their suitability for on-land disposal," New Hampshire Agr. Exp. Sta. Bull. 503, 19 pp. (1975).

97. Bittell, J. E. and R. J. Miller. "Lead, cadmium, and calcium selectivity coefficients on a montmorillonite, illite, and kaolinite," *J. Environ. Quality* 3:250–253 (1974).

98. Dowdy, R. H. and W. E. Larson. "Metal uptake by barley seedlings grown on soils amended with sewage sludges," *J. Environ. Quality* 4:229–233 (1975).

99. Hahne, H. C. H. and W. Kroontje. "Significance of pH and chloride concentration on behavior of heavy metal pollutants: mercury (II), cadmium (II), zinc (II), and lead (II)," *J. Environ. Quality* 2:444–450 (1973).

100. Lagerwerff, J. V., G. T. Biersdorf and D. L. Brower. "Retention of metals in sewage sludge. I: Constituent heavy metals," *J. Environ. Quality* 5:19–23 (1976).

101. Lagerwerff, J. V., G. T. Biersdorf and D. L. Brower. "Retention of metals in sewage sludge. II. Incorporated radioisotopes," *J. Environ. Quality* 5:23–25 ((1976).

102. MacLean, A. J., B. Stone and W. E. Cordukes. "Amounts of mercury in soil of some golf course sites," *Can. J. Soil Sci.* 53:130–132 (1973).

103. Santillan-Medrano, J. and J. J. Jurinak. "The chemistry of lead and cadmium in soil: solid phase formation," *Soil Sci. Soc. Amer. Proc.* 39:851–856 (1975).

104. Singh, B. R. "Migration of ions in soils. I. Movement of Zn^{65} from surface application of zinc sulphate in soil columns," *Plant and Soil* 41:619–628 (1974).

105. Singh, B. R. "Migration of ions in soils. II. Movement of Mn^{54} from surface application of manganese sulphate in soil columns," *Plant and Soil* 41:629–636 (1974).

106. Zimdahl, R. L. and J. M. Foster. "The influence of applied phosphorus, manure, or lime on uptake of lead from soil," *J. Environ. Quality* 5:31–34 (1976).

107. Korte, N. E., J. Skopp, E. E. Nieble and W. H. Fuller. "A baseline study on trace metal elution from diverse soil types," *Water, Air, and Soil Pollution* 5:149–156 (1975).

108. Haghiri, F. "Plant uptake of cadmium as influenced by cation exchange capacity, organic matter, zinc, and soil temperature," *J. Environ. Quality* 3:180–183 (1974).

109. Cunningham, J. D., D. R. Kenney and J. A. Ryan. "Phytoxicity and

uptake of metals added to soils as inorganic salts of in sewage sludge," *J. Environ. Quality* 4:460–462 (1975).

110. Dijkshoorn, W. and J. E. M. Lampe. "Availability for ryegrass of cadmium and zinc from dressings of sewage sludge," *Neth. J. Agr. Sci.* 23:338–341 (1975).

111. Chaney, R. L. "Crop and food chain effects of toxic elements in sludges and effluent," pp. 129–141, In *Recycling Municipal Sludges and Effluents on Land*, Nat. Assoc. of State Univ. and Land-Grant Coll., Washington, D.C. (1973).

112. Miller, R. H. "Factors affecting the decomposition of an anaerobically digested sewage sludge in soil," *J. Environ. Quality* 3:376–380 (1974).

113. El-Bassam, Von N., P. Poelstra and M. J. Frissel. "Chrom und Quecksilber in einem seit 80 Janren mit stadtischem Abwasser berieselten Boden," *Z. Pflanzenern. Bodenk.* 3:309–316 (1975).

114. Johnson, R. D., R. L. Jones, T. D. Hinesly and D. J. David. "Selected chemical characteristics of soils, forages, and drainage water from the sewage farm serving Melbourne, Australia," Dept. of the Army, Corps of Engineers, 54 pp. (1974).

115. Zwerman, P. J. and F. A. M. de Haan. "Significance of the soil in enviromental quality improvement—a review," Institute for Land and Water Management Research Technical Bull. 88, pp. 121–155, Wageningen, The Netherlands (1973).

116. Sandstead, H. H., W. H. Allaway, R. G. Burau, W. Fulkerson, H. A. Laitinen, P. M. Newberne, J. O. Pierce and B. G. Wixson. "Cadmium, zinc, and lead," pp. 43–56, In *Geochemistry and the Environment, Vol. I. The Relation of Selected Trace Elements to Health and Disease*, Nat. Acad. Sci., Washington, D.C. (1974).

117. Gilkes, R. J., R. C. Young and J. P. Quirk. "Leaching of copper and zinc from trace element superphosphate," *Aust. J. Soil Res.* 13:33–41 (1975).

118. Lee, K. W. and D. R. Keeney. "Cadmium and zinc additions of Wisconsin soils by commercial fertilizers and wastewater sludge application," *Water, Air and Soil Pollution* 5:109–112 (1975).

119. Stenström, T. and M. Vahter. "Cadmium and lead in Swedish commercial fertilizers," *Ambio* 3(2):91–92 (1974).

120. John, M. K., C. J. Van Laerhoven, V. E. Osborne and I. Cotic. "Mercury in soils of British Columbia, a mercuriferous region," *Water, Air and Soil Pollution* 5:213–220 (1975).

121. Kirkham, M. B. "Value of phosphorus in sewage sludge spread on agricultural land," In G. V. Levin, Ed., *Biological Waste Treatment*, CRC Press, Cleveland, Ohio (1976).

122. Biesinger, K. E., A. E. Lemke, W. E. Smith and R. M. Tyo. "Comparative toxicity of polyelectrolytes to selected aquatic animals," *J. Water Pollution Control Fed.* 48:183–187 (1976).

123. Oliver, B. G. and E. G. Cosgrove. "Metal concentrations in the sewage, effluents, and sludges of some southern Ontario wastewater treatment plants," *Environ. Letters* 9:75–90 (1975).

124. Olver, J. W., W. C. Kreye and P. H. King. "Heavy metal release by chlorine oxidation of sludges," *J. Water Pollution Control Fed.* 47:2490–2497 (1975).

125. Van Hook, R. I. "Cadmium, lead, and zinc distributions between earthworms and soils: potentials for biological accumulation," *Bull. Environ. Contamination and Toxicology* 12:509–512 (1974).

126. Lynam, B. T. (General Superintendent). U.S. EPA notice of intent to issue a policy statement on acceptable methods for the utilization or

disposal of sludge from publicly-owned wastewater treatment plants. Comments and recommendations of The Metropolitan Sanitary District of Greater Chicago. The Metropolitan Sanitary District of Greater Chicago, Chicago, Illinois, 7 sections plus 8 appendices (1974).

127. Bloomfield, C. and G. Pruden. "The effects of aerobic and anaerobic incubation on the extractabilities of heavy metals in digested sewage sludge," *Environ. Pollution* 8:217–232 (1975).
128. Engler, R. M. and W. H. Patrick, Jr. "Stability of sulfides of manganese, iron, zinc, copper, and mercury in flooded and nonflooded soil," *Soil Sci.* 119:217–221 (1975).
129. Nambier, E. K. S. "Mobility and plant uptake of micronutrients in relation to soil water content," pp. 151–163, in D. J. D. Nicholas and A. R. Egan, Ed., *Trace Elements in Soil-Plant-Animal Systems,* Academic Press, New York (1975).
130. Kirkham, M. B., D. R. Keeney and W. R. Gardner. "Uptake of water and labelled nitrate at different depths in the root zone of potato plants grown on a sandy soil," *Agro-Ecosystems* 1:31–44 (1974).
131. Raimond, Y. "Distribution of phosphate and potassium fertilizers in arable stratum-applied to surface or plowed into soil," *Revue de l'Agriculture* 28:1085–1115 (1975).
132. Tarr, J. A. "From city to farm: urban wastes and the American farmer," *Agr. Hist.* 69:598–612 (1975).
133. Sedgwick, W. T. "Sewage and the farmer. A problem in the conservation of waste," *Sci. Amer.* 107(1):38 (1912).
134. Duedall, I. W., H. B. O'Connors and B. Irwin. "Fate of wastewater sludges in the New York Right apex," *J. Water Pollution Control Fed.* 47:2702–2706 (1975).
135. Guarino, C. F., M. D. Nelson, S. A. Townsend, T. E. Wilson and E. F. Ballotti. "Land and sea solids management alternatives in Philadelphia," *J. Water Pollution Control Fed.* 47:2551–2564 (1975).
136. Sanjour, W. "Cadmium and environmental policy," Office of Water and Hazardous Materials, U.S. EPA, Washington, D.C., 22 pp. (1974).
137. Ember, L. R. "Ocean dumping: Philadelphia's story," *Environ. Sci. Tech.* 9:916–917 (1975).
138. Page, A. L. and A. C. Chang. "Trace element and plant nutrient constraints of recycling sewage sludges on agricultural land," paper presented before the Second Nat. Conf. on Water Reuse: Water's interface with energy, air, and solids. Chicago, Illinois, May 4–8, 1975. Sponsored by Amer. Inst. Chem. Eng. and Environ. Protection Agency: Tech. Transfer (1975).
139. Lund, L. J., A. L. Page and C. O. Nelson. "Movement of heavy metals and nitrogen beneath sewage sludge and effluent disposal lagoons," *Agron. Abstr.* 66:33 (1974).
140. Bouwer, H. "Developing drainage design criteria," pp. 67–79, in J. van Schilfgaarde, Ed., *Drainage for Agriculture,* Amer. Soc. Agron., Madison, Wisconsin (1974).
141. Raats, P. A. C. and W. R. Gardner. "Movement of water in the unsaturated zone near a water table," pp. 311–357, in J. van Schilfgaarde, Ed., *Drainage for Agriculture,* Amer. Soc. Agron., Madison, Wisconsin (1974).
142. Hem, J. D. "Chemistry and occurrence of cadmium and zinc in surface water and groundwater," *Water Resources Res.* 8:661–679 (1972).
143. United States Environmental Protection Agency. "Dumps may contaminate ground water," *Environ. News,* Friday, March 14, 1975, Washington, D.C., 2 pp. (1975).

144. Zanoni, A. E. "Potential for ground water pollution from the land disposal of solid wates," *CRC Critical Reviews in Environ. Control* 3:225–206 (1973).

145. Weist, W. G., Jr. and R. A. Pettijohn. "Investigating ground water pollution from Indianapolis' landfills—the lessons learned," *Ground Water* 13(2):191–196 (1975).

146. Lieber, M. and W. F. Welsch. "Contamination of ground water by cadmium," *J. Amer. Water Works Assoc.* 46:541–547 (1954).

147. Josephson, J. "Quality assurance for groundwater," *Environ. Sci. Tech.* 10:226–227 (1976).

148. MacKenzie, A. J. and F. G. Viets, Jr. "Nutrients and other chemicals in agricultural drainage waters," pp. 489–508, in J. van Schilfgaarde, Ed., *Drainage for Agriculture*, Amer. Soc. Agron, Madison, Wisconsin (1974).

14

HEAVY METAL RELATIONSHIPS IN THE PENN STATE "LIVING FILTER" SYSTEM

R. C. Sidle and L. T. Kardos
Department of Agronomy, The Pennsylvania State University,
University Park, Pennsylvania

INTRODUCTION

The "living-filter" concept involves the controlled application of wastewater to the land to efficiently utilize the soil-plant system as a renovation media. Research conducted by Parizek *et al.* (1) emphasized the fertility benefits of the applied wastewater as well as the renovation of potential recharge percolate.

In recent years some of the research emphasis has shifted toward heavy metals associated with the wastewater and their ultimate distribution in a land disposal system. Copper (Cu), zinc (Zn), and cadmium (Cd) are the heavy metals of primary concern in most municipal wastewater applied to agricultural land. The major areas of interest regarding heavy metals are plant toxicities caused by excessive metal loadings, ultimate food chain toxicities resulting from crop uptake of metals, and subsequent transfer of these metals through foods to animals and humans. Copper and zinc do not usually approach toxic levels in the ultimate food chain as a result of wastewater disposal since plant toxicities occur before these metals reach levels that would be harmful to animals and humans. Relatively low levels of Cu in forage tissue have been known to be toxic to sheep (2). Cadmium poses the greatest concern to the ultimate food chain since it is readily taken up by plants and accumulates in the vital organs of animals and humans eating these plants. Due to their similar chemistry, Zn and Cd can compete for plant uptake.

Some research (3, 4) has shown that at low Cd levels in growth media, increasing the relative abundance of Zn with respect to Cd can result in suppressed Cd uptake by plants. Because of the interaction of Cd and Zn, Chaney (5) suggested that a Cd:Zn ratio of less than 1 percent in the applied wastewater be maintained to guarantee protection of the ultimate food chain.

Data presented by Konrad and Kleinert (6) for municipal sewage treatment plants in Wisconsin showed the following ranges of heavy metals in effluents: Cu, 0.02–0.8 ppm; Zn, 0.04–1.0 ppm; Cd 0.02–0.04 ppm. Total concentrations of heavy metals commonly found in soils and plants are reported in a literature survey by Allaway (7). Data on crop uptake of heavy metals has been reported by Sidle et al. (4) for a land disposal system. Levels of heavy metals in soils treated with sludge have been reported by Hinesly (8) and Boswell (9). Very little data have been presented regarding the distribution of heavy metals applied via wastewater to forested areas. Sidle and Sopper (10) found there was no increase in Cd uptake due to wastewater irrigation in selected forest species in two different areas.

The purpose of this paper is to evaluate the long-term effects of wastewater disposal with regard to heavy metals in selected areas of the Penn State "living-filter" project. The areas to be considered include a reed canarygrass area, a corn area, and an abandoned agricultural field which had been planted with white spruce. Plant uptake over time and soil accumulation of metals was examined.

MATERIALS AND METHODS

The soil in the three research areas was a Hublersburg clay loam. Soil pH at irrigated sites was approximately 6.3 in the upper 30 cm of the reed canarygrass and corn areas and 5.3 in the same depth of the old field area. Cation exchange capacity averaged 16 meq per 100 g of soil for this depth interval.

Wastewater used in the project's spray irrigation system was obtained from the Pennsylvania State University Sewage Treatment Plant. Sewage effluent received secondary treatment by trickling filter and activated sludge processes and was subsequently chlorinated. Since 1971 anaerobically digested liquid sludge was injected into the irrigated effluent at a ratio of 1 part sludge to 13 parts effluent during selected irrigation cycles in the reed canarygrass area only. Average concentrations of heavy metals found in the effluent and effluent with sludge injected are shown in Table 1.

The reed canarygrass research area has been irrigated since

Table 1. Average concentrations of heavy metals in applied wastewater.

	Effluent (ppb)	Effluent with Sludge Injected (ppb)
Cu	68	501
Zn	197	730
Cd	2.7	5.0

1964 with 5 cm of wastewater per week on a year-round basis. Prior to 1971 irrigation was with sewage effluent only, but beginning in 1971 most of the wastewater applied contained injected sludge during the middle 4 hr of a 12-hr cycle. A nearby unirrigated reed canarygrass plot, which was not harvested until 1972, was used as a control. The reed canarygrass was harvested three times each year, and from 1972 through 1975 samples from each cutting were analyzed for heavy metals. Soil samples of the 0–30-cm depth zone were collected in 1969, 1971, 1973 and 1975 in the treated area and in 1973 and 1975 in the control area.

The corn area has been irrigated weekly with 5.0 cm of sewage effluent during the growing season since 1963. During 1972 and 1973 this irrigation rate was increased to 7.5 cm per week. No sludge was applied in this area at any time. A corn-oats-hay-hay strip crop rotation was practiced up until 1968. Since then the area was managed in continuous corn silage. An adjacent control area received fertilizer, but no irrigation. Corn silage samples from 1972 to 1975 were analyzed for heavy metals. Soil samples from the 0–30-cm depth zone were collected in 1966, 1971, 1973 and 1975 in both the treated and control areas.

An abandoned old field planted with white spruce seedlings in 1955 has been irrigated April through November since 1963 with sewage effluent containing no sludge. The irrigation rate was 5.0 cm per week. An adjacent old field plot has been maintained as an unirrigated control. White spruce needles, wild strawberry and goldenrod were sampled in 1974 for heavy metals analysis. Soil samples were taken (1974) at depths of 0–15 cm and 15–30 cm for heavy metals analysis.

Plant tissue samples collected from all research areas were prepared for analysis by a digestion procedure similar to that outlined by Perkin-Elmer (11). A 50-ml sample of wastewater was digested by the same procedure. Soils were extracted with 0.1 N HCl and analyzed for heavy metals as outlined by Sidle et al. (12).

The digested or extracted samples were ultimately analyzed

for Cu and Zn using standard atomic absorption spectroscopy. For Cd analysis, a graphite-furnace attachment together with a deuterium arc background corrector were used in conjunction with an atomic absorption spectrophotometer to obtain lower detection limits and eliminate background interferences. Percolate samples collected in the reed canarygrass area in December 1975 were acidified and directly analyzed for heavy metals as outlined above. Only four percolate samples were obtained from the treated area and one from the control area.

RESULTS AND DISCUSSION

Reed Canarygrass Area

Cumulative applications of Cu in the reed canarygrass receiving 5 cm per week area are presented in Figure 1 along with

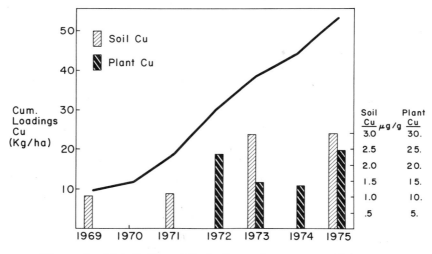

Figure 1.   Distribution of Cu in the reed canarygrass 5-cm area.

extractable soil copper values and plant copper concentrations. The Cu loading level in 1969 (9.6 kg/ha) reflects the estimated applications of Cu from 1964 to 1969 based on concentrations in 1972–1974 and known hydraulic loadings. The rate of Cu application increased substantially from 1971 on, due to the introduction of sludge injection into the irrigated wastewater. Extractable soil Cu (0–30-cm depth) increased significantly from 1971 to 1973 as a result of sludge injection. There was no significant increase in soil Cu from 1973 to 1975. Concentrations of Cu in reed canarygrass were more dependent upon the load-ings for a given year rather than upon the cumulative Cu load-ings. For example, the lowest plant Cu level between 1972 and 1975 occurred in 1974 which corresponded to the lowest Cu load-ing (5.45 kg/ha) during that period. Copper concentrations in forage harvested from an unirrigated control area (7.17–8.87 μg/g) were significantly lower than those in the treated area for all four years sampled. Percolate samples taken at the 30-cm depth in December 1975 had slightly higher Cu concentra-tions in the treated area (8.0 ppb) than in the control (2.0 ppb).

Distribution of Zn in the irrigated reed canarygrass area from 1969 to 1975 is presented in Figure 2. The rate of Zn load-ing increased after sludge injection in 1971. Extractable soil Zn (0–30-cm depth) increased substantially from 1971 to 1973 fol-lowing sludge injection. A slight decrease in extractable soil Zn was noted from 1973 to 1975. Plant uptake of Zn remained rela-

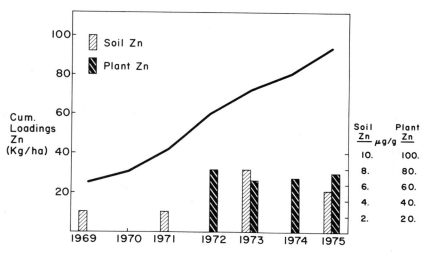

Figure 2.   Distribution of Zn in the reed canarygrass 5-cm area.

tively constant from 1972 to 1975. The highest plant Zn concen-tration (79.73 μg/g) was recorded in 1972, which was the year of highest Zn loading (18.20 kg/ha). The control area had sig-nificantly lower forage Zn concentrations (39.21–51.71 μg/g) than the treated area for all years tested. Zinc levels in the soil water sampled in December 1975 at the 30-cm depth showed slightly higher concentrations in the treated area (49 ppb) than in the control area (26 ppb).

Cadmium loadings from 1969 to 1975, as well as Cd concentrations in soil extracts and plant tissue, are shown in Figure 3.

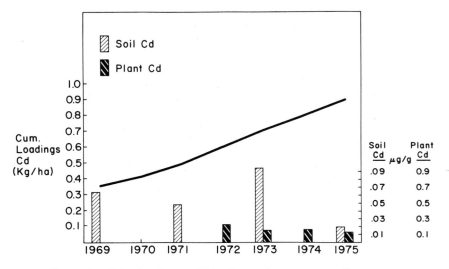

Figure 3.   Distribution of Cd in the reed canarygrass 5-cm area.

Application of Cd increased at a slightly greater rate following sludge injection in 1971, but not to the extent that Cu and Zn did. Extractable soil Cd increased slightly following sludge injection in 1971, but then dropped off markedly in 1975. Much sampling variability in extractable Cd levels in the surface soil has been encountered at these very low Cd levels ($<1\ \mu g/g$). Cadmium concentrations in reed canarygrass forage remained low ($<0.3\ \mu g/g$) and did not increase with time. Cadmium levels in percolate samples at the 30-cm depth were below the lowest detection limit (<0.25 ppb) in both treated and control areas.

Cornfield

Cumulative Cu loadings from 1966 to 1975 in the 5.0 cm per week corn area along with extractable soil Cu and plant Cu are presented in Figure 4. Copper loadings were at a fairly constant low rate over time. Extractable soil Cu in the 0–30-cm zone showed a slight increase over time. The highest level of extractable soil Cu ($1.25\ \mu g/g$) recorded in 1975 was less than half the concentration in the treated reed canarygrass soil, while the total Cu loadings in the reed canarygrass area were more than six times those in the cornfield. Concentrations of Cu in corn silage

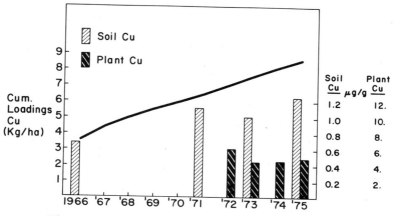

Figure 4. Distribution of Cu in the corn 5-cm area.

did not show any increase with time and were all well below
levels that would cause food chain problems.

Zinc loadings and concentrations in the soil and silage are
shown in Figure 5. Although there was a very slight increase
in extractable soil Zn in the 0–30-cm zone time, the concentration in 1975 (3.52 μg/g) was less than the 1975 concentration
in the reed canarygrass area (5.14 μg/g). Cumulative Zn loadings through 1975 were 2.7 times higher in the reed canarygrass
area than in the corn area. The highest concentration of Zn in

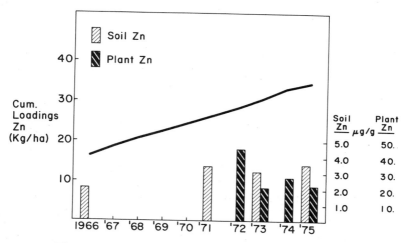

Figure 5. Distribution of Zn in the corn 5-cm area.

corn silage was in 1972 (45.15 μg/g). Silage Zn concentrations in 1973 through 1975 were not substantially different from those in the fertilized but unirrigated control area.

Loadings of Cd (Figure 6) are very low due to the low aver-

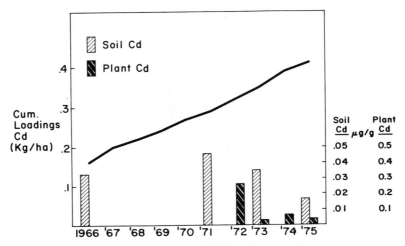

Figure 6.   Distribution of Cd in the corn 5-cm area.

age concentration of Cd in the irrigated effluent (2.7 ppb). Total Cd loadings through 1975 were less than half the corresponding loadings in the reed canarygrass area. There was no significant difference in extractable soil Cd between 1966 and 1973 in the corn area. As with Zn, the highest Cd levels in silage were found in 1972. Concentrations of Cd in silage sampled in 1973 through 1975 were lower in the effluent irrigated area than in the fertilized but unirrigated control area for each year.

Cadmium-Zinc Relations in Crop Areas

The average Cd:Zn ratios in the applied effluent and sludge-injected effluent were 1.38 percent and 0.68 percent, respectively. The Cd:Zn ratios in the extractable portion of the soil from the irrigated reed canarygrass area decreased with time from 2.34 percent in 1969 to 0.37 percent in 1975. Also, soil Cd:Zn ratios were higher in the reed canarygrass control in 1973 (1.42 percent) and 1975 (0.65 percent) than in the irrigated area in 1973 (1.18 percent) and 1975 (0.37 percent). Plant Cd:Zn ratios in the reed canarygrass area were lower during all years than the corresponding Cd:Zn ratios in the applied wastewater and extractable soil fraction. Cadmium:Zn ratios in forage did not increase with time; in fact, the lowest Cd:Zn ratio in the irrigated reed canarygrass forage was recorded in 1975 (0.17 percent).

As in the reed canarygrass area, Cd:Zn ratios in the cornfield soils (0–30-cm) decreased with time from 1.59 percent in 1966 to 0.48 percent in 1975. In all years Cd:Zn ratios in the treated soil extracts were lower than in the unirrigated control soils. Cadmium:Zn ratios in corn silage did not increase with time and ranged from 0.59 percent in 1972 to 0.16 percent in 1973, all of which are below the Cd:Zn ratio of applied effluent (1.38 percent).

Old Field—White Spruce Area

Total applications of Cu, Zn and Cd in the old field area from 1963 to the end of 1974 were 12.2 kg/ha, 36.2 kg/ha and 0.47 kg/ha, respectively. Levels of 0.1 N HCl extractable Cu, Zn and Cd in soils sampled at the 0–15-cm and 15–30-cm depths in 1974 are presented in Table 2. Higher levels of Cu and Zn were found in treated areas at both depths, although the differences between treated and control plots were not significant at the 0–15-cm depth because of sampling variability. The increase in the treated area at the 15–30-cm depth suggests that there may have been some movement of these metals into this depth zone associated with the long-term effluent irrigation. Cadmium research (10) in this area showed no increase in soil Cd due to effluent irrigation at either depth. Copper, zinc and cadmium concentrations

Table 2.   Heavy metal concentration in soil from old field (1974).

	Copper	Zinc (mg/g)	Cadmium [a]
		0–15-cm depth	
Treated	1.637 a[b]	9.733 a	0.070 a
Control	1.178 a	7.292 a	0.083 a
		15–30-cm depth	
Treated	0.880 a	1.333 a	0.013 a
Control	0.707 b	0.898 b	0.015 a

[a] Cd Data from Sidle and Sopper (10).

[b] Means of a particular metal for a given depth followed by the same letter are not significantly different at the 0.05 level.

in goldenrod, white spruce and wild strawberry foliar samples (1974) are shown in Table 3. Zinc concentrations in both goldenrod and white spruce were significantly higher in the unirrigated control than in the treated area. Higher levels of Cu and Cd were also found in the control area for goldenrod. These lower concentrations of heavy metals in the effluent-treated area are probably the result of a dilution of heavy metals in the larger biomass of vegetation on the treated area.

SUMMARY AND CONCLUSIONS

The distribution of heavy metals applied via wastewater irrigation to three research areas—a reed canarygrass area, a corn area and an old field—has been evaluated. Heavy metal levels in the applied wastewater were generally in the lower portion of the ranges for heavy metals in effluents.

Significant increases in 0.1 N HCl extractable soil Cu and Zn were found in the reed canarygrass area (0–30-cm depth) after sludge was injected into the irrigated wastewater in 1971. Much

Table 3. Heavy metal concentration in vegetation sampled in the old field (1974).

	Goldenrod	White Spruce $(\mu g/g)$	Wild Strawberry
		Zinc	
Treated	47.9 b[b]	14.4 b	33.6 a
Control	206.1 a	57.1 a	79.8 a
		Copper	
Treated	6.0 b	3.5 a	10.8 a
Control	10.5 a	4.2 a	12.5 a
		Cadmium[a]	
Treated	0.42 b	0.12 a	0.18 a
Control	0.82 a	0.15 a	0.31 a

[a] Cd Data from Sidle and Sopper (10).

[b] Means of a particular metal for a given species followed by the same letter are not significantly different at the 0.05 level.

lower increases in extractable soil Cu and Zn were found in the surface 30 cm of soil in the effluent-irrigated corn area. The Cd status of soils in both areas did not show any sustained accumulation with time as a result of wastewater irrigation. Soil Cd:Zn ratios in both the reed canarygrass and corn areas showed no increase with time. Percolate samples collected during the dormant season in the reed canarygrass area suggested that some Cu and Zn may be leaching out of the upper 30 cm of soil. Extractable Cu and Zn levels were slightly higher at both the 0–15 and 15–30-cm depths in the treated old field area. In the soils sampled from all three research areas, there appeared to be no serious heavy metals contamination due to wastewater loadings.

Concentrations of Cu, Zn and Cd in reed canarygrass forage and corn silage did not increase substantially over the time period tested (1972–1975). Levels of heavy metals in the treated plant samples tended to reflect the metal loadings in the applied wastewater for a particular year rather than being influenced by cumulative metal loadings over a long period of time. Levels of Cu and Zn in the reed canarygrass forage were consistently higher in the treated area than in the unirrigated control. The higher forage Cu levels found, 24.92 $\mu g/g$ in 1975 and 23.61 $\mu g/g$ in 1972, may be marginal in terms of toxicity to sheep. The low levels of heavy metals in corn silage were often as low as or lower than corresponding heavy metals concentrations from a normally fertilized field. The lower concentrations of Cu, Zn and Cd in white spruce and goldenrod foliage sampled in the old field area indicate that under relatively low metal loading rates the actual concentration of Cu, Zn and Cd may be reduced in selected species.

ACKNOWLEDGMENTS

This paper is a contribution from the Pennsylvania State University Agricultural Experiment Station and the Institute for Research on Land and Water Resources, the Pennsylvania State University, University Park, Pennsylvania.

REFERENCES

1. Parizek, R. R., L. T. Kardos, W. E. Sopper, E. A. Myers, D. E. Davis, M. A. Farrell and J. B. Nesbitt. "Wastewater Renovation and Conservation," Penn State Studies No. 23, The Pennsylvania State University, University Park, Pennsylvania, 71 pp. (1967).
2. Todd, J. R. "Trace elements in animal nutrition," New Lecture Series No. 66. The Queen's University, Wayne, Boyd and Son Ltd., Belfast, p. 9–13 (1972).
3. Lagerwerff, J. V. and G. T. Biersdorf. "Interaction of zinc with uptake

and translocation of cadmium in radish," *Proc. 5th Ann. Conf. on Trace Substances and Environmental Health*, University of Missouri, Columbia, pp. 515–522 (1972).

4. Sidle, R. C., J. E. Hook and L. T. Kardos. "Heavy Metals Application and Plant Uptake in a Land Disposal System for Wastewater," *J. Env. Quality* 5:97–102 (1976).

5. Chaney, R. L. "Recommendations for Management of Potentially Toxic Elements in Agricultural and Municipal Wastes," *In:* Factors Involved in Land Application of Agricultural and Municipal Wastes, USDA, ARS, National Program Staff Pub., Beltsville, Maryland, pp. 97–120 (1974).

6. Konrad, J. G. and S. J. Kleinert. "Survey of toxic metals in Wisconsin," Tech. Bull. No. 74. Dept. of Natural Resources, Madison, Wisconsin, pp. 1–7 (1974).

7. Allaway, W. H. "Environmental Cycling of Trace Elements," *Adv. Agron.* 20:235–271 (1968).

8. Hinesly, T. D., R. L. Jones and E. L. Ziegler. "Effects on corn by applications of heated anaerobically digested sludge," *Compost Sci.* 13:26–30 (1972).

9. Boswell, F. C. "Municipal Sewage Sludge and Selected Element Applications to Soil: Effect on Soil and Fescue," *J. Env. Quality* 4:267–272 (1975).

10. Sidle, R. C., and W. E. Sopper. "Cadmium Distribution in Forest Ecosystems Irrigated with Treated Municipal Wastewater and Sludge," *J. Env. Quality* (in press).

11. Perkin-Elmer Corp. "Analytical methods for atomic adsorption spectrophotometry," Perkin-Elmer Corp., Norwalk, Connecticut Sect. AY-5, pp. 1–2 (1971).

12. Sidle, R. C., J. E. Hook and L. T. Kardos. "Accumulation of Heavy Metals in Soils as Influenced by Extended Wastewater Irrigation," *J. Water Poll. Cont. Fed.* (in press).

15

MONITORING SEWAGE SLUDGES, SOILS AND CROPS FOR ZINC AND CADMIUM

D. E. Baker, M. C. Amacher and W. T. Doty
Department of Agronomy, The Pennsylvania State University, University Park, Pennsylvania

INTRODUCTION

Monitoring the potential for accumulation of any chemical element within the food chain from soils involves an evaluation of the following soil effects: variable ionic activities or intensities at the plant root surface, variable relationships among labile quantities in relation to both total composition and solution activities, and relative intensity effects in which the availability of one ion is affected by that of other ions. Plant uptake of an ion from soils is affected by inherent differences among species and varieties within species, ion interactions and soil-plant interactions. Interpretations of experimental results from studies of element accumulation in the food chain involving soil testing, soil chemical equilibria, plant analysis and studies of plant physiological processes have been subjects of several review articles (1–9) beginning with the excellent article on trace element cycling by Allaway in 1968 (10).

While soil and plant data have revealed the complexities of the problem of monitoring the uptake of trace metals from soils by plants, the application of sewage sludge to land further complicates the monitoring problem. Like soils and especially the humus component of soils, sewage sludges vary in organic and inorganic composition and other properties (2, 11). The trace metals are complexed by sewage sludge during the sewage digestion process; like soil humus, sludge is not totally inert but does resist further decomposition when incorporated in soil. Miller (12) reported that a maximum of 20 percent of the carbon added as sewage sludge evolved from soil during a six-month incubation under controlled conditions. Therefore, it is difficult to predict long-term effects of trace metal additions when the metals are added to different soils as components of sewage sludges. The objective of this presentation is to provide perspective to problems of monitoring sewage sludges, soils and crops for zinc and cadmium and to present research results that aid in establishing approaches and standards for protecting the food chain from unsafe accumulations of cadmium.

REVIEW OF LITERATURE

While zinc is an essential element for both plants and animals, cadmium is nonessential and potentially toxic to both plants and animals. Turner (13) in 1973 reported that most plant species could tolerate a solution concentration of cadmium of up to 0.1 ppm. His results compared with others (1) suggest that plant roots tolerate approximately equal ionic activities of cadmium and copper. Page *et al.* (14), working with nutrient solution concentrations of cadmium ranging from 0.1 to 10 ppm, reported that a 50 percent growth reduction was associated with nutrient solution concentrations of 0.2 to 0.9 ppm Cd for the sensitive plants—turnips, field beans, red beets and lettuce—while less sensitive crops including sweet corn, pepper, barley, tomato and cabbage required 1.2 to 9 ppm Cd for a 50 percent reduction in growth. Bingham *et al.* (15) observed that cadmium applied to soil at a rate of 160 ppm depressed zinc in leaf tissue of field beans from 47 ppm to 24 ppm. They reported that a 25 percent plant yield decrement was associated with soil cadmium additions of 4 to 640 ppm, a diagnostic leaf concentration of 3 to 160 ppm Cd, and concentrations within the edible portion of various crops ranging from 2.0 to 80.0 ppm Cd. Results reported by Haghiri (16) indicated that uptake and translocation of [115M]Cd by soybean plants was more efficient when cadmium was supplied to roots than when supplied to leaves. Compared to concentrations of cadmium considered harmful to animals and man, the relatively high concentrations of cadmium found toxic to plants make the results for plant toxicity studies seem irrelevant to the problem of cadmium in the environment.

From soil-plant relationships and the chemistry of cadmium, even small additions of cadmium with sewage sludge would be expected to increase plant accumulations. Hakne and Kroontje

(17) showed that at the normal pH range of soils and at chloride solution concentrations of less than 10^{-3} molar, cadmium exists almost entirely as Cd^{++}. Applications of sewage sludge relatively low in cadmium (7.4 ppm) and zinc (1070 ppm) by Dowdy and Larson (18) increased cadmium and zinc accumulation by potatoes. For sewage sludge applications of 0, 112, 225, and 450 metric tons per hectare, cadmium in potatoes was 0.12, 0.11, 0.21, and 0.23 ppm and zinc was 24, 31, 41, and 53 ppm for the respective rates of sludge. Lettuce leaves accumulated 2.67 ppm Cd when the highest rate of sludge was applied. Giordano et al. (19) applied up to 10 ppm Cd to soil from sewage sludge containing 50 ppm Cd, and they found corn forage ranging from 0.7 to 5.3 ppm Cd compared with 0.2 to 1.2 ppm Cd in grain. Boswell (20) found 28 ppm Cd in fescue tissue from additions of 5.6 metric tons per hectare of sludge containing 165 ppm Cd. In a greenhouse study, Jones et al. (21) observed cadmium accumulations of up to 18.5 ppm in tissue of soybean plants from additions of 87.1 metric tons per hectare of sewage sludge containing 129 ppm Cd. Kirkham (22) found up to 24 ppm Cd in corn leaves and up to 1 ppm Cd in corn grain from a site used for long-term sludge disposal. While variability among plant species and varieties with respect to tissue accumulations of zinc and cadmium makes it difficult to use plant analysis for monitoring soil levels of these elements, it is evident that plant composition results are required to monitor the concentrations of cadmium entering the food chain.

Soil testing with respect to macro elements, trace elements and nonessential, potentially toxic elements is required for monitoring soil accumulations of zinc, cadmium and other elements. Extracting cadmium by the familiar double-acid method (23) did not reflect cadmium additions to soils studied by Kirkham (22). In addition to the double-acid method, extraction with 0.1 N HCl has been used to extract trace elements from soils (23). The chelator, DTPA (diethylenetriaminepentaacetic acid), has been used to extract soils for several trace elements (24). Formation constants with DTPA suggest that it could be used to reflect soil availabilities of iron, copper, cadmium, nickel, zinc, and manganese (1, 2, 8, 9, 23, 24).

For animals, toxicity symptoms are evident within a short time when rations contain 50 to 100 ppm Cd (2, 25). Cadmium tends to accumulate in the human body with time, and although threshold doses for long-term toxic effects are not known with certainty, some recommended limits on cadmium intake suggest that long-time exposure to relatively low levels of cadmium within diets could be hazardous to human health. Friberg et al.

(25) suggest that a 50-year cadmium ingestion by man of 132 μg/day could result in kidney damage if 5 percent of the cadmium were retained. The F.Y. 1973 Total Diet Studies of the Bureau of Foods (26) indicated that the dietary intake of cadmium was approximately 80 percent of the WHO/FAO limit of 57.1 to 71.4 μg/day per person. For toddlers, whole milk was predicted as the major source of cadmium, amounting to 23 percent of the total intake. For adults, cereal grain products were the major source of cadmium, amounting to 28 percent of the total intake. For example, shredded wheat samples had an average level of cadmium more than ten times greater than the average level in milk samples. A recent review by Sandstead *et al.* (27) estimated normal intake of cadmium at 40–190 μg/day. The threshold for safety was placed at 170–500 μg/day, and intake for people suffering from over-exposure to cadmium in Japan was estimated to be 600–1,000 μg/day. Normal accumulation of cadmium in man was estimated at 0.1–1.8 μg/day, which would indicate a retention of less than 5 percent.

In a report by Schroeder and Nason (28) it was shown that for rats there was no net accumulation of cadmium when the dietary concentration was 0.07 ppm. Other studies have indicated that cadmium accumulation within the kidney and liver was substantial within relatively short periods of time at dietary levels of 3 to 5 ppm Cd in rations (29, 30). It was on the basis of these results that Baker and Chesnin (2) suggested a maximum concentration of 1 ppm Cd in leaves of soybeans, cereal grains and forages as a means of protecting animal health and preventing a buildup of cadmium within the food chain. The Bureau of Foods' survey results (26) suggest that 1 ppm Cd in plant leaves is too high for safe dietary levels over a period of years.

Zinc and cadmium are associated within the earth's crust, and this association carries over into biological systems. A review of the literature in geochemistry (2) indicates that soil parent materials range from 4 to 300 ppm Zn and 0.1 to 0.3 ppm Cd. The zinc/cadmium ratio averages about 600:1 for the lithosphere and about 100:1 for soils. The average composition of soils has been reported as 50 ppm Zn and 0.5 ppm Cd. Accumulation of cadmium in soils relative to that for Zn suggests that there will be an increase in cadmium within the food chain over time from additions of cadmium to soils regardless of the zinc/cadmium ratio. Therefore, the suggested beneficial effects to man of high zinc on the toxicity of cadmium (25, 27) might not be achievable through soil additions of the two elements.

Animal manure is a source of trace elements, ranging from

43 to 247 ppm Zn with an average of 96 ppm (2). Animal diets range from approximately 20 to 60 ppm Zn; therefore, the zinc concentrations in manure suggest a two- to three-fold increase in trace element concentrations in manure over the dietary levels. Land applications of manure from animals receiving supplemental amounts of trace elements will increase dietary levels of those elements over time. Additions of manure to a Tripp soil in Nebraska for 31 years increased the soil test level of zinc from 0.23 ppm to 2.81 ppm (2). If dietary levels of cadmium range from 0.27 to 0.62 ppm (27), then manure should range from 0.054 to a maximum of 1.8 ppm Cd. If the dietary levels of cadmium were increased to a maximum of 1 ppm, as suggested for a maximum tolerable concentration of cadmium in plant leaves (2), then manure concentrations of cadmium would increase as high as 3 ppm. Even using the upper limit of 3 ppm Cd in plant leaves as proposed by Melsted (7), manure concentrations of cadmium should not exceed 9 ppm. Thus, it is possible to postulate that wastes containing as little as 5 to 10 ppm Cd would eventually cause an increase in the concentrations of cadmium within the food chain.

MATERIALS AND METHODS

To compare their compositions with respect to fertilizer value and potential toxic effects on the food chain, fresh manure samples were collected from dairy cattle, horses, swine, chickens and deer in Centre County, Pennsylvania, and sewage sludge samples were collected biweekly from six treatment plants considered representative of cities and towns in Pennsylvania. The samples were analyzed for components as indicated in Table 1. The digestion procedure was the nitric and perchloric acids technique of Olson (31). The metallic ions were determined by flame atomic absorption, with the exception of cadmium, which was determined by flameless atomic absorption using a graphite furnace.

Sewage sludge from plant number 6, Tables 2 and 3, was used in a field experiment in 1974 and 1975 to study soil-plant-food chain interrelationships. Since the plant compositions were affected most by zinc and cadmium in the sludge, and since cadmium accumulation by plants has a potential for serious adverse effects on human health, only the monitoring results for zinc and cadmium are presented. Yield, plant growth, soil water potentials, and other parameters of agronomic importance will be published elsewhere.

The field plots were established on a land boundary between

Table 1.   Composition[a] of fresh animal feces collected in Centre County, Pennsylvania, in 1975.

Determination	Dairy Cattle				Horses		Swine		Chickens		Deer
	A	B	C	D	A	B	A	B	A	B	
pH	6.45	6.00	6.85	6.70	7.30	7.10	6.30	6.30	6.40	6.60	7.3
Solids	15.7	17.7	22.3	17.8	24.1	23.6	31.6	28.8	19.2	21.5	40.0
Potassium	1.02	0.93	1.03	0.94	1.10	1.22	0.91	1.09	2.41	2.08	0.22
Calcium	1.13	1.19	–	2.78	0.71	1.02	1.63	2.19	8.40	7.88	2.38
Magnesium	0.42	0.42	–	0.62	0.19	0.28	0.60	0.88	0.95	1.26	0.34
Nitrogen	2.75	2.83	2.40	1.95	3.53	3.44	3.75	3.45	6.10	6.11	2.71
Phosphorus	0.74	1.15	0.90	1.01	0.89	0.86	1.55	2.36	2.72	1.59	0.16
Sodium	0.21	0.18	0.34	0.38	0.30	0.30	0.21	0.23	0.15	0.12	0.01
Iron	0.15	0.17	0.25	0.15	0.05	0.06	0.07	0.08	0.19	0.28	0.04
Aluminum	0.21	0.24	0.15	0.10	0.05	0,06	0.02	0.03	0.18	0.23	0.13
Manganese	111	127	338	186	87	140	409	503	237	318	1156
Zinc	304	346	192	198	112	144	550	663	152	198	108
Cadmium	0.38	0.14	0.12	0.14	0.08	0.12	0.23	0.29	0.54	0.38	–
Zn/Cd	800	2471	1600	1414	1400	1200	2391	2286	281	521	–
Copper	27	29	35	40	21	30	50	62	20	20	11
Nickel	30	<1	12	12	<1	24	14	17	26	12	12
Cobalt	14	9	23	10	<1	2	14	17	4	8	6

[a] Elements K, Ca, Mg, N, P, Na, Fe and Al are expressed as a percentage of dry matter, and other elements are expressed as ppm in dry matter.

Hagerstown silt loam and Murrill silt loam. The plots were replicated four times in a randomized block design with each plot split to include two crops. Both field corn (Pioneer 3773 in 1974 and Pioneer 3780 in 1975) and grain sorghum (Northrup King hybrid 180A for both years) were planted in each plot. Both plantings in all plots received row fertilizer to supply 12 + 36 + 12 in 1974 and 8 + 34 + 13 in 1975 as pounds per acre of N, P_2O_5 and K_2O, respectively. Variable treatments included the soil incorporation of no treatment or check; commercial fertilizer (100 + 25 + 0); dairy cattle manure at 5 tons dry matter per acre (from herds C and D of Table 1); and sewage sludge at 5, 10 and 20 tons of dry matter per acre.

Plant leaf samples were collected from both crops on August 7, 1974, and again on August 5, 1975. Grain samples were collected for both crops in mid-September each year. Soil samples were collected after harvest year from the plow layer of each plot. In 1975, core samples were taken from the plots to determine changes in carbon content. The percentage of carbon in the soil

Table 2. Summary of zinc concentrations (ppm) and statistical parameters[a] obtained for sewage sludge samples from six sewage treatment plants in 1974–75.

						Standard Deviation as Percentage of Mean			
Location	Times Sampled	Range (ppm)	Mean (ppm)	F-Ratios for time	E.M.S.	Duplicates	Time	Error	All
1	23	426-2036	1086	76.2[d]	2,316	6.9	38.7	4.4	27
2	19	1325-3155	2479	11.9[d]	30,315	5.0	24.3	7.0	18
3	18	4090-10051	6787	3.1[b]	874,012	11.7	24.3	13.8	20
4	23	921-1434	1174	22.2[d]	1,889	5.6	11.4	3.7	12
5	23	2476-13461	5018	114.8[d]	78,366	0.01	59.8	5.6	42
6 (dry)	19	3376-5215	4557	21.0[d]	23,252	1.6	15.3	3.4	11
6 (cake)	5	2529-3502	2937	65.7[c]	3,008	5.8	15.1	1.9	10

[a] E.M.S. refers to error mean square used to calculate F-ratios. The standard deviation is expressed as a percentage of the mean for results for duplicate determinations, times sampled, unexplained error, and all results combined, respectively.

[b, c, d] Indicate significance at the 0.05, 0.01, and 0.001 levels, respectively.

averaged over the four replications was 1.57, 1.59, 1.64, 1.76, 1.84, and 2.32 for the respective treatments indicating that the sludge was resistant to decomposition. The 40 tons of sludge per

Table 3. Summary of cadmium concentrations (ppm) and statistical parameters[a] obtained for sewage sludge samples from six sewage plants in 1974–75.

						Standard Deviation as Percentage of Mean			
Location	Times Sampled	Range (ppm)	Mean (ppm)	F-Ratios for time	E.M.S.	Duplicates	Time	Error	All
1	23	0-16	10.2	6.77[b]	3.58	1.0	48.5	18.6	36.3
2	19	13-49	32.4	34.2[b]	3.21	1.7	32.3	5.5	22.9
3	18	69-167	123	10.1[b]	102	5.1	26.2	8.2	19.1
4	23	52-327	80.4	5.7[b]	552	29.2	69.9	29.2	53.2
5	23	42-551	127	105[b]	195	0.3	112	11.0	78.8
6 (dry)	19	172-446	257	80.3[b]	71.5	2.6	29.5	3.3	20.7
6 (cake)	5	150-189	167	484[b]	1.1	1.2	13.6	0.6	9.1

[a] E.M.S. refers to the error mean square used to calculate F-ratios. The standard deviation is expressed as a percent of the mean for results for duplicate determinations, times sampled, unexplained error and all results combined, respectively.

[b] Indicates significance at the 0.001 level.

acre added an equivalent of about 2 percent carbon to the soil. Crop yields for the two higher rates of sludge were about equal to that for commercial fertilizer, and the yields for the manure and lowest rate of sludge were intermediate between the check and the other treatments.

The plant leaf and grain samples were dried at 70°C in a steam-heated forced air dryer and ground in a stainless steel Wiley mill to pass a 20 mesh screen. The plant materials were prepared for analyses by a dry ashing-acid digestion procedure. All glassware and crucibles were soaked in a dilute nitric acid bath and rinsed with deionized water before use. A 2.000-g portion of each sample was weighed into a fused silica crucible and ashed in a muffle furnace that increased in temperature at a rate of 1°C per minute to a final temperature of 450°C. After dry ashing overnight, the cooled samples were treated with 10 ml of 1 N HNO_3 and 5 ml of 6 N HCl, covered and allowed to stand overnight. The contents of each crucible were thoroughly mixed, transferred and diluted to a final volume of 50 ml without filtering. An aliquot of each sample was transferred to a plastic vial and stored for cadmium analysis, while the remainder of each sample was filtered into storage bottles for zinc and other analyses by flame atomic absorption.

Standards were prepared from Zn metal and CdO using the same amounts of the acids as in the samples. For cadmium analysis the Perkin Elmer Model 403 atomic absorption spectrophotometer was equipped with a model 2100 hydrogen graphite analyzer and a deuterium arc background corrector. The reliability of the methods was checked by analyzing NBS standard reference material 1571—orchard leaves. Analyses of five replicated samples gave a mean value of 0.10 ppm Cd with a coefficient of variation of 8 percent. Values for zinc averaged 28 ppm with a coefficient of variation of 7 percent. The NBS values were given as 0.11 ± 0.02 ppm for cadmium and 25 ± 3 ppm for zinc. The procedure as developed and used on plant materials was found reliable and precise.

To compare soil testing methods for zinc and cadmium, the soil samples from each plot were extracted with 0.1 N HCl, DTPA and solution A of the Baker method as presented in the *Handbook on Reference Methods for Soil Testing* (23). The results for each test were expressed as ppm of each element in the air dry soil. Results obtained from the new approach to soil testing (23, 24, 32) were used in a computer program to calculate pZn and pCd in the samples, which are analogous to pH. The calculations are from a computer program presented in *Agronomy Abstracts* in 1975 by Richard Eshelman. The method of calculation is a converging iteration on the free DTPA ligand concentration, which accounts for the metal ions and ligand mass balance and formation constants of the metal ligand species.

Each set of plant analysis and soil test results was subjected to an analysis of variance. As a statistical measure of differences among means, F-ratios were calculated from the analysis of variance using the mean square for error as the lesser mean square for comparing major effects and treatment times year interactions. The use of the fixed model means that the levels of significance apply only to the field plots studied and the results are not statistically applicable to other locations. Only from theoretical considerations are the results made applicable to other soils.

RESULTS AND DISCUSSION

Composition of Animal Manure

The composition results for fresh manure samples collected in Centre County, Pennsylvania, in 1975 are reported in Table 1. Supplementation of dairy cattle and swine rations with zinc and copper is evident from the data obtained. While the average zinc composition of feces was expected to be near 100 ppm, the average for all samples was 270 ppm Zn. For the horses and deer, zinc in feces ranged from 108 to 144 ppm Zn. Cadmium ranged from 0.08 for one sample of horse feces to 0.54 ppm in a sample of chicken feces. The zinc/cadmium ratio was higher than expected from results published for feed grains and forages (2, 33, 34). The addition of zinc to rations explains the high zinc/cadmium ratio in feces of dairy cattle and swine; however, the results for horses indicate that the abundance of zinc in farm animal feed in Centre County might be more than 1,000 times that of cadmium.

A feeding trial with broiler chicks (29) involving dietary additions of 0, 3, 12 and 48 ppm Cd resulted in manure containing 0.46, 5.55, 37.87 and 111.41 ppm Cd on a dry matter basis. While body accumulations of cadmium were substantial, especially in liver and kidney, the percentage of body accumulation from the diet was relatively small, so the concentrations of cadmium in the feces were 1.7, 3.1 and 2.3 times the concentrations added in the ration. The manure concentrations of zinc and other elements were not affected by dietary cadmium. Thus, the two- to three-fold increase in trace element concentrations in manure over dietary levels appears realistic.

Cadmium and Zinc in Sewage Sludge

Guidelines (4, 7, 35–37) used to establish limits on metal concentrations in sewage sludge applied to cropland require that sludge from each treatment plant be analyzed. A stated upper limit for concentrations of cadmium and zinc (2, 34) should not be considered as being equal to the desired upper limit to be tolerated. The true amount of an element plus the resultant sum of errors in sample collection, preservation, preparation and chemical analysis should not exceed the desired upper limit. Baker and Chesnin (2) indicated that a coefficient of variation over time of 50 percent for sludge samples from a given plant appeared to be a realistic goal.

Sludge samples from six treatment plants in Pennsylvania were collected biweekly from July 1, 1974, to June 30, 1975. The results for zinc and cadmium are presented in Tables 2 and 3. The variation in composition over time was random for both elements, with no seasonal trends being observed. While the samples from every plant were significantly different over time, the standard deviation when expressed as a percentage of the mean ranged from 11–60 percent for zinc and 14–112 percent for cadmium. While the precision of analyses was variable over time among plants, it is evident that plants were more different than samples from within plants over time. For example, randomly selecting three sampling times and averaging the results for cadmium revealed that the averages for the six plants would have been estimated as 10.6, 32, 122, 75, 257 and 260 ppm Cd, respectively. Only the results for plant number 5 were extremely different when only three random samples were used. The three random values of 76.8, 519.2 and 174.3 included the highest value obtained from treatment plant 5, so one would not be satisfied to average these three values. Addition of the results of the fourth random sample from plant 5 of 105.1 ppm Cd reduced the average to 219 ppm Cd; elimination of the highest value and averaging the other three values provides an estimate of 119 ppm Cd for plant 5. The relatively large amount of variation in sludge cadmium for plant 5 was reflected by a large standard deviation over time.

It was difficult to obtain representative subsamples from sludges produced by plants 3 and 4. The variations between duplicate samples are reflected by higher standard deviations. Even though the precision is less than that considered acceptable for monitoring biological systems, it is possible to conclude that from three or four samples of sludge collected over time from a treatment plant one can predict their average zinc and cadmium concentrations over that period of time.

Cadmium and Zinc in Plant Leaves and Grain

The sludge used in the field study had a relatively high concentration of zinc (4557 ppm) and cadmium (257 ppm) and a narrow zinc/cadmium ratio (17:7). Both zinc and cadmium concentrations were increased in corn leaves from the initial application of the lower rate of sludge, Table 4. The zinc content in

Table 4.   Zinc, cadmium and Zn/Cd ratio in corn leaves grown in field plots in 1974 and 1975.

Soil Treatments	Zinc (ppm) 1974	Zinc (ppm) 1975	Cadmium (ppm) 1974	Cadmium (ppm) 1975	Zn/Cd 1974	Zn/Cd 1975
Check	19.1	20.7	0.08	0.14	257	528
Fertilizer	22.7	25.6	0.15	0.15	161	399
Manure – 5	21.8	25.0	0.15	0.17	160	291
Sludge – 5	40.2	74.0	1.09	0.83	39.6	119
Sludge – 10	51.5	82.9	1.66	0.95	25.1	91.5
Sludge – 20	76.3	136.2	2.47	3.40	31.8	41.0
F-Ratios for:						
Treatments (T)	55.8^a		29.0^a		3.6^c	
Years (Y)	35.4^a		0.00		2.8	
T x Y	6.7^b		1.72		1.3	

a,b,c Indicates significance at the 0.001, 0.01 and 0.05 levels, respectively.

corn leaves reflected zinc additions within years and over years, while the uptake of cadmium increased with rates of sludge but did not change over years. These trends in corn leaf zinc and cadmium were reflected by their respective concentrations in corn grain, Table 5. The much wider zinc/cadmium ratio in corn grain than in corn leaves, however, reflects a substantial degree of discrimination between these elements with respect to their translocation from leaves to grain. Considerable work with another pair of elements, cadmium and strontium, did not indicate a discrimination within corn genotypes (38).

The trends for accumulations of zinc and cadmium by sorghum leaves were not greatly different from those observed for corn leaves, Table 6. For lower levels of zinc, uptake by sorghum leaves was greater than for corn leaves; however, at higher sludge rates, the sorghum leaf zinc tended to be lower than for corn leaves. Therefore, the zinc/cadmium ratio was lower for sorghum leaves than for corn leaves.

The accumulations of zinc and cadmium in sorghum grain

Table 5. Zinc, cadmium, and Zn/Cd ratio in corn grain grown in field plots in 1974 and 1975.

Soil Treatments	Zinc (ppm) 1974	1975	Cadmium (ppm) 1974	1975	Zn/Cd 1974	1975
Check	20.7	28.5	0.005	0.015	3907	5526
Fertilizer	21.2	24.2	0.015	0.003	1445	9719
Manure – 5	21.2	26.4	0.010	0.005	2027	9012
Sludge – 5	22.2	30.8	0.068	0.025	357	1422
Sludge – 10	27.0	36.3	0.154	0.065	192	577
Sludge – 20	25.7	39.9	0.180	0.120	152	335
F-Ratios for:						
Treatments (T)	29.1^a		41.7^a		6.7^c	
Years (Y)	158.0^a		19.0^b		13.7^c	
T x Y	6.1^c		3.9		3.1	

a,b,c Indicates significance at the 0.001, 0.01 and 0.05 levels, respectively.

showed almost no discrimination, Table 7. The zinc/cadmium ratios in sorghum grain reflected those in the leaves. For both corn and sorghum, grain accumulations of cadmium were different over years. The grain cadmium decreased in 1975 for corn and increased for sorghum. However, for corn, varieties were different for the two years, and the varieties have not been characterized with respect to elemental accumulation.

Table 6. Zinc, cadmium, and Zn/Cd ratio in sorghum leaves grown in field plots in 1974 and 1975.

Soil Treatment	Zinc (ppm) 1974	1975	Cadmium (ppm) 1974	1975	Zn/Cd 1974	1975
Check	23.4	20.6	0.29	0.25	175	274
Fertilizer	25.0	28.8	0.43	0.10	113	290
Manure – 5	25.4	24.1	0.18	0.26	153	211
Sludge – 5	37.9	43.2	1.58	1.08	25.0	41.6
Sludge – 10	44.6	53.5	1.86	1.72	26.2	33.3
Sludge – 20	47.2	56.3	2.27	2.29	23.1	26.2
F-Ratios for:						
Treatments (T)	490.0^a		37.7^a		22.8^a	
Years (Y)	63.1^a		1.6		13.3^b	
T x Y	18.4^a		0.6		2.8	

a,b Indicates significance at the 0.001 and 0.01 levels, respectively.

Table 7. Zinc, cadmium, and Zn/Cd ratio in sorghum grain grown in field plots in 1974 and 1975.

Soil Treatment	Zinc (ppm)		Cadmium (ppm)		Zn/Cd	
	1974	1975	1974	1975	1974	1975
Check	11.1	21.7	0.031	0.035	375	633
Fertilizer	13.5	22.1	0.047	0.108	294	295
Manure – 5	13.1	23.7	0.040	0.063	348	422
Sludge – 5	15.6	26.7	0.237	0.470	70.8	58.3
Sludge – 10	16.3	30.6	0.309	0.876	53.9	35.9
Sludge – 20	17.9	32.7	0.364	0.855	49.4	38.0
F-Ratios for:						
Treatments (T)	8.3^b		80.2^a		47.4^a	
Years (Y)	138.2^a		87.1^a		4.2	
T x Y	0.9		16.7^a		3.5^c	

a,b,c Indicates significance at 0.001, 0.01 and 0.05 levels, respectively.

Baker and Chesnin (2) suggested 1 ppm Cd in plant leaves of major food crops like corn, sorghum and soybeans as the "safe" upper limit to be tolerated until more is known relative to its movement within the food chain. If this guideline were applied to the conditions of this experiment, the five tons per acre rate of this sludge would be the maximum to be tolerated. The comparatively high grain accumulation of cadmium by sorghum has also been reported for soybeans (29). When soybean leaves contained 0.71 ppm Cd on this site, the soybean grain contained 0.72 ppm Cd. Relatively high accumulations of cadmium by soybean grain have been reported previously by Jones, Hinesly and Ziegler (21).

Soil Testing for Zinc and Cadmium

Considering the variation of results caused by the lack of uniform mixing of the sludge treatments with the surface soil, the results presented in Tables 8, 9 and 10 indicate that extraction with 0.1 N HCl, the DTPA extractant, or the Baker method will reflect soil additions of zinc and cadmium. Each five tons of sludge added approximately 22.8 ppm Zn and 1.3 ppm Cd to the surface soil. The 0.1 N HCl removed essentially all of the cadmium and more than 50 percent of the zinc.

The DTPA extractant as proposed by Lindsay and Norvell (5, 8, 9) was expected to provide an approximation to the labile concentrations of both zinc and cadmium within the soils. From

Table 8. Zinc, cadmium and Zn/Cd ratios in soil receiving various treatments in 1974 and 1975 as measured by an extraction with 0.1 N HCl (23). Each value represents a mean for four replications.

Soil Treatment	Zinc (ppm) 1974	Zinc (ppm) 1975	Cadmium (ppm) 1974	Cadmium (ppm) 1975	Zn/Cd 1974	Zn/Cd 1975
Check	3.05	3.12	0.14	0.12	21.4	26.2
Fertilizer	3.28	3.81	0.12	0.13	28.6	28.8
Manure – 5	3.37	3.75	0.13	0.11	26.8	34.0
Sludge – 5	10.66	30.87	1.20	2.37	8.9	13.0
Sludge – 10	32.91	62.94	2.52	5.62	12.0	11.6
Sludge – 20	67.12	121.78	6.91	11.08	9.3	11.3
F-Ratios for Comparing						
Treatments (T)	33.6[a]		34.9[a]		109.3[a]	
Years (Y)	11.7[b]		8.3[c]		16.8[b]	
T x Y	3.0[c]		2.3		2.6	

[a,b,c] Indicates significance at 0.001, 0.01 and 0.05 levels, respectively.

the 1975 soil test data of Table 9, it was calculated that 34, 35 and 32 percent of the added zinc and 59, 62 and 58 percent of the added cadmium applied with the 5, 10 and 20 ton rates of sludge, respectively, was removed by the DTPA extractant from

Table 9. Zinc, cadmium and Zn/Cd ratios in soils receiving various treatments in 1974 and 1975 as measured by an extraction with DTPA by the method of Lindsay and Norvell (23). Each value represents a mean for four replications.

Soil Treatment	Zinc (ppm) 1974	Zinc (ppm) 1975	Cadmium (ppm) 1974	Cadmium (ppm) 1975	Zn/Cd 1974	Zn/Cd 1975
Check	1.12	1.31	0.08	0.08	14.7	16.0
Fertilizer	1.34	1.89	0.09	0.09	15.0	22.0
Manure – 5	1.26	1.70	0.08	0.08	15.6	22.3
Sludge – 5	4.84	15.44	0.67	1.51	7.1	10.4
Sludge – 10	14.96	31.96	2.05	3.18	7.1	10.1
Sludge – 20	32.08	58.75	4.09	5.98	7.3	9.8
F-Ratios for Comparing						
Treatments (T)	27.7[a]		35.5[a]		33.6[a]	
Years (Y)	11.3[b]		5.5[c]		31.0[a]	
T x Y	2.6		1.4		1.8	

[a,b,c] Indicates significance at 0.001, 0.01 and 0.05 levels, respectively.

Table 10. Zinc, cadmium and Zn/Cd ratios in soils receiving various treatments in 1974 and 1975 as measured by the method of Baker (23). Each value represents a mean for four replications.

Soil Treatment	Zinc (ppm) 1974	1975	Cadmium (ppm) 1974	1975	Zn/Cd 1974	1975
Check	1.12	2.75	0.09	0.06	12.5	46.7
Fertilizer	1.48	4.00	0.10	0.08	14.5	53.6
Manure – 5	1.42	3.56	0.09	0.06	16.6	65.0
Sludge – 5	5.08	55.37	2.35	1.69	4.1	34.3
Sludge – 10	14.75	59.62	2.08	2.66	7.1	22.6
Sludge – 20	28.72	47.67	8.48	3.66	5.5	12.8
F-Ratios for Comparing						
Treatments (T)	9.7^a		5.2^b		4.5^b	
Years (Y)	17.7^a		1.0		38.8^a	
T x Y	3.7^c		0.9		1.8	

a,b,c Indicates significance at 0.001, 0.01 and 0.05 levels, respectively.

the soil. These data suggest that use of the DTPA extractant would be excellent for use on soil-sludge combinations to determine which component regulates the labile amounts of each element. For this soil it is postulated that a labile concentration of approximately 1 ppm will yield leaf concentrations of 1 ppm Cd in corn, sorghum and soybean leaves. In an experiment involving Cd additions as a soluble salt over a two-year period, 1 ppm Cd added to soils of this site resulted in leaf concentrations of 0.36 ppm and 0.65 ppm Cd for corn and soybeans, respectively; 2 ppm soil Cd resulted in leaf concentrations of 0.82 and 0.71 for leaves of the respective crops. Thus, the percentage of cadmium that remains labile over time appears to be a function of soil adsorption or fixation.

Results from the Baker method lack the degree of precision obtained by the DTPA extraction. However, the method was designed to reflect both labile amounts of trace elements and, in addition, to allow the calculation of ionic activities within the testing solution that reflect those of the soil system (23, 24). The Baker method removed 66, 52, and 36 percent of the cadmium applied with the 5, 10 and 20 ton rates of sludge, respectively. While the DTPA extraction removed a constant percentage of the added cadmium, the Baker method removed proportionately less as the rates of sludge increased, which was also observed for plant uptake.

For soil systems in equilibrium, it has been established that plant uptake of an ion is a direct function of its activity or relative partial molar free energy at the root surface (39). For one soil site where other factors governing uptake of an ion are equal, the availability of an element should be reflected by the negative logarithm of the activity, which is analogous to the conventional pH. For different soils, the relationship of ion activity to labile amounts adsorbed on the solid phase is also different (24). For relatively abundant elements in soils, ion uptake is more a function of the activity or relative partial molar free energy, but for less abundant elements, like cadmium, the labile concentration is expected to be more important because of its effect on diffusion of ions to plant roots (24, 39).

Results for pZn and pCd by the Baker method are presented in Table 11. Plots of the 1975 results for pZn and pCd related to zinc and cadmium in corn leaves are presented in Figure 1. The results were essentially linear for the sludge treatments; however, the check, fertilizer and manure treatments were associated with pZn and pCd values, which indicated a high degree of fixation within the soil. The results for pZn–pCd of Table 11, when compared with the similar changes in slope of curves for both ions in Figure 1, indicate the similarity of bonding for the two ions in the soil. While plants tolerate solution concentrations of 0.1 ppm Cd (pCd − 6.0), corn leaf concentrations of

Table 11. Soil test results by the Baker method expressed as estimates of the negative logarithm of the zinc activity (pZn) and cadmium activity (pCd) in test solutions at equilibrium. Each value represents a mean for four replications.

Soil Treatment	pZn		pCd		pZn − pCd	
	1974	1975	1974	1975	1975	1975
Check	11.7	11.5	13.5	13.8	-1.8	-2.3
Fertilizer	11.1	11.2	12.9	13.6	-1.8	-2.4
Manure – 5	11.3	11.4	13.2	13.8	-1.9	-2.4
Sludge – 5	10.9	10.0	12.1	12.1	-1.2	-2.1
Sludge 10	10.4	9.9	11.9	11.9	-1.5	-2.0
Sludge – 20	9.6	9.8	11.0	11.6	-1.4	-1.8
F-Ratios for Comparing						
Treatments (T)	14.8[a]		31.0[a]			
Years (Y)	1.1		6.7[b]			
T x Y	1.3		0.7			

[a,b] Indicates significance at 0.001 and 0.05 levels, respectively.

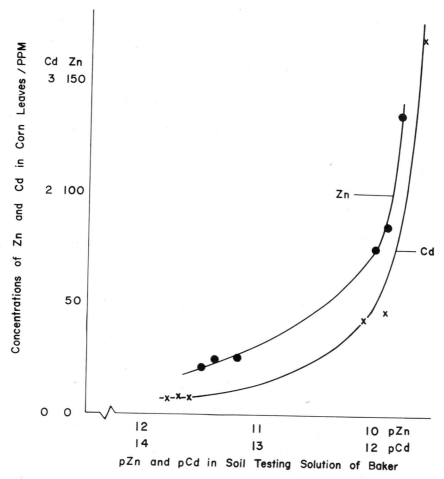

Figure 1. Plots of the negative logarithms of calculated activities for zinc and cadmium in the Baker soil testing solution against concentrations of the respective elements in corn leaves for 1975. Each point represents the mean for four replications.

1 ppm Cd could be expected in any system when the activity of cadmium at the plant root surface remains at approximately 1×10^{-12} molar or when pCd is equal to 12. Results of Tables 10 and 11 and Figure 1 suggest that zinc and cadmium uptake are regulated by soil properties, sludge properties and the rates of the elements applied. Uptake of cadmium by plant leaves indicates that the sludge initially regulates pCd and pZn, and with

time the soil and/or stable humus from sludge tends to fix a portion of the trace elements.

If the initial availability of cadmium to plants is regulated by the sludge, then sludge composition should be monitored. From the historical work on ion activities associated with ions bound by an exchanger (32), the availability of cadmium or zinc should be a function of their mole fraction on the adsorption sites. For the trace elements, zinc and especially cadmium, the mole fraction is a direct function of the concentration of cadmium within the sludge. From the results presented here and those presented previously, soil applications of 0.5 ppm Cd (1 lb/acre) per year for three years are considered safe with respect to cadmium within the food chain. The rate is equivalent to 50 ppm Cd in sludge on a dry matter basis when 10 tons per acre are applied as a source of nitrogen for corn.

The availability of cadmium increased relatively more than did zinc from applications of sludge. Since both elements are eventually bound within the hydrous oxide fraction of the soil, it might be that the ion activities reflect their solubilities. The negative logarithms for the solubility products for $Zn(OH)_2$ and $Cd(OH)_2$ were calculated at 16.2 and 14.3, respectively (40). The fact that $Cd(OH)_2$ is about 100 times as soluble as $Zn(OH)_2$ could be important in explaining the relatively greater activity and biological availability of cadmium than zinc in soils. For high pH soils it could be significant to note, however, that $ZnCO_3$ is more soluble than $CdCO_3$ ($pK_{ZnCO_3} = 9.9$; $pK_{CdCO_3} = 11.2$). Additional work on phase equilibria is needed to explain better the retention of zinc and cadmium in relation to suspension activities of the ions (5, 8, 41).

SUMMARY AND CONCLUSIONS

The results presented and those of other investigators when related to concepts of surface adsorption of ions as it regulates ion activity and availability to plants lead to the following conclusions:

1. Additions of wastes containing only 5 to 10 ppm Cd can be expected to increase the concentrations of Cd within the food chain over time.
2. Manure concentrations of cadmium are about 2.3 times as high as those of the animal ration.
3. Sludge composition from different treatment plants over time varies with respect to zinc and cadmium, from 15 to 100 percent of the mean. A variation of 50 percent should be expected. Therefore, if an upper limit of 50 ppm Cd in

sludge is desired, one seeks a sludge containing less than 33 ± 17 ppm Cd.

4. For maintaining less than 1 ppm Cd in leaves of corn, sorghum and soybeans, the loading capacity of the site studied over a two-year period was 0.5 ppm per year or 1 ppm Cd within the soil over the two-year period.

5. The 0.1 N HCl extractant removed essentially all of the cadmium and more than 50 percent of the zinc added with sludge to the soil. DTPA removed about 30 percent of the zinc and 60 percent of the cadmium added over two growing seasons.

6. The Baker method extracts zinc and cadmium in relation to their availability, and their activities in the testing solutions reflect the relatively high activity and biological availability of cadmium compared with zinc.

7. For controlling cadmium within the food chain, sludge applied as a source of nitrogen for corn on this site should not contain more than 33 ± 17 ppm Cd; the labile cadmium within the soil should not exceed 1 ppm, and pCd by the Baker method should not be lower than 12.0.

8. Soil levels of zinc should not exceed levels associated with a zinc toxicity in plants.

9. While mole fractions of zinc on the adsorption complex are expected to reflect levels of zinc that are phytotoxic, safe levels of cadmium within the food chain will be reflected more by pCd and labile cadmium and their relationships within the soil.

ACKNOWLEDGMENTS

This paper is a contribution of the Pennsylvania Agricultural Experiment Station, University Park, Pennsylvania 16802 and was authorized for publication on March 23, 1976, as Paper No. 5044 in the Journal Series. Partial support of this research from Fair Funds administered by the Pennsylvania Department of Agriculture is gratefully acknowledged. The authors wish to express their appreciation to Raymond Shipp for assistance in the collection of manure and sludge samples and to Leon Marshall for assistance with the field work.

REFERENCES

1. Baker, D. E. "Copper: Soil, water plant relationships," *Federation Proc.* 33:1188–1193 (1974).
2. Baker, D. E. and L. Chesnin. "Chemical monitoring of soils for environ-

mental quality and animal and human health," *Adv. Agron.* 27:305–374 (1975).

3. Bouwer, H. and R. L. Chaney. "Land treatment of wastewater," *Adv. Agron.* 26:133–176 (1974).

4. Chaney, R. L. "Metals in plants—Absorption mechanisms, accumulation, and tolerance. Metals in the biosphere, Univ. of Guelph, Guelph, Ontario, Canada, pp. 79–99 (1975).

5. Lindsay, W. L. "The chemistry of metals in soils. Metals in the biosphere," Univ. of Guelph, Guelph, Ontario, Canada, pp. 47–62 (1975).

6. Lisk, D. J. "Trace metals in soils, plants and animals," *Adv. Agron.* 24:267–325 (1972).

7. Melsted, S. W. "Soil-plant relationships (Some practical considerations in waste management)," *Proc. Joint Conf. Recycling Municipal Sludges and Effluents on Land*, pp. 121–128 (1973).

8. Norvell, W. A. "Equilibria of metal chelates in soil solution," in: *Micronutrients in Agriculture.* Soil Sci. Soc. Amer., Madison, Wisconsin pp. 115–138 (1972).

9. Viets, Jr., F. G. and W. L. Lindsay. "Testing soils for zinc, copper, manganese and iron." In: *Soil Testing and Plant Analysis*, L. M. Walsh and J. D. Beaton, Eds. Soil Sci. Soc. Amer, Madison, Wisconsin, pp. 153–172 (1973).

10. Allaway, N. W. "Agronomic controls over the environmental cycling of trace elements," *Adv. Agron.* 19:235–274 (1968).

11. Page, A. L. "Fate and effects of trace elements in sewage sludge when applied to agricultural land," U.S. EPA, Nat. Environ. Res. Center, Cincinnati, Ohio, p. 97 (1974).

12. Miller, R. H. "Factors affecting the decomposition of an anaerobically digested sewage sludge in soil," *J. Environ. Quality* 3:376–380 (1974).

13. Turner, M. A. "Effect of cadmium treatment on cadmium and zinc uptake by selected vegetable species," *J. Environ. Quality* 2:118–119 (1973).

14. Page, A. L., F. T. Bingham, and C. Nelson. "Cadmium absorption and growth of various plant species as influenced by solution cadmium concentration," *J. Environ. Quality* 1:288 (1972).

15. Bingham, F. T., A. L. Page, R. J. Mahler, and T. J. Ganje. "Growth and cadmium accumulation of plant growth on a soil treated with a cadmium-enriched sewage sludge," *J. Environ. Quality* 4:207–211 (1975).

16. Haghiri, F. "Cadmium uptake by plants," *J. Environ. Quality* 2:93–96 (1973).

17. Hakne, H. C. H. and W. Kroontje. "Significance of pH and chloride concentration on behavior of heavy metal pollutants: Mercury (II), Cadmium (II), Zinc (II), and Lead (II)," *J. Environ. Quality* 2:444–450 (1973).

18. Dowdy, R. H. and W. E. Larson. "The availability of sludge-borne metals to various vegetable crops," *J. Environ. Quality* 4:278–282 (1975).

19. Giordano, P. M., J. J. Mortvedt, and D. A. Mays. "Effects of municipal waste on crop yields and uptake of heavy metals," *J. Environ. Quality* 4:394–399 (1975).

20. Boswell, F. C. "Municipal sewage sludge and selected element application to soil: effect on soil and fescue," *J. Environ. Quality* 4:267–273 (1975).

21. Jones, R. L., T. D. Hinesly, and E. L. Ziegler. "Cadmium content of soybeans grown in sewage sludge amended soil," *J. Environ. Quality* 2:351–353 (1973).

22. Kirkham, M. B. "Trace elements in corn grown on long-term sludge disposal site," *Environ. Sci. Technol.* 9:765–768 (1975).

23. Council on Soil Testing and Plant Analysis. *Handbook on Reference Methods of Soil Testing.* Athens, Georgia (1974).

24. Baker, D. E. "A new approach to soil testing: II. Ionic equilibria involving H, K, Ca, Mg, Mn, Fe, Cu, Zn, Na, P, and S," *Soil Sci. Soc. Am. Proc.* 37:537–541 (1973).

25. Friberg, L. M., M. Piscator, and G. Nordberg. *Cadmium in the Environment.* Chemical Rubber Co. Press, Cleveland, Ohio (1971).

26. Bureau of Foods. "Compliance program evaluation. Total diet studies: FY 1973 (7320.08)." Food and Drug Admin. H.E.W., Washington, D.C., 21 pp. (1975).

27. Sandstead, H. H., *et al.* "Cadmium, zinc and lead." In: *Geochemistry and the Environment*, Vol. 1. National Acadamy of Science, Washington, D.C., pp. 43–56 (1974).

28. Schroeder, H. A. and A. P. Nason. "Interactions of trace metals in rat tissues. Cadmium and nickel with zinc, chromium, copper and manganese," *J. Nutr.* 104:167–178 (1974).

29. Baker, D. E., R. M. Eshelman and R. M. Leach. "Cadmium in sludge potentially harmful when applied to crops," *Sci. Agric.* XXII:14–15 (1975).

30. Doyle, J. J. and W. H. Pfander. "Cadmium Toxicity in Sheep," *J. Nutr.* 104:160–166 (1974).

31. Olson, O. E. "Fluorometric analysis of selenium in plants," *Assoc. Off. Anal. Chem.* 52:627–634 (1969).

32. Baker, D. E. "A new approach to soil testing," *Soil Sci.* 112:381–391 (1971).

33. Huffman, Jr., E. W. D. and J. F. Hodgson. "Distribution of cadmium and zinc/cadmium ratios in crops from 19 states east of the Rocky Mountains," *J. Environ. Qual.* 2:289–291 (1973).

34. Shipp, R. F. and D. E. Baker. "Pennsylvania's sewage sludge research and extension program," *Compost Sci.* 16:6–8 (1975).

35. Chaney, R. L. "Crop and food chain effects of toxic elements in sludges and effluents," *Proc. Joint Conf. on Recycling Municipal Sludges and Effluents on Land*," pp. 129–141 (1973).

36. Franz, M. "Survey of sewage sludge research at agricultural experimental stations," *Compost Sci.* 15(2):16–19 (1974).

37. Kenney, D. R., K. W. Lee and L. M. Walsh. "Guidelines for the application of wastewater sludge to agricultural land in Wisconsin," Dept. of Nat. Res. Tech. Bull. 88, 36 pp. (1975).

38. Bradford, R. R. and D. E. Baker. "Variable selectivity for Sr by corn hybrids which accumulate different concentrations of Sr, Ca, Mg and K," *Agron. J.* 61:766–768 (1969).

39. Baker, D. E. and P. F. Low. "Effect of the sol-gel transformation in clay-water systems on biological activity: 2. Sodium uptake by corn seedlings," *Soil Sci. Soc. Am. Proc.* 34:49–56 (1970).

40. National Bureau of Standards. "Selected values of Chemical Thermodynamic Properties," *Technical Notes #270*, 3:233–261, U.S. Department of Commerce (1968).

41. Santillan, Medrano, J. and J. J. Jurinak. "The chemistry of lead and cadmium in soil: Solid phase formation," *Soil Sci. Soc. Amer. Proc.* 39:851–858 (1975).

42. Kelling, K. A., A. E. Peterson, J. A. Ryan, L. M. Walsh and D. R. Kenney. "Crop responses to liquid digested sewage sludge," Proc. Int. Conf. on Land for Waste Mgt. Ottawa, Canada, pp. 243–252 (1973).

16

HEAVY METAL RELATIONSHIPS DURING LAND UTILIZATION OF SEWAGE SLUDGE IN THE NORTHEAST

R. L. Chaney, S. B. Hornick, and P. W. Simon
Agricultural Environmental Quality Institute
Agricultural Research Service
Biological Waste Management and Soil Nitrogen Laboratory
U.S. Department of Agriculture
Beltsville, Maryland

INTRODUCTION

Application of sewage sludges on farmland is increasing for several reasons: (a) use on farmland is usually the lowest cost option for ultimate disposal of sludge; (b) there is a desire to utilize "wastes" beneficially where possible as sludge can add needed macronutrients and micronutrients, improve soil physical properties, and support low cost revegetation of problem soils; and (c) sludge available for use on land is increasing sharply as sewage treatment is improved, as ocean disposal and incineration become less environmentally acceptable, and as new sludge processing technology makes sludge use on land more convenient and less expensive [for example, the raw sludge composting developed at Beltsville (1)].

Potential problems from heavy metals in municipal sludges have been identified by a number of recent reviews: Page (2), Chaney (3), Chaney and Giordano (4), Braude, Jelinek and Corneliussen (5), Webber (6), Leeper (7). Although phytotoxicity from zinc, copper and nickel have been identified as one potential problem when sludges are utilized on farmland, the greatest concern has been focused on increased crop uptake of cadmium and its subsequent movement to the human food chain. Uncontrolled sludge use on land has been identified as a major potential cause of increased dietary Cd (5,8) partially because food comprises the bulk of human exposure to Cd.

Recently, laboratory, greenhouse and field studies have begun to characterize sludge, soil and crop factors involved in heavy metal (particularly Cd) aspects of sludge use on farmland. Sludge Cd ranges from < 1 to 3400 ppm, is seldom above 25 ppm without abatable Cd sources, and is often below 10 ppm at suburban community treatment plants. Different Cd compounds may occur in sludges depending on industrial sources, and may have different availability to plants (9). Soil properties affect plant availability of added Cd. John, VanLaerhoven and Chuch (10) have found that increased soil capacity to absorb Cd reduces plant Cd uptake. This capacity was related to CEC. Many have reported reduced Cd uptake at higher soil pH (10-12). Crops differ both in relative Cd uptake to diagnostic leaf tissues when growing on the same soil, and in Cd transport to grain or other edible tissues relative to Cd uptake to diagnostic leaves (4,12-14). Soil temperature and the presence of other microelements influence Cd uptake to plant tops and transport to edible tissues (4). Thus, opinion about sludge Cd has varied from (a) low Cd sludges are not safe enough for crops (15) to (b) high Cd sludges are safe to grow corn grain with good management (16). Studies on crop uptake of sludge Cd versus soluble inorganic Cd salts indicate further complications in judging experiments on safety of Cd added to soils (17).

A sludge may cause only slight increases in corn grain Cd while soybean grain Cd may be increased to very high levels (16, 18). This relationship can be reversed depending on soil pH and sludge Cd/Zn (12). Some animal studies of the availability of Cd indicate large increases in kidney and liver with no effect on muscle, while others report that muscle and egg Cd can be increased by dietary Cd levels present in crops grown on soils treated with high Cd sludges (19). Thus, the plant availability of sludge Cd additions to farmland is a complicated, controversial and important topic. An understanding of long-term impact on food Cd which may result from utilizing sewage sludge as an agricultural resource is necessary to set policy on ultimate disposal of sewage sludge.

Why Study Long-Term Sludge Use Sites?

For years an underlying question of newly begun research projects concerning the plant availability of sludge-applied heavy metals has remained unanswerable: "These results for 2 or 3 or 5 years look very interesting, but what will happen after 10 or 25 years?" The wisdom of particular guideline schemes (limit Zn, Cu and Ni; limit Cd; limit Cd:Zn, limit N; limit Cd:NH$_4$-N) is always tested against current research plot results, whereas long-term safety is the goal of the guidelines.

The studies reported in this paper were designed to try to answer some of the questions that new field plots cannot answer. The selection of both long-term continuous sludge use sites, and of sites where sludge use had ceased, and the study of many combinations of sludges, soils and crops, all widen the knowledge base on factors which control the safety of sludge use in the long-term. Fields where sludge had been used many years are certainly nearer to equilibrium with regard to metal reactions than are new field plots. It seems clear that society cannot wait for scientists to establish field plots for each type of sludge and soil and study many crops grown thereon for 25 continuous years before some safe method for ultimate disposal of municipal sewage sludge is selected.

Reports on Long-Term Sludge or Sewage Application Farms

A few researchers have reported studies of existing practice. Evans (20) and Manson and Merritt (21,22) extolled the virtues of sludge use on farmland, yet made no objective studies on metals in sludges, soils or crops. They observed apparent crop response to sludge N, heard farmers say they wanted all the sludge they could work into their cropping pattern, and heard treatment plant personnel tell about how inexpensive sludge disposal on farmland was.

Some of the early reports which alerted researchers to the need to study heavy metal aspects of sludge and sewage use were reports on long-term use sites. Rohde (23) described phytotoxicity (apparently due to Cu, Zn, lowered pH and sensitive crops) at the Berlin sewage farm. LeRiche (24) found that Zn and other metals added over many years remained plant-available. Neither of these reports considered the continuing availability of applied Cd.

More recently, several reports have considered Cd as well as other elements. Most of these reports covered low Cd wastes and soils, soil pH values high enough to minimize plant uptake, and/or crops which accumulate only low levels of Cd and other elements. Often soil pH, organic matter, CEC, available P and texture were not characterized. From comparisons with a control field and a field where low-Cd sludge was applied for 15 years, Andersson and Nilsson (25) found that rape fodder was enriched

Table 1.   Treatment practices, mean flows and sludge handling at treatment plants studied.

City	Flow Yearly Mean (MGD)	Sewage Treatment 1°	Sewage Treatment 2°	Sludge Processing	Form Hauled [a]	Crops	Hauler [b]	Land Ownership [c]
1	15	Yes	Act.	AnDig $1°$-H,$2°$	L,C	Any	M	P
2	20	Yes	None	AnDig $1°$-H,$2°$	L,C,D	Any	M	P
3	2.7	Yes	T.Filt	AnDig $1°$-H,$2°$	L	Any	M	M,P
4	0.9	Yes	T.Filt	AnDig $1°$-H,$2°$	L	Any	M	M,P
5	1.5	Yes	T.Filt	AnDig $1°$-H	L,SB	Any	M	P
6	3.9	Yes	Act.	AnDig $1°$-H	C	Grass	C	P
7	3.8	Yes	Act.	AnDig $1°$-H	C	Grass	C	P
8	0.5	Yes	T.Filt	AnDig $1°$-H	L	Any	C	P
9	1.7	Yes	T.Filt	AnDig $1°$-H,$2°$	L,SB	Any	M	P
10A	9.0	Yes	Act.	AnDig $1°$-H	L,SB	Any	M	P
10B	9.0	Yes	Act.	Heat Treat.	C	Any	M	P
11	3.7	None	Act.	Aerobic	L	Any	C	P
12	0.4	Yes	T.Filt	AnDig $1°$-H	L,SB	Any	C	P
13	5.5	Yes	Act.	AnDig $1°$-H [d]	L,SB	Any	C	P
14	1.7	Yes	T.Filt	AnDig $1°$-H [d]	L,SB	Any	M	P
15	3.7	Yes	T.Filt	AnDig $1°$-H,$2°$	L	Any	M	P
16	0.2	Yes	T.Filt	AnDig $1°$	L,SB	Any	M,P	P
17	1.5	Yes	T.Filt	AnDig $1°$-H,$2°$	L	Any	M	P,M
18	6.0	Yes	Act.	AnDig $1°$-H [d],$2°$	L	Any	M	P,M
19	1.2	Yes	T.Filt	AnDig $1°$-H	L	Grass	M	M
20	1.2	None	Act.	Aerobic; Lagoon	L	Any	C	P
21	0.7	Yes	Act.	AnDig $1°$-H [d]	L,SB	Any	M	P
22	33	Yes	T.Filt	AnDig $1°$-H,$2°$	C	Any	M	P,M
23A	15	Yes	Act.	AnDig $1°$-H	C	Any	C	P,M
23B	10	Yes	Act.	AnDig $1°$-H	C	Any	C	P,M
23C	5	Yes	T.Filt	AnDig $1°$-H	SB	Any	C	P,M
25	0.8	Yes	Act.	AnDig $1°$-H,$2°$	L,SB	Any	M	P
26	309	Yes	Act.	AnDig $1°$-H	C	Any	C	P,M
27	10	Yes		AnDig $1°$-H	C	Any	C	P
29A	50	Yes	Act.; T.Filt	AnDig $1°$-H	C	Any	C	P
29B	25	Yes	No	AnDig $1°$-H	C	Any	C	P
30	2.5	Yes	T.Filt	AnDig $1°$	SB	Any	M	P
31	1.4	Yes	T.Filt	AnDig $1°$-H	SB	Any	P	P
32	25	Yes	Act.	AnDig $1°$-H,$2°$	D	Any	M	P
33	0.5	Yes	Act.	Aerobic	SB	Any	M	P
34	0.4	Yes	Act.	AnDig $1°$-H	L	Any	M	P
35	0.9	Yes	T.Filt	AnDig $1°$-H	L,SB	Any	M	P
36	0.4	Yes	T.Filt	AnDig $1°$-H	L,SB	Any	M	P
37	0.1	Yes		AnDig $1°$-H,$2°$	L,SB	Any	M	P
38		Yes	T.Filt	AnDig $1°$-H,$2°$	L,SB	Any	C	P
39	5.0	Yes	Act.	AnDig $1°$-H,$2°$	L	Any	M	M,P
40	1.9	Yes	Act.	AnDig $1°$-H,$2°$	SB	Any	P	P
41	3.0	Yes	T.Filt	AnDig $1°$-H,$2°$	SB	Grass	M	M

[a] Type of sewage sludge applied to land where L = liquid, SB = sandbed, C = cake, and D = dried.

[b] Main spreaders of sewage sludge where M = municipality, C = private contractor, and P = local townspeople.

[c] Owners of land on which sewage sludge was applied where P = private, and M = municipality.

[d] Digester is poorly heated in winter.

in Zn, Cu and As, slightly higher in Cr, Mo, Hg and Pb, and not higher in Cd. Chaney, White and Simon (12) reported no increase in Cd, Pb or Cu in pasture grasses unless the crop was physically contaminated with sludge. This was on a farm at Hagerstown, Maryland, where sludge was used as the only fertilizer since 1951. Johnson *et al.* (26) found no increase in Cd in pasture crops growing on the Werribee Sewage Farm near Melbourne, Australia, although soil levels of sewage metals had increased substantially during the 80 years of sewage farming. Pike, Graham and Fogden (27,28) reported studies on microelement accumulation in soils and availability to vegetable crops at a disused sewage farm at Beaumont Leys, Leicester, England, many years after sewage and sludge spreading had ceased. Although they did not report soil pH and other properties other than metal content, and they had no real control plots, their results showed that: (a) added Zn, Ni and Cd remained extractable by acetic acid, (b) Zn and Ni remained plant-available, (c) Pb, Cr and Cu in crops were relatively unchanged, and (d) crop Cd levels were high (approximately 2–10 ppm Cd in dry lettuce) but were not appreciably different than their "ordinary garden" crops grown in the same community. Their results were not statistically evaluated.

Recently, Kirkham (29) reported the metal content of sludge, soil and corn tissue at a 35-year-old sludge disposal farm at Dayton, Ohio. The treated soils were very high in total metals: 2065 ppm Zn, 843 ppm Cu, 70.5 ppm Cd. Control soils contained 158 ppm Zn, 51 ppm Cu, and 2.0 ppm Cd. The corn leaves grown on the control and treated soil contained 67 and 196 ppm Zn, 10 and 40 ppm Cu, and 2.1 and 13.9 ppm Cd, respectively, while the corn grain contained 12 and 79 ppm Zn, 8 and 12 ppm Cu, and 0.8 and 0.9 ppm Cd, respectively. These data appear very encouraging regarding the long-term low availability to corn of sludge-borne Cd. However, insufficient study of the sludge-spreading practice, of site soil properties, and Cd availability to other crops were reported to assess the full impact of these high soil Cd levels. Further study of the Dayton site by J. A. Ryan (EPA, Cincinnati, Ohio, personal communication) has confirmed the high foliar Cd increase and lack of grain Cd increase (control and treated soil-grown grain contained about 0.1 ppm Cd). Exclusion of Cd from corn grain also has been observed by Jones *et al.* (16) and Bingham *et al.* (30).

Reported studies of long-term sludge use farms are, unfortunately, equivocal regarding the long-term plant availability of applied Cd. Often inadequate data were reported to assess the crop results, or only crops known to exclude Cd were sampled. However, the experimental technique of evaluating the long-term plant availability of Cd on these sites appears to have remained meaningful and needed.

Thus, a program was started in our laboratory to: (a) locate long-term sludge use sites with a wide range of soil Cd contents, (b) grow and analyze a number of crops representing a range of Cd accumulation characteristics on these sites and matched control fields, and (c) characterize existing sludge use practice regarding heavy metals in sludges, soils and crops. This chapter is a preliminary report of the results of these studies, showing sludge composition, metal accumulation in treated soils, and crop uptake of heavy metals from treated and control soils. These studies have shown that use of high Cd sludges lead to long-term increases in crop Cd, but that proper soil pH management and crop selection can minimize the impact of sludge use on farmland on Cd movement into the food chain.

METHODS AND MATERIALS

Finding Long-Term Sludge Fields

Based on research experience at Hagerstown, several criteria for sludged fields worth the expense of further scientific evaluation were set: (a) the longer or longer ago sludge was applied, the better; (b) the greater the cumulative loading of sludge and/or sludge-borne metals, the better; and (c) the sludge should have been tilled into the soil at least intermittently. Both fields which had received recommended sludges (Cd/Zn $\leqslant$ 1.5 percent) and fields which had received nonrecommended sludges were sought in order to evaluate the long-term availability of heavy metals. A few fields which had received sludges very high in Cd or Cd/Zn were especially sought. It was hoped that some locations where good soil pH management (pH $\geqslant$ 6.5) had occurred simultaneously with sludge utilization would also be found.

Sewage treatment plants listed by Carroll et al. (31) as having land-applied liquid sludge for more than 5 years were contacted first. The Departments of Health, Environmental Protection, or Water Pollution Control in the states near Beltsville were contacted in order to learn which cities were known to apply any form of sludge on land. Also contacted were water pollution control groups, environmental groups, consulting engineers, and other researchers. Rodale Press helped provide in the search by asking their readers who knew of sludged locations to write the laboratory.

Treatment plant representatives responsible for sludge disappearance and/or the farmers receiving sludge were telephoned to obtain an oral history of how and where sludge was applied.

If a site looked promising, either a visit was scheduled immediately, or a box with sampling supplies was sent by mail to allow the treatment plant personnel to return samples of sludge and, if possible, treated soil. Samples returned were analyzed for heavy metals, and site visits were scheduled where application history and sludge composition were of particular interest. On initial site visits samples included: (a) current sludges—those available for land application; (b) any other sludge—those found in sandbeds or old stockpiles; (c) sludged soils on farms—topsoil and subsoil of treated fields, preferably the part of the farm reportedly highest in cumulative sludge loading; (d) untreated soil— such as neighbors' fields, fence rows, etc., of the same soil type as the treated soil; and (e) vegetation—plants growing on sludge and control locations if available during the sampling visit. Both bucket auger and composite (20 cores, 1-inch-diameter) soil samples were collected in these fields. Analysis of these samples provided objective data about long-term sludge composition, application rates, pH management, depth of tilling sludge into the soil, etc. If these results and the oral history confirmed that the site met the criteria for further study, sites were revisited to find the best locations for field plots.

Determining Locations of Field Plots

Prior to a site visit, all persons who could provide information were contacted: treatment plant personnel (superintendent, sludge application foremen; truck drivers); farmers who received sludge many years; and County Agricultural Extension Agents and Soil Conservation Service staff familiar with soils in the area. Soil maps for the farms were obtained if possible.

Based on the preliminary sampling pattern and interview information, sludged fields were selected for intensive sampling. A grid pattern (every 10–20 yd) was sampled. The soil pH was measured in the field, and areas with soil pH of interest were sampled more intensively, often on a 5-yd grid pattern (bucket auger samples) with 20 core composite samples taken within the grid pattern.

When the field pH of the selected available plot land was established, paired control fields were sought. Soil maps and the kind assistance of County Agents and SCS soil scientists helped to find nearby fields with very similar soil properties. If the landowner permitted sampling and certified sludge had never been applied to the land, the area was sampled and field pH values determined. If the soil pH, type, color, texture, slope, etc., were near that of the long-term sludged field, a small grid was sampled (points and composites) to define a possible control plot area.

Installation of Field Plots

Based on the analyses of soil and plant samples from potential field plot areas, specific locations within sludge-treated fields
were selected for plot use. Both DTPA-extractable metals and
soil pH were used to select relatively uniformily treated areas.
Fertilizer and lime requirements of the plot areas were determined
by further analysis of the samples from the selected area.

Where the initial soil pH was low, the experimental design
included paired unlimed and limed plots. Three replications of
these two pH treatments were laid out at both sludged and control fields. Plot areas were 15 x 20 ft with 6-ft treated borders.
Where the initial soil pH was $\geq$ 6.5, four replications were laid
out at each sludged and control field; no lime amendments were
made—only the existing field pH was studied.

On the low pH fields, the soil pH was raised to $\geq$ 6.5 for the
limed plots by application of 2000 lb $CaCO_3$ equivalent to a hy-
drated lime plus dolomitic limestone as required so that 110 percent of the lime requirement was applied. NPK fertilizers were
applied as recommended. Since none of the soils were deficient
in P, K or Mg, only crop requirements of the macrofertilizer
elements were applied. N at 200 lb/acre was applied to the chard
and orchardgrass area of each plot; P and K at 100 lb P_2O_5 and
100 lb K_2O/acre were applied to the entire plot.

The crops grown on the plot were selected because they represent different metal accumulation patterns and should allow overall assessment of the range of potential impacts sludge will have
on the food chain. Inoculated "Williams" soybeans, Fordhood
Giant Swiss chard, and common orchardgrass were seeded in
1975; Clintford oat was seeded in Spring 1975, while soybean,
chard and "Paris White" Romaine lettuce are to be grown in
1976. A satisfactory stand of orchardgrass was not obtained until
Fall 1975.

The plots were fenced to reduce wildlife problems, and blood
meal was used as a deer repellent. Although several measures
were taken to reduce groundhog populations near the plot areas
in 1975, the sludged plots at city 4 were grazed so heavily and
repeatedly that no soybean samples were obtained.

Sludge, Soil and Plant Analyses

Sludges

Weighed mixed samples were dried at 105°C to determine
percent solids, and ashed at 500°C overnight to determine percent ash. The ash was treated with 5 ml conc. HNO_3, heated

to dryness, then dissolved with 40 ml of 3 N HCl and refluxed 2 hr with intermittent stirring. After filtering, the samples (about 2 g dry sludge) were diluted to 100 ml ($\sim$ 1 N HCl).

Soils

Soil Total Metal samples were prepared similarly to sludge samples except that the extract from 5 g of soil was diluted to 50 ml 1 N HCl. *DTPA-Extractable Metals* were extracted according to Follett and Lindsay (32). Fifteen g soil and 30 ml extracting solution (5 mM DTPA, 10 mM $CaCl_2$, 100 mM triethanolamine, 1 ppm phenylmercuric acetate, pH 7.30) were shaken to 2 hr in a 125-ml polyethylene flask on a wrist action shaker. The samples were filtered through pre-folded Whatman 111-V filter paper. *Soil pH* was measured after 1 hr on a 1:1 soil:water (volume) slurry. Lime requirements were determined by a buffer method.

Plant samples were dry ashed overnight at 480°C, treated with 2 ml conc. HNO_3 to dryness, dissolved in 10 ml 3 N HCl and refulxed for 2 hr. After filtration through an acid-rinsed Whatman 40 paper, the samples were diluted to 25 ml. Foliar samples were 2.00 g; seed and fruit samples were 4.00 g.

Metal analysis was by atomic absorption spectrophotometry. Background correction was used for Cd, Pb and Ni but not for Zn, Cu, Mn and Fe. Because of the g tissue/ml analyte and the sensitivity of the Varian-Techtron AA-6 used, no samples required graphite furnace analysis. Some soil samples were sent to the Ohio Soil Testing Laboratory for routine fertility, cation exchange capacity and organic matter analysis.

RESULTS

Sewage Treatment and Site Handling

Table 1 summarizes the general information about the sewage treatment plants visited. Sludge had been land-applied for at least 5 years and as many as 37 years by the different plants. Size ranged from under 1 MGD to 390 MGD; several treatment plants hauling predominantely liquid sludge were 20–30 MGD, although most were smaller. Usually treatment plants over 15 MGD applied some vacuum filter cake. Nearly every sludge was anaerobically digested; many plants had heated-stirred 1° digesters and unheated 2° digesters/holding-storage tanks. One treatment plant with aerobic digestion had a holding lagoon. Odor problems were noted only where sludges from unheated or poorly heated digesters were land-applied and not tilled in immediately.

Although sandbeds served as winter or wet weather back-ups for some treatment plants, most sandbeds observed (even glass-house-covered) were not being used. The decision to not use available sandbeds was usually based on convenience and dirtiness rather than on economic analysis. Large treatment plants made vacuum filter cake and had some provision for cake storage.

Sludge was applied to/for any crop normally grown in the area—predominantly corn, grass and legume hay, wheat, oat, soybean, orchards, turfgrass sod, golf courses and cemeteries, gardens, tobacco, etc. Liquid sludge was applied from 1 to 30 times per year, with some sprayed into standing wheat, oat and corn. Liquid sludge application to pasture (hay crops) was an important component of most liquid-only systems; cows and spreader trucks were seen in the same field, and some farmers reported that cows grazed freshly sludged pastures. Most farmers asked for sludge to be applied only after hay harvest or after grazing rather than to mature hay crops.

Most municipalities hauled their own sludge, and truck drivers did other work during bad weather; several contracted disappearance of their liquid sludge and especially of their cake sludge. Sludge was generally applied to private farmland. It was also applied to municipality-owned land purchased for sludge application and farmed tenant farmers or left as wasteland. One city leased private farmland for summer application. Some sludge was used on disturbed land and leveled gravel-mines for revegetation.

Most sludge was hauled and applied to private land free of charge. However, some treatment plants charged for all deliveries beyond a set distance, or charged all who do not elect to accept filter cake or sandbed sludge. One farmer with a well drained field at the gate of a treatment plant actually was paid because "the trucks made it harder to plow;" many who got free sludge felt it made plowing easier. Sludge delivery foremen played a very signigicant role in finding private farms where sludge could be applied and in maintaining good relations with farmers. Without this expertise many systems would require much more expenditure for bad weather back-ups.

Permit Status of Sludge Use

Although several states have had some permit programs over the years, and all have relied on public nuisance regulations, none (Maryland, Virginia, Delaware, District of Columbia, Pennsylvania) had specific sludge application regulations or guidelines. City 18 has applied all its sludge for 26 years with no regulatory interest until late 1975. Some counties and even townships prohibited importation of sludge but allowed any local sludges without permit. The only factor which caused imposition of permits was an odor problem, although many plants operated under an "understanding" with their county health department that they must prevent contamination of surface waters and public nuisance conditions. No location required any records or monitoring until 1975.

Agronomic Aspects of Sludge Utilization Practice

Considerable variation exists among cities and farmers in the agricultural aspects of planning sludge use. Most people saw sludge as similar to animal manures. A few farmers planned the rate of sludge and additional fertilizer needed by their crops based on the N fertilizer value of the sludge; others presumed it would replace all fertilizers and limestone (one for 25 continuous years) ; others ignored any fertilizer value sludge might have. The treatment plant provided sludge total N contents in only a few cases.

Generally, no treatment plant regularly analyzed their sludge for nutrients and especially not for heavy metals. Although several treatment plants had obtained heavy metal analyses, until very recently it was only done in relation to digester upsets. Recently Shipp and Baker (33), the Maryland Department of Agriculture, the Virginia Water Pollution Control Board, the U.S. EPA, and our laboratory have provided sludge analyses to a number of treatment plants. Because the Shipp and Baker program charges the farmer $100 for a complete sludge analysis, few have requested this valuable service; one county-wide survey and a number of individual treatment plants in Pennsylvania have been trying to provide some meaningful sludge analyses to private farmers wanting to use sludge since this analytical service became available.

A few farmers who used sludge sampled their soils regularly and followed the recommendation of their state agricultural experiment station or consultant doing the analyses. Because some farmers limed regularly, recommended pH management long-term sludge use sites were found. However, very few farmers sought or followed agronomic (pH; N, P, K, Mg fertility) advice, and some provided improper soil samples to their advisors. Some farmers received bad advice: one was told to apply 200 lb/acre of urea-N supplemental to more than enough sludge-N because his consultant found a very high Bray-1 P value (city 4). None of the sludge-use locations had been sampled and analyzed for soil or crop heavy metals before researchers came. No farmer utilized foliar analysis as a management tool. Until recently no farmer or city had asked their county agent for advice on how to use sludge; in 1975 one treatment plant held a public meeting with its sludge users, county agents, and Ray Shipp.

Had any agricultural *problems* been observed? Even though some locations were found with quite low soil pH and high cumulative loadings of sludge metals, there was no visible evidence of plant injury. Few crops sensitive to Zn, Cu or Ni are grown in the area examined. Swiss chard grown on the field plots showed a substantial yield response to liming on the strongly acid sludges fields, more so than on any of the control plots; however, no chlorosis was observed. Slower seedling growth was observed in several fields, but mature crops showed no symptoms. One farmer reported a severe yield reduction for several years where several truckloads of high metal sandbed sludge were applied; the yield reduction disappeared before we visited.

Lodging of wheat, oats and barley occurred on several farms. This problem was related to: (a) excessive sludge-N application, and (b) higher-than-needed application of poorly digested sludges where a higher proportion of the sludge-N could become crop available. Several long-term users felt that the new wheat varieties were less susceptible to sludge-induced lodging than older varieties. On the other hand, adequate-to-heavy fluid sludge top dressing on wheat caused remarkable yield increases—greater tillering, bigger plants, and longer heads; these visual observations may have been biased due to inadequate fertilization of the rest of the field.

Usually too little sludge was applied to satisfy crop N requirements. We were unable to demonstrate sludge metal or phosphorus residues in several sludged fields.

Soil pH was generally too low for good farming ($\leqslant 5.6$). The tendency of sludge to cause lowered soil pH is unappreciated and often the highest cumulative application area of a field was 0.5–1.5 pH units lower in pH than the rest of the field. Occasionally very high sludge and soil Cd and very low soil pH coincided (city 25).

What could the usually poor pH management of sludge fields be attributed to? It was felt that the following factors can each contribute: (a) farmers and treatment plant personnel are seldom informed of the need to lime in order to minimize heavy metal uptake; further they are not advised about the fact that sludge use causes pH lowering; (b) sludge is often used on rented or leased land where tenant farmers often do not maintain proper pH management even if the soil is tested; (c) some farmers using sludge employ very few good agronomic practices and have a low pH since they do not apply lime as a general practice; and (d) uneven sludge application coupled with even soil sampling could obscure localized extremely low pH (4.5) areas due to high localized sludge application.

Composition of Sludges Studied

Table 2 shows the mean metal contents of the sludges collected at the 43 treatment plant sources to date. The minimum, maximum, mean and median values for each element are shown in Table 3 along with maximum values for sludges from "domestic" sources. Each element and Cd/Zn varied widely showing the influence of both metal sources, ratio of source flow to total flow, and dilution by a large proportion of food processing wastewater (city 20). Metal plating (Cu, Zn, Ni, Cd, Cr), metal pigment use (CdS at city 25), and acid stripping of metal oxides (Pb, Cu) were known metal sources which contributed to the higher than "domestic" sludges.

The results are biased to some extent by some high metal sludges because we asked for leads to this type of sludge (see footnote b in Table 2). However, the overall results show that the median sludge content for each metal and Cd/Zn (Table 3) were well below the domestic maximum levels of Chaney and Giordano (4). Table 4 shows that the "domestic" levels were exceeded by a low percentage of the sludges when one element is considered at a time: Cd $\geq$ 1.5 percent was exceeded by only 26 percent of the sludges (if biased sources deleted, only by 16 percent of the sludges). However, if any one of the elements or Cd/Zn is high, sludge can fail the "domestic" test. A number of evaluations of these sludges are listed in Table 4.

The variation and effect of pretreatment enforcement on sludge metal levels are shown in Tables 5 and 6. At city 9 a plater discharged a poorly treated wastewater high in at least Zn, Cd, Cu and Ni. The digested sludge from city 9 had been consistently high in these metals. However, in 1975 the plater was forced (by a temporary shut-off of sewer service) to operate his pretreatment facility more carefully. Although the 1350 ppm Cd in raw sludges on August 8 may indicate his occasional failure (pumping Cd wastes down a commode during the temporary shut-off of his main sewer outlet), the digested sludge Cd level fell from 200 to 50 ppm Cd. Zinc and Ni were also reduced. In April 1976, the plating wastewater Cd of 0.034 mg/l was still higher than the remaining trunk sewers in the town (about 0.005 mg/l).

The influence of a CdS discharger on digested and raw sludge Cd levels is shown in Table 6. A manufacturer of "Bug Lights" discharges waste CdS to the sewer; the production is high during winter and spring so the bulbs can be ready to ship. The

Table 2.   Microelement content[a] of municipal sewage sludges available for use on land at 43 treatment plants in the northeast.

City[b]	Zn	Cd	Cu	Ni	Pb	$\frac{Cd}{Zn}$
			(mg/kg dry sludge)			(%)
1[c]	3450	100	1010	185	960	2.8
2[c]	3290	160	670	260	1230	4.9
3[c]	3420	7.3	3490	1260	500	0.21
4	1720	22.1	1100	34	540	1.28
5	1200	16.4	940	50	640	1.34
6	690	3.8	520	10	240	0.56
7	1140	61.4	1270	40	320	5.4
8	1190	9.7	900	54	490	0.82
9	5050	169.0	1510	980	350	3.3
10A[c]	1450	16.9	1450	77	440	1.16
10B[c]	5320	176.0	630	180	630	3.3
11	660	38.4	790	240	240	5.8
12	935	7.4	1080	36	640	0.79
13[c]	6430	683.0	1810	530	660	10.6
14	1410	8.9	1980	30	340	0.63
15	2380	11.1	690	38	505	0.47
16	1160	8.4	590	18	195	0.73
17	1350	19.2	2740	69	650	1.42
18	870	7.0	690	37	275	0.80
19	1430	6.6	600	24	500	0.46
20	228	0.6	2600	41	52	0.26
21	920	15.7	710	28	2080	1.71
22	1260	10.8	870	108	1420	0.85
23A	600	5.0	350	35	300	0.83
23B	960	10.0	265	40	140	1.04
23C	1320	16.0	495	50	700	1.21
25	1250	970	1170	88	270	0.78
26	1780	15.9	500	42	590	0.84
27[c]	2150	269.0	1340	120	1770	12.5
29A	4800	18.6	2590	210	620	0.38
29B	1200	9.4	860	160	470	0.78
30	1190	10.0	240	39	490	0.84
31	2590	9.0	620	25	2660	0.35
32	2000	13.3	820	44	520	0.66
33	1240	9.1	500	30	270	0.73
34	6120	22.6	660	49	4900	0.37
35	1560	17.6	570	40	340	1.12
36	1690	9.1	650	30	680	0.54
37	2100	12.6	2940	37	420	0.60
38	1080	13.0	770	25	470	1.20
39	2460	95.0	670	95	470	3.90
40	1540	8.0	1220	46	1300	0.52
41	1670	12.0	540	25	330	0.71

[a] Each value is the mean content of all samples of sludge obtained from a treatment plant to date.

[b] Numbers represent different municipalities; letters represent different treatment plants.

[c] Visits to treatment plants based on advance information about heavy metal release to the sewer, metal-caused digester upsets, etc.

Table 3. Summary of mean compositions of sludges ready for use on land at 43 treatment plants in the northeast.

Element	Minimum	Maximum	Mean	Median	Maximum Domestic [a]
Zn (ppm)	228	6430	2010	1430.0	2500
Cd (ppm)	0.6	970	72.2	13.0	25
Cu (ppm)	240	3490	1080.0	790.0	1000
Ni (ppm)	10	1260	129.0	42.0	200
Pb (ppm)	52	4900	735.0	500.0	1000
Cd/Zn (%)	0.26	78.0	3.64	0.84	1.00

[a] Chaney & Giordano (4).

April 1976 raw sludge was clearly the yellow color of this pigment and was heavily contaminated. This discharge could easily be alleviated by recovery of the Cd wastes for direct reuse or reprocessing.

A pattern found typical of "domestic" sludges in general and suburban sludges in particular is the small variability of sludge metal levels. One such sludge is used repeatedly at Beltsville in experiments conducted by USDA and the University of Maryland Department of Agronomy and Veterinary Science. Analyses of 12 samples taken from August to December 1975 are sum-

Table 4. Portion of treatment plants producing sludges which exceeded "domestic" limits.

Element	Maximum Domestic Sludge	Percentage Exceeding
Zn (ppm)	2500	21
Cu (ppm)	1000	37
Ni (ppm)	200	14
Pb (ppm)	1000	16
Cd (ppm)	25	23
Cd/Zn (%)	1.0	44
Cd/Zn (%)	1.5	26
Any [a]	–	56
Any [b]	–	44

[a] Sludge which exceeded any of the Zn, Cd, Cu, Ni or Pb concentration limits or 1.5% Cd/Zn.

[b] Sludge which exceeded 150% of Zn, Cd, Cu, Ni or Pb concentration limits or 1.5% Cd/Zn.

Table 5.   Composition of plating shop wastewater, raw and digested sludge grab samples at city 9; city officials began to enforce pretreatment in July 1975.

Element	Plating Shop Wastewater (mg/1)		Raw Sludge (mg/kg dry sludge)		Digested Sludge (mg/kg dry sludge)			
	7/9/75	4/76	7/23/75	8/6/75	4-5/75	7/75	11/75	4/76
Zn	880	4.2	4020	7340	5470	5060	3650	3030
Cd	29.6	0.034	60	1350	190	178	90	50
Cu	17.2	0.095	1070	1450	1500	1590	1440	1370
Ni	138.0	2.08	402	935	1120	1080	460	410
Pb	9.2	0.071	345	410	340	360	310	305
Cr	1.85	–	1040	2660	–	2590	1030	–

marized in Table 7. Sludge Cd varied little contrary to reports of others. Expression on a liquid basis would show greater variability because sludge percent solids vary more than sludge metals.

Salmonellae and *ascaris* ova present in the sludges ready for use on land were measured by Dr. W. D. Burge of our laboratory and will be reported separately. *Salmonellae* were present in some sludges, and *ascaris* ova were present in most of the sludges.

Effect of Sludge Use on Soil Composition

Analysis of soils for heavy metals is a powerful tool in locating where sludge has been land-applied because added metals remain in the surface soil. This information allows one to characterize the uniformity of the application process. The general results for DTPA-extractable Zn, Cd, Cu and Ni on a number of cities

Table 6.   Composition of sludges sampled at city 25 where CdS is discharged to the sewer.

Sample	Zn	Cd	Cu	Ni	Pb	$\frac{Cd}{Zn}$	Date	Type
			(mg/kg dry sludge)					
F-0360	1120	730	1210	70	270	0.65	6/75	Sand Bed
F-0359	1520	1230	1160	100	270	0.81	6/75	2° Digested
F-1310	1090	640	960	78	205	0.59	8/75	Sand Bed 2
F-1311	1200	720	1100	88	210	0.60	8/75	Sand Bed 3
F-1308	685	1820	740	63	130	2.66	4/76	Raw
F-1309	1300	1520	1420	105	280	1.17	4/76	2° Digested

Table 7. Variability of sludge composition at a suburban treatment plant, 23B.[a]

	Minimum	Maximum	Mean	Median
Solids (%)	3.65	5.93	4.99	5.26
Ash (%)	52.6	62.0	59.1	60.1
Zn (ppm)	680	1022.0	891.0	915.0
Cu (ppm)	230	394.0	321.0	340.0
Ni (ppm)	19	31.0	26.0	25.0
Cd (ppm)	7.72	8.65	8.23	8.22
Pb (ppm)	210.0	430.0	325.0	327.0
Fe (%)	9.1	12.8	11.6	12.3
Cd/Zn (%)	0.80	1.19	0.94	0.90
Zn (eq)*[b]	1020	1710	1440	1510

[a] Based on 12 samples collected in August through December 1975.

[b] Zn (equivalent)* = 1 x ppm Zn + 2 x ppm Cu + 4 x ppm Ni – 200.

and farms are shown in Table 8. Analyses of the specific soils sampled in conjunction with plant samples are shown in Tables 9, 10 and 11. Analyses of the plot area soils are shown in Table 12. It was beyond the scope of this chapter to describe each field sampled at each city. Up to 150 soil samples have been taken from some fields in the general search, in the plot location search and in the sampling of farmer-grown crops.

The kinds of equipment used and attitudes of the persons who spread sludge were both factors in how uniform the sludge was spread. "Disposal" attitudes appeared to cause nonuniform application practices with any type of equipment; entire loads were discharged without moving the trucks. High loading spots occurred where the driver opened a valve at the rear of the truck and slowly returned to the driver's seat. Some cities had arranged innovative equipment to control opening and closing the valve from the driver's seat or door. Spreader bars were arranged at several cities to allow even spreading, while others spread the sludge fairly evenly with splash plates; each of these practices requires that the truck drive everywhere it spreads. Many of the cities in Pennsylvania had side-delivery tank trucks. With one type of nozzle a path about 3–5 ft wide was covered, while with another type of nozzle a path 30 ft wide was covered. These trucks could spread from all-weather roads if needed. Where this practice occurred, sludge metals were much higher beside the road than anywhere else in the field. More sludge was often applied nearer to field entrances than further from the field entrances.

Table 8. Microelement content of long-term sludge treated and control soils at selected cities.[a]

| City | Farm | Sludged | pH | DTPA-Extractable Metals (ppm) | | | | |
				Zn	Cd	Cu	Ni	
1	1A	1962–76	5.1	91	2.73	43	2.0	edge near road
	1A	1962–76	5.7	38.0	1.39	15	1.0	interior
	1C	1962–76	6.2	29.6	1.41	10.7	0.5	
	2A	1967–75	4.5	30.2	1.19	7.7	1.6	edge of field
	2A	1967–75	5.0	21.9	0.96	6.9	1.1	interior
	2A	1967–75	5.5	7.5	0.16	3.0	0.5	control
	2C	1967–75	6.9	22.4	1.23	11.6	0.6	garden
	3A	1965–76	5.8	15.0	0.68	6.8	2.2	hay strip
	3A	1965–76	5.7	11.1	0.56	6.8	1.2	wheat strip
	3A	1965–76	6.5	7.6	0.15	2.7	0.4	fence row
	4A	1965–76	6.6	1.3	0.87	8.0	0.7	edge of field
	4A	1965–76	5.9	74.0	3.02	30	1.0	interior of field
	4A	1965–76	5.8	5.3	0.10	1.4	0.4	corner near house
2	1	1959–68	5.7	95.0	4.6	26.0	3.2	
	1	–	6.1	7.0	0.16	2.0	0.2	
5	1	1966–73	6.0	88	1.27	50	1.3	edge
	1	1966–73	6.1	24	0.57	26	0.3	interior
	2	–	6.2	7	0.12	4	0.5	
6,7	1	1967–75	6.5	26	0.76	36	0.8	Area A
	1	1967–75	6.7	76	0.76	100	1.6	Area B
8,11,12,13	1A	1968–75	6.6	92	9.0	18	8.8	near road
8,11,12	1B	1968–75	6.6	17	0.87	6	0.9	interior
	1C	–	6.7	4	0.09	2	0.3	
19	1A	1960–76	7.1	19.1	0.19	10.6	0.4	near edge 1
	1A	1960–76	5.85	63.2	0.42	27.8	1.4	on edge 1
	2A	1960–76	6.60	40.9	0.35	14.5	0.5	on edge 2
	2A	1960–76	5.70	84.0	0.61	37.7	1.6	near edge 2
21	1A	1960–75	5.7	14.2	0.26	6.8	0.5	edge of field
	1A	1960–75	7.3	6.5	0.16	4.8	0.2	interior
	1B	1970–75	6.1	30.2	0.56	18.0	0.6	disposal field-leased
	1C	–	5.6	6.6	0.09	1.0	0.6	
25	1A	1964–75	5.0	13.8	7.2	23.9	1.8	corn
	1A	–	3.8	3.7	0.12	3.2	0.2	fence row
	2A	1964–75	5.0	19.8	27.0	23.0	1.2	beside road
	2A	1964–75	5.2	10.2	10.9	10.9	0.9	near road
	2A	1964–75	5.3	5.7	4.6	6.7	0.6	interior
	2B	–	3.8	2.7	0.08	1.0	0.2	fence row
	3	1970,1974	4.9	16.3	17.1	22.0	3.9	pasture

[a] See soil data for cities 4, 9, 13, 21 and 23 in Tables 9, 10, 11, 12 and 13.

Farmers left some strips in hay to provide all-season application sites; these strips were often more heavily spread than strips where row crops were grown yearly. Well-drained fields and areas were more heavily spread than those slowly drained. At city 21 a field was rented to allow land application year around.

Table 9. Microelement content of oat grain, red clover, and associated soils (0–15 cm) at city 13.

Location	Soil pH	DTPA-Extractable				Oat Grain, 1975				
		Zn	Cd	Cu	Ni	Zn	Cd	Pb	Cu	Ni
		(ppm, air dry soil)				(ppm dry crop)				
F–0416	6.9	8.0	1.36	4.6	0.8	30.3	0.22	0.71	4.0	2.1
F–0418	6.6	6.8	1.06	5.1	0.9	32.8	0.10	0.61	4.2	1.5
F–0420	6.4	61.8	8.4	25.3	4.1	49.8	2.13	0.56	4.7	7.5
F–0422	6.2	36.6	5.8	15.6	3.0	46.0	1.32	0.56	4.3	5.6
F–0424	6.1	25.8	3.8	10.1	2.0	45.6	1.68	0.50	4.8	5.8
		DTPA-Extractable				Red Clover, 1976				
F–1681	6.6	62.8	8.9	29.0	4.2	43.4	0.88	2.3	7.7	2.2
F–1682	6.6	30.1	4.5	10.8	2.3	57.6	0.92	2.4	10.8	1.6
F–1683	6.1	24.6	3.4	8.6	2.2	45.5	0.60	1.6	7.9	1.8
F–1684	5.9	24.6	3.0	7.4	2.2	50.6	0.66	2.6	7.2	2.3

Field B had higher accumulated metals in 6 years than Field A after 15 years of sludge use at fertilizer rates.

The presence of greater fluid sludge storage capacity at the treatment plant led to less reliance on road-side spraying. Several plants could store up to 3 months' sludge production. The storage lagoon at city 20 was helpful in matching spreading needs to weather demands. Larger treatment plants applying fluid sludge usually needed to apply sludge every day the

Table 10. Zn and Cd in soil and corn grain in city 1.

Location	Soil Total		pH	Soil DTPA		Corn Grain	
	Zn	Cd		Zn	Cd	Zn	Cd
	(ppm)			(ppm)		(ppm)	
F–0555 [a]	147	6.7	5.3	43	2.15	34	0.42
F–0557	124	4.9	5.5	37	1.67	21	0.21
F–0558	238	10.8	5.2	74	2.92	26	0.31
F–0565 [a,b]	134	5.6	5.1	37	1.68	23	0.72
F–0566	156	6.9	5.5	44	2.03	24	0.75
F–0567	125	5.9	5.2	39	2.08	34	0.85
F–0568	—	9.1	6.1	34	2.42	23	0.50
F–0569	77	3.1	5.2	22	1.18	23	0.39
F–0570	90	3.9	5.1	27	1.32	21	0.62
F–0577 [a,c]	156	5.7	7.4	33	1.67	21	0.062
F–0578	75	2.3	6.9	14	0.70	17	0.048
F–0579	131	9.0	6.1	33	2.94	20	0.64

[a] Liquid sludge applied since 1965.

[b] Limed filter cake sludge applied 1975.

[c] Windblown limestone dust.

Table 11. Microelements in soils and crops at city 23 where high lime filter cake sludge was utilized at 200–400 Dry T/A.[a]

		pH	Zn	Cd	Pb	Cu	Ni	Mn	Fe
				(ppm dry weight)					
Soil	Sludged Total	7.8	128.0	1.01	42.0	45.6	12.4	339	–
	Control	5.3	28.0	0.08	10.4	3.1	4.8	186	–
Soil	Sludged DTPA	7.8	19.3	0.28	5.5	15.8	0.44	15	174
	Control	5.3	2.3	0.07	3.2	0.9	0.41	20	81
Grass	Sludged	7.8	61	0.24	3.5	7.7	1.0	116	410
	Control fertilized	6.2	23	0.18	6.1	4.7	1.2	87	112
	Control unfertilized	6.1	13	0.11	4.6	4.1	0.18	110	194

[a] Estimated from sludge and soil analysis and known 14-in. application depth.

weather permitted. Some plants shifted to filter cake production part of the year because they had been unable to make the arrangements to apply fluid sludge. The spreading equipment used, the ability of sludge-spreading personnel to use soil maps, the cooperativeness that farmers displayed in allowing some compaction in their fields in order to receive free fertilizer, and the political difficulty of constructing storage capacity were all important factors in the shift to dewatering. Land use of dewatered sludges generally led to uneven application. At one city, sludge cake is dumped in piles in the field when it is produced, and later spread using front-end loaders. At another, the farmer spreads the sludge cake with his manure spreader (sometimes intermixed with manure from a common storage area).

Table 12. Microelement content of soybean and corn grain and associated soils (0–15 cm) at city 21.

Location	Soil pH	DTPA-Extractable				Soybean Grain, 1975				
		Zn	Cd	Cu	Ni	Zn	Cd	Pb	Cu	Ni
		(ppm, air dry soil)				(ppm, dry crop)				
F–0749	6.4	8.8	0.22	5.3	0.8	48	0.30	0.50	12.0	1.8
F–0750	5.5	13.6	0.33	8.1	1.2	67	0.19	0.84	14.3	2.4
F–0751	5.9	9.6	0.23	5.2	0.6	56	0.23	0.74	12.8	2.8
F–0752	6.2	5.9	0.17	4.2	0.5	51	0.23	0.60	13.9	2.0
F–0753	6.3	3.7	0.11	2.3	0.4	45	0.18	0.89	11.8	1.5
F–0754	6.4	3.9	0.11	3.0	0.5	44	0.37	0.57	12.7	1.4
		DTPA-Extractable				Corn Grain, 1975				
F–0758	5.2	7.1	0.13	3.8	1.2	19	0.23	0.87	1.8	0.7
F–0759	5.0	6.4	0.17	4.1	0.9	18	0.56	0.78	2.0	0.4
F–0760	5.0	10.2	0.22	7.1	0.8	16	0.43	0.77	1.7	0.4

Extreme variability of soil pH was found among fields where sludge was applied. Fields at or near pH were common, while several fields had become calcareous from high cumulative applications of limed filter cake. Within fields, the pH often varied in a manner related to sludge-spreading practice; a general pH of 4.5 occurred along the all-weather road at city 4, farm 2. Soil pH was dependent mostly on the farmers attitudes and training.

In a few cases, data are shown for both total and DTPA-extractable metals. DTPA usually extracted half of the sludge-applied metals. However, the percent extractability of added Zn was lower in fields with consistently high soil pH. No clear pattern of pH or time after application effect on Cu or Cd extractability was noted, perhaps because of soil or sludge effects on this equilibrium.

Table 12 shows that available soil P was increased in long-term sludge use sites. Organic matter and C.E.C. appeared to have increased on some sites where sludge was continually applied at or above fertilizer rates. Available K and Mg were unrelated to sludge spreading unless the farmer chose to apply nothing but sludge. Sludge did not correct B deficiency in alfalfa at city 18.

Effect of Sludge Use on Crop Composition

As described in the introduction, crop uptake of metals added with sludge is both very important and very complicated. The results in Tables 9, 10, 11, 12 and 14 show the effects of crop differences, soil pH, and other soil factors on crop uptake of metals. The data reported emphasize plant Cd uptake on long-term sludge use sites.

Table 9 shows the Cd content of oat grain and red clover at city 13. Several locations were extremely high in DTPA-Cd compared to control soils (Table 13). Oat grain Cd reached 2.1 ppm; Kjellstrom et al. (34) reported that normal oat grain contains only 0.01 ppm Cd. On the same general area of the field, red clover (whole plants) growing the next spring was much lower in Cd than the oats. One would have expected a grain crop to be much lower in Cd than a foliar crop based on the plot studies of Bingham et al. (13,30).

Table 10 reports the Zn and Cd content of corn grain at city 1. These results generally corroborate those of Jones et al. (16) except that pH had a greater effect at city 1. Corn grain grown on the calcareous area in field 3 had a much lower grain Cd content. The recent application of sludge filter cake caused higher Cd than the repeated application of fluid sludge. Over all loca-

Table 13.   Composition of soils in field plots before treatments were applied.[a]

City	Sludged		pH	C.E.C		O.M. (%)	P (ppm)	DTPA-Extractable				Pb
				meq/100 g				Zn	Cd	Cu	Ni	
								(ppm, air dry soil)				
4	—	Hagerstown	5.7	9		3.0	77	6.4	0.14	2.3	0.3	7.1
4	1962-75	sil	5.2	13		4.9	176	42.3	0.55	22.4	0.9	6.2
9	—	Lansdale	5.3	8		2.5	20	3.1	0.12	1.8	0.8	4.8
9	1961-73		5.1	8		2.8	98	17.3	1.36	6.6	1.8	4.8
13	—	Readington	5.3	9.5		2.6	52	4.2	0.10	2.1	0.6	5.0
13	1967-74	sil	5.6	11		3.8	124	47.0	6.30	16.3	3.6	5.8
1	—	Lansdale	5.6	—		—	—	3.7	0.12	1.2	0.6	—
1	1967-75	loam	5.4	—		—	—	53.9	2.20	20.4	1.2	—
1	1967-75		6.5	—		—	—	40.9	1.88	16.3	0.6	—
19	—	Genesee	6.3	—		—	—	7.0	0.20	2.3	1.0	—
19	1960-76	sil	6.6	—		—	—	23.0	0.40	18.2	0.8	—
39	—	Hagerstown	6.6	—		—	—	1.6	0.05	0.8	—	—
39	1960-71		6.7	—		—	—	70.3	3.51	86.5	0.7	—

[a] Plots established at cities 4, 9 and 13 in 1975; plots established at cities 1, 19 and 39 in 1976; limestone applied to appropriate plots at cities 4, 9, 13 and 1.

tions, DTPA-Cd was a poor predictor of corn grain Cd because it does not discriminate between previously and recently applied Cd and is insensitive to soil pH. Corn grain is normally about 0.03–0.10 ppm Cd; the use of a nonrecommended sludge substantially increased the Cd content of this crop.

Table 11 reports on a site where a huge amount of low-metal sludges were applied. The soil pH was high. Tall fescue was somewhat higher in Zn and Cu than control fescue, but Cd and Pb were not really different. The crop Cd is a little higher than control grasses at some other locations (0.03–0.15 ppm).

Table 12 shows a complicated situation. The cumulative sludge application is not very high, and the soybean field is near pH 6 while the corn field is near pH 5. The corn grain was substantially enriched above normal Cd levels by use of a domestic sludge; however, the soil pH is so low that a matched pH control crop would be needed to make any reasonable comparison. Numerous pH 5 sludged fields grow corn grain and silage in the northeast.

Crop Pb and Cu were usually unaffected by long-term sludge utilization. Crop Mn was usually influenced by only soil pH; the fescue Mn was considerably higher than expected at city 23 (Table 11). Foliar Zn was usually increased in every crop grown on sludged soils. Levels of Zn generally described as potentially phytotoxic (> 500 ppm) were observed only in chard.

Table 14 shows the Cd results for the chard and soybean tissues grown on the established field plots in 1975. The chard Cd results at city 4 show no effect of sludge-applied Cd. At city 13, chard Cd was increased at the low soil pH, but not at the high pH. At city 9, chard Cd (73 ppm) was much higher than expected at the low soil pH, but dropped to 5.5 ppm at the high soil pH. The chard bore no symptoms; chard yield on the low pH plot was about one-third the yield on the high pH plot. The control chard at city 9 was already considerably higher than the control chard at the other cities.

Soybean leaves and grain were substantially increased in Cd at cities 9 and 13. Again, there were differences among cities. At city 9, raising the soil pH lowered grain Cd about 60 percent, while at city 13, raising the soil pH lowered grain Cd about 75 percent. Soybean grain normally contains $\leqslant$ 0.2 ppm Cd; soybean grain grown on field plots at Beltsville containing 105 T of a domestic sludge/acre are only 0.2 ppm Cd.

Thus, the high Cd and high Cd/Zn sludges used at cities 9 and 13 led to substantial increases in soybean Cd. At city 9, sludges had not been applied for nearly 3 years at harvest time, yet soybean grain contained 3.7 ppm Cd at the low pH and 1.5 pp, at the high pH. Soybeans grown on high Cd/Zn sludge-amended plots at Beltsville also have high Cd content (up to 1.7 ppm), while domestic sludges have not led to high soybean grain Cd.

Thus the crops grown on long-term sludge use sites show (a) soil pH significantly affects crop contents of Cd; (b) crops differ widely in Cd uptake, in the relative transport of Cd to grain, and in the influence of pH on Cd uptake; (c) Cd remained avail-

Table 14.   Cd content of crop tissues grown on the field plots in 1975.

City	DTPA Cd	pH	Chard Leaf	Soybean Leaf (ppm, dry crop)	Grain
4 Control	0.13	5.7	0.6	0.27	0.17
4 Control	0.14	6.7	0.5	0.26	0.15
4 Sludged	0.53	5.2	1.9	—	—
4 Sludged	0.57	6.2	0.6	—	—
9 Control	0.13	5.3	3.6	1.04	0.36
9 Control	0.10	6.7	1.2	0.55	0.28
9 Sludged	1.13	4.8	73.0	10.7	3.70
9 Sludged	1.19	6.6	5.5	1.87	1.51
13 Control	0.09	5.3	0.9	0.24	0.16
13 Control	0.10	6.4	0.5	0.17	0.13
13 Sludged	7.15	5.6	7.0	5.70	2.64
13 Sludged	5.45	6.6	1.8	2.38	0.65

able to crops even though sludge application had ceased; (d) soil characteristics other than pH influence Cd uptake; and (e) high Cd and Cd/Zn sludges produced higher crop Cd levels than do domestic sludges, even when both are applied at fertilizer rates.

Adherence of Spray-Applied Sludge to Tall Fescue

As described earlier, application of fluid sludges to pastures and hay fields is an important component of most all-season, all-weather fluid sludge-only systems. When fluid sludges are spray-applied to establish forage crops, the sludge solids could adhere to the crop. Because several of the microelements (Cu, Pb, Hg, Cr, Fe) in sludge are translocated to the tops of forage crops in very limited amounts, sludge contamination of the forage could increase these metals to levels not attainable when sludge is only mixed with the soil.

Chaney, Lloyd and Simon (unpublished) conducted an experiment to test the adherence of fluid sludge from treatment plant 23-B to tall fescue. Tall fescue is currently the recommended grass for grazing pastures in the states we studied because of its agronomic qualities. Thus sludge and crop are currently being studied to determine the efficiency and safety of sludge use on grazed pastures in a cooperative study of the Washington Suburban Sanitary Commission, the University of Maryland Departments of Agronomy and Veterinary Sciences, and our laboratory at Beltsville.

Part of the results from the sludge adherence study are shown in Table 15. The percentage of sludge in the harvested forage was calculated from the increased levels of Zn, Cu, Cd, Pb and Fe in contaminated fescue *vs.* untreated fescue. Sludge comprised as high as 32 percent of the harvestable forage; rainfall was ineffective in removing the adhering sludge, but subsequent growth diluted the sludge content of the standing crop. The results of this study were corroborated by weekly field measurements of the sludge content in 4-week rotation paddocks: in treatment 1, sludge was applied after mowing, then the crop grew 3 weeks before the cattle rotated onto treatment 1; in treatment 2, the sludge was applied after the 3-week growth period, the day before the cattle rotated into treatment 2. The apparent percent sludge on the first day of grazing is about 5 percent for treatment 1 and 15 percent for treatment 2.

Whether sludge would adhere to other forage crop species as much or as long as it did to tall fescue, or whether other fluid sludges would adhere as much as that of treatment plant 23-B

Table 15. Calculated sludge content on tall fescue forage after fluid sludge application.[a]

Treatment (cm)	0	Days After Application 14 (% Sludge In/On Forage)	80
Unmowed			
0.51	26.2	15.9	5.4
1.03	31.5	21.8	5.5
Mowed			
0.51	22.7	16.5	5.7
1.03	31.5	22.5	2.4
Rainfall (cm)	0	27	41

[a] Fluid sludge from treatment plant 23-B was spray-applied on September 16, 1975. Sludge contained 5.93% solids and 901 mg Zn, 342 mg Cu, 8.41 mg Cd, 323 mg Pb, and 111 g Fe/kg dry solids. The 0.51-cm application (5,500 gal/acre) supplied 60 lb available N/A and 1.13 dry T solids/acre. The mowed fescue was about 7 cm high and the unmowed was about 35 cm high.

remains unclear. This sludge is quite high in Fe because of Fe used in phosphate removal. Other crops have different surface characteristics and shapes. Because surface contamination can increase forage levels of some microelements and persistent organic compounds far above levels normally attainable, sludge use on pastures may require special management or equipment (as injection into sod). Cows and sludge trucks have been observed in the same field of mature forage. The recent observation of PCB in milk above FDA tolerances from sludge use on pasture at Bloomington, Indiana, points out the risks other than heavy metals. Whether sludge-borne metals and other compounds adhering to forage crops are unsafe can only be settled by grazing trials such as those in progress in Beltsville.

DISCUSSION AND SUMMARY

The studies reported in this paper were started because the information obtainable in studies of long-term sludge use farms appeared to be too valuable to neglect. The general areas of information to be studied were: (a) the role and actions of each party in sludge use; (b) the long-term plant availability of sludge-applied heavy metals in practice; and (c) the research, regulatory and advisory needs to allow safe, beneficial landspreading of sewage sludge as an agricultural resource. Information on the long-term availability of heavy metals is needed to allow design of sludge use systems, to form a scientific basis for regulatory guidelines, and to rationally prioritize research funding. Currently, all these are progressing on a very restricted information base since too few sludges, soils, pH values, crops, and climates have been studied.

There are some limitations in the interpretability of the data gained in studies of these old sites. In contrast to new plots where a weighed amount of a material which can be analyzed is applied on a known date, the exact composition and yearly amount of sludge applications at old sites cannot be learned. Although in-depth interviews with farmers, sludge truck drivers, and plant superintendents were conducted, the limitations in their record keeping, knowledge of application rate, frequency of application, specific location of application, and the lack of previous sludge analyses or samples became limitations in the research information.

However, the intensive program of sampling the analysis of sludges and soils provided an objective baseline to evaluate verbal accounts and the few written records. A pattern of sludge use should have a particular increase in soil Zn, Cd, Cu or the specific metal in question. Usually, all parties agree as to when sludge use on a particular farm or field started or stopped. If the accumulated total metal content of the treated soil agrees well with the reported histories and the analyses, then the plant availability results should have a meaning. Usually, more precise histories were obtained after the soil and sludge samples of the first site visit were analyzed. These built-in limitations were not so great that one could ignore the plant uptake results. If the information about sludge composition and frequency, rate and location of sludge use did not agree fairly well with the objective soil metal results, the site received much less consideration for the field plot program.

What has this research program found to date?

Sludge Metals

1. Although sludge metal contents vary widely, many sludges are free of abatable metals.

2. Information about high metal levels in sludges was frequently enough for treatment plant personnel to identify specific polluters.

3. Simple source-searching strategies can identify major contributors of metals.

4. Digested sludge is a good tool for monitoring metal pollution among cities although it seldom can identify low-level metal discharges.

5. Adequate sludge analysis is a serious deficiency of nearly every sludge use program.

6. Once informed of the evidence about presence of excessive sludge levels of a metal, most treatment plant superintendents got in touch with suspected contributors to help work out source controls.

7. In the absence of a national program to control metal release to the sewer, many sources will threaten to leave town in order to avoid local enforcement of pretreatment.

Sludge Spreading Practice

1. Generally, sludge is spread at rates and in ways that are convenient for the treatment plant personnel.

2. Until very recently treatment plant personnel and farmers could not obtain meaningful information about how much sludge was needed to fertilize a particular crop.

3. Uneven application rates, application near all-weather roads and strip crop farming lead to very uneven and unpredictable levels of soil metals.

4. Because sludge composition, rate, frequency of application, cropping pattern, and fertilizer and lime practices affect soil pH, soil pH values from 4.5–7.8 were found. Proper sampling of soils and subsequent addition of limestone to raise the pH to 6.5 before any sludge application is permitted would obviate the low Ph situations.

5. The lack of inexpensive all-weather sludge-spreading equipment has contributed to uneven applications.

6. The absence of sludge analysis for fertilizer elements and adequate advisory information to calculate sludge application rate from sludge composition and crop requirements and soil test information have contributed to both inadequate and excessive sludge applications.

7. In the absence of appropriate advisory and regulatory information, very high soil Cd (27 ppm DTPA-extractable Cd) concentrations can coincide with very low soil pH (5.0).

8. Currently, any sludge is being applied to any soil to grow any crop; no restrictions other than odor have been consistently imposed.

9. Utilization of domestic sludges at fertilizer rates leads to a slow build-up of metals in the soil.

Crop Availability of Metals

1. Crops were not influenced by additions of sludge Pb.

2. With the exception of chard, crops were not higher in Cu from sludge use; a large proportion of added Cu remained DTPA-extractable.

3. Where sludges high in Ni were used, plant levels of Ni were increased only at low soil pH; soybean seed accumulates Ni where Ni is available for crop uptake.

4. Zn remained available to plants over many years of sludge use and for many years after sludge use stopped. Crops differed widely in Zn uptake from the same sludge-treated soil; grain Zn levels increased less than foliar Zn. Zinc phytotoxicity was not indicated except for chard. Maintaining a high soil pH appears to reduce DTPA-extractability of sludge-applied Zn, but only at a higher pH than is ordinarily maintained in the Northeast (pH 7).

5. Cd addition to Northeastern soils from sludge use drastically increased available soil Cd; control levels were ordinarily 0.05–0.15 ppm DTPA-Cd while sludge use at fertilizer rates increased DTPA-Cd to 0.5–25 ppm. *Sludge-applied Cd remained available to crops* even after sludge use ceased; soybean grain Cd of 3.7 and 1.5 ppm, and chard Cd of 73 and 5.5 (low and high pH, respectively) were ovserved at city 9 where sludge application had ceased at least two crop seasons before the plots were installed. Research work has suggested strongly that increases in available Cd (DTPA or 0.1 N HCl extractable) from any Cd application will increase Cd levels in crops; air pollution, P-fertilizer use, and use of both polluted and domestic sludges each increase DTPA-extractable Cd in soils. Use of domestic sludges even at high cumulative rates had little influence on crop Cd levels while use of high Cd and high Cd/Zn sludge even at low rates increased crop Cd levels. Maintaining high soil pH reduced crop-available Cd only somewhat while it reduced crop-available Zn considerably; high soil pH appeared to reduce transport of Cd to corn grain.

6. Crops differed in Cd uptake. Although the low transport of available Cd to corn grain was confirmed, oat was observed to reach high grain Cd levels. At city 13, oat grain contained up to 2.1 ppm Cd while red clover hay contained only 0.8 ppm Cd. Chard reached extremely high leaf Cd levels with no symptoms; although the Cd content of chard at city 9 dropped from 73 to 5.5 ppm upon appropriate limestone addition. Soy leaves and grain, and grass Cd levels decreased to only one-half or one-third of these crops grown on the unlimed plots.

Agronomic Aspects of Sludge Use

1. No phytotoxicity was observed in crops ordinarily grown on farms where sludge is used.

2. Low soil pH was commonly observed; although pH 4.5 was

observed, the most common range was 5.0–5.7 which is not unusual for the Northeast.

3. Even application of sludge was occasional and then only on one strip or one small field. Customarily sludge application was uneven with very heavy accumulation near all-weather roads.

4. In the absence of regulatory controls or advisory information, any sludge was applied at uncalculated rates for growth of any crop.

5. Both farmers and treatment plant personnel desire the information, training and equipment to use sludge as an agricultural resource.

Regulatory and Advisory Needs Indicated

1. Farmers and treatment plant personnel are unable to calculate sludge N which will become available to crops. Research must define the mineralization of sludge N, and define the N loss from volatilization of both the NH_4-N in the fluid sludge and the NH_4-N formed on the soil surface. Does sludge NH_4-N applied to pasture crops, cover crops, and the like volatilize as much as when applied to bare soil?

2. Farmers need special agronomic information to interpret routine soil tests and determine lime and fertilizer needs when they use sludge. The sludge use advisory network could involve County Agricultural Extension Staff, sludge utilization coordinators employed by the treatment plants, or officials in state health, environmental, or agricultural agencies.

3. Metal and nutrient analyses of sludges must be provided to sludge users. Most small treatment plants cannot afford the equipment and trained personnel needed to make the metal analyses; however, they could determine total N and NH_4-N, percent solids, volatile solids, and phosphorus contents. Some public unit should provide the needed sludge analyses, and/or test the reliability of treatment plant analyses. EPA has not defined an approved method of sludge metal analysis, nor provided sludge standard materials.

4. The real world "ordinary practice" of pH management, crop selection, and uniformity of sludge spreading must be accounted for in regulatory controls on sludge use. Low soil pH will likely occur at some time after sludge has been used. Nongrain crops will likely be grown. Crops other than corn will likely be grown. Farmers can raise high Cd crops easily using normal agronomic practices if sludges containing abatable Cd are used on privately owned farmland. Exclusion of high Cd and high Cd/Zn sludges from private farmland would avoid most of this potential hazard to the food-chain.

Although this report shows a number of situations where high crop Cd levels have been reached, it seems clear that (a) proper pH management could alleviate the current problems, and (b) proper restrictions on acceptable sludges (Cd $\leqslant$ 1.5 percent of Zn; Cd $\leqslant$ 25 ppm) coupled with reasonable soil pH management could prevent these situations from developing again. Many educational needs of farmers must be satisfied so that they can choose safe sludges and optimal management practices for safe and beneficial sludge use.

ACKNOWLEDGMENTS

The authors gratefully acknowledge the kind assistance and cooperation of all treatment plant personnel and farmers who allowed us to study their practices and use their land, and informed us of past histories of sludge application in confidence; the financial and scientific support for this research program through Interagency Agreements with the Food and Drug Administration and the Environmental Protection Agency; the laboratory and field technical support of Jacqueline Janzegers, Cheryl Lloyd, Robert Cowherd, Michael White, and Milagros Morella.

This chapter is a contribution from the U.S. Department of Agriculture, Agricultural Environmental Quality Institute, Agricultural Research Service, Biological Waste Management and Soil Nitrogen Laboratory, Beltsville, Maryland 20705, and the Maryland Environmental Service, Annapolis, Maryland 21401.

REFERENCES

1. Epstein, E., G. B. Willson, W. D. Burge, D. C. Mullen and N. K. Enkiri. "A forced aeration system for composting wastewater sludge, *J. Water Pollution Control Fed.* 48:688–694 (1976).
2. Page, A. L. "Fate and effects of trace elements in sewage sludge when applied to agricultural lands. A literature review study," U.S. Environ. Prot. Agency Rept. No. EPA-670/2-74-005, 108 pp. (1974).
3. Chaney, R. L. "Crop and food chain effects of toxic elements in sludges and effluents," In: *Recycling Municipal Sludges and Effluents on Land,* Nat. Assoc. St. Univ. and Land-Grant Coll., Washington, D.C., pp. 129–141 (1973).
4. Chaney, R. L. and P. M. Giordano. "Microelements as related to plant deficiencies and toxicities," In: *Soils for Management and Utilization of Organic Wastes and Wastewaters,* L. F. Elliott and F. J. Stevenson, Eds. Soil Sci. Soc. Amer., Inc., Madison, Wisconsin (in Press, 1976).
5. Braude, G. L., C. F. Jelinek and P. Corneliussen. "FDA's overview of the potential health hazards associated with the land application of municipal wastewater sludges," *Proc. 2nd Natl. Conf. Municipal Sludge Management,* Information Transfer, Inc., Rockville, Maryland, pp. 214–217 (1975).

6. Webber, J. "Effects of toxic metals in sewage on crops," *Water Poll. Contr.* 71:404–413 (1972).

7. Leeper, G. W. "Reactions of heavy metals with soil with special regard to their application in sewage wastes," Dept. of Army, Corps of Engineers, under contract No. DACW 73–73–C–0026, 70 pp. (1972).

8. Fulkerson, W. and H. E. Coeller, Eds. "Cadmium, the dissipated element," ORNL Rept. No. ORNL–NSF–EP–21, Oak Ridge National Laboratory, Oak Ridge, Tennessee, 473 pp. (1973).

9. Cunningham, J. D., D. R. Keeney and J. A. Ryan. "Yield and metal composition of corn and rye grown on sewage sludge-amended soil," *J. Environ. Quality* 4:449–454 (1975).

10. John, M. K., C. J. VanLaerhoven and H. H. Chuch. "Factors affecting plant uptake and phytotoxicity of cadmium added to soils," *Environ. Sci. Technol.* 6:1005–1009 (1972).

11. Andersson, A. and K. O. Nilsson. "Influence of lime and soil pH on Cd availability to plants," *Ambio.* 3:198–200 (1974).

12. Chaney, R. L., M. C. White and P. W. Simon. "Plant uptake of heavy metals from sewage sludge applied to land," In: *Proc. 2nd Nat. Conf. Municipal Sludge Management*, Information Transfer, Inc., Rockville, Maryland. pp. 169–178 (1975).

13. Bingham, F. T., A. L. Page, R. J. Mahler and T. J. Ganje. "Yield and Cadmium accumulation of forage species in relation to cadmium content of sludge-amended soil," *J. Environ. Quality* 5:57–60 (1976).

14. Dowdy, R. H. and W. E. Larson. "The availability of sludge-borne metals to various vegetable crops," *J. Environ. Quality* 4:278–282 (1975).

15. Linnman, L., A. Andersson, K. O. Nilsson, B. Lind, T. Kjellstrom and L. Friberg. "Cadmium uptake by wheat from sewage sludge used as a plant nutrient source," *Arch. Environ. Health.* 27:45–47 (1973).

16. Jones, R. L., T. D. Hinesly, E. L. Ziegler and J. J. Tyler. "Cadmium and zinc contents of corn leaf and grain produced by sludge-amended soil," *J. Environ. Quality.* 4:509–514 (1975).

17. Cunningham, J. D., D. R. Keeney and J. A. Ryan. "Phytotoxicity and uptake of metals added to soils as inorganic salts or in sewage sludge," *J. Environ. Quality* 4:460–462 (1975).

18. Jones, R. L., T. D. Hinesly and E. L. Ziegler. "Cadmium content of soybeans grown in sewage-sludge amended soil," *J. Environ. Quality* 2:351–353 (1973).

19. Baker, E. D., R. E. Eshelman and R. M. Leach. "Cadmium in sludge potentially harmful when applied to crops," *Science in Agriculture* 22(4):14–15 (1975).

20. Evans, J. O. "Using sewage sludge on farmland," *Compost Sci.* 9(2):16–17 (1968).

21. Manson, R. . and C. A. Merritt. "Land application of liquid municipal wastewater sludges," *J. Water Pollution Control Fed.* 47:20–29 (1975).

22. Manson, R. J. and C. A. Merritt. "Farming and municipal sludge: They're compatible," *Compost. Sci.* 16:16–19 (1975).

23. Rohde, G. "The effects of trace elements on the exhaustion of sewage-irrigated land," *J. Inst. Sew. Purif.* 1962:581–585 (1962).

24. LeRiche, H. H. "Metal contamination of soil in the Woburn Market-Garden experiment resulting from the application of sewage sludge," *J. Agr. Sci.* 71:205–208 (1968).

25. Andersson, A. and K. O. Nilsson. "Enrichment of trace elements from sewage sludge fertilizer in soils and plants," *Ambio.* 1:176–179 (1972).

26. Johnson, R. D., R. L. Jones, T. D. Hinesly and D. J. David. "Selected chemical characteristics of soils, forages, and drainage water from the

sewage farm serving Melbourne, Australia," Dept. of the Army, Corps of Engineers, 54 pp. (1974).

27. Pike, R. E., L. C. Graham and M. W. Fogden. "An appraisal of toxic metal residue in the soils of a disused sewage farm. I. Zinc, copper, and nickel," *J. Assoc. Publ. Anal.* 13:19–33 (1975).

28. Pike, E. R., L. C. Graham and M. W. Fogden. "An appraisal of toxic metal residue in the soils of a disused sewage farm. II. Lead, cadmium, and arsenic, with notes on chromium," *J. Assoc. Publ. Anal.* 13:48–63 (1975).

29. Kirkham, M. B. "Trace elements in corn grown on long-term sludge disposal site," *Environ. Sci. Technol.* 9:765–768 (1975); and comments by G. W. Leeper and M. B. Kirkham, *Environ. Sci. Technol.* 10:284–285 (1976).

30. Bingham, F. T., A. L. Page, R. J. Mahler and T. J. Ganje. "Growth and cadmium accumulation of plants grown on a soil treated with a cadmium-enriched sewage sludge," *J. Environ. Quality* 4:207–211 (1975).

31. Carroll, T. E., D. L. Maase, J. M. Genco and C. N. Ifeadi. "Review of landspreading of liquid municipal sewage sludge," Environ. Prot. Tech. Series, EPA–670/2–75–049, 95 pp. (1975).

32. Follett, R. H. and W. L. Lindsay. "Changes in DTPA-extractable zinc, iron, manganese, and copper in soils following fertilization," *Soil Sci. Soc. Amer. Proc.* 35:600–602 (1971).

33. Shipp, R. F. and D. E. Baker. "Pennsylvania's sewage sludge research and extension program," *Compost Sci.* 16(2):6–8 (1975).

34. Kjellstrom, T., B. Lind, L. Linnman and G. Nordberg. "A comparative study on methods for cadmium analysis of grain with an application to pollution evaluation," *Environ. Res.* 8:92–106 (1974).

SECTION III

HEALTH
ASPECTS

17

HEALTH HAZARDS OF AGRICULTURAL, INDUSTRIAL AND MUNICIPAL WASTES APPLIED TO LAND

D. Strauch
Institute of Animal Medicine and Animal Hygiene, University of Hohenheim, Stuttgart, Federal Republic of Germany

AGRICULTURAL WASTES

Hygienic problems involved in the application of animal waste to land must be considered under two aspects: a) major problems in epidemiology of infectious diseases; and hygienic problems involved in animal husbandry, many of which are closely associated with environmental hygiene.

Infectious Diseases

Epidemiological problems center upon the identification, the utilization of factors inhibiting infections within the agricultural establishment and the prevention of contributing factors, the protection of rural animal husbandry against the danger of infection resulting from large animal confinement, and the protection of the general public against zoonoses or health hazards due to residues of drugs, active agents, and additives as well as therapy-resistant microorganisms.

Epidemiological problems involved in large animal feedlots are closely associated with those of animal waste disposal. Incidence of latent infections increases when animals of homogeneous populations are concentrated in confinement. Most infected animals eliminate the pathogenic agent by way of feces, urine or others so that germs, ultimately, come into contact with the floor of the buildings. Many pathogens are excreted as has been shown by the results of several studies in Romania (1) in pig confinements with a yearly production of 100,000 to 150,000 fattening pigs. From seven confinements 159 samples of effluent and sludge were examined. The total bacterial count varied between 7×10^6 and 3×10^{11}/ml, the total $E.$ $coli$ count from 15×10^6 and 6×10^{11}/kg. One hundred and forty-two (142) strains of $E.$ $coli$ with pathogenicity for man and animals were isolated. These strains belonged to 23 different sero-groups and -types. One hundred and six (106) strains of salmonella were isolated which belonged to 12 sero-types. The strains are noted in the decreasing frequency of their isolation: $S.$ $derby,$ $S.$ $anatum,$ $S.$ $münchen,$ $S.$ $heidelberg,$ $S.$ $meleagridis,$ $S.$ $panama,$ $S.$ $senftenberg,$ $S.$ $manhattan,$ $S.$ $enteritidis,$ $S.$ $morbificans,$ $S.$ $panama$ type Z, $S.$ $infantis.$ Thirty-two (32) strains of leptospira of the type $L.$ $pomona$ and $L.$ $tarassovi$ were also isolated. These results show that the raw and partially treated waste from animal confinements possess microorganisms which are pathogenic for man and animals. These findings concur with the results of other authors all over the world.

Conventional Waste Handling

Conventional livestock units where bedding is used do not cause a special epidemiological problem because if proper management procedures are carried out, dung heaps develop such high temperatures as to destroy pathogens that may be present. The safety of this procedure is demonstrated by its being stipulated in many countries in the official provisions of dung disinfection for the control of infectious diseases, in terms of the so-called dung packing. After three weeks the dung is considered disinfected and can be used for agricultural purposes.

Liquid Waste Handling

Both large animal confinement and modern rural livestock management have introduced new housing systems. Straw for bedding is no longer being used. Animal excreta are generally collected jointly in liquid form, the so-called "liquid manure." This mixture of urine, dung, forage remains and splashed water also is called slurry. It is either collected and stored within the animal building or it is drained from the building and kept in under- or aboveground reservoirs until it is used. No matter what handling and storing methods are used, spontaneous generation of heat that could entail the destruction of pathogens will not occur in this medium either in summer or in winter.

Self-disinfection

A possible measure would consist in awaiting a sort of self-disinfection of the slurry during storage. Pertinent laboratory experiments and others performed in practice in farm liquid manure pits have shown, however, that these expectations are not founded. In fact, salmonella or stable forms of parasites remain alive in liquid manure both in summer and in winter over many months. Figure 1 shows the findings of laboratory tests which indicate that the viability of salmonella can be nearly one year in winter (8° C) and approximately six months in summer (17°C). Experiments carried out in farm liquid manure pits corroborate these findings. Survival times obtained in praxi

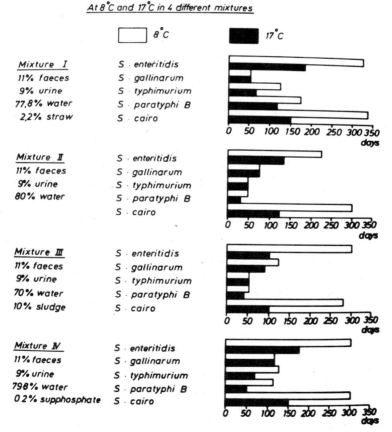

Figure 1. Viability of salmonellae in slurry in a laboratory experiment.

were slightly reduced to 286 days. The marked differences between summer and winter storage periods in this test were not as distinct as in the laboratory one (2).

Other pathogens with a viability of several months' duration in this medium include the viruses, especially if they are enclosed in tissue or in fecal matter. Most farms do not have sufficient storage capacity for slurry over many months. The average storage time for slurry in Europe is usually 1 to 3 months.

Figure 2 illustrates the seven direct and indirect ways by

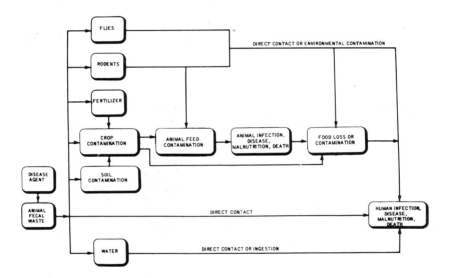

Figure 2. Animal fecal waste—disease relationships (3).

which contaminants in animal excreta pass not only to man but also to other animal production units. Successful prevention of infections is possible only if these ways can be blocked efficiently. In view of the many ways of disease transmission, a great number of measures would be necessary to block all ways of transmission unless epidemiological measures are launched before germs are released into the environment. This latter approach should be done at the animal production facility.

That a hygienic treatment of infected slurry is necessary can be proved by a number of cases in which transmission of infections has been observed after spreading of slurry to pastures. In Germany a severe salmonella outbreak in a dairy herd with 50 cows has been reported (4). Six sero-types of salmonella were isolated: *S. stanley, S. typhimurium, S. heidelberg, S. braend-rup, S. london* and *S. uganda.* The source of infection was meat-meal from a rendering plant. The same salmonella could be found in the slurry pits as well as on samples of soil and grass of the pastures which were manured with infected slurry during winter two months before the samples were taken. Thus a cycle of infection in this farm was built up. Even after the feeding of the infected meat-meal was stopped and all infected animals were eradicated, salmonella still could be isolated from the slurry.

In England an outbreak of *S. typhimurium* infection in cattle grazing on a pasture which had been irrigated with slurry three weeks before has been described (5). *S. typhimurium* was isolated from the slurry system and from four carrier cows after the outbreak.

These observations are in line with investigations in England, Northern Ireland and New Zealand (6, 7, 8, 9). Slurry, contaminated with either *E. coli* or *S. dublin*, was applied under varying climatic conditions to growing pasture in England. Survival was measured by daily assessment of the number of viable organisms per gram of grass or soil. Under the conditions of the test, *E. coli* survived for seven or eight days. *S. dublin* persisted for 18 days on the lower levels of grass and up to 12 weeks in the soil. When the grass was cropped, however, no recoveries were made after seven days. An assessment was made of the palatability for calves of slurry-polluted pasture. Calves were allowed to graze a pasture, half of which had been heavily polluted with slurry. For two days they rejected the polluted area but by seven days had grazed it well. Calves that grazed pasture to which 10^6 *S. dublin*/ml of slurry had been applied on the previous day became infected, but no cases resulted when the contamination rate was reduced to 10^3/ml.

In Northern Ireland *S. typhimurium* survived in three different soils for 159 days and in another soil for 110 days. These results indicate that salmonella can survive in the surface layers of the soil for over three months even when exposed to winter climatic conditions. It is evident from this preliminary work that soils constitute a large potential hazard as vectors of salmonella from animal to animal on the farm.

In New Zealand enumeration of salmonella on inoculated plots containing soil or sheep feces under various climatic and environmental conditions suggested that survival is greatest (18 weeks) when the organisms are in contact with organic material. Surprisingly the survival time in the summer period was longer than during winter.

If cases are added in which animals were infected by sewage

and human excretions, then it is clear that all infected excretions possess a high infective potential (10, 11, 12).

In one case *S. aberdeen* was established as a cause of illness in 30 out of a herd of 90 milking cows. The illness was only moderately severe and all animals responded to treatment. The source of infection was human sewage effluent overflowing onto grazing land. In another case a cow aborted as a result of *S. para typhi B* transmitted by an infected woman who took care of the cow. In England another case of paratyphoid B-infection in cattle has been described. *S. paratyphi B* caused generalized subclinical infection of cows and cases of human enteric fever at a dairy farm. The cattle were infected from a stream receiving the sewage effluent of a village in which a chronic carrier lived. At the same time as the farm outbreak a waterborne outbreak of paratyphoid fever occurred in villages several miles away. The water supply was chlorinated and no failure of its treatment had been detected. *S. paratyphi B* was isolated from the septic tank of a cottage near the water source and from soil over a break in the effluent pipe. Several inhabitants of the cottage worked at the infected dairy farm and one showed serological evidence of the infection.

Another severe case of transmission of salmonella through slurry is reported in Sweden (13). In a dairy farm more than 50 out of 96 cows became ill within three days during grazing. Twenty cows died within two days. Bacteriological examinations proved *S. choleraesuis* to be the cause of the disease outbreak. The effluent of an infected pig farm (*S. choleraesuis*) had flowed into a dirty pond. This pond had connections with an open ditch system in the pastures of the dairy farm. Heavy rain caused the contents of the pond to enter the ditches of the pastures to which the cows had access.

These examples indicate that animal fecal wastes as liquid manure have an epidemiological significance for land disposal which should not be underestimated. The above information is realistic as shown by the fact that as many as 100,000,000 leptospires have been reported to be shed per ml of cattle urine (14). In acute cases in calves 10,000,000 salmonella organisms per gram of feces are discharged (15).

In the United States infections with leptospirosis in cattle and swine are common. The most common sero-type infecting cattle is *L. pomona*. Leptospires survive for days to weeks outside the live animal. In recent research leptospires survived up to 138 days in the manure of a field-simulated model of an oxidation ditch (16). These authors also have observed human cases of leptospirosis associated with aerosol-borne transmission from the urine of infected cattle in farmers, veterinarians, packing house workers and hunters. Their studies indicated that salmonella died (decimal reduction) and leptospires lived or multiplied in the manure slurry environment of the laboratory oxidation ditch. In aerosol studies in the laboratory, leptospires were detected in the air on one occasion and salmonella several times. Leptospires were not transmitted to hamsters by recycled feed that was gathered from the leptospiral-contaminated manure slurry nor were the hamsters infected via aerosols. Salmonella was transmitted to turkey poults by feeding salmonella-contaminated feed and via aerosols. These comprehensive studies are very convincing.

In the U.S. salmonellosis is one of the major communicable disease problems. There are an estimated two million human cases per year. Salmonellosis causes substantial losses to the livestock and food industries with the cost estimated at $300,000,000 annually. Salmonellosis is a threat to everyone as a foodborne disease (16).

In summary it can be stated that the land disposal of infected animal wastes can result in transmission of pathogenic microorganisms to man and animals. Therefore, it has to be required that spreading of germs via infected slurry must be prevented by on-site treatment of liquid manure by chemical or physical methods.

COMMON HYGIENIC PROBLEMS

Soil Contamination

On a great number of farms, the number of animals is excessive compared with the agricultural acreage available. Therefore, excrements cannot be spread on available agricultural fields without impairing soil and plants. Large amounts of liquid manure and nutrients applied over several years make it impossible for the over-stressed soil filter to retain the nutrients absorbed along with the liquid manure. The application of very heavy loadings of liquid manure into soils with relatively high water saturation allows liquid manure to sink into lower depths or to run off. The consequence is pollution of ground and surface waters (17).

Table 1 shows data of some examined fields. The soils tested were sandy with a mean carbon content of 2.8 percent and a mean nitrogen content of 0.1 percent. The table indicates the roughly estimated amounts of liquid manure and nutrients spread per hectare over an average of 15 years.

Table 1. Amounts of liquid manure and nutrients on the examined fields (17).

	In 15 Years	Average per Year
m^3 liquid manure	2,000	133
kg N	15,000	1,000
kg CaO	13,500	900
kg P$_2$O$_5$	12,000	800
kg K$_2$O	7,000	470

Table 2 indicates the concentration of nutrients in different depths of the soil. Concentration of phosphoric acid is particularly high. This is due to the application of great amounts of slurry from pigs and poultry.

Table 2. Accumulation of nutrients (kg/ha) in the soil due to liquid manure from pigs and poultry (17).

Depth of Soil:	0–30 (cm)	30–60 (cm)	60–90 (cm)	0–90 (cm)
P$_2$O$_5$, total	+ 7,000	+ 2,900	+ 1,300	11,200
K$_2$O	+ 540	+ 360	+ 180	1,080
Mg	+ 178	+ 68	+ 36	282
N, total	+ 280	+ 90	+ 338	708
NO$_3$	+ 36	+ 14	+ 9	59

Phosphoric acid contained in slurry diffuses more rapidly in the soil than that of mineral fertilizers because organic matter contained in slurry favors phosphate diffusion. When spreading slurry on highly water-saturated soils, phosphoric acid in relatively high quantity will be transported immediately after application to lower depths.

Although phosphate accumulation occurs in deeper layers it has not yet been identified in ground water (Table 3). Pollution of ground water by phosphates can be avoided by intermittent application of large amounts of slurry into relatively dry soil.

Permanent application of highly water-diluted slurry or frequent rainfall after spreading accelerates the diffusion of phos-

Table 3. Nitrate and phosphate contents (mg/l) of leakage waters and ground waters in 1.5–3.0 m depth after manuring the fields with 160 m³/ha/year of slurry (17).

	No. of Fields	NO_3-N	NH_3-N	P_2O_5
Without slurry	5	5	3	0.23
With slurry	12	53	7	0.23
With slurry in autumn	5	94	5	0.12

phates and all other nutrients into the subsoil and thus possibly into ground water.

Another substance that must be considered within the scope of environmental protection is nitrate. Table 3 shows nitrate content found in ground water under fields treated with very high amounts of liquid manure over several years. Nitrate-N content under these fields averaging 53 mg was 10 times as high as under nontreated fields. Nitrate-N content of the fields manured in autumn (rainfalls, wet soil) a few weeks before the water samples were taken averaged 94 mg and was substantially higher than under the other fields manured longer before. The 10- to 20-fold higher content of nitrate found in ground water is injurious to health (methemoglobinemia of babies) according to the recommendations of WHO and EC.

In some parts of Northwest Germany the number of animals kept per hectare is extremely high and the danger for soils and ground waters is also large. Legislation is in preparation which will allow only the "usual rate" of animals per hectare defined by the ratio of three "manure livestock units" (MLU). One MLU is equivalent to a yearly excretion of 80 kg N and 70 kg P_2O_5.

The heavy manure load on the fields in some parts of the Netherlands is the reason why farmers ship their surplus of manure to other parts of the country. The organizational basis of this transfer, the so-called "manure-banks," was established to act as a brokerage between farms with manure surplus and those which need organic fertilizers. Thus in Holland in 1973/74, 75,000 tons of slurry were transported with tank wagons—75 percent over a distance between 8 and 20 km and 25 percent between 20 and 50 km (18). In 1975 the slurry transactions increased to 395,000 tons.

In Northwest Germany (Figure 3) there already are difficulties for the farms with manure surplus. These farms are not able to ship their slurry to other parts of the area since most of the communities already have an animal population of between

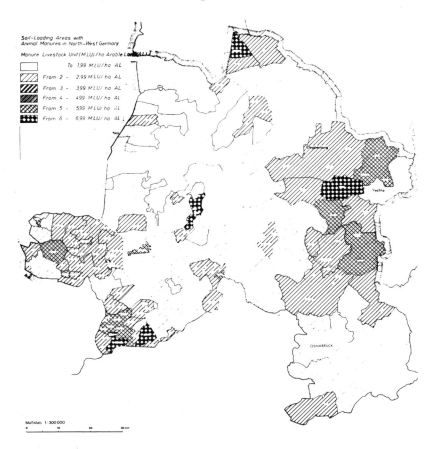

Soil-Loading Areas with
Animal Manures in North-West Germany

Manure Livestock Unit (MLU) / ha Arable Land

	To 1.99 MLU/ ha AL
	From 2 - 2.99 MLU/ ha AL
	From 3 - 3.99 MLU/ ha AL
	From 4 - 4.99 MLU/ ha AL
	From 5 - 5.99 MLU/ ha AL
	From 6 - 6.99 MLU/ ha AL

Maßstab 1 : 300 000

Figure 3.　Soil-loading areas with slurry in a part of Northwest Germany
(19).

2 and 3 MLU/ha. Thus the chance to ship manure to other areas
is very limited. If the number of animals to be kept per hectare
is not limited by law in these areas, the health hazards for the
public will steadily increase (19).

Water Contamination

Even when infected animal slurries are treated by settling
out solids, aeration, or sludge drying, pathogens are not eradicated. It can be expected, in case of infections of animals in a
large feedlot, that large numbers of pathogens would be discharged into rivers. In nearly all animal confinements, some ani-

mals may discharge pathogens without showing signs of infection (20). Therefore it is expected that feedlot wastewaters discharged into river waters will have to be disinfected. Chlorine disinfection is the most common method. When properly applied and controlled, chlorination of wastewaters from animal feedlots for disinfection is an effective measure for improving the bacteriological quality of wastewater and protecting the human and animal population against transmission of enteric diseases via the water route.

Problems with Antibiotics

In the early 1950s, the feeds containing low levels of antibiotics were found to increase the growth of farm animals. This was a major breakthrough for commercial animal feedlots. Soon, however, it also was found that the widespread use of antibiotics for nontherapeutic purposes resulted in the development of microorganisms which became resistant to antibiotics. The situation was made worse by the fact that some bacteria became capable of transferring their antibiotic resistance to organisms which had not had contact with these antibiotics.

Although scientists recognize that transmission of drug resistance poses a real threat to disease control in both animals and man, no one seriously recommends that use or production of antibiotics be discontinued. However, transferable resistance cannot be taken lightly. Everybody facing this problem should be aware of the potential hazards of transferable resistance because of the public health implications and because understanding the phenomenon of resistance helps explain why some patients do not respond to the therapy. If transferable resistance proves to be more common than has been shown so far, and there is much evidence that this is so, the discharges of feedlot wastes have to be given special consideration as a source of resistant bacterial strains.

Investigations in Czechoslovakia have shown that the surroundings of large animal confinements have a heavy contamination of all kinds of intestinal microorganisms including 100 percent resistant *E. coli* strains. The self-cleaning process of the soil in the surroundings of these farms is considerably retarded so that health hazards to the population living in the vicinity have to be taken into consideration.

It has been feared that the transfer of resistance capability from coliform bacteria to salmonella might be a setback for the control of certain infectious diseases in human medicine. Further tests have shown, however, that transfer *in vivo* does not occur as frequently as was to be expected according to tests *in vitro*. In addition, resistant coli bacteria of animal origin have proved not to resist for long the intestinal flora in humans. Therefore, it has been inferred that animals cannot be a major source of resistant coli bacteria for man since he is the carrier of antibiotic-resistant coli bacteria of human origin. Only in the personnel of agriculture or slaughterhouses who are professionally in close contact with the animals may resistant strains of coli bacteria of animal origin be found sporadically (21). New substances on the market are reported not to be resorbed, hence will not get into food of animal origin. Manufacturers claim that these antibiotics do not produce transferable resistance. If this can be confirmed in long-term experiments the problem comes closer to a solution.

INDUSTRIAL WASTES

Introduction

In the struggle by modern industry to eliminate or alleviate the effects of pollution, greater priority has generally been given to the limitation of toxic and other hazardous discharges to the atmosphere, sewers and waterways than to the limitation of the disposal of toxic and other hazardous waste to land.

The discharge of such waste to land is now beginning to cause concern. If a waste has a degree of solubility or is potentially mobile in any way, there is a risk that it will affect the water supply of the district. If the disposal site is on impermeable land, there is a probability that there will be runoff to surface water, and if the tip is on permeable land the toxic material may reach the underground water.

The fact that, in the absence of fissures, this movement may be slow, perhaps taking decades, is not necessarily an advantage. Until the pollution shows itself in the water there is a tendency to regard the disposal site as safe and a large quantity of the waste may be disposed of so that the site may be eventually polluted in a permanent or semipermanent manner, whereas if pollution quickly becomes evident, disposal could cease and pollution may be no more than temporary.

Nevertheless, slow movement may be an advantage with waste which is amenable to biological or chemical purification, provided these agencies in the soil are not overloaded. Chemical purification, however, usually involves the insolubilization of metals by a base exchange action so that the toxic metal is not eliminated, but merely "fixed."

Water pollution is not the only risk from hazardous waste. There is also the possibility of fire, explosion and effects (such as poisoning, infection and skin irritation) on trespassers. Figure 4 shows the possible pathways between chemical waste and human disease.

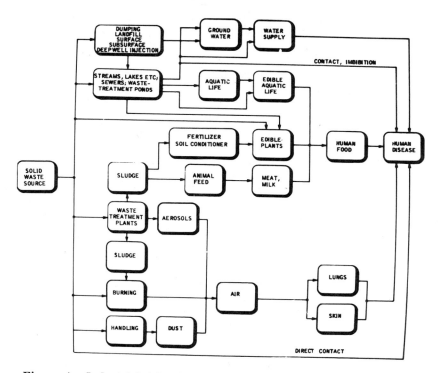

Figure 4. Industrial (chemical) waste—human disease pathways (3).

These problems are not new, but they have increased in magnitude in recent years. At the same time, because of greater land use and demand for water caused by population and industrial pressures, suitable safe waste disposal sites are becoming more difficult to find.

This presupposes that land disposal is a proper method for the disposal of persistant toxic or hazardous substances. There is considerable doubt whether any disposal site can be regarded as truly safe for such waste or that such disposal represents a proper use of land. However, it seems certain that, in conjunction with incineration and other presently available means of disposal, substantial land disposal will continue to be used until sufficient alternative facilities become available.

To meet this problem of increased amounts of waste and limited disposal sites, strict controls over disposal and a search for alternative methods are required. Before considering the problem further, it should be noted that discussion will not be restricted to solid toxic or hazardous waste. Many sludges and even liquid wastes which cannot be disposed of to sewers, streams, etc., present similar problems and need to be disposed of in similar ways. Since they form part of the same problem it is logical to consider them. Certain wastes such as radioactive waste, are not included in this context because of the very special considerations involved

One of the problems of controlling the disposal of toxic and hazardous waste is the adequate definition of which waste is to be included. The difficulty arises because of the variety of such waste, because the composition is not always known and because the potential danger will depend on the quantity or concentration present. Possibly the most helpful procedure is to have a system of notification (except for clearly defined "safe waste" which is not considered a special hazard in any quantity) to a controlling waste disposal authority, one of whose duties will be to authorize how and where such waste is to be disposed.

An example of a list of safe wastes is given in the original publication (22). Wastes not shown on such a list will cover a wide range of hazards. Many types may not present a high hazard and proper disposal may be relatively simple, but other types may be dangerous even in small quantities. To assist the controlling authorities, industry, other waste producers, disposal contractors, etc., to ensure proper disposal with the least difficulty, a list of the major types of industrial toxic and other hazardous waste should be produced. It would be the intention to introduce codes of practice for such waste as soon as possible, but these should not be framed in a manner which would discourage innovation in disposal.

Types and Amounts of Toxic and Other Hazardous Waste

The number of possible types of hazardous waste is very high and, for some of the less important wastes, decision as to the best disposal method(s) will probably rest largely with the area waste disposal authority. However, it is probable that most wastes will fit into convenient classification groups and that for each major type of waste, national or international codes of practice can be produced.

Although the major types of hazardous waste may differ from community to community, a typical list might include the following: cyanide waste; metal finishing waste including acid wastes; polychlorinated biphenyls and analogs; mineral oils; solvents; mercury waste; other metals and their compounds including, for example, manganese, lead, zinc, vanadium, boron, tellurium, nickel, chromium, cadmium, barium and beryllium; arsenical waste, including antimony and selenium; pharmaceutical waste; medical waste and infected material; biocedes; tarry waste; asbestos; plastics waste, including monomers; phenols, cresols and simple derivatives.

The above list is not intended to be exhaustive, but simply indicative of the more common types of hazardous waste. Many organic and inorganic substances might well be added.

Many communities produce very little information on the types and amounts of hazardous waste generated in their areas. This is unfortunate as such information is essential to the formulation of a proper disposal scheme. Some information is available and a selection is included in the original publication (22). The tables in the publication indicate the variety of ways used to express such information.

Environmental and Health Risks

Waste can produce harmful effects of a biological, chemical, physical, mechanical or psychological nature. It is not easy to clearly distinguish the effects of waste in general from those of the more toxic or otherwise hazardous waste. For example, human pathogens in feces provide a biological threat; some industrial waste poses chemical hazards; flammable materials involve physical danger of fires or explosions, and broken glass causes mechanical hazards. Many other effects, some not clearly defined, also merit consideration, including psychological and behavioral disturbances and repercussions on both the material and the aesthetic environment.

All health hazards not only require consideration under correct or normal conditions of handling and disposal; they should also take account of possible incorrect procedures, accidents (particularly during transportation), reactions between wastes, synergistic actions, the capacity for and speed of degradation of the wastes, including possible deleterious effects of the degradation products.

In human populations there usually exist certain critical groups which are more exposed to specific risks than the rest of the population, for whom the consequences of exposure are more severe. Examples include pregnant women, children, and some disabled groups. The suceptibilities of such groups must be borne in mind when assessing risk.

Hazards from waste may add to those from other sources. For example, danger from heavy metals may arise from air, water, food, drugs, occupational exposure, etc., and such metals may have cumulative effects on specific organs or tissues with other substances having the same target.

The disposal of hazardous waste to land not only may result in soil pollution but may result in water and air pollution as well. Of these, water pollution is the most important. Reference has been made to this in the introduction. Such contamination may threaten water intended for drinking purposes, for agricultural purposes, for the food industry and even for recreational purposes. More insidious effects on human health may occur through adverse effects on aquatic and marine life, including phytoplankton and zooplankton and through concentration of toxic substances in the food chain. Some waste, such as nitrates and phosphates, although not particularly toxic in themselves, may damage the ecological equilibrium through eutrophication of waters.

Air pollution arising from the deposition of hazardous waste is usually less important and often transient. However, a distinction must be drawn between the dangers of toxic gaseous substances or particulates arising from natural chemical or biological processes after disposal and the dangers arising from treatment of waste (such as hydrochloric acid arising from the burning of polyvinyl chloride) where special precautions are almost always necessary.

The elimination of all possibilities of risk to human health and the general environment must be the object of all those responsible for hazardous waste disposal. In this respect, the responsible authorities should take into account the international conventions such as the London and Oslo Conventions, and the various international recommendations such as those of the United Nations bodies IMCO and GESAMP.

For cases of land use of industrial wastes and other wastes that may contain undesired elements, an expert committee of the Federal Health Office in Berlin has prepared a proposal for the tolerable concentrations of various elements in soil that are agriculturally used (Table 4).

Excessive amounts of trace nutrient elements and other elements including heavy metals are harmful to crops and to the

Table 4. Tolerable concentrations of various elements in soil with reference to the permissible limits for plants (1974).

Element	Content in ppm	
	Normal Soils	Tolerable in Soils (Proposal)
Be	0.1–10	10
F	10–500	500
B	2–100	100
Cr	1–100	100
Ni	1–100	100
Co	1–50	50
Cu	2–100	100
Zn	10–300	300
As	1–50	50
Se	0.1–10	10
Mo	0.2–10	10
Cd	0.01–1	5
Hg	0.01–1	5
Pb	0.1–10	100

food chain. The composition of the wastes varies considerably. Control of their composition by continuous chemical analyses is not feasible. Restriction is necessary to a random sample survey, to a waste source control, and to a regular soil testing in order to prevent the concentration of hazardous elements from exceeding the tolerable level in the soils (23).

Infective Agents

The causative agents of nearly all infectious diseases of man are excreted via the intestinal or urinary tract or they come to the sewage with other excretions of the body. For a long time it was believed that the sewage treatment processes would destroy the pathogenic microorganisms. However, this opinion is not always correct (Tables 5 and 6).

In Sweden the following parasitological results were obtained in 22 samples of the effluent of a municipal sewage treatment plant: *Diphyllobothrium latum* 100 percent, *Taenia* 45 percent, *Ascaris* 27 percent, *Toxocara* 27 percent. The salmonella content in 279 samples of effluent of various plants was between 30 percent and 56 percent, and in the effluent of each plant investigated salmonella could be found. Eighty-four (84) strains of salmonella were isolated which belonged to 25 different serotypes (25).

Table 5. Isolation of salmonella from sewage sludge in 44 treatment plants in Switzerland (24).

Kind of Sludge	No. of Samples	Salmonella-Positive (absolute)	(%)
Raw sludge	49	45	91.8
Stabilized sludge (aerobically)	32	25	78.1
Digested sludge	138	113	81.9
Total	219	183	83.6

Table 6. Examples of removal of microorganisms by primary and biological sewage treatment per 100 ml (26).

Organisms	Primary Treatment Influent	Effluent	Biological Treatment Influent	Effluent
E. coli	10^6–10^7	0.3–0.6×10^6–10^7	0.9–22×10^6	5.4–170×10^3
Salmonella				
summer	125	56	240	0.22
winter	0.5	4	0.2	0.1
Clostridium perfringens	73,000	46,000	46,000	27,000
Mycobacteria	20–100	20–100	–	–
	100–10,000	100–1,000	150,000	1,000
Protozoa	not available		not available	
Parasite eggs	0.4–3.0	0.5–1.0	not available	
Enteric viruses	not available		autumn 10,000	100
			winter 0.1	0.1

All these results show that the hygienic effect of sewage treatment plants must be estimated as only very small.

As already was shown in the discussion of agricultural wastes, salmonella have a high survival time in the environment. The results of a literature survey are shown in Figure 5. Under practical conditions in a large area with agricultural land disposal of sewage sludge disposal, the following results were obtained (27). Viable salmonella were found in 26 percent of grass samples for 5 weeks, in 59 percent of topsoil samples for 10 weeks and in 84 percent of sludge-crust samples for 16 weeks after the last spreading of sludge to the fields. These results were supplemented with findings in Switzerland (28). They found viable

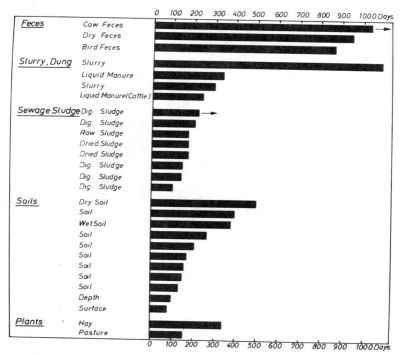

Figure 5. Tenacity of salmonella in the environment (24).

salmonella on grass and in topsoil of sludge-treated pastures under natural climatic conditions for 47 weeks.

These observations indicate a threat to the health of animals which come into contact with sewage- and sludge-treated pastures or other feed. That this proves to be correct can be shown by a review of reports during the last 10 years in international literature (6, 24, 28, 29, 30, 31). These authors describe their own observations of animal infections caused by land disposal of sewage and/or sewage sludge as well as summarizing other literature. Only two of these observations shall be mentioned here.

In a dairy farm in the Netherlands with regular land disposal of sewage sludge, 4.7 percent of the cows were salmonella-infected, whereas the average of infected cows in the whole country was only 0.3–0.5 percent (31).

Epidemiological relations between agricultural sludge disposal and salmonella infections in dairy farms can be shown (Figure 6) by the statistics of the Institute for Veterinary Bacteriology

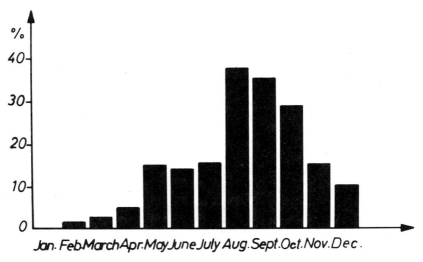

Figure 6. Seasonal incidence of salmonella isolation from adult cattle in Switzerland 1969–1974 (13,877 samples) (28).

in Zürich/Switzerland (28). The analysis of 13,877 seasonal equally distributed investigations of adult cattle showed an unequivocal accumulation of infections with salmonella during the green fodder period. The first increase of infections was traced to the distribution of sludge during the winter and the culmination of cases in August/September to the massive spreading of sludge after hay-making and the following short intervals in cutting the grass.

According to a report of the US Department of Agriculture (32), not all animals which are infected with salmonella show symptoms of the disease. However, these carriers represent an important danger for public health because the food from them is not recognized as infected and therefore not treated according to veterinary legislation. The same report describes a case of sludge disposal on the lawn of a hospital (aerobically stabilized and dried sludge), after which a salmonella epidemic occurred among the patients. Identical salmonella types were isolated from the sludge and the feces of the patients.

All these examples indicate that it is a perversion of hygienic principles if more and more sewage is collected and treated in plants whereas the sludge—which mainly consists of human feces—afterwards is distributed over large areas without being disinfected. Figure 7 identifies the possible pathways of human fecal wastes (sewage and sludge) to animals and man.

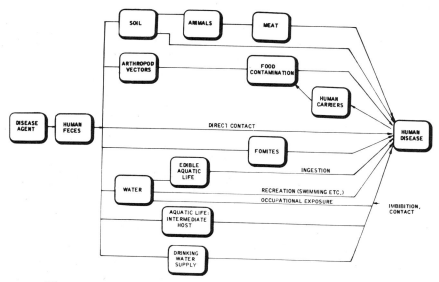

Figure 7. Human fecal waste—human disease pathways (3).

The foregoing discussions also are applicable for parasitic eggs. As far as pathogenic viruses are concerned, because of complex test procedures little research has been done on the possibility of hazardous build-up of pathogens in soil and ground water during land disposal of sewage and sewage sludge. In a study initiated by the City of St. Petersburg, Florida and performed by the Epidemiology Research Center, the following results with viruses were obtained (Table 7).

According to these authors, several laboratory studies over the years by various investigators have indicated that virus is removed in the upper few inches of soil. However, no extensive field studies have been done to prove or disprove these data. The findings in St. Petersburg showed that viruses can and do sur-

Table 7. Virus load in various wastewaters (33).

Concentration Technique	Influent (PFU/gal)	Unchlorinated Effluent	Chlorinated Effluent (PFU/gal) Contact time		
			5 min	7 min	15 min
PEG	122.9	210.4	88.2	207.9	ND
MA	62.7	74.0	10.9	23.7	0.2

vive percolation through the soil. Although virus isolations from the ground waters occurred infrequently, they cannot be discounted. In view of the inefficiency of the concentration technique and the absence of continuous monitoring, it is possible that higher virus concentrations were present in the ground waters. Since each virus PFU has the potential of causing disease in man or animals, secondary wastewater spray irrigation on sandy soil should be recognized as a potential hazard.

Noninfective Agents

Sewage sludge also contains a variety of heavy metals and trace elements which influence growth and yield of plants and may influence the food chain from plant via animal to man. A part of these minerals may have a limiting influence on the waste disposal over a long time. Investigations (34) have shown (Tables 8, 9 and 10) the maximum time for which sludges can be used in agriculture. The data in Table 10 indicate that salt, phosphorus, zinc and iron may limit the time for the use of municipal (and industrial) sludges. When these periods are exceeded, additional health hazards for animals and man may arise.

SUMMARY

The number of latent infections in animal production has steadily increased all over the world caused by concentration of animals in confinements. The infective agents occur in the liquid manure and survive the usual storage time. By land disposal of slurry, especially on pastures and other green fodder, infections of animals can be transmitted. For an effective interruption of the infection cycles, hygienic on-site treatment of infectious slurry may be needed before land disposal.

Additional dangers for public health arise when too large amounts of slurry are distributed on land. Thus infective agents and undesired chemical compounds such as P_2O_5 and NO_3 can be transmitted to surface and ground waters.

Another hygienic risk for man and animals is the occurence of transmissible resistance of microorganisms caused by feeding antibiotics to improve growth in animal production. By the use of newly developed substances which are not resorbed and are said not to cause resistance, this problem possibly can be solved.

The problems of land use of industrial wastes include the occurrence of toxic organic and inorganic substances. High concentrations in plants and food derived from animals can be a

Table 8. Supply of minerals in kg/ha by application of different amounts of sludges with and without regard to extraction by plants and leaching processes (DM = Dry Matter) (34).

Mineral	10 t DM without kg/ha	10 t DM with kg/ha	20 t DM without kg/ha	20 t DM with kg/ha	25 t DM without kg/ha	25 t DM with kg/ha	50 t DM without kg/ha	50 t DM with kg/ha	75 t DM without kg/ha	75 t DM with kg/ha	100 t DM without kg/ha	100 t DM with kg/ha
P_2O_5	140.1	+ 59	280	+ 199	350	+ 269	700	+ 619	1050	+ 969	1401	+ 1320
K_2O	14.9	- 325	29.8	- 310	37	- 303	74	- 266	111	- 229	149	- 191
Mg	.3.9	- 46	7.8	- 42	10	- 40	20	- 30	30	- 20	39	- 11
Ca	590	+ 270	1080	+ 760	1475	+ 1155	2950	+ 2630	4425	+ 4105	5900	+ 5580
Cu	3.6	+ 3.5	7.2	+ 7.1	9.0	+ 8.9	18.0	+ 17.9	27.0	+ 26.9	36.1	+ 36.0
Zn	21.9	+ 21.5	43.8	+ 43.4	54.8	+ 54.6	109.6	+ 109.2	164.4	+ 163.8	219.2	+ 218.4
Mn	3.6	+ 2.6	7.2	+ 6.2	9.0	+ 8	18.0	+ 17	27.0	+ 26	36.1	+ 34.7
Fe	106.3	+ 104.7	212.6	+ 211.0	265.8	+ 264.4	531.6	+ 530.4	797.4	+ 795.4	1063.2	+ 1061
Pb	3.7	+ 3.6	7.4	+ 7.6	9.2	+ 8.9	18.4	+ 17.9	27.6	+ 26.9	37.0	+ 36.9
B	0.09	- 0.26	0.18	- 0.17	0.23	- 0.12	0.46	+ 0.11	0.69	+ 0.34	0.9	+ 0.55
N	150	- 20	300	+ 130	375	+ 205	750	+ 580	1125	+ 955	1500	+ 1330
Salt	180		360		450		900		1350		1800	
C	2190	+ 1190	4380	+ 3380	5475	+ 4475	10950	+ 9950	16425	+ 15425	21900	+ 20900
C:N-Rest		:	26	: 1	22	: 1	17	: 1	16	: 1	16	: 1

Table 9. Average figures for the loss of various minerals by extraction from plants and leaching in the zone of 0–20 cy per ha and year (34).

Mineral	Dimension	Extraction by Plants	Leaching	Loss
N	kg/ha	130 (50–220)	40 (10–90)	170
P_2O_5	kg/ha	80 (30–120)	1 ($<$1–10)	81
K_2O	kg/ha	300 (100–500)	40 (10–70)	340
Mg	kg/ha	30 (10–70)	20 (5–70)	50
Ca	kg/ha	70 (20–110)	250 (70–140)	320
Cu	g/ha	80 (30–140)	30 (15–50)	110
Zn	g/ha	250 (200–300)	100 (50–100)	350
Fe	g/ha	1500 (500–10000)	100 ($<$100–500)	1600
Mn	g/ha	700 (400–1500)	250 (150–1100)	950
Pb	g/ha	50 (20–150)	$<$10	60
B	g/ha	150 (40–400)	200 (150–250)	350
C	kg/ha	1000 (0–5900)	$<$1	1000
Organic Substance	kg/ha	2000 (0–12000)	$<$1	2000

health hazard for animals and man. Therefore the maximum concentrations in soil and plants should be determined and routinely controlled.

Infections of animals and man can occur as a result of land disposal of infected sewage and sewage sludge. To avoid this hazard, sewage and sludge should be disinfected before use. If this is not feasible they should be used only on arable land and not on pastures and other green fodder areas.

Since sewage and sludge also contain undesired chemical compounds which may cause hazards for animals and man, the same control measures as for the land use of industrial wastes should be employed.

Table 10. Maximum time of application (in years) of sludge with regard to the losses indicated in Table 8 (34).

Mineral (g)	10t/ha (yr)	20t/ha (yr)	25t/ha (yr)	50t/ha (yr)	75t/ha (yr)	100t/ha (yr)
Salt	17	8	7	3	2	1
P_2O_5	50	15	11	5	3	2
Cu	83	42	33	17	11	8
Zn	11	6	4	2	1	1
Mn	417	208	167	83	56	42
Fe	29	14	11	6	4	3
Pb	83	42	33	17	11	8
B	50	25	15	10	8	5

REFERENCES

1. Tomascu, V., F. Marschang, R. Mogo Minzat, O. Rusu, E. Crainiceanu and J. Grozav. "Einige bakteriologische Charakteristika der Abwaesser aus Schweinegrossanlagen," *Tieraerztl. Umschau* 29:210–212 (1974).
2. Strauch, D. "Hygienische Probleme bei der Gewinnung, Behandlung und Verwertung tierischer Exkremente." In: Strauch-Baader-Tietjan, *Abfaelle aus der Tierproduktion*, Verl. Eugen Ulmer, Stuttgart (1976).
3. Hanks, T. G. Solid waste/disease relationships. PHS. Publ. No. 999–UIH–6 (1967).
4. Münker, W. Personal communication (1965).
5. Jack, E. J. and P. T. Hepper. "An outbreak of *Salmonella typhimurium* infection in cattle associated with the spreading of slurry," *Vet. Rec.* 84:196–199 (1969).
6. Taylor, R. J. and M. R. Burrows. "The survival of *Escherichia coli* and *Salmonella dublin* in slurry on pasture and the infectivity of *S. dublin* for grazing calves," *Br. Vet. J.* 127:536–543 (1971).
7. Taylor, R. J. "A further assessment of the potential hazard for calves allowed to graze pasture contaminated with *Salmonella dublin* in slurry," *Brit. Vet. J.* 129:354–359 (1973).
8. Stewart, D. J. "The survival of *Salmonella typhimurium* in soils under natural climatic conditions," *Res. Exper. Rec. Min. Agric. N. Ireland* 11:53–56 (1961).
9. Tannock, G. W. and J. M. B. Smith. "Studies on the survival of *Salmonella typhimurium* and *Salmonella bovis-morbificans* on soil and sheep feces," *Res. Vet. Sci.* 13:150–153 (1972).
10. Bicknell, S. R. "*Salmonella aberdeen* infection in cattle associated with human sewage," *J. Hyg., Camb.* 70:121–126 (1972).
11. Hensel, L. and H. Frerking. "Eine vom Menschen stammende Paratyphus-B-Infektion als Abortusursache bei einer Kuh," *Der Landarzt* 40:1518–1520 (1964).
12. George, J. T. A., J. G. Wallace, H. R. Morrison and J. F. Harbourne. "Paratyphoid in man and cattle," *Brit. Med. J.* 3:208–211 (1972).
13. Hoflund, S. "Wasserverunreinigungen als Krankheitsursache in Rinder- und Schafbestaenden," *Dt. tieraerztl. Wschr.* 68:97–100 (1961).
14. Gillespie, R. W. H. and J. Ryno. "Epidemiology of Leptospirosis," *Am. J. Publ. Hlth.* 53:1950–1955 (1963).

15. Loken, K. I. Personal communication (1967).

16. Diesch, S. L., P. R. Goodrich, B. S. Pomeroy and L. A. Will. "Survival of pathogens in animal manure disposal," EPA Technology Series, EPA-660/2-75-102 (June 1975).

17. Vetter, H. and A. Klasink. "Grenzen fuer die Anwendung hoher Fluessigmistgaben." In: Vetter, H., Mist und Guelle, DLG-Verlag. Frankfurt/Main (1973).

18. Jongebreur, A. A. "Transport von Fluessigmist und die Funktion von Mistbanken in Holland," *Dt. Gefl.wirtsch. u. Schweineprod.* 27:469-470 (1975).

19. Best, E. and D. Strauch. "Zur Frage einer moeglichen Umweltbelastung durch Wirtschaftsduenger, dargestellt am Beispiel des Regierungsbezirks Osnabruech sowie der Kreise Cloppenburg und Vechta des Verwaltungsbezirks Oldenburg," *Tieraerztl. Umschau* 30:118-128 (1975).

20. Strauch, D. "Hygienic problems in large animal confinements," WHO seminar on agricultural wastes (28 Sept-4 Oct), Bratislava/CSSR (1975).

21. Braeuchle, B. "Hygienische Probleme der Resistenzbildung durch Beifuetterung von Antibiotika," Ldw. Diplomarbeit, Universitaet Hohenheim, FRG (1971).

22. Fish, R. A. "Toxic and other hazardous wastes," In: *Manual on Solid Waste Management*, World Health Organization, EURO 3404-1, (1973).

23. Tietjen, C. "The admissible rate of waste (residue) application to land with regard to high efficiency in crop production and soil pollution abatement," *Proc., Waste Management Conf.*, Cornell University (1976).

24. Hess, E., G. Lott and C. Breer. "Sewage-sludge and transmission cycles of salmonellae," *Zbl. Bakt. Hyg.*, I. Abt. Orig. B 158:446-455 (1974).

25. Strauch, D. "Gesetzliche und hygienische Grundlagen der Klaerschlammbehandlung," *Giessener Berichte zum Umweltschutz* 1/1972: 3-27 (1972).

26. Lund, E. "Public health aspects of wastewater treatment," In: *Radiation for a Clean Environment*, pp. 45-60, Int. Atomic Energy Agency, Vienna (1975).

27. Köser, A. "Die tierhygienische Seite der landwirtschaftlichen Abwasser- und Schlammverwertung," *Schriftenr. d. Kurat. f. Kulturbauwesen, Heft* 16:25-42, Verl. Wasser und Boden, Hamburg (1967).

28. Hess, E. and C. Breer. "Epidemiology of salmonellae and fertilizing grassland with sewage sludge," *Zbl. Bakt. Hyg.*, I. Abt. Orig. B 161: 54-60 (1975).

29. Strauch, D. and K. H. Knoll. "Zum Thema Klaerschlammproblem," Internationale Arbeitsgemeinschaft fuer Muellforschung, Informationsbl. No. 24:12-13 (1965).

30. Strauch, D. and E. Parráková. "Gefaehrdung des Weideviehs bei der landwirtschaftlichen Verwertung von Klaerschlamm," *Staedtehygiene* 21:195-198 (1970).

31. Kampelmacher, E. H. Personal communication No. 11 (1973).

32. U.S. Department of Agriculture and the Food and Drug Administration. "An evaluation of the salmonella problem," National Academy of Science (1969).

33. Wellings, F. M., A. L. Lewis and C. W. Mountain. "Pathogenic viruses may thwart land disposal," *Water Wastes Eng.* 12:70-74 (1975).

34. Gaul, D., W. Scholl, D. Strauch, W. Mueller and K. Miersch. "Kompost aus Klaerschlamm," Hohenheimer Arbeiten, Pflanzliche Produktion, Heft 78, Verl. Eugen Ulmer, Stuttgart (1975).

18

MICROBIAL CONCERNS WHEN WASTES
ARE APPLIED TO LAND

J. W. Doran, J. R. Ellis, and T. M. McCalla
U.S. Department of Agriculture
Lincoln, Nebraska

INTRODUCTION

As a result of implementation of the National Environmental Policy Act of 1969 and, more specifically, the respective amendments for the Clean Air Act and Federal Water Pollution Control Act of 1970 and 1972, land application seems the most feasible means for disposal of animal and domestic sewage wastes. The magnitude of these wastes is not small. In 1975, 1.8 billion metric tons of animal wastes were produced, 1.0 billion tons of which were available for application to land. Between 7 and 16 million metric tons of sewage sludge were produced in 1975 and, by 1980, the quantity of sludge disposed of on land is expected to increase to 35 million tons per year (1, 2). Approximately 530 million hectares of cropland, pasture, and rangeland—two-thirds of the land area of the continental United States—are potentially available for accepting wastes (3). Uniform distribution of all the animal wastes and sewage sludges produced yearly over this available agricultural land would result in a loading rate of 2.0 metric tons of waste per hectare. But the actual rates of waste application to land are much higher because transportation costs and land accessibility keep waste disposal near points of origin. Because of the magnitude of this disposal problem, it is important that we understand the effects of waste application on the environment and determine the safe loading capacities of soils.

Microbial concerns associated with land applications of waste can be placed into three general categories. The first is the potential for disease transmission to humans and domestic animals, resulting from the contamination of water supplies, feedstuffs, or air by pathogenic microorganisms in waste materials. Second is the effect of waste application upon microbial populations and their activities in decomposition of organic materials and nutrient cycling. The third is the accelerated production or mobilization by microorganisms of compounds toxic to either plants or animals.

The type, chemical composition, and mode and rate of application of wastes are important in evaluating microbial effects of land-applied wastes. Discussion here will be limited to evaluation of the microbial concerns associated with land-disposal of fecal wastes, with emphasis on animal manures and sewage sludge.

Although the compositions of animal and human fecal material differ widely, similarities exist. Wastes of fecal origin contain large numbers of microorganisms; are sources of carbon, energy, and nitrogen for microbial growth; and sometimes contain toxic compounds. Marked shifts in microbial activities in soil, resulting from waste applications, greatly influence the soil-plant environment and environmental quality in general. When manures and sludges are applied to land at appropriate rates, they can be considered a resource rather than a waste. However, when these materials are applied at rates that exceed the soil's capacity to assimilate them, environmental contamination and disease may result.

WASTES AS CARRIERS OF PATHOGENS

The social concern for disease potential and the unaesthetic aspects of wastes have created problems in their utilization and disposal. Historically, animal wastes have been used as a source of crop nutrients. In contrast, the use of municipal wastes for crop production, although prevalent in many countries, has not been accepted in the U.S. In addition to the pathogen hazard of waste recycling, needs for waste disposal and nutrient cycling should be recognized. Fecal "wastes" are a resource, not a useless material. Many benefits may be derived from efficient use of animal and municipal wastes for crop produciton.

The disease-transmission potential in disposal of animal and human wastes has long been of concern and with some justification. Memories of past epidemics, before man discovered modes of disease transmission and developed sanitation controls, have left a residue of fear. However, improved sanitation and massive immunization programs have reduced the incidence of disease. In the U.S., both tuberculosis and brucellosis had target eradication dates of 1976. From 1961 to 1970, there was an average of only 2.3 cases of waterborne illnesses per 100,000 people reported annually in the U.S. (4).

Of several hundred diseases that can be transmitted among animals of different species, more than a hundred also can be transmitted to man (5). However, a smaller number of pathogens are significant in animal or municipal wastes. The common pathogens found in these wastes are the bacterial pathogens: *Salmonella, Shigella, Mycobacterium, Erysipelothrix, Leptospira;* the enteroviruses, adenoviruses, and hepatitis viruses; and among the intestinal parasites, *Entamoeba histolytica* and *Ascaris lumbricoides.*

One consideration in using wastes on the land is pretreatment of the waste material. Pathogens have survived activated sludge, trickling filter oxidation, and disinfection treatments of municipal wastes; it has been reported that even after the final disinfection, wastes can be a potential health hazard (6). However, a 10^3–10^5 reduction of certain pathogens can be expected if municipal wastes receive primary and secondary treatments and disinfection (Table 1). When wastes treated in this manner are applied to the land at a rate of 5 cm effluent/week, about five pathogens would be applied per m^2 (7). Salmonella generally can be found in wastes or other materials (soil, water, cattle feeds) whenever it is sought; the mere presence of a pathogen, however, does not necessarily represent a disease hazard.

To survive or remain virulent, pathogens usually depend on the favorable conditions of the host. When a microbe encounters

Table 1. Estimated wastewater pathogens applied to soil.[a]

| Pathogen | No. Organisms/Million Liters | | Organisms Applied /ha/day[b] |
	Raw Wastewater	Disinfected Secondary Effluent	
Salmonella	5.3×10^9	1.3×10^5	1,580
Mycobacterium	5.3×10^7	4.0×10^3	49
E. histolytica	4.0×10^6	3.2×10^3	38
Helminth ova	6.6×10^7	1.3×10^3	16
Virus	1.1×10^{10}	5.3×10^5	6,480

[a] After Foster and Engelbrecht (7).

[b] Applied at rate of 5 cm/wk.

a situation in which it cannot function normally, growth usually stops and the cell dies. Numerous environmental conditions affect the cell after it leaves the natural host. Although organic matter acts as a protective agent, cells that have been stressed by waste treatment encounter unfavorable moisture conditions, pH, temperatures, sunlight, and nutrient levels when applied to land. Toxic substances in the waste, soil antibiotics, and antagonistic organisms also present obstacles to pathogen survival. Most pathogens lose the battle.

In soils receiving sewage sludge, most pathogens will perish or be reduced to low numbers in two to three months. Although some pathogens have long survival times in the soil (Table 2),

Table 2.   Reported pathogen survival.[a]

| Organism | Survival Time | |
	Plant Surface	Soil
Bacteria		
Salmonella	1–42 days	7–168 days
Shigella	1– 7 days	
Mycobacterium	10–49 days	3–15 months
Leptospira		15–43 days
Erysipelothrix		21 days
Viruses		
Enterovirus	4 days	25–170 days
Poliovirus		32 days
Parasites		
Entamoeba histolytica	3 days	8 days
Ascaris lumbricoides ova	27–35 days	2–6 years

[a] After Burge (10), Dunlop (6), Foster and Engelbrecht (7), Gerba *et al.*, (11), and Sepp (9).

most do not survive long on plant surfaces. When long survival times have been reported, initial inoculation levels were high, most pathogens were detected in low numbers, and no indication was given of the actual disease potential. Salmonella, although reported to survive as long as six weeks on vegetable surfaces, usually did not survive longer than one week (8). It has been recommended that application of sewage wastes to fruit or vegetable crops be stopped one month before harvest (9). Rudolfs *et al.* (8) recommended a year's lapse after application before growing vegetables that are eaten raw.

Another area of concern in waste utilization and disposal is the danger of infective aerosols from sprinkler application of effluents. It has been reported that workers in municipal treatment plants are exposed to a viable pathogen every other breath when working within 1.6 m of the downwind side of an activated sludge aeration tank, but have fewer sick days than the general population (10). However, Ledbetter (12) showed that, while wastewater treatment workers in Texas have a lower incidence of pneumonia, they had 28 and 58 percent increases in colds and flu, respectively, when compared with water treatment workers. Sewage and laboratory-created sprays leave no aerosol buildup seconds after sprays cease (10). Therefore, the effect of aerosols can be kept to a minimum if effluents are disposed of by spray irrigation on warm, sunny days, with a low wind velocity and a relative humidity between 40 and 60 percent.

Although the disease potential should not be overlooked, the fact remains that raw and processed sewages have been used in agricultural production in Europe and Australia for nearly 100 years with no known adverse effects on animal or human health (1). Disease outbreaks associated with disposal of raw wastes have usually involved accidental contamination of surface water and land. Salmonella outbreaks and worm infections have been reported when fresh sewage has overflowed a cow pasture or vegetables irrigated with sewage have been eaten raw (8, 13).

If used in a manner that protects surface and ground water and human and animal health, animal wastes and sewage sludges can be valuable resources in the production of food and fiber for human and animal consumption (14). However, treated effluents should not be sprinkled on human food crops because of possible treatment failure. Treated and lagoon effluents can be used on grain, forage, and other animal feed crops, but cattle should be restricted from grazing pastures until two to three weeks after sewage sludge and effluent have been applied. Areas that receive raw sewage sludge and effluent applications should be restricted to feed crop production and closed to public access.

Continuing research is needed in the area of pathogens in wastes, and water standards should be changed to reflect new knowledge. The presence of a pathogen and the use of indicator organisms often do not define the source of pollution or the disease potential. The animal host tolerance and resistance to pathogens, as well as virulence factors, still need to be defined to set rational guidelines for waste disposal.

MICROORGANISMS—AGENTS OF DECOMPOSITION

Microorganisms are extremely important "middlemen" between the producers and consumers in biological ecosystems. All organic substances of living organisms are eventually decomposed into elements and simpler compounds that can be recycled. Microorganisms can metabolize a wide array of carbonaceous substances, most of which are used as carbon and energy sources for growth. The role of microorganisms as decomposers is essential to the success of land use for assimilation of organic wastes. When manures or sludges are added to land, degradation and decomposition by soil microorganisms continue until the wastes are converted to gaseous carbon and nitrogen compounds, minerals, water, and microbial cells. The ability of microorganisms to decompose organic materials is enhanced by complexity and metabolic diversity of the microbial populations in soil. Different groups of microorganisms often reach large numbers in the surface of agricultural soils; estimates of 10^8 bacteria, 10^6 actinomycetes, 10^5 fungi, and 10^5 protozoa per gram of soil are representative values (15).

An important concern in assessment of land application of wastes is the effects of disposal practices on microbial processes of decomposition and nutrient cycling. For both crop production and environmental quality, four of the most important microbial processes in soil are mineralization, nitrification, denitrification, and immobilization. Mineralization is the conversion of organic forms of an element to inorganic forms and, for carbon, represents the conversion of organic carbon compounds to carbon dioxide. Mineralization of organic nitrogen compounds results in the formation of ammonium-N or ammonia and, consequently, this transformation also is termed ammonification (Figure 1). The oxidation of ammonium-N to nitrite-N and nitrate-N is called nitrification and is extremely important in the regeneration of nitrate, the preferred form of nitrogen taken up by plants. The nitrification process is largely mediated by autotrophic organisms, which utilize the oxidation of amonium-N and nitrite-N as sole sources of energy. Microbial denitrification can modify the impact of the nitrifying process because nitrite and nitrate forms of nitrogen are converted to nitrous oxide (N_2O) and dinitrogen (N_2). Both these forms of nitrogen are gaseous and can be lost from the soil to the atmosphere. Denitrification requires an organic energy source and the absence of oxygen. Nitrate or nitrite, rather than oxygen, acts as the terminal electron acceptor in the energy-yielding electron transport system of de-

Ammonification

$$\text{ORGANIC N} \longrightarrow NH_4^+ \text{ or } NH_3 \uparrow$$

Nitrification

$$NH_4^+ \longrightarrow NO_2^- \longrightarrow NO_3^-$$

Denitrification

$$NO_3^- \longrightarrow NO_2^- \longrightarrow N_2O \uparrow \longrightarrow N_2 \uparrow$$

Immobilization

$$\text{INORGANIC N} \longrightarrow \text{ORGANIC N}$$

Figure 1.   Microbial transformations of nitrogen important to land disposal of wastes.

nitrifying microorganisms. Immobilization is the process by which inorganic elements are reincorporated into organic cellular components.

Chemical composition and rates of land application of manures and sludges are important considerations in evaluating waste decomposition and transformation into beneficial or hazardous products. The contents of available carbon, nitrogen, salts, and possibly toxic elements will determine the rate at which wastes can be applied to land. Chemical composition of animal wastes and sludges can vary considerably.

A fresh cow manure slurry of feces and urine can have a BOD value as high as 100,000 mg O_2/liter, whereas the BOD of runoff from cattle feedlots is considerably lower and ranges from 100 to 10,000 mg O_2/liter (16). As can be inferred from the data shown in Table 3, the chemical composition of animal manures and sewage sludges is determined by waste type, treatment, and moisture content. Drying and aging of animal manures results in release of ammonia, and more than one-half of

Table 3.   Chemical composition of animal manures and sewage sludges.

	N	P	K	Ca	Mg	Na	Dry Matter % Wet Wt	References
			% Dry Wt					
Animal Manures[a]	2.0	1.2	1.8	–	–	–	–	(1)
23 Feedlot manures	1.3	0.5	1.5	1.3	0.5	0.7	46–79	(2)
Liquid feedlot manure	4.5	1.1	3.1	2.5	–	1.6	11	(5)
Dry dairy manure	1.6	0.6	3.3	2.9	–	0.8	69	(5)
Poultry	3.8	1.6	1.7	3.8	0.5	0.4	–	(19)
Swine	5.1	1.2	2.5	–	–	–	–	(19)
Sewage Sludges	2.0	1.7	0.4	–	–	–	5–20	(1)
7–State average	3.2	1.9	0.3	5.0	0.5	0.4	–	(2)
7–State range	0–17.6	0–6.1	0.02–1.9	0.1–25.0	0.03–2.0	0.01–2.7	–	(2)

[a] Values are expressed as average composition.

the original nitrogen can be lost through volatilization (17). Miller (18) found that activated sewage sludge contained 5.6 percent nitrogen, whereas anaerobically digested sludge contained only 2.4 percent nitrogen.

The treatment of sludges and manures before disposal also affects decomposition rate. Miller (20) found that anaerobically digested sludge did not decompose rapidly in the soil and that a maximum of 17–20 percent of the carbon for soils amended with 90 or 224 metric tons/ha of sludge was evolved as CO_2 after six months. Other studies have shown that the addition of undigested or primary sludge to soils resulted in a 27 percent loss of carbon after 54 days' incubation, whereas a parallel addition of digested sludge resulted in less than a 10 percent loss of carbon during this same time period (21). Because of the differences in degree of digestion of sludges, the nitrogen availability in the soil resulting from mineralization can range from 5 percent per year for stabilized sludge to 70 percent per year for undigested sludges (1). Like sewage sludges, the decomposition rate for animal manures depends greatly on both chemical composition and previous management. Pratt (17) commented that the nitrogen in dairy corral and feedlot manures, which contained 1.6–4.5 percent N on a dry-weight basis, respectively, appears to be mineralized at rates of from 45 to 75 percent the first year, 10 to 20 percent the second year, and 5 percent the third year in irrigated soils and climate of southern California.

Waste applications often cause changes in the indigenous soil microflora. Because of the large and heterogeneous population of soil microorganisms and the slow decomposition rate of soil organic matter (2–10 percent each year), competition is intense for available substrates (22). Addition of manures or other wastes to soil greatly increases available carbon and nitrogen levels, resulting in a burst of microbial activity and stimulation of certain populations. Miller (20) found there was a change from a bacterial population dominated by gram-positive bacteria in an unamended soil to a population where 50 percent or more of the isolates were gram-negative bacteria in sludge-amended soils. Also, the number of spore-forming microorganisms and the average cell size had decreased for sludge-amended soils. Physiological changes in organisms resulting from sludge application included faster relative growth rates, increased tolerance to high salt concentrations, and enhanced ability to grow better at lower temperatures. Biochemical changes for organisms isolated from sludge-amended soils included changes in enzyme activities and substrate use and increased resistance to antibiotics.

Changes in microbial populations resulting from waste application will invariably influence nutrient transformations and cycling. Premi and Cornfield (23) found that moderate additions of sewage sludge (less than 114 kg NH_4–N/ha equivalent in sludge) stimulated nitrification and mineralization of soil organic nitrogen, but higher loading rates (greater than 28 kg NH_4–N/ha equivalent in sludge) delayed nitrification and temporarily immobilized ammonium-N. They attributed the inhibitory effect of high sludge applications on nitrification to the presence of toxic organic materials. In a review of nitrification and denitrification processes as related to wastewater treatment, Focht and Chang (24) commented that organic nitrogen compounds, such as peptone, aniline, guanadines, urethanes, ureas, and pyridines, have been shown to inhibit nitrifiers. Molina *et al.* (25) observed that bactericidal properties of the liquid portion of anaerobically digested sewage sludge were not related to parasitic relationships, protein content, antibiotics, or nutrient competition. Toxic metals also may affect nitrification. Tomlinson *et al.* (26) demonstrated that concentrations of chromium and copper of 120 and 150 ppm, respectively, resulted in a 75 percent inhibition of nitrification in activated sludge. The reduction of nitrification resulting from sludge additions to soil may be due to oxygen limitations brought on by heterotrophic activity.

The oxygen status under which manures and sludges are decomposed will determine the nature of the decomposition processes and products formed more than any other single factor. The rate and completeness of microbial decomposition of wastes and mineralization of organic nitrogen are determined by oxygen supply. Where the availability of oxygen is high, organic wastes are decomposed by the aerobic fungi, actinomycetes, and bacteria; the oxygen is used as a terminal electron acceptor in metabolic processes that result in production of CO_2 and water. Under anaerobic conditions, fungi and actinomycetes are excluded from the metabolizing population and decomposition is taken over by the slower-growing and less efficient anaerobic and facultative anaerobic bacteria. Anaerobic degradation is almost completely a fermentation process, which results in the accumulation of compounds that normally require oxygen for further breakdown.

As can be seen in Table 4, the products formed under anaero-

Table 4. Products formed from manures and sludges subjected to aerobic or anaerobic decomposition.

| Oxygen Status | Compounds of | | |
	Carbon (& Hydrogen)	Nitrogen	Sulfur
Aerobic	CO_2 (60%) microbial cells (40%) H_2O	$NH_4^+ \rightarrow NO_3^-$	$H_2S \rightarrow SO_4^=$
Anaerobic	CO_2 (20%) microbial cells (5%) Organic intermediates[a] (70%) CH_4 (5%) H_2	NH_3 N_2 (from NO_3^-) organic N intermediates[b]	H_2S mercaptans

[a] Alcohols and organic acids.

[b] Pyridines, amines, indoles, skatoles, etc.

bic conditions are quite different than those formed in the presence of oxygen. The aerobic decomposition of readily available wastes results in the production of CO_2, microbial cells, water, and inorganic compounds of nitrogen and sulfur. With anaerobic decomposition, yields of CO_2 and microbial cells are much lower, and most of the carbon products formed consist of residual organic intermediates, such as alcohols and organic acids. Undesirable consequences of anaerobic decomposition of wastes are the production of odorous compounds, such as amines, mercaptans, indoles, skatoles, and hydrogen sulfide, and the production of phytotoxic compounds, such as ammonia and hydrogen sulfide.

Loading rates for waste disposal on land are important because both the availability of carbon and the high moisture of some wastes will affect the oxygen status in soils. Soil moisture content alone can be an important consideration because, in wet soils, decomposition activity is dominated by bacteria, whereas in drier soils, fungi and actinomycetes are much more active. Parr (16) suggested that by applying wastes at acceptable loading rates (20 to 70 metric tons/ha, dry-weight basis), problems associated with waste decomposition, anaerobiosis, and undesirable end-products could be minimized. However, excessive loading of soils with waste materials (greater than 100 metric tons/ha) will result in oxygen depletion, slower decomposition, and accumulation of odorous, phytotoxic end products that could reduce soil productivity.

MICROBIAL PRODUCTION AND MOBILIZATION OF POLLUTANTS AND TOXIC SUBSTANCES

Microorganisms play a central role in some chemical transformations that produce or mobilize compounds toxic to plants or animals or degradative to the environment. Aside from surface runoff and crop removal, nutrients and toxic substances are most commonly lost from soil by leaching or volatilization. Microorganisms are essential catalysts in both solubilization of elements and production of volatile compounds. Generally, undesirable microbial products can be formed in two ways when wastes are applied to land. First, the stimulation of microbial activity can result in a direct transformation of elements or substances contained in the soil itself. Second, microbial decomposition of the waste can result in release of substances contained in the waste that are either directly toxic or are transformed to toxic products.

Often, when large quantities of animal manures or sewage sludge are applied to land, nitrification of ammonium-N released from these wastes can produce nitrate levels that far exceed plant needs, which may result in leaching and contamination of ground water with nitrate. Excess nitrate production on waste-amended soils has been associated with toxicity to livestock grazing high-nitrate forage, eutrophication of inland waters, and public health hazards when potable water supplies contain nitrate-N concentrations exceeding 10 ppm (27). Problems associated with nitrogen in wastes generally can be forecast by determining the nitrogen content of the waste applied to the land.

During the decomposition of manures and sludges, the C/N ratio of these materials either increases or decreases until it approaches 10:1, the ratio of native soil organic matter. For wastes low in nitrogen, the C/N ratio decreases with time as carbon is lost as CO_2, but nitrogen is retained through assimilation into microbial cellular material. For wastes high in nitrogen, with a C/N ratio of less than 10:1, nitrogen is mineralized and released to the soil with waste decomposition because more nitrogen is present than the decomposing microorganisms need.

Since most manures, sludges, and plant residues contain about 40 to 50 percent organic carbon, immobilization or mineralization of nitrogen can be estimated from nitrogen content. The nitrogen contents of several organic wastes and residues are shown in Table 5. The addition of materials containing more

Table 5.   Total nitrogen content of some organic wastes and residues.[a]

	Nitrogen (% of dry matter)
Cattle feedlot manure	1.2 - 2.0
Fresh cow manure, litter-free	2.4
Poultry manure	3.5 - 5.0
Municipal waste compost	1.0
Sewage sludges	2.0 - 6.0
Corn stover	0.9
Wheat straw	0.5

[a] After Parr (16).

than 1.2 to 1.5 percent nitrogen to soil (C:N < 10:1) will result in the mineralization and release of nitrogen to the soil where it can be taken up by plants, lost as a volatile product, fixed in the soil, or leached into ground water. The addition to soil of materials containing less than 1.2 percent nitrogen will result in immobilization of soil nitrogen. It is well known, for example, that large quantities of wheat straw or corn stover added to soil will induce stunting and nitrogen deficiency in crop plants unless nitrogen fertilizer is added with the residues.

The effects of nitrification on nitrate levels in soils can be greatly modified by microbial denitrification. Ellis et al. (28) found that anaerobic conditions in the profile of stocked Nebraska feedlots resulted in denitrification of nitrate produced at feedlot surfaces, thus circumventing nitrate pollution of ground water. However, when feedlots were abandoned, the oxygen levels of profiles increased, resulting in stimulation of microbial nitrification and contamination of ground water with levels of nitrate-N as high as 77 ppm (Table 6). In the treatment of nitrogenous wastes, oxygen levels can be altered to remove nitrogen. Adequate oxygen levels during initial treatment stages result in microbial nitrification of ammonium-N to nitrite-N and nitrate-N, both of which can be removed as gaseous nitrogen compounds through microbial denitrification if anaerobic conditions are induced during final stages of treatment (29).

Table 6. Nitrate- and ammonium-nitrogen in ground water samples of feedlots and cropland.[a]

Site Description	NH_4^+ -N (ppm)		NO_3^- -N (ppm)	
	High	Mean	High	Mean
Upland feedlots	22.9	1.7	38.6	7.2
Flat (river valley) feedlots	10.2	2.6	11.4	2.6
Abandoned feedlots	0.5	0.4	77.2	40.5
Cropland-corn	2.3	0.5	14.9	6.2
Grassland	2.7	1.4	1.7	1.1

[a] From Ellis, *et al.*, (28).

The volatilization of ammonia from organic wastes represents both the loss of a valuable plant nutrient and potential for eutrophication of surface waters with readsorption. Ammonia can be produced by conversion of ammonium-N to ammonia-N at high pH, through the hydrolysis of urea, or as a result of microbial breakdown of uric acid. Beef cattle feedlots and dry-lot dairies release large amounts of ammonia to the atmosphere (30, 31) and volatilization of ammonia may represent a significant pathway by which nitrogen is transported to surface waters in areas receiving animal wastes (32).

The microbial production of phytotoxic substances may be an important consideration in land disposal of organic wastes. Cropping systems are frequently used to aid in assimilation and recylcing of nutrients contained in waste materials. Most of the research on microbial production of phytotoxic substances has been related to use of crop residues. McCalla and Norstadt (33) stated that the microbial activities stimulated by residue application could result in the production of many phytotoxic substances. The amount of plant growth inhibited by phytotoxicity depended greatly on how much the residue had decomposed in the soil (34). Little information is available with respect to the production of phytotoxic substances from sludges and manures applied to soil, but phytotoxins probably will be produced. The addition of animal manure to soil does result in the production of ethylene, a gas that retards both root growth and plant development (35). The anaerobic decomposition of manures and sludges, as stimulated by excessive application rates, will lead to the production of ammonia, hydrogen sulfide, and alcohol, all of which can be toxic to plant germination and growth.

Trace element toxicities are an important consideration in the land application of wastes. It has been stated that a major factor limiting the long-range use of sewage sludge on agricultural land is the potential for buildup of zinc, copper, nickel, cadmium and other toxic elements in the soil (1). Much of the concern associated with heavy metal content of sludges has been related to crop productivity and quality. Biomagnification of many toxic elements is often halted by toxicity to one of the food chain components, but with selenium, cadmium, molybdenum and possibly lead, food plants can grow at normal or near-normal rates and still contain sufficient quantities of these elements to cause toxicity or imbalance in animals consuming these plants (36). Concentrations of trace elements in sewage sludges vary widely and ranges reported by Page (37) for sludges from approximately 300 treatment plants in the United States, Canada, Sweden, England and Wales were as follows (in $\mu g/g$ dry matter): silver (5–150), boron (5–1,000), cadmium (1–1,500), chromium (20–40,615), copper (52–11,700), molybdenum (2–1,000), nickel (10–5,300), lead (15–2,600), and zinc (72–49,000). Excessive concentrations of heavy metals in sewage sludge usually result from industrial activities. Animal wastes can also contain toxic elements. Manure concentrations of copper and zinc are often elevated because these elements frequently are added to feed (1). Inorganic and organic arsenic is fed to both poultry and swine (19), and recently selenium was cleared as a feed additive. As with sewage sludges, the trace element content of animal waste varies greatly.

Soil microorganisms continuously produce compounds that can act as chelates in forming soluble organic complexes with metallic cations. The microbial formation of mineral acids (nitric and sulfuric) can lead to solubilization of oxide, hydroxide, and phosphate precipitates of many metals. Microbial metabolism of organic materials also can cause solubilization of metals indirectly through production of reducing conditions.

The microbial alkylation of trace elements may be a major concern in the land application of wastes, as this transformation often results in both mobilization and change in toxicity of many elements. Methylation is an essential metabolic process of all biological systems. Consequently, it is not surprising that microorganisms can methylate many elements. The microbial metabolism of inorganic and organic forms of many elements results in the formation of many different methylated compounds, including at least one volatile form for each element (Table 7).

Table 7.   Microbial formation of methylated compounds.

Element	Methylated Product	Reference
Hydrogen	$CH_4 \uparrow$ [a]	41
Lead	$(CH_3)_4Pb \uparrow$	40
Mercury	$(CH_3)_2Hg \uparrow$, CH_3Hg^+	42
Arsenic	$(CH_3)_2AsH \uparrow$, $(CH_3)_3As \uparrow$	
	$CH_3AsO(OH)_2$, $(CH_3)_2AsO(OH)$	38, 41
Sulfur	$(CH_3)_2S \uparrow$, CH_3SH	
	$(CH_3)_2S_2$	39
Selenium	$(CH_3)_2Se \uparrow$, CH_3SeH, $(CH_3)_2Se_2$	39, 43
Tellurium	$(CH_3)_2Te \uparrow$	39

[a] $\uparrow$ = volatile compounds

The methylated forms of lead, mercury, arsenic and tellurium are more toxic than their inorganic precursors (38–40). Aside from toxicity, the methylated compounds of sulfur, selenium, tellurium and arsenic are extremely malodorous at very low concentrations.

Methylated compounds of mercury, arsenic, sulfur and selenium can be formed microbially under either aerobic or anaerobic conditions in soil and sediments (42, 44, 45). Banwart and Bremner (46) have shown that methylated compounds of sulfur are produced from manure under both anaerobic and aerobic conditions. Many organic selenium compounds can be rapidly metabolized to form methylated products. The metabolism of trimethylselenonium, a compound formed in the urine of animals metabolizing selenium, by soil microorganisms can result in release of up to 90 percent of the selenium in this compound as volatile dimethylselenide within 32 days (45).

The methylation process is energy-requiring in that energy is needed for methyl group synthesis and transfer or for the reduction of many inorganic elements that must first be reduced before they can be methylated. Consequently, the addition of manures and sludges to land will stimulate microbial methylation by increasing carbon availability and microbial activity. Adding wastes high in lead, mercury, arsenic, and selenium to soils undoubtedly will result in the formation of methylated compounds of these elements. The addition of an available carbon source to soil will markedly stimulate methylation and volatilization of both added and indigenous selenium (43). Many soils in the Western United States produce vegetation which contains selenium at levels toxic to animals. The land disposal of wastes could modify such toxicities, as the addition of large quantities of organic matter to seleniferous soils will result in volatilization and loss of selenium from soil (45).

The total number of elements that can be methylated by microorganisms is uncertain. Arguments based on chemical properties have suggested that cadmium, zinc, lead, tin, and germanium would not be methylated biologically because either intermediates for methylation are unstable or the methylated compounds have never been successfully synthesized (47). However, as illustrated by the recent demonstration of the microbial methylation of lead (40), the use of chemical properties to predict biological methylation potentials is not always reliable. Further research is needed to determine the effects of land application of wastes on microbial methylations.

SUMMARY

Land application appears to be the most feasible means for disposal of both animal and domestic sewage wastes. Potential for transmission of disease as a result of waste application to land is negligible when fecal materials receive adequate pretreatment and are applied to land at acceptable rates and in a manner in which rapid percolation and runoff are minimized. However, surface and ground waters should be monitored frequently for indications of fecal contamination in areas where wastes are applied to land.

Important concerns in land application will be rate of waste decomposition and the production of undesirable microbial products. Excessive additions of organic materials to soil will result in the production of anaerobic conditions and the formation of both odorous and phytotoxic compounds. Under such conditions, waste decomposition is decreased and soil buildup of organic matter can limit the capacity of soil to accept wastes. Proper loading of organic wastes in soil can result in optimal balance between removal of nitrogen and reduction in pollution. The stimulation of microbial methylation by wastes added to soil may be important in that this transformation can result in mobilization of many toxic trace elements through volatilization.

ACKNOWLEDGMENTS

This chapter is a contribution from the Soil, Water, and Animal Waste Management Research Unit, North Central Region, Agricultural Research Service, U.S. Department of Agriculture, in cooperation with the Nebraska Agricultural Experiment Station, Lincoln.

REFERENCES

1. CAST Report. "Utilization of animal manures and sewage sludges in food and fiber production," Report No. 41, Council for Agricultural Science and Technology (P. F. Pratt, Task Force Chairman), Headquarters Office: Dept. of Agronomy, Iowa State Univ., Ames, Iowa (1975).
2. McCalla, T. M., J. R. Peterson, and C. Lue-Hing. "Properties of agricultural and municipal wastes," in: *Proc., Symp. on Soils for Management and Utilization of Organic Wastes and Waste Waters*, Muscle Shoals, Alabama, March 11–13, 1975, Soil Science Society of America, Madison, Wisconsin (1976).
3. USDA. *Agricultural Statistics*, U.S. Govt. Printing Office, Washington, D.C. (1975).
4 Cran, G. F., and L. J. McCabe. "Review of the causes of waterborne disease outbreaks," *J. Amer. Water Works Assoc.* 65:74–84 (1973).
5. Diesch, S. L. "Survival of leptospires in cattle manure," *J. Amer. Vet. Med. Assoc.* 159:1513–1517 (1971).
6. Dunlop, S. G. "Survival of pathogens and related disease hazards," pp. 107–122. in: *Proc. Symp. on Municipal Sewage Effluent for Irrigation*, C. W. Wilson and F. E. Beckett, Eds., Louisiana Polytechnic Inst., Ruston, Louisiana (1968).
7. Foster, D. H., and R. S. Engelbrecht. "Microbial hazards in disposing of waste water on soil," pp. 247–270. in: *Recycling Treated Municipal Wastewater and Sludge Through Forests and Cropland*. W. E. Sopper and L. T. Kardos, The Pennsylvania State Press, University Park, Pennsylvania (1973).
8. Rudolfs, W., L. L. Falk, and R. A. Ragotzkie. "Contamination of vegetables grown in polluted soil, VI. Application of results," *Sewage Ind. Wastes* 23:992–1000 (1951).
9. Sepp, E. "The use of sewage for irrigation: a literature review," California State Dept. Publ. Health, Bur. Sanitation Eng., Berkeley, California (1971).
10. Burge, W. D. "Pathogen considerations," pp. 37–50, in: *Factors Involved in Land Application of Agricultural and Municipal Wastes*, USDA–ARS National Program Staff, Soil, Water, and Air Sciences, Beltsville, Maryland (1974).
11. Gerba, C. P., C. Wallis, and J. L. Melnick. "Fate of wastewater bacteria and viruses in soil," *Amer. Soc. Civil Eng. Proc., Irr. Drain. Div.* IR3: 157–174 (1975).

12. Ledbetter, J. O. "Health hazards from wastewater treatment practices," *Environ. Lett.* 4:225–232 (1973).

13. Sorber, C. A., and K. J. Guter. "Health and hygiene aspects of spray irrigation," *Amer. J. Publ. Health* 65:47–52 (1975).

14. Menzies, J. D. "Pathogen considerations in land application of organic wastes," in *Proc. Symposium on Soils for Management and Utilization of Organic Wastes and Waste Waters*, Muscle Shoals, Alabama, March 11–13, 1975, Soil Science Society of America, Madison, Wisconsin (1976).

15. Alexander, M. *Introduction to Soil Microbiology.* John Wiley & Sons, Inc., New York (1961), 472 p.

16. Parr, J. F. "Organic matter decomposition and oxygen relationships," pp. 129–139, in: *Factors Involved in Land Application of Agricultural and Municipal Wastes*, USDA–ARS Spec. Publ. (1974).

17. Pratt, P. F. Personal communication (March 1976).

18. Miller, R. H. "The soil as a biological filter," pp. 71–94. in: *Recycling Treated Municipal Wastewater and Sludge Through Forest and Cropland*, W. E. Sopper and L. T. Kardos, Eds., The Pennsylvania State University Press, University Park, Pennsylvania (1973).

19. Azevedo, J., and P. R. Stout. "Farm manures: an overview of their role in the agricultural environment," Univ. of California, Div. Agri. Sci. Manual 44, Agricultural Publication, Univ. of California, Berkeley (1974).

20. Miller, R. H. "Soil microbiological aspects of recycling sewage sludges and waste effluents on land," pp. 79–90, in: *Proc. Joint Conf. on Recycling Municipal Sludges and Effluents on Land*, Champaign, Illinois, July 9–13, 1973, Natl. Assoc. State Univ. and Land-Grant Colleges, Washington, D.C. (1973).

21. ARS, USDA. First progress report, "Incorporation of sewage sludge in soil to maximize benefits and minimize hazards to the environment," Beltsville, Maryland (1972).

22. Broadbent, F. E. "Organics," pp. 97–101, in *Proc. Joint Conf. on Recycling Municipal Sludges and Effluents on Land*, July 9–13, 1973, Champaign, Illinois, Natl. Assoc. State Universities and Land-Grant Colleges, Washington, D.C. (1973).

23. Premi, P. R., and A. H. Cornfield. "Incubation study of nitrification of digested sewage sludge added to soil," *Soil Biol. Biochem.* 1:1–4 (1969).

24. Focht, D. C., and A. C. Chang. "Nitrification and denitrification processes related to wastewater treatment," *Advan. Appl. Microbiol.* 19:153–186 (1975).

25. Molina, J.-A. E., O. C. Braids, and T. D. Hinesly. "Observations on bactericidal properties of digested sewage sludge," *Environ. Sci. Technol.* 6:448–450 (1972).

26. Tomlinson, T. G., A. G. Boon, and C. N. A. Trotman. "Inhibition of nitrification in the activated sludge process of sewage disposal," *J. Appl. Bacteriol.* 29:266–291 (1966).

27. McCalla, T. M. "Animal waste recycling," presented, Amer. Natl. Cattlemen's Assoc., Environmental Sciences Comm., Phoenix, Arizona (January 26, 1976).

28. Ellis, J. R., L. N. Mielke, and G. E. Schuman. "The nitrogen status beneath beef cattle feedlots in eastern Nebraska," *Soil Sci. Soc. Amer. Proc.* 39:107–111 (1975).

29. Prakasam, T. B. S., and R. C. Loehr. "Microbial nitrification and denitrification in concentrated wastes," *Water Res.* 6:859–869 (1972).

30. Elliott, L. F., G. E. Schuman, and F. G. Viets, Jr. "Volatilization of

nitrogen-containing compounds from beef cattle areas," *Soil Sci. Soc. Amer. Proc.* 35:752–755 (1971).

31. Hutchinson, G. L., and F. G. Viets, Jr. "Nitrogen enrichment of surface water by absorption of ammonia volatilized from cattle feedlots," *Science* 166:514–515 (1969).

32. Luebs, R. E., K. R. Davis and A. E. Laag. "Enrichment of the atmosphere with nitrogen compounds volatilized from a large dairy area," *J. Environ. Quality* 2:137–141 (1973).

33. McCalla, T. M., and F. A. Norstadt, "Toxicity problems in mulch tillage," *Agric. Environ.* 1:153–174 (1974).

34. Guenzi, W. D., T. M. McCalla and F. A. Norstadt. "Presence and persistence of phytotoxic substances in wheat, oat, corn, and sorghum residues," *Agron. J.* 59:163–165 (1967).

35. Burford, J. R. "Ethylene in grassland soil treated with animal excreta," *J. Environ. Quality* 4:55–57 (1975).

36. Allaway, W. H. "Agronomic controls over environmental cycling of trace elements," *Advan. Agron.* 20:235–274 (1968).

37. Page, A. L. "Fate and effects of trace elements in sewage sludge when applied to agricultural lands," EPA 670/2–74–005, U.S. Environmental Protection Agency, Office of Research and Development, Cincinnati, Ohio (1974).

38. Alexander, M. "Microorganisms and chemical pollution," *BioScience* 23:509–515 (1973).

39. Challenger, F. "Biological methylation," *Advan. Enzymol.* 12:429–491 (1951).

40. Wong, P. T. S., Y. K. Chau, and P. L. Luxon. "Methylation of lead in the environment," *Nature* (London) 253:263–264 (1975).

41. McBride, B. C., and R. S. Wolfe. "Biosynthesis of dimethylarsine by methanobacterium," *Biochemistry* 10:4312–4317 (1971).

42. Wood, J. M. "Biological cycles for toxic elements in the environment," *Science* 183:1049–1052 (1974).

43. Doran, J. W., and M. Alexander. "Microbial formation of dimethyl selenide," paper presented before Ann. Mtg., Amer. Soc. Microbiol., New York (April 29, 1975).

44. Becker, W. F., A. A. Moghissi, F. H. F. Au, E. W. Bretthauer, and J. C. McFarlane. "Formation of methylmercury in a terrestrial environment," *Nature* (London) 249:674–675 (1974).

45. Doran, J. W. "Microbial transformations of selenium in soil and culture," Ph.D. dissertation. Cornell University (January 1976).

46. Banwart, W. L., and J. M. Bremner. "Identification of sulfur gases evolved from animal manures," *J. Environ. Quality* 4:363–366 (1975).

47. Thayer, J. S. "Biological methylation—its nature and scope," *J. Chem. Educ.* 50:390–391 (1973).

19

POTENTIAL FOR ADVERSE HEALTH EFFECTS ASSOCIATED WITH THE APPLICATION OF WASTEWATERS OR SLUDGES TO AGRICULTURAL LANDS

D. E. Weaver and J. L. Mang
SCS Engineers, Long Beach, California
W. A. Galke and G. J. Love
U.S. Environmental Protection Agency
Research Triangle Park, North Carolina

Of the multitude of problems routinely facing urban areas today, one of the most urgent is that concerning the disposal of large volumes of liquid and solid waste generated in the metropolitan area. The ecological problems associated with direct discharge into surface waters, the constant dwindling of fresh water resources, and the difficulty in locating suitable landfill sites for disposal of sludges have forced various governmental agencies to examine alternate disposal practices. One alternate disposal practice that is currently receiving renewed interest is the application of wastewaters and sludges to agricultural lands. The utilization of agricultural lands as a depository for wastewaters and wastewater sludges for both disposal and agricultural enrichment purposes is not a new concept by any means. It is one that has been exploited for centuries.

This solution, however, has the attendant disadvantage of providing a closed-loop system for the propagation of various pathogenic microorganisms that can cause adverse health effects in man. This problem can be viewed in terms of a biological analogy to an electrical circuit. A human infected with a pathogenic organism provides a gain in the system through multiplication of the microorganism within the intestinal tract of that human being. Wastewater treatment operations will provide certain system losses as will other factors in the transmission of the microorganism through the environment to man. Whenever the overall system gain exceeds one, the population of pathogenic microorganisms will exponentially increase, thereby creating an unstable system. Figure 1 schematically illustrates such a situation which would apply to a pathogen with an infective dose of one organism. Under these conditions, clearly an epidemic can result.

The obvious advantage of the application of wastewaters and sludges to agricultural lands is the inherent recycling of nutrients that were derived from soil by plants and utilized at various points in the human food chain. In agrarian societies, the majority of constituents found in waste being returned to agricultural lands came originally from these lands in the form of plant matter. The total system, therefore, could be thought of as a closed-loop system. In modern industrial societies, this simplistic system is significantly complicated by the addition of large amounts of materials to wastewater streams that originate from sources other than the food supply. In particular, these materials include trace element contaminants, synthetic/organic chemicals, and biocidal materials. These additional materials along with pathogenic microorganisms pose potential adverse health consequences to mankind when wastewaters and sludges are applied to agricultural lands.

The questions concerning the potential health significance of the application of wastewaters and sludges to agricultural lands cannot be looked at in a vacuum, however, but must be considered in the same context as other alternative methodologies. This concept was utilized as part of an Environmental Protection Agency-sponsored study entitled "State-of-the-Art Evaluation of Potential Health Effects Associated with Wastewater

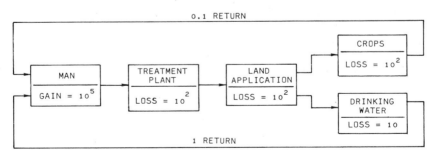

$$\text{OVERALL SYSTEM GAIN} = 10^5 \left(\frac{1}{10^5} + \frac{1}{10^6} \right) = 1.1$$

Figure 1. Example of a potential epidemic situation where the overall system has a gain greater than one.

Treatment and Disposal Systems." Under the auspices of this study, existing knowledge has been evaluated to determine potential public health problem areas which will require future research. The most important conclusion from this study is that for the majority of contaminants found in wastewater streams, only limited and incomplete knowledge is available concerning their pathways to mankind, the gains or losses in these pathways, and their roles in producing adverse health effects upon reaching the public.

Various environmental pathways returning contaminants to man were evaluated in this study. Figure 2 shows a simplified schematic of the pathways considered. With respect to this conference, however, the applicable pathways lie within the area of the dashed box. The contaminants considered in this study were classified into the five following categories:

Elemental Contaminants—includes elements and their various chemical forms.

Synthetic/Organic Contaminants—includes a wide variety of chemicals which have been introduced either as part of industrialization, naturally-occurring organic chemicals, or organic chemicals produced by the treatment processes.

Biocidal Contaminants—includes those chemicals that have been specifically designed for biocidal purposes.

Biological Contaminants—includes viruses, bacteria, fungi, and other microorganisms.

General Contaminants—includes those constituents found in wastewaters that do not belong to any of the above categories. The constituents include the most commonly measured parameters such as BOD, COD, nitrogen species, phosphorous species, etc.

The pathways through the biosphere by which these contaminants may travel from agricultural land application operations to mankind can be seen in Figure 3. This illustration represents the major environmental routes from this source to man. In relation to the application of wastewaters and sludges to agricultural lands, three separate recipients, as shown in Figure 3, of contaminants must be considered because each of these can receive different contaminant dosages from the same original source. Those local workers who are directly exposed to wastewaters and sludges will receive different dosages of contaminants than the general public who may be consuming crops grown in these land application situations. With a wide variety of pathways affecting different segments of the populace, the potential adverse health effects and methodology for controlling these problems vary greatly.

In further considering the transfer through the biosphere, it becomes clear that different contaminants will travel through

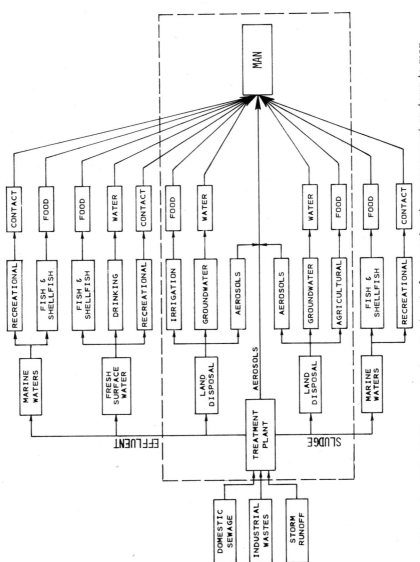

Figure 2. Potential transport of contaminants from wastewater management systems to man.

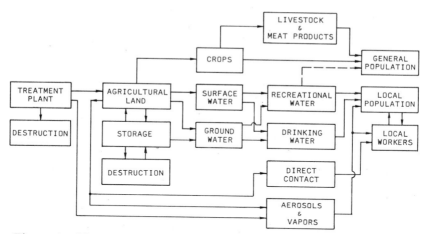

Figure 3. Major pathways for contaminant transfer back to man from land application of wastewater or sludge.

different pathways. In discussing the potential health effects on the various segments of the populace, it is convenient to divide the contaminants into two groups, conservative or nonconservative materials.

The conservative materials can be thought of as those which are not destroyed or degraded by any of the pathways in question. On the opposite side of the spectrum, the nonconservative materials, such as the biologicals and certain synthetic/organic chemicals, may actually be destroyed or degraded at several points along the environmental pathways leading to man.

Considering the conservative materials, the pathways through the biosphere can be analyzed in terms of conventional mass balance considerations. The pathways chart has been modified in Figure 4 to conform to the situation that can be expected for various conservative materials such as the elemental contaminants. One of the most important concepts in the land application of wastewaters and sludges is that of the storage function provided by soils for the conservative materials.

The storage function in any flow system of this type creates an effective time constant for system responses. The actual time constants depend upon each individual contaminant, the mass flow rate through the soil, and the interactions among the contaminants themselves and with various components of the soil system. Contaminants that have weak bonding properties, such as sodium (Na), will have limited storage capacity and would consequently appear in the ground water after short periods of

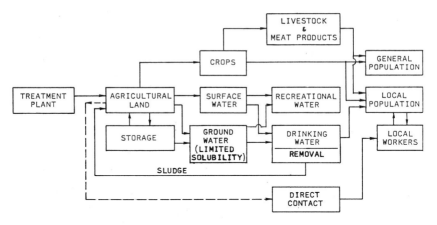

Figure 4. Major pathways for conservative contaminant transfer back to man from land application of wastewater or sludge.

time. Conversely, a contaminant which is strongly absorbed by soil, such as some of the multivalent heavy-metal cations, may have time constants measured in centuries or eons. These long time periods provide an excellent "short-term" solution to the disposal problem. However, if wastewaters or sludges are continually applied for a time period comparable to this time constant, the situation will be reached where the output concentration in the ground water is determined by the solubility limit of each individual contaminant rather than by soil adsorption. From the preceding discussion, it is clear that on a geological time scale, the conservative contaminants that reach soils from land applications will eventually be mobilized with the oceans being the final repositories.

Concerning the nonconservative materials a flow diagram depicting the environmental pathways to mankind is presented in Figure 5. In the case of nonconservative materials, the actual destruction or degradation of these contaminants at various points in the pathways becomes exceedingly important. The storage function of soils in land application practices allows time for degradation, and the relative time constants for the various materials become functions of both the degradation rates and storage capabilities.

For a majority of the nonconservative materials, the storage capacity of soil combined with a degradation or destruction rate will limit transmission of these contaminants by way of the ground water pathway. Clearly then, pathways which have limited storage capabilities, such as aerosols, can become exceed-

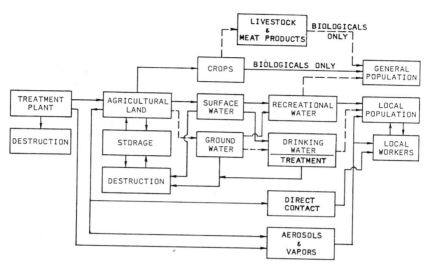

Figure 5. Major pathways for nonconservative contaminant transfer back to man from land application of wastewater or sludge.

ingly important for the nonconservative biological contaminants even though the mass transfer to man via this pathway is very small. Considering nonconservative, nonbiological contaminants, the primary pathway to man would be through runoff from application of wastes to agricultural lands to surface waters which can be a water supply for the general populace. This drinking water pathway can potentially transfer significant amounts of contaminants, including biologicals.

The case of fixed nitrogen forms provides an exception to the preceding categorization of conservative and nonconservative materials. In this situation, the reduced nitrogen forms, such as NH_3, can be converted to oxidized forms, such as NO_3. The reduction of the oxidized nitrogen forms, however, results in nitrogen gas. Under this condition, nitrogen is still conserved, but consideration must be given to the fact that it exists in an inert form as N_2 gas. The only species of nitrogen that have any health significance are those in an oxidized state. The pathway chart for the nitrogen species is presented in Figure 6. It must be noted that agronomic rates of application result in nitrogen leaving the soil in the form of protein contained in crops rather than as a water-borne contaminant.

Beyond the simple classification of contaminants into conservative and nonconservative materials, each contaminant must be examined individually and synergistically to determine which

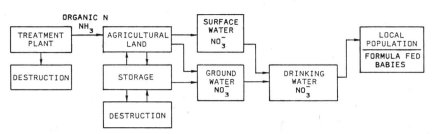

Figure 6. Major pathways for nitrogen forms of health significance returning to man from land application of wastewater or sludge.

pathways through the biosphere will return what quantities to the populace and whether this return rate is of any potential health significance. Clearly, the application of wastewaters and sludges to agricultural lands must not be looked at in a vacuum, but should be considered along with other disposal practices. When the environmental pathways from disposal operations have been clearly defined, then objective decisions can be made as to which system can be applied with public health concerns being of prime consideration.

20

PATHOGEN SURVIVAL IN SOILS RECEIVING WASTE

S. M. Morrison and K. L. Martin
Department of Microbiology, Colorado State University,
Fort Collins, Colorado

INTRODUCTION

One of the important factors that allowed the human race to survive into modern times was the development of the concept of soil replenishment by the return of animal and human wastes to agricultural land. With sparse population densities and the lack of knowledge and control of the transmission of infectious disease agents, the dangers of pathogenic organism survival in soils and transmission of disease by the soil route was only a minor epidemiological factor. At the present time, however, we have a completely different set of circumstances. The high densities of human and animal population in close proximity to waste disposal sites and our extreme concern for all modes of transmission of disease-producing agents makes it imperative that we better understand the survival patterns of pathogens when they are put into a soil environment.

In choosing our topic for review, we naively felt that assembling the existing literature would not be difficult and that, hopefully, a pattern or scheme of information might emerge which could be presented as well documented scientific truths or conclusions. For many reasons the conclusions are elusive and we only hope that we do not add to the information turmoil that exists on this topic.

Because soil disposal has been such a long established procedure, a great many reports in the technical literature discuss and describe it as an economical, effective and safe means of handling wastes. Very few of these reports indicate any scientific investigation of the dangers or the control of viable pathogenic organisms that may be present. There are reports of pathogen survival in animal manures, in human fecal materials, in slurries and sludges of these wastes, in fluid systems such as lagoons, ponds or irrigation systems and in compost and sanitary landfills. Also, there are numerous studies of the longevity of nonpathogenic agents such as the index organisms of environmental pollution and bacteriophages in wastes, waters and soils. Also, many tests have been performed under artificial experimental conditions of the laboratory. As valuable as many of these reports are, they do not directly provide the needed information on the behavior of disease-producing agents in the soil disposal system so as to minimize the health hazards. A typical example of the difficulty in evaluating the health implications is a fine scientific paper recently presented, entitled "Virus Migration through Soils." In the conclusions the authors (1) state, "The implications of these data for terrestrial waste disposal projects are rather limited since water was used as eluent instead of treated sewage effluent and much larger quantities of virus were used than would be found under the conditions of irrigation projects."

EXISTING INFORMATION

A great deal of variability exists in the reported information on survival of organisms when they are applied to soil or when laboratory simulations of this process are tested. A great number of variables control the results and make for some difficulty in comparing results; these hard-to-control variables also may have had a limiting effect on the research that has been done. Among these important variables are: a) the organism and its physiologic state, b) the physical and chemical nature of the receiving soil, such as pH, porosity, and organic and elemental content, c) atmospheric conditions, particularly moisture, temperature and, possibly, sunlight, d) the mode of application of wastes to the soil as well as organism loading and frequency of application, e) the biological interactions of organisms in the waste and organisms in the waste-soil environment, and f) the methods of detecting and enumerating organisms in soils and their limitations.

One additional factor, out of the pervue of this report, is the host-parasite relationships, the ability of the specific organism to cause disease and of the host to resist infection.

A group of bacteria, the spore formers, are very resistant to die-off if their spores are incorporated into soils. Pasteur (2) reported survival of *Bacillus anthracis* in soil for 12 years and Koch (3) isolated the organism from earthworms. Wilson and Russell (4) opened a 60-yr-old vial of soil and found viable anthrax organisms. The aerobic *Bacillus anthracis* could be a serious health problem if anthrax were not a relatively rare disease in our animal and human population and if we were lax in controlling disposal of infected dead animals, their wastes and contacts.

The spores of pathogenic anaerobes as *Clostridium tetani,* *C. perfringens* and *C. botulinum* also persist for long periods and are known to be abundant in manured soils (5). Large numbers of vegetative cells of these clostridia are also present in soils (6) and these may survive very well. Garcia and McKay (7) showed that *C. septicum* remained viable for 30 days in both sterile and nonsterile soil. Strains low in "sporulating potency" seem to be highly toxigenic (8) and several have been isolated from soils. Aeration of soil is detrimental to these vegetative organisms and the exotoxin of *botulinum* is destroyed by aerobic soils (9) but heavy, water-logged soils encourage them. Soils, particularly those with waste additions, have always presented a serious health threat from tetanus, enteritis, wound infections and other diseases. Our defense has and will continue to be general sanitation, immunization, treatment of wounds and proper handling and storage of food products.

The difficulty in arriving at a simple formula for predicting the survival of a pathogenic bacterium in soil is illustrated very clearly by a historical review of the work done by many workers over an 85-yr period with *Salmonella typhi*. Table 1 summarizes this information. Typhoid fever is not a disease of, nor is it shed by, farm animals and it has fortunately declined in humans in the United States so the danger of disseminating it through contaminated soil is very low. In 1889, Grancher and Deschamps (10) isolated the organisms from soil 5.5 months after contamination, while Karlinski (11) reported 3 months. While Martin (12) found briefer survival in sterilized soil, Smith (13) believed die-off would be faster in unsterile soil. Beard (14), in 1940, tried to correlate the results of much of the earlier work and carried out extensive studies. All workers showed the importance of moisture to survival. Beard felt that sunlight might affect the surface organisms and that inocula of typhoid bacilli survived longer in fresh compared to sewage-contaminated soils.

Table 1.   Selected chronology of reported survival of *Salmonella typhi* in soils.

Researchers	Year	Application or Soil Type	Survival Time
Grancher & Deschamps (10)	1889	AC[a]	5.5 mo
Karlinski (11)	1889–1891	AC	3 mo
Robertson (15)	1898	AC	86-315 days
Martin (12)	1898–1901	AC-unsterile soil	404 days
Rullman (16)	1901	organically applied unsterile soil	100 days
Demster (17)	1902	AC-various soils	12–42 days
Firth & Horrocks (7)	1902	AC	32 and 55 days
Sedgwick & Winslow (18)	1902	dry soil	2 weeks
		moist soil	>2 weeks
Claudits (19)	1904	AC	70 days
Mair (20)	1908	sterile soil	9 days
		unsterile soil	42 and 74 days
Melick (21)	1917	infected feces	74 days
Murillo (22)	1919	garden soil	30–36 days
		sterile sand	55 days
Kligler (23)	1921	AC	70 days
Grandi (24)	1930	AC	20 days
Beard (14)	1940	various soils-wet	30–120 days
Wade (25)	1950	sand	2 weeks
		organic muck	6 weeks
Mallmann & Litsky (26)	1951	dry soil	14->42 days
		wet soil	5–19 days
Pikovskaya *et al.* (27)	1956	AC	2–110 days

[a] AC = Artificial Contamination.

He also found that fecal suspensions gave faster death than saline suspensions and found no antagonistic action of other bacteria in soil. Adverse effects on longevity were seen on either side of pH 7.5. Frozen soils could contribute to extended survival.

In 1950 (28) Rudolf *et al.* reviewed the information to date and their general conclusions concerning the fate of the *typhi* in soil are worth presenting as they probably are representative of what happens to many, if not all, of the nonsporulating pathogenic bacteria, with the exception of the very resistant *Mycobacterium tuberculosis*.

> From the results presented, certain general conclusions suggested themselves concerning the fate of *E. typhosa* (*Salmonella typhi*) in soil.
> 1. The viability was variable, having been reported ranging from less than 24 hr in peat to more than two years in freezing moist soils, but generally less than 100 days.
> 2. The moisture content of the soil appeared important, the trend being for greater survival in the moister soils.
> 3. The moisture holding capacity was an influence—that is, survival

was less in sandy soils than in soils of higher water-holding capacity.

4. Colder temperatures increased the survival of *E. typhosa*.
5. Longer survival time in soil, as compared with feces, had not been definitely demonstrated.
6. The low pH values of peat and acid soils were unfavorable to *E. typhosa*.
7. Sunlight was unfavorable to *E. typhosa* in soils, but this may have been more of a heating and drying effect, than of lethal effects of sunlight.
8. The antagonistic or beneficial effects of other microorganisms in soil on *E. typhosa* had not been conclusively demonstrated. Greater survival of the organism after inoculation into sterilized than into unsterilized soil indicates antagonistic effects. However, whether this was due to effects of other bacteria or to the predatory action of higher forms of life, such as protozoa, had not been established.
9. The presence of pollution in soil in the form of sewage or feces either had no effect or reduced the survival of the typhoid bacteria. Whether this was due to the presence of added organic matter or to the antagonistic effect of sewage organisms was not demonstrated.

Krone (29) pointed out the importance of a pathogenic organism adapting to a soil environment, away from its natural host, for survival. He stated that "generalizations are difficult, but it appears that longer survival is encouraged by reduced temperatures, near neutral pH, and absence from antagonistic organisms."

In writing about pathogens applied to soil, Foster and Engelbrecht (30) emphasized that continuous additions of wastes could lead to an accumulation of the pathogenic agent and an equilibrium between die-off or removal and fresh additions. The continuing additions of survival-stimulating moisture would raise the equilibrium to higher levels.

Waksman (31) and Waksman and Woodruff (32, 33), in the period of development of antibiotics, presented considerable material indicating the hostile nature of the soil environment for pathogenic organisms. The following paragraph reflects Waksman's views of the soil as receiver of pathogenic materials:

Considering the millions of years that animals and plants have existed on this planet, one can only surmise the great numbers of microbes causing the numerous diseases of all forms of life that must have found their way into the soil or into streams and rivers. What has become of all these pathogenic bacteria? This question was first raised by medical bacteriologists in the eighties of the last century. The soil was searched for bacterial agents of infectious diseases.

It was soon found that, with very few exceptions, organisms pathogenic to man and animals do not survive very long. This was at first believed to be due to the filtration effect of the soil upon the bacteria (34). It came to be recognized, however, that certain biological agents are responsible for the destruction of the pathogenic organisms. These investigations led to the conclusion that the soil can hardly be considered as a carrier of most of the infectious diseases of man and animals. The fact that many pathogens can grow readily in sterilized soil but do not survive long in normal fresh soil tends to add weight to the theory of the destructive effect upon pathogens of the microbiological population in normal soil.

He also summarized the factors that caused rapid disappearance of disease organisms from soil as: a) unfavorable environment; b) lack of sufficient or proper food supply; c) destruction by predacious agents such as protozoa and other animals; d) destruction by various saprophytic bacteria and fungi considered as antagonists; e) formation by these antagonists of specific toxic or antibiotic substances destructive to the pathogens; f) in the case of some organisms at least, increase of the bacteriophage content of the soil resulting in lysis.

Waksman (31) cites the work of Briscoe (35) who mixed bovine tuberculosis organisms in manure and spread a 2-in. layer on the ground. The organisms remained virulent for 2 months in sunlight and longer in the shade. In garden soil the tubercle bacilli remained alive for over 213 days. In a buried infected guinea pig the organisms were viable for 71 days but dead by 99. The seasonal effect on *tuberculosis* was shown by spreading the organisms in cow feces on pasture land (36) in England. They remained alive for 5 months in winter, 2 during spring, 4 months in fall and less than 2 in summer. Maddock (37) showed that manure addition to soil favored survival of *Mycobacterium tuberculosis*. Survival was for 178 days; on grass survival was 49 days.

Mom (38) demonstrated that animals grazing near a sanitorium that was discharging viable tubercle bacilli into drainage ditches became infected. Six months after closing the sanitorium no viable organisms were found in the ditch.

Table 2 presents a few selected survival times observed for some of the bacterial pathogens applied to soils in various forms.

Some early work reviewed by Waksman (31) showed that *Brucella melitensis* survived 25 days in soil, 69 days in dry sterile sand and 20 in manured soil (46, 47). Also, Houston (56) showed the rapid destruction of the organisms of cholera and diphtheria in soil.

Cameron (40) showed that *Brucella abortus* survived 66 days in wet soil at room temperature and 29 days at 77°F. When near freezing, it survived for up to 800 days. An interesting study was done by Kuzdas and Morse (39) with dead guinea pigs infected with a strain of *B. abortus*. The pigs were put on the sur-

Table 2.   Selected references to survival of bacteria in soils.

Organism	Application	Survival Time	Year	Researchers
Bacillus anthracis	AC[a]	12 yr	1881	Pasteur (2)
	Stored soil	60 yr	1964	Wilson & Russell (4)
Clostridium septicum	AC	30 days	1969	Garcia & McKay (8)
Brucella abortus	AC-sterile soil	188 days	1954	Kuzdas & Morse (39)
	AC-frozen soil	670 days		
	Infected guinea pig surface			
	winter	44 days		
	summer	1 day		
	subsurface			
	winter	29 days		
	AC-wet soil (room temp.)	66 days	1957	Cameron (40)
	AC-manure and soil (26°C)	29 days		
	AC-manure and soil (frozen)	800 days		
Listeria monocytogenes	AC-moist clay soil	6–7 mo	1960	Welsheimer (41)
	AC-fertile moist garden soil	295 days		
	AC-dry soil	7 days		
Sphaerophorus necrophorus (*Fusobacterium necrophorum*)	Swamp pasture	10 mo	1934	Marsh & Tunnicliff (42)
Leptospira	AC	15–43 days	1955	Smith & Self (43)
Mycobacterium tuberculosis	Manured garden soil	213 days	1912	Briscoe (35)
	Manured pasture		1930	Williams & Hoy (36)
	fall	4 mo		
	winter	5 mo		
	spring	2 mo		
	summer	2 mo		
	AC	6 mo	1933	Maddock (37)
	?	6 mo	1955	Dedie (44)
	Sewage and soil	15 mo	1957	Greenberg & Kupka (45)
Brucella melitensis	AC-dry sand	69 days	1905–1906	Horrocks (46); Gilmour (47)
	AC-manure and soil	20 days		
	AC-soil			
Salmonella (other than *typhi*)	AC	15–70 days	1956	Bergner-Rabinowitz (48)
	Sprinkled domestic sewage	4 days	1957	Müller (49)
	Semiliquid manure	27–281 days	1971	Best *et al.* (50)
	Sewage studies	56–112 days	1971	Kenner *et al.* (51)
	AC	14 weeks	1971	Tannock and Smith (52)
	Bovine manure	24 weeks	1972	Findlay (53)
	AC	6 weeks	1972	Tannock and Smith (54)
	Slurry	76 days	1973	Taylor (55)

[a] AC = Artificial Contamination.

face of the ground or buried 2 ft below the surface in winter and summer periods. On the surface the spleens showed organisms for 44 days in winter and only 1 in summer. Subsurface tissue showed organisms for 29 days in winter. They also did longevity studies in soils. At 25 or 37° C survival was up to 188 days, with sterilized soils favoring survival. In frozen soil, survival was as long as 670 days.

In 1960, Welsheimer (41) inoculated *Listeria monocytogenes* into fertile garden and clay soils, both kept moist. In clay survival was 6–7 months, while in the rich soil the organisms were alive at 295 days. In dry soil, death was almost complete in 67 days.

A factor associated with survival of bacteria in soil is the migration of the agents through the solid medium. McCoy (57) worked with indicator organisms and concluded that they were removed by adsorption in percolation through soil. She felt there was no great concern that they would move any great distance from the point of application of manure or wastewater. This conclusion may not account for all the variables of soil type and condition, waste application procedures, organism size and type, etc.; however, this topic is outside the realm of this report and deserves separate in-depth review.

Greenberg and Thomas (58) reviewed the results of percolating sewage and concluded that, "a bacteriologically safe water can be produced from settled or more highly treated sewage if the liquid passes through at least 4 feet of soil." However, they also stated, "Further investigation is needed of sewage percolation in different soils and of phenomena associated with the movement of water into such soils to generalize the conclusions reached as a result of this study with Hanford fine sandy loam."

A brief reference is in order to the work of Stotzky and co-workers (59–61) who have been demonstrating the controlling influence of specific geochemical elements in soils on the survival of bacteria, actinomycetes and fungi. More work is needed before its meaning on our subject can be understood.

One of the forms in which wastes are applied to pasture land is as a slurry or as irrigation water. If cattle graze the pasture too soon after application, a health risk is involved. Jack and Hepper (62) showed that, for up to 8 weeks, animals with *S. typhimurium* were carriers, shedding the organism. Gibson (63, 64) felt that the carrier state for *S. dublin* could last many years, constantly adding infectious inocula to the soil. Miner *et al.* (65) observed this asymptomatic carrier state in feedlot runoff studies. Recently, Taylor (55) contaminated a pasture with a known quantity of *S. dublin*. Only one calf in twelve showed symptoms, but the organism could be isolated from grass for a week and from soil for 76 days.

An important factor demonstrated by this work is that disease transmission by bacterial agents through the enteric route is dose-dependent (66). The danger of infection is reduced if dosages of the organisms are kept low. Another example of dose threshhold is shown by the grass sprayed with 4 x 10^6 tubercle bacilli per ft^2 that did not infect fed guinea pigs (45). Heavier concentrations of the organism caused infection in the pigs.

Tannock and Smith (52, 54) worked with *Salmonella*-infected sheep manure put into soil plots. In summer the survival was shorter than in winter and the fecal material protected the organism for a longer period of time than in plain soil. They ascribed the protection by feces to the greater action of sunlight, dehydration and higher temperature in low-organic soil.

A health danger associated with applying pathogen-laden wastes to crop land is the potential for contaminating food materials with bacteria and protozoa and other parasitic organisms. The danger of transmitting phytopathogenic agents also exists but has not been reviewed; also viruses may be spread but almost no literature evaluating this problem exists. Rudolfs and associates (28, 67–74) did an extensive literature search on the subject and reported on their own research. With tomatoes they were able to show surface contamination, particularly of imperfect fruit, with coliforms (28). They felt that if no fecal material were applied for 30 days before harvest, the contamination was reduced to safe levels. *Salmonella* and *Shigella* were undetected on the tomatoes in 7 days after contamination.

In a study of *Endamoeba histolytica* cysts on tomatoes and lettuce (70), it was shown that the cysts are very sensitive to drying and even after vegetables are contaminated, death is rapid in dry weather. Crops can be easily contaminated by irrigation or fertilization with fecal matter, by direct contact with soil and even by splashing or droplets of rain. Studies with *Ascaris* eggs on lettuce and tomatoes showed the eggs to be viable for at least a month. However, the eggs did not develop into the infectious motile embryo stage (71). Deterioration of the eggs in approximately a month was contrary to their literature search that showed *Ascaris* and *Trichuris* eggs to be very resistant in soils (75). Spindler (76) studied *Ascaris* eggs and concluded that they "die much more rapidly in soil than commonly believed."

Orenstein's work in South Africa indicated that sewage-irrigated farms provided vegetables that did not seem to increase intestinal infections (77) ; most of the vegetables were cooked before eating. Dunlop *et al.* (78–81) extensively studied row crops irrigated with sewage-contaminated water containing high levels of *Salmonella*. They almost never were able to recover the organism from vegetables. However, they definitely warned of the dangers of contaminated irrigation waters, not only for *Salmonella* but other bacteria and parasites (78).

In 1957, Müller (49) reported on crops watered with sewage containing *Salmonella*. With potatoes, the organism was recovered from soil and tubers for 40 days, from carrots for 10 days and from cabbage leaves and gooseberries for 5 days. In Darmstadt, Germany (82) roundworm infections were high because of improperly treated sewage on vegetable farms. In early studies in China, Winfield and Chin (83, 84) were concerned with the dangers of *Endamoeba histolytica* transmission by night soil in cropped areas. Yoshida (85) showed that *Ascaris lumbriocoides* survived 5–6 months in winter soil. Strong sunlight was very lethal (86).

Some of the survival observations on parasitic agents in soil are given in Table 3.

Robinson (67) felt that grazing cattle on sewage farms was not a dangerous practice. An extensive literature review of sewage effluents and land disposal was written recently by Benarde (91). In a review by Dotson (92) he stated, "although no record exists of diseases having been caused by using digested sludge as soil conditioner or fertilizer, this is still a concern of many people."

A few comments on microorganism survival in soil waste disposal systems are in order. Much of the investigative work is somewhat old and the results are difficult to evaluate because of varying methodology, conditions of the research and objec-

Table 3. Selected references to survival of protozoans and parasites in soil.

Organism	Application	Survival Time	Year	Researchers
Ascaris eggs	?	up to 7 yr	1953	Müller (87)
	Sewage	2.5 yr	1959	Gudzhabidze (88)
	Winter soil	5–6 mo	1920	Yoshida (86)
Endamoeba histolytica cysts	AC[a]	8 days	1949	Beaver & Deschamps (89)
Ancylostoma				
larvae	Infected feces	6 weeks	1955	Smith & Self (43)
ova	Drying sludge	60–80 days	1943	Cram (90)

[a]AC = Artificial Contamination.

tives of the work. The entire problem needs extensive scientific study in light of today's situation. It may seem surprising that little information was found directly relating to *Shigella*, the causative agent of bacillary dysentery. Because it is a pathogen affecting humans almost exclusively in this country, it would not be found in animal wastes. In domestic wastes the organism does not compete well with the other intestinal types outside of the human body (93) and probably is rarely present in material put on or into the soil.

Our knowledge of the behavior of viruses in the soil waste environment is less developed than the bacteriological information and is going to be more difficult and expensive to develop. The state-of-the art is exemplified by some opening remarks by Berg (94) at a recent symposium:

> Land spreading and spray irrigation are more and more entertained as means for disposing of treatment sludges and effluents. Productive disposal of such wastes is not new, having long been practiced in many areas. But the risk of respiratory transmission of viruses by airborne droplets that may carry for long distances is yet to be determined. The risk of virus transmission by intrusion into ground waters or by absorption into plants consumed by man also needs to be studied. In this country, the possible risks associated with land spreading of sludges and spray irrigation with effluents are under intensive discussion. The degree to which sludge digestion destroys viruses has been questioned. Pasteurization is just now under consideration as a solution.

In 1973, Foster and Engelbrecht (30) stated that, "Little definitive information on the survival of virus in soil exists, but one could expect it to be of the same order of magnitude as survival in wastewater where persistence as long as 100 days has been reported (95)." Certainly better experimental information is critically needed.

The majority of virus studies related to soil disposal have been done with human sewage rather than with animal wastes. Well over 100 separate virus entities are recognized as infectious agents by the fecal-oral route and it is certain that very many more will be identified as research proceeds. Much of work with sewage has been focused on poliomyelitis virus—early, because of the interest in the search for the causative agent of this disease and then because a large amount of the virus is enterically shed by vaccinated individuals. A major interest has been the transport of virus through solid media as soils rather than our direct concern in this paper with survival of the infectious agent. A good review of virus transport in soils from a health standpoint has been written by Romero (96). Many experimental studies have used bacteriophages or even synthetic particles. These are of the same size range as viruses, and phages have some of the same properties as animal viruses. However, there are differences and limitations to the use of phages as models (97). These authors worked with phages and concluded that virus retention on soil particles is an adsorption process, pH affects adsorption because of the amphoteric nature of the virus coat and that increased cation concentration of the liquid phase of a soil-water system enhances adsorption. Also of interest was their observation that adsorption of virus particles cannot be judged on the basis of various tests which normally are used to characterize a soil. In general they found that adsorption increased with clay content, silt content, ion-exchange and glycerol retention capacity.

In a study of virus irrigation on sandy soils in Florida (98) the importance of the soil/water ratio was shown. Once virus passed the thin organic layer it moved down rapidly if the water content of the lower depths were high. The results of Young and Burbank (99) were very interesting. While tests with Hawaiian soils were effective in adsorbing coliphage, the same soils did not retain polio II virus.

Bagdasaryan (100) showed that survival of enterovirus in soil is affected by pH, temperature and moisture. Under adverse conditions of dry sandy soil, pH 5 and about 20° C, the virus survived 6 to 12 days. Lower temperatures, moist soil and higher pH extended the life of enterovirus considerably. Lefler and Kott (101) studied survival of polio virus in sand columns. When added in oxidation pond effluents the virus was still detectable in 91 days.

SUMMARY AND RECOMMENDATIONS

From the technical literature reviewed in this presentation it is evident that pathogenic microorganisms, potential agents of disease transmission to man, animals and probably plants, are capable of surviving in soils that have received wastes. Many factors regulate the longevity and pathogenicity of these organisms. Many die rapidly when put into the biologically competitive and the physically and chemically hostile environment of the soil. It is particularly hostile to organisms that have become specialized and adapted to the temperature and nutrition of the host cells and organs in which they produce the abnormal condition of disease. Moisture is probably the largest single factor in determining the length of survival. Lowered temperature such as winter conditions usually leads to lengthened survival of the pathogens. Spore-forming cells, some parasites and protected cells such as the mycobacteria may be extremely long-lived. Other factors such as the nature of the receiving soil, the organism inoculum size, the type and condition of the cells, sunlight, and antagonistic biological and chemical agents of the soil-waste mix all play a role in the survival of the microorganisms. The variations in survivability observed certainly accentuate the lack of adequately controlled experiments so that one factor at a time might be evaluated.

In all too many of the very fine engineering and agricultural studies adding agricultural and municipal waste materials to soils, the microbiological testing and observations are either lacking or done in an inadequate manner. Even when biological testing is done it is usually focused on indicator organisms which may not give data comparable to the functions of the pathogens. It probably is generally true that the indicator organisms are more resistant than the pathogens so that we are erring on the side of public health safety.

One additional circumstance has served as a safety factor in the disposal of waste to soils, aesthetics. Even though our knowledge has come a long way since malaria was named for foul smelling air, the emanation of unpleasant odors tends to limit the time, place and methods of disposal and the types and quantity of waste materials that are applied to a particular disposal system.

As microbiologists, with strong interests in public health and the environment, we must take a firm stand on the potential dangers of distributing large quantities of pathogen-containing waste materials on or into the land. We cannot ignore the observations of the many highly skilled workers who have shown the presence of highly dangerous, viable bacteria, protozoa and other parasites, fungi and viruses in the receiving soils or waters that have drained these soils.

However, the scientific observations of almost 100 years has shown very few, if any, instances in which adequately treated and handled animal or human waste have caused disease when applied to appropriate soils in a cautious manner. The limiting terms that have been used in the last statement must be emphasized. If leachate reaches ground water, if runoff reaches drinking or recreational water resources, if pathogen-laden wastes are put on edible food or feed crops which are consumed by man or animal too soon after application, then waste application to soil is a dangerous health hazard. If soil disposal is compared to the several alternatives available for destroying or utilizing the vast amounts of animal wastes and, at times, an almost equal quantity of treated municipal and commercial biological wastes, the choice of soil disposal ranks very favorably.

Among the general precautions advisable to reduce the possible transmission of disease agents are the following:

1. Human wastes must be adequately treated to reduce pathogen levels to as low numbers as possible.

2. Do not apply manures, sludges or slurries to crops that are to be eaten or grazed unless adequate time is allowed for die-off or the produce is to be thoroughly cleaned and sanitized.

3. Limit the quantities of wastes added to a single site to reduce probability of pathogen build-up.

4. Have adequate knowledge of the geology and hydrology of a receiving field so that ground waters, particularly nearby wells, are protected and runoff to surface waters is minimized.

5. Avoid high-density human population centers (and perhaps even high-density animal areas) where vectors such as wind, insects, and rodents may serve as the carriers of the infectious agents.

6. Maintain a high level of immunity, where it is possible, in the human and animal population to reduce the epidemiologic hazards.

7. Health care in the area, both animal and human, is essential to treat the infected and to reduce the potential for high levels of infectious organisms in the wastes and the soils receiving them.

The criteria just presented are quite general and certainly are not easy to attain. They certainly should be goals that all concerned with soil disposal of wastes should strive for. One additional point must be strongly emphasized. We do not know all we should about the fate and dangers of pathogenic organisms put into the soil environment. We must strive to establish methods for directly studying infectious agents and their behavior in the soil.

REFERENCES

1. Duboise, S. M., B. P. Sagik, B. E. D. Moore and J. F. Malina, Jr. "Virus migration through soils," pp. 233–240. *In: J. F. Malina, Jr. and B. P. Sagik (eds.). Virus Survival in Water and Wastewater Systems,* Center for Research in Water Resources, The University of Texas at Austin (1974).
2. Pasteur, L. *C. R. Acad. Sci.* 92:209 (1881). *Cited in:* Wilson, G. S. and A. A. Miles (102).
3. Koch, R. *Mitt. Reichsgesundh. Amt.* 1:49 (1881). *Cited in:* Wilson, G. S. and A. A. Miles (102).
4. Wilson, J. B. and K. E. Russell. "Isolation of *Bacillus anthracis* from soil stored 60 years," *J. Bacteriol.* 87:237–248 (1964).

5. Burrows, W. *Text book of Microbiology.* W. B. Saunders, Co., Philadelphia (1968).

6. Bruner, D. W. and J. H. Gillespie. *Hagan's Infectious Diseases of Domestic Animals.* Comstock Publishing Associates, Ithaca (1973).

7. Firth, R. H. and W. H. Horrocks, "An inquiry into the influence of soil, fabrics and flies in the dissemination of enteric infection," *Br. Med. J.* 2:936 (1902). *Cited in:* Rudolfs, W., *et al.* (28).

8. Garcia, M. M. and K. A. McKay. "On the growth and survival of *Clostridium septicum* in soil," *J. Appl. Bacteriol.* 32:362–370 (1969).

9. Garcia, M. M. and K. A. McKay. "Pathogenic microorganisms in soil: an old problem in a new perspective," *Can. J. Comp. Med.* 34:105–110 (1970).

10. Grancher, J. and E. Deschamps. *Arch. de Med. Exper. et d'anat. Path.* 1:33 (1889). *Cited in:* Beard, P. J. (14).

12. Martin, S. *Rep. Med. Oc., Local Gov. Bd.* 99:382 (1898). *Cited in:* Beard, P. J. (14).

13. Smith, J. L. *J. Hyg.* 4:418 (1904). *Cited in:* Beard, P. J. (14).

14. Beard, P. J. "Longevity of *Eberthella typhosa* in various soils," *Am. J. Public Health* 30:1077–1082 (1940).

15. Robertson, J. *Brit. Med. J.* 1:69 (1898). *Cited in:* Beard, P. J. (14).

16. Rullmann, W. *Centralbl. f. Bakt.*, 1, Abt. 30:321 (1901). *Cited in:* Beard, P. J. (14).

17. Demster. *Brit. Med. J.* 1:1126 (1894). *Cited in:* Rudolfs, W., *et al.* (28).

18. Sedgwick, W. T. and C. E. A. Winslow. *Mem. Am. Acad. Arts and Sci.* 12:508 (1902). *Cited in:* Beard, P. J. (14).

19. Clauditz, H. "Typhus and plants," *Hyg. Rundschau* 14:865 (1904). *Cited in:* Dunlop, S. G. (79).

20. Mair, W. "Experiments on the survival of *B. typhosa* in sterilized and unsterilized soils," *J. Hyg.* 8:37 (1908). *Cited in:* Dunlop, S. G. (197).

21. Melick, C. O. "The possibility of typhoid infection through vegetables," *J. Infect. Dis.* 21:28 (1971). *Cited in:* Dunlop, S. G. (79).

22. Murillo, F. "Cholera, typhoid and vegetables," *Plus Ultra* (Madrid) 2:115 (1919). *Cited in:* Rudolfs, W. *et al.* (28).

23. Kligler, I. J. "Investigation on soil pollution and the relation of the various types of privies to the spread of intestinal infections," *Rockefeller Inst. of Med. Res. Monograph,* No. 15 (1921). *Cited in:* Rudolfs, W. *et al.* (28).

24. Grandi, D. "Importance of green vegetables in transmission of typhoid fever," *Iniene Moderna* 23:65 (1930). *Cited in:* Dunlop, S. G. (79).

25. Wade, S. T. "The persistence of the VI antigen of *S. typhosa*," M.S. Thesis, Michigan State University, East Lansing (1950). *Cited in:* Krone, R. B. (29).

26. Mallmann, W. L. and W. Litsky. "Survival of selected enteric organisms in various types of soil," *Am. J. Public Health* 41:38–44 (1951).

27. Pikovskaya, R. E., S. J. Rtskhiladze and M. G. Gelashvili. "Concerning the self-purifying properties of basic soil types found in the Soviet State of Georgia," *Gig. i. san.* 1:15 (1956). *Cited in:* Dunlop, S. G. (79).

28. Rudolfs, W., L. L. Falk and R. A. Ragotzkie. "Literature review on the organisms in soil, water, sewage and sludges, and on vegetation, I. Bacterial and virus diseases," *Sewage Ind. Wastes* 22:1261–1281 (1950).

29. Krone, R. B. "The movement of disease producing organisms through

soil," pp. 75–105. *In:* C. W. Wilson and F. E. Beckett (eds.). *Municipal Sewage Effluent for Irrigation.* Louisiana Polytechnic Inst., Ruston (1968).

30. Foster, D. H. and R. S. Engelbrecht. Microbial hazards in disposing of wastewater on soil, pp. 247–270. *In:* W. E. Sopper and L. T. Kardos (eds.). *Recycling Municipal Wastewater and Sludge through Forest and Cropland.* The Pennsylvania State University Press, University Park (1973).

31. Waksman, S. A. *Microbial antagonisms and antibiotic substances,* The Commonwealth Fund, N.Y. (1945).

32. Waksman, S. A. and H. B. Woodruff. "Survival of bacteria added to soil and the resultant modification of soil population," *Soil Sci.* 50: 421–427 (1940).

33. Waksman, S. A. and H. B. Woodruff. "The soil as a source of microorganisms antagonistic to disease-producing bacteria," *J. Bacteriol.* 40:581–600 (1940).

34. Bail, J. and F. Breinl. "Versuche über das seitliche Verdringen von Verunreinigungen im Boden," *Arch. f. Hyg.* 82:33–56 (1914). *Cited in:* Waksman, S. A. (31).

35. Briscoe, C. F. "Fate of tubercle bacilli outside the animal body," *Ill. Agr. Exper. Sta. Bull.* 161:279–375 (1912). *Cited in:* Waksman, S. A. (31).

36. Williams, R. S. and W. A. Hoy. "The viability of *B. tuberculosis* (bovinus) on pasture land, in stored faeces and in liquid manure," *J. Hyg.* 30:413–419 (1930).

37. Maddock, E. C. G. "Studies on the survival time of the bovine tubercle bacillus in soil, soil and dung, in dung, and on grass with experiments on the preliminary treatment of infected organic matter and on the cultivation of the organism," *J. Hyg.* 33:103–117 (1933).

38. Mom, C. P. Institute for sewage treatment-annual report (Netherlands) (1948–49). *Cited in:* Rudolfs, W. (67).

39. Kuzdas, C. D. and E. V. Morse. "The survival of *Brucella abortus,* U.S.D.A. strain 2308, under controlled conditions in nature," *Cornell Vet.* 44:216–228 (1954).

40. Cameron, H. S. "The viability of *Brucella abortus,*" *Cornell Vet.* 22: 212–224 (1932).

41. Welsheimer, H. S. "Survival of *Listeria monocytogenes* in soil," *J. Bacteriol.* 80:316–321 (1960).

42. Marsh, H. and E. A. Tunicliff. "Experimental studies of foot-rot in sheep," *Montana Agr. Expt. Sta. Bull.* 285:3–16 (1934). *Cited in:* Garcia, M. M. and K. A. McKay (9).

43. Smith, D. J. W. and H. R. M. Self. "Observations on the survival of *Leptospira australis* A in soils and water," *J. Hyg. Camb.* 53:436 (1955). *Cited in:* Dunlop, S. G. (79).

44. Dedie, K. "Organisms in sewage pathogenic to animals," *Stadtehyg.* 6:177 (1955). *Cited in:* Dunlop, S. G. (79).

45. Greenberg, A. E. and E. Kupka. "Tuberculosis transmission by wastewaters—a review," *Sewage Ind. Wastes* 29:524–537 (1957).

46. Horrocks, W. H. "Further studies on the saprophytic existence of *Micrococcus melitensis,*" Reports of the commission for the investigation of Mediteranean fever. Part I. pp. 5–20 (1905) Part IV, pp. 27–31 (1906). *Cited in:* Waksman, S. A. (31).

47. Gilmour, R. T. "Further notes on the isolation of the *Micrococcus melitensis* from peripheral blood; and experiments on the duration of life of ths microbe in earth and in water," Reports of the commission

for the investigation of Mediterranean fever. Part IV, pp. 3–7 (1906). *Cited in:* Waksman, S. A. (31).

48. Bergner-Rabinowitz, S. "The survival of coliforms, *Streptococcus faecalis* and *Salmonella tennessee* in the soil and climate of Israel," *Appl. Microbiol.* 4:101–106 (1956).

49. Müller, G. "The infection of growing vegetables by spraying with domestic drainage," *Stadtehyg.* 8:30–32 (1957). *Cited in:* Dunlop, S. G. (79).

50. Best, E., W. Müller and D. Strauch. "Untersuchungen über die Tenazität vom Krankheitserregern in tierschen Fäkalein. IV. Mitteilung: Tenazitätsversuche mit *Salmonellen* in natürlich gleagerten Flüssigmisten vom Rindern and Kälbern," *Berl. Munch. Tierarztl. Wschr.* 84: 184–188 (1971).

51. Kenner, B. A., G. K. Dotson and J. E. Smith. "Simultaneous quantitation of *Salmonella* species and *Pseudomonas aeruginosa*," Environmental Protection Agency, National Environmental Research Center, Cincinnati (1971).

52. Tannock, G. W. and J. M. B. Smith. "Studies on the survival of *Salmonella typhimurium* and *Salmonella morbificans* on pasture and in water," *Aust. Vet. J.* 47:557–559 (1971).

53. Findlay, C. R. "The persistence of *Salmonella dublin* in slurry, in tanks, and on pasture," *Vet. Rec.* 91:233–235 (1972).

54. Tannock, G. W. and J. M. B. Smith. "Studies on the survival of *Salmonella typhimurium* and *Salmonella bovis morbificans* on soil and sheep faeces," *Res. Vet. Sci.* 13:150–153 (1972).

55. Taylor, R. J. "A further assessment of the potential hazard for calves allowed to graze pasture contaminated with *Salmonella dublin* in slurry," *Br. Vet. J.* 129:354–358 (1973).

56. Houston, A. C. "Report on inoculation of soil with particular microbes, pathogenic and other," Local Govt. Board, *Rep. Med. Officer* 28:413–438 (1898–1899). *Cited in:* Waksman, S. A. (31).

57. McCoy, E. "Removal of pollution bacteria from animal waste by soil percolation," Annual Meeting, Am. Soc. Agric. Eng., Paper 69–430 (1969).

58. Greenberg, A. E. and J. F. Thomas. "Sewage effluent reclamation for industrial and agricultural use," *Sewage Ind. Wastes* 26:761–770 (1954).

59. Stotzky, G. "Clay mineral and microbiol ecology," *Trans. N.Y. Acad. Sci.* 30:11–21 (1967).

60. Stotzky, G. and A. H. Post. "Soil minerology as a possible factor in geographic distribution of *Histoplasma capsulatum*," *Can. J. Microbiol.* 13:1–7 (1967).

61. Stotzky, G. and L. T. Rem. "Influence of clay minerals on microorganisms, I. Montmorillonite and kaolinite on bacteria," *Can. J. Microbiol.* 12:547–563 (1966).

62. Jack, E. J. and P. T. Hepper. "An outbreak of *Salmonella typhimurium* infection in cattle associated with the spreading of slurry," *Vet. Rec.* 84:196–199 (1969).

63. Gibson, E. A. "Reviews of the progress of dairy science. Section E. Diseases of dairy cattle. *Salmonella* infections in cattle," *J. Dairy Res.* 32:97–134 (1965).

64. Gibson, E. A. "Disposal of farm efflent," *Agriculture* 74:183–188 (1967).

65. Miner, J. R., L. R. Fina and C. Piatt. "*Salmonella infantis* in cattle feedlot runoff," *Appl. Microbiol.* 15:627–628 (1967).

66. McCullough, N. B. and C. W. Eisele. "Experimental human salmonellosis," *I. J. Infect. Dis.* 88:278–289; II. *J. Immunol* 66:595–608; III. *J. Infect. Dis.* 89:209–213; IV. *J. Infect. Dis.* 89:259–265 (1951).

67. Rudolfs, W. (chairman) "A critical review of the literature of 1950 on sewage, waste treatment, and water pollution," *Sewage Ind. Wastes* 23:555–642 (1951).

68. Rudolfs, W., L. L. Falk and R. A. Ragotzkie. "Literature review on the occurrence and survival of enteric, pathogenic, and related organisms in soil, water, sewage, and sludges and on vegetation, II. Animal parasites," *Sewage Ind. Wastes* 22:1417–1427 (1950).

69. Rudolfs, W., L. L. Falk and R. A. Ragotzkie. "Contamination of vegetables grown in polluted soil, I. Bacterial contamination," *Sewage Ind. Wastes* 23:253–268 (1951).

70. Rudolfs, W., L. L. Falk and R. A. Ragotzkie. "Contamination of vegetables grown in pollution soils, II. Field and laboratory studies on *Endomoeba* cysts," *Sewage Ind. Wastes* 23:478–485 (1951).

71. Rudolfs, W., L. L. Falk and R. A. Ragotzkie. "Contamination of vegetables grown in polluted soil, III. Field studies on *Ascaris* eggs," *Sewage Ind. Wastes* 23:656–660 (1951).

72. Rudolfs, W., L. L. Falk and R. A. Ragotzkie. "Contamination of vegetables grown in polluted soil, IV. Bacterial decontamination," *Sewage Ind. Wastes* 23:739–751 (1951).

73. Rudolfs, W., L. L. Falk and R. A. Ragotzkie. "Contamination of vegetables grown in polluted soil, V. Helminithic decontamination," *Sewage Ind. Wastes* 23:853–860 (1951).

74. Rudolfs, W., L. L. Falk and R. A. Ragotzkie. "Contamination of vegetables grown in polluted soil, VI. Application of results," *Sewage Ind. Wastes* 23:992–1000 (1951).

76. Spindler, L. A. "On the use of a method for the isolation of *Ascaris* eggs from soil," *Am. J. Hyg.* 10:157 (1929). *Cited in:* Rudolfs, W. (67).

75. Brown, H. W. "Studies on the rate of development of viability of the eggs of *Ascaris lumbricoides* and *Trichuris trichura* under field conditions," *J. Parasitol.* 14:1 (1927). *Cited in:* Dudolfs, W. (68).

78. Dunlop, S. G. "The irrigation of truck crops with sewage contaminated water," *Sanitarian* 15:107–110 (1952).

77. Orenstein, A. J. "A contribution to the discussion on the hazard of *Ascaris* infestation from sewage," *Public Health* 14:1 (1950). *Cited in:* Rudolfs, W. (67).

79. Dunlop, S. G. "Survival of pathogens and related disease hazards," pp. 107–122. *In:* C. W. Wilson and F. E. Beckett (eds.) *Municipal Sewage Effluent for Irrigation,* Louisiana Polytechnic Inst., Ruston (1968).

80. Dunlop, S. G., R. M. Twedt and W. L. Wang. "*Salmonella* in irrigation water," *Sewage Ind. Wastes* 23:1118–1122 (1951).

81. Dunlop, S. G., R. M. Twedt and Wang. "Quantitative estimation of *Salmonella* in irrigation water," *Sewage Ind. Wastes* 24:1015–1020 (1952).

82. Reinhold, F. "Further experiences with the roundworm plague," *Ber. Abwassertechnischen Verein.* I:145 (1949). *Cited in:* Rudolfs, W. (67).

83. Winfield, G. F. and T. Chin. "Studies on the control of fecal-borne diseases in North China, VI. The epidemiology of *Ascaris lumbricoides* in an urban population," *Chinese Med. J.* 54:233 (1938). *Cited in:* Rudolfs, W. *et al.* (68).

84. Winfield, G. F. and T. Chin. "Studies on the control of fecal-borne diseases in North China, VII. The epidemiology of the parasitic

amoebae," *Chinese Med. J.* 56:265 (1939). *Cited in:* Rudolfs, W. *et al.* (68).

85. Yoshida, S. "On the development of *Ascaris lumbricoides* L.," *J. Parasitol.* 5:105 (1919). *Cited in:* Rudolfs, W. *et al.* (68).
86. Yoshida, S. "On the resistance of *Ascaris* eggs," *J. Parasitol.* 6:132 (1920). *Cited in:* Rudolfs, W. *et al.* (68).
87. Müller, G. "Investigations on the survival of *Ascaris* eggs in garden soil," *Zentralbl. Bakteriol.* 159:377 (1953). *Cited in:* Dunlop S. G. (79).
88. Gudzhabidze, G. A. "Experimental observations on the development and survivial of *Ascaris lumbricoides* eggs in soil of irrigated agricultural fields," *Med. Parazit.* 28:578 (1959); *Abst. Soviet Med.* 4:979 (1960).
89. Beaver, P. C. and G. Deschamps. "The viability of *Entamoeba histolytic cysts* in soil," *Am. J. Trop. Med.* 29:189 (1949). *Cited in:* Dunlop, S. G. (79).
90. Cram, E. B. "The effect of various treatment processes on the survival of helminth ova and protozoan cysts in sewage," *Sewage Works J.* 15:1119–1138 (1943).
91. Benarde, M. A. "Land disposal and sewage effluents: appraisal of health effects of pathogenic organisms," *J. Am. Water Works Assoc.* 65:432–440 (1973).
92. Dotson, G. K. "Some constraints of spreading sewage sludge on cropland," *Compost Sci.* 14:12–15 (1973).
93. Wang, W. L., S. G. Dunlop and R. G. Deboer. "The survival of *Shigella* in sewage, I. An infect of sewage and fecal suspensions on *Shigella flexneri*," *Apply. Microbiol.* 4:34–38 (1956).
94. Berg, G. "The virus hazard—a panorama of the past, a pressage of things to come," p. xiii-xviii. *In:* J. F. Malina, Jr. and B. P. Sagik (eds.). *Virus Survival in Water and Wastewater Systems*, Center for Research in Water Resources, The University of Texas at Austin (1974).
95. Akin, E. W., W. H. Benton and W. F. Hill, Jr. "Enteric viruses in ground and surface water: a review of their occurrence and survival," pp. 59–74, 13th Water Quality Conf., University of Illinois (1971).
96. Romero, J. C. "The movement of bacteria and viruses through porous media," *Ground Water* 8:37–48 (1970).
97. Drewry, W. A. and R. Eliassen. "Virus movement in ground water," *J. Water Pollut. Control Fed.* 40:R257–R271 (1968).
98. Wellings, F. M., A. L. Lewis and C. W. Mountain. "Virus survival following wastewater spray irrigation of sandy soils," pp. 254–360. *In:* J. E. Malina, Jr. and B. P. Sagik (eds.). *Virus Survival in Water and Wastewater Systems*. Center for Research in Water Resources, The University of Texas at Austin (1974).
99. Young, R. H. F. and N. C. Burbank, Jr. "Virus removal in Hawaiian soils," *J. Am. Water Works Assoc.* 65:598–604 (1973).
100. Bagdasaryan, G. A. "Survival of viruses of the enterovirus group (poliomyelitis, Echo, Coxackie) in soil and on vegetables," *J. Hyg. Epedemiol., Microbiol. and Immunol.* (Prague) (1964). 8:497–505 *Cited in:* Lefler, E. and Y. Kott (101).
101. Lefler, E. and Y. Kott. "Virus retention and survival in sand," pp. 84–91. *In:* J. F. Malina, Jr. and B. P. Sagik (eds.). *Virus Survival in Water and Wastewater*, Center for Research in Water Resources, The University of Texas at Austin (1974).
102. Wilson, G. S. and A. A. Miles. *Principles of Bacteriology and Immunity*, 5th Ed., The Williams and Wilkins Co., Baltimore (1964).

21

PUBLIC HEALTH ASPECTS OF DIGESTED SLUDGE UTILIZATION

S. J. Sedita, P. O'Brien, J. J. Bertucci, C. Lue-Hing, and D. R. Zenz
The Metropolitan Sanitary District of Greater Chicago,
Chicago, Illinois

INTRODUCTION

Traditionally and practically the most difficult problem facing municipal and private treatment facilities has been the disposal of solid residues from wastewater treatment. As the technology for removing various pollutants from wastewater becomes available, the disposal of the resulting solids becomes a more difficult technical and social problem.

There is no precise national inventory of the quantities of municipal sludges produced in the U.S. although there are some estimates available. These estimates are based primarily on domestic population, and the typical amounts of sludge generated per capita. These estimates do not include the quantities of sludge resulting from the discharge of industrial wastewaters in municipal systems.

Dean (1) has estimated that wastewaters from the national population of 200 million results in 20,000 tons/day of sludges.

Fuhrman (2) has estimated that the actual national quantity of sludge in 1972 is approximately 3.6 million tons/year (or about 10,000 tons/day). This estimate is lower because it is based on only the sewered population and includes the effect of process variations such as digestion and incineration. By 1990, he estimates that the quantity of sludge will increase to 4.7 million tons/year (or about 13,000 tons/day). These estimates do not include industrial users or municipal plants. The authors believe that both these estimates are low. For example, the Metropolitan Sanitary District of Greater Chicago (MSDGC) produces approximately 700 dry tons of sludge solids per day for ultimate disposal from a population of 5.5 million people plus the industrial contribution. This rate of sludge production translates to about 25,500 tons/day or 9.3 million tons per year for a population of 200 million. It is clear from these estimates that by 1990, the municipal sludge production in the U.S. could range from approximately 5.0 to 11.5 million tons per year, assuming a U.S. population of 246* million by 1990.

Regardless of which estimate is accepted, it is evident that the problem of ultimate sludge disposal is not decreasing but rather increasing.

On a worldwide basis, higher living standards and environmental concerns will promote the demand for more sophisticated and advanced wastewater treatment processes. Older waste treatment practices, producing little or no sludge, will be supplanted by processes which will generate larger quantities. Other currently acceptable sludge disposal techniques will gradually give way, in the face of social pressures, to new and more innovative solids disposal techniques.

Superimposed upon this picture is the increasing urbanization in many nations, which will increase the difficulty of finding locations in and around large cities with space adequate for solids disposal.

DISPOSAL ALTERNATIVE

The basic processes available to most municipalities as sludge disposal alternatives may be reduced to the following: incineration; ocean disposal; landfill; and land spreading.

Among these alternatives, land spreading is becoming increasingly popular with many municipalities and indeed throughout the world. It offers the advantage of recycling nutrients to the land, and of reclaiming land spoiled by strip mining or other damaging practices. Sludge, stabilized by anaerobic digestion or other means, is applied to the land. The stabilization processes eliminate noxious odors and fly problems. Grain and forage crop yields are increased by the nutrients and the water supplied by irrigating with digested sludge. Sludge imparts favorable characteristics to soils because of its high humus content. Land spreading of sludge allows urban areas to recycle the fertilizer value of sludge to farming areas from which much of the organic materials and nutrients originate. Considering the current shor-

---

*U.S. Census Bureau estimate.

tage of fuels, and the high energy and fuel requirements for inorganic fertilizer production, sludge recycle for agricultural use offers a real opportunity to reduce energy requirements in this sector. Also, many municipalities can offer their sludge at very low cost to the farmer, thus effecting significant savings both for the farmer and the municipality producing the sludge.

SLUDGE APPLICATION TO LAND

The MSDGC has embarked upon a program to utilize the digested sludge from its treatment plants on land in the central portion of Illinois. Currently, the MSDGC owns over 15,000 acres (6,075 ha) of land approximately 200 miles (320 km) south of its 900-sq mi (2340-sq km) drainage area. Liquid digested sludge containing approximately 4 percent solids is barged down the Illinois River and then pumped through a 10.8-mile (17.3 km) pipeline to holding basins. During the application season, principally April through October, liquid sludge from the holding basins is applied to the land either by spraying or soil incorporation. To date, rates up to 30 dry tons per acre (67.2 MT/ha) per year have been applied during the growing season to fields which were principally planted in corn.

Approximately 6,000 wet tons (5,400 metric wet tons) of 4 percent liquid digested sludge is currently being transported daily to the site in Illinois. In 1974, about 475 dry tons (431 MT) per day were being removed from the MSDGC's west-southwest secondary treatment facility which currently has a design capacity of 1200 million gallons ($3.51 \times 10^6 M^3$) per day.

ENVIRONMENTAL MONITORING OF SLUDGE APPLICATION TO LAND

Even though sludge is beneficial as a fertilizer and soil amendment, its application to land must not result in significant degradation of local waters and the soil ecosystem.

As defined by Blakeslee (3) monitoring implies observance of the performance of the system, checking the quality of affected natural systems such as surface and ground waters and evaluating environmental impacts as changes in quality occur.

The type of monitoring program is defined by the size and purpose of the project. Large-scale land application programs would require extensive monitoring at high frequency, while with smaller projects less extensive and frequent monitoring would suffice.

The MSDGC developed a comprehensive environmental monitoring system for its 15,000-acre (6,075-ha) sludge recycle project in West Central Illinois. This monitoring system was developed and instituted in 1971 and is being used to continually evaluate the surface and ground water quality of the site. Water quality data are collected at monthly intervals. This includes sampling of major streams entering and leaving the property, wells, and strip-mined reservoirs. Chemical, physical and biological parameters are included in the sampling program. Among the biological parameters are fish population and condition estimates, algal quality, indicator bacteria, total bacteria and total virus levels. Animal parasite studies are conducted on a continuing basis for the district by the University of Illinois Department of Veterinary Medicine at Urbana-Champaign.

PUBLIC HEALTH ASPECTS

The major emphasis of this report will be on the public health aspects of the MSDGC sludge recycle project, with specific reference to the bacteriological, viral and animal parasite studies. In an operation of this nature, there will be public concern regarding the possibility of disease transmission to the surrounding populace. No exhaustive epidemiological studies have been carried out on a municipal sludge application project; even so, there exists a wealth of indirect information which indicates that such public health problems are not significant (4).

Several mechanisms might be operational for the direct or indirect contamination of food and/or domestic animals by sewage sludge. These are (5) :

1. Direct physical contact from spraying.
2. Contact between low-growing crops and sludge.
3. Development of a cycle of ingestion, infection, discharge excretion and recontamination.
4. Survival of *Salmonella, Mycobacteria* and other bacteria through the treatment processes.
5. Survival of viruses through the treatment processes and transportation or uptake by plants.

Several studies have attempted to assess the hazards involved in the application of wastewater to land. These studies have been mostly concerned with aerosol hazards, crop contamination, and contamination of water from runoff and percolation (4, 6).

In general terms the majority of studies present considerable data relating to the possible or potential hazards involved in the application of sewage or sludge to land. Actual cases of disease resulting from wastewater and, more particularly, sludge application to land are extremely scarce and indicate that few instances have been related to either practice. This is not to imply that because such instances are rare safeguards can be ignored. It is obvious that with appropriate processing safeguards, adequate site preparation, continuous site monitoring and sensible application techniques the risks will remain insignificant and full advantage can be taken of sludge as a resource.

PROCEDURES

Environmental Monitoring

The MSDGC monitoring program at its Fulton County land application site was instituted in 1971 and is carried out on a monthly basis.

The chemical, physical and biological quality of the surface waters, strip-mine reservoirs and ground water is under constant surveillance. With respect to the biological parameters the following is a listing of the sampling sites (Figure 1):

1. There are 14 sampling sites for bacteriological analyses, 6 for soils, 2 stream sites and 6 wells (3 in sludged and 3 in unsludged areas).
2. Virological analyses are run on a major stream which drains the MSDGC property and at the discharge point of a reservoir (R-3) which drains a major sludge application area.
3. Bacteriological analyses are run on the sludge from the barges and on the lagooned sludge prior to application.
4. Animal parasite studies are carried out for the MSDGC by the University of Illinois on two sites, one sludged, the other unsludged.

From June 1971 to the present samples have been collected at the stations shown in Figure 1. In general, 50 to 70 samples are collected throughout each year at each station. Wells were sampled once per month. Samples for viral analyses were collected once per month.

Bacteriological Analyses

Samples for bacteriological analyses were collected in sterile 4-oz reagent bottles treated with enough sodium thiosulfate to neutralize 15 mg/l chlorine. Stream and lake samples were shipped on ice to the MSDGC Research and Development Laboratory in Stickney, Illinois for analysis. Total coliform (TC), fecal coliform (FC), fecal streptococci (FS), total plate count (TPC) and total virus (TV) analyses were performed on samples from R-3, S-1, and S-2. The indicator organism analyses were performed according to the membrane filter technique outlined in *Standard Methods for Water and Wastewater Analyses* (7). Total plate counts were similarly determined using 0.45-μm Milli-

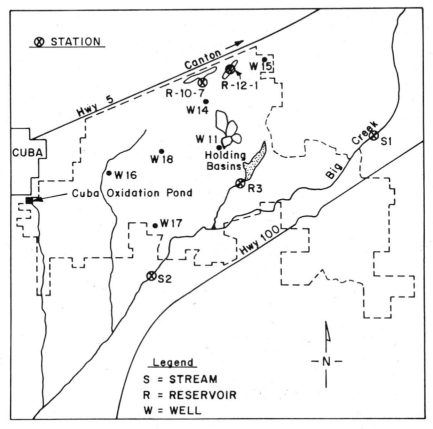

Figure 1. Fulton County bacteriological and viral sampling stations.

pore membrane filters overlaid on plate count agar and incubated at 35° C·for 48 hr.

Runoff basins and wells were analyzed for fecal coliforms only. These analyses were originally performed at the MSDGC labs as previously described. Since mid-1973 these analyses have been run at the site laboratory. Samples from reservoir 10 and reservoir 12 were analyzed for TC, FC and FS at the R&D laboratory.

Analysis of liquid fertilizer was begun in 1975. TPC, TC, FC and FS analyses were performed at the R&D laboratory using the procedures described. *Salmonella* species were estimated using a procedure developed by Kenner and Clark (8). *Staphylo-*

coccus aureus and *Pseudomonas* species were estimated using the procedure outlined in *Standard Methods* (7).

Virological Analyses

Virus concentrations at the three sampling sites chosen (S-1, S-2 and R-3) were too low to detect without prior concentration. Two virus concentration techniques were employed. Viruses from 4-liter volumes of Big Creek water (S-1 and S-2) samples were concentrated approximately 1000-fold by means of a modified aluminum hydroxide technique (9). This technique was found to be practical for water volumes up to 4 liters. Viruses in R-3 samples occurred at levels that required the processing of larger volumes. The PE 60 method (10), was employed to concentrate viruses from 20-liter volumes of water from R-3 samples. By this procedure viruses were concentrated approximately 4000-fold.

Until March 1974 virus concentrates were assayed by the tube dilution method in primary cultures of Rhesus monkey kidney cells. From March 1974 to the present, virus concentrations have been estimated by means of the plaque assay technique in cultures of the B.G.M. cell line.

Statistical Analyses

Stepwise regression analysis was employed to study correlations among levels of viruses and indicator bacteria at stations S-1, S-2, and R-3.

RESULTS AND DISCUSSION

Bacteriological and Viral Analyses of Sludge

In 1975 bacteriological analyses were begun on barged and lagooned sludge. The results of one such analysis are shown in Table 1. Sample sizes for *Salmonella* analyses, run as most prob-

Table 1. Bacterial analyses of barged and lagooned sludge.

Sample Source	Total Coliform	Fecal Coliform	Fecal Streptococci	Total Plate Count	Salmonella
			(#/100 ml)		
Barged	3.0×10^5	5.0×10^3	2.6×10^4	3.3×10^{10}	1.6
Lagooned	9.6×10^4	3.6×10^4	2.6×10^4	7.1×10^9	NR[a]

[a] NR = not recovered.

able number, were 50 ml (one tube) and five tubes each at 10, 1.0 and 0.1 ml. The data are presented to give an indication of the relative magnitude of these parameters in the MSDGC sludge being shipped to and applied at the Fulton County land disposal site.

In addition to bacteria, virus levels in sludge are of concern from the public health viewpoint. Due to the difficulty of estimating virus concentrations in materials containing high levels of solids there are no definitive studies of virus die-off with respect to the indigenous viruses in sludge. There is, however, evidence available regarding the die-away of viruses seeded into digested sludge.

The MSDGC has conducted two studies of the reduction of virus levels in anaerobic digesters. In the first of these studies (11) MS-2 coliphage was inoculated into pilot digesters. It was found that 87.7–96.3 percent of the viruses were inactivated in 24 hr and from 99.0–99.6 percent were inactivated in 48 hr (Figure 2).

In the second study (12) the inactivation rates of four enteric picornaviruses were estimated in seeding experiments. Figure 2 shows a composite graph with die-away curves for the five viruses studied (13). After 24 hr of digestion, average inactivation was 94.4, 99.11, 90.04 and 60.60 percent respectively for poliovirus Type 1, coxsackie virus Type A-9, coxsackie virus Type B-4, and echovirus Type 11. After 48 hr the respective inactivations were 98.80, 99.93, 98.65 and 92.90 percent.

These studies taken with others in the literature, such as the Meyer *et al.* (14) study with swine enterovirus, indicate that the anaerobic digestion process is very effective in reducing virus levels in sludge. More work is needed on this aspect of sludge digestion particularly in the area of quantities and types of indigenous virus and digester die-off rates.

Bacteriological and Viral Studies in Streams and Reservoirs in Fulton County

As noted above and as can be seen in Figure 1, S-1 and S-2 represent samples upstream and downstream of the MSDGC sludge application site. The stream, Big Creek, drains nearly the entire site. R-3 is a reservoir which drains one of the major application areas, and which discharges into Big Creek prior to its leaving the site boundary. Any influence which the sludge application areas may have on the water quality of Big Creek should be reflected at R-3 and S-2, with S-1 being a control sample.

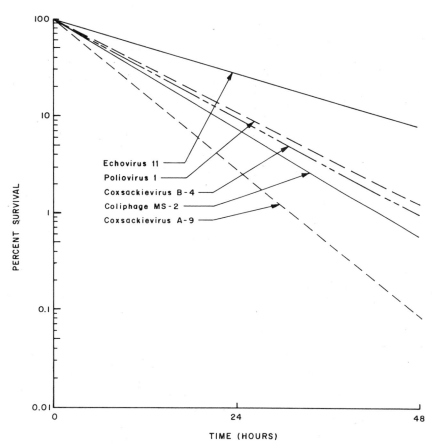

Figure 2. Inactivation of viruses during anaerobic digestion; regression-derived curves (12).

Figures 3, 4 and 5 show TC, FC, FS, TP and TV counts for S-1, R-3, and S-2 respectively.

TPC increases progressively in all sampling sites as seen by the increase in the annual geometric mean of counts from 5.80, 5.64 and 3.90 respectively for S-1, S-2 and R-3 in 1972 to respective values of 8.91, 8.96 and 7.95 for 1975.

It can be stated that runoff from the site has increased the heterotrophic bacterial population of R-3. It is not possible to establish any relationship between sludge application on the site and TPC increases in S-1 and S-2. R-3 could be expected to have only a small influence on S-2 since the discharge from this reservoir represents less than 10 percent of the flow in Big Creek.

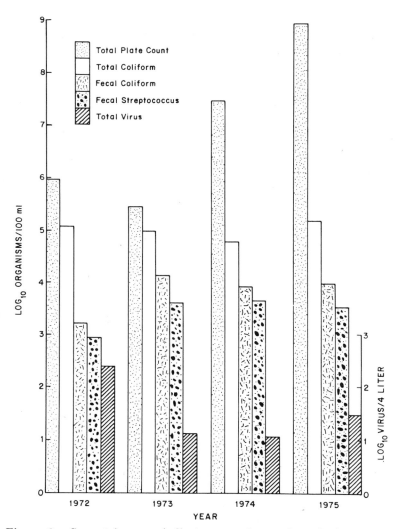

Figure 3. Geometric mean indicator organism and total virus counts 1972–1975, stream site S-1, Fulton County, Illinois.

Increases in TPC at S-1 would be expected to correspond most directly with increases in S-2 since the flow above S-1 is mainly from the unchlorinated effluent of the wastewater treatment plant at Canton, Illinois, a facility which provides secondary treatment only.

Total coliform, fecal coliform and fecal streptococci data (Fig-

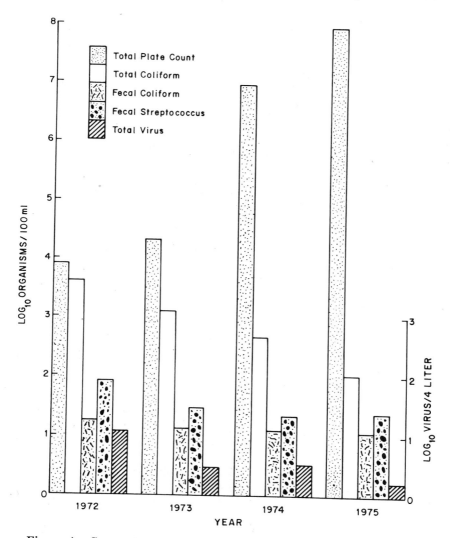

Figure 4. Geometric mean indicator organism and total virus counts 1972–1975, reservoir site R-3, Fulton County, Illinois.

ures 3, 4 and 5) are much more amenable to interpretation than the TPC data. The data indicate that in S-1, S-2 and R-3 the indicator organisms have remained at relatively constant levels for the period of the study reported (1972–1975). In every case the levels of TC, FC and FS in S-1 and S-2 are 1 to 1.5 orders of magnitude higher than the corresponding counts in R-3. This is to be expected due to the influence of the unchlorinated efflu-

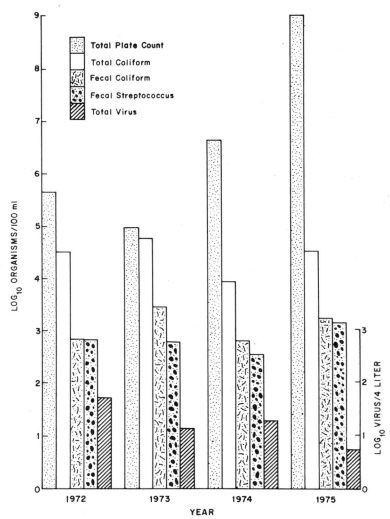

Figure 5. Geometric mean indicator organism and total virus counts 1972–1975, stream site S-2, Fulton County, Illinois.

ent entering Big Creek at Canton, Illinois. These data indicate that the sludge application site has relatively little impact on the water quality of Big Creek in terms of indicator organisms.

Total virus counts can be seen to decrease throughout the course of the study. The highest virus levels in S-1 and S-2 (Fig-

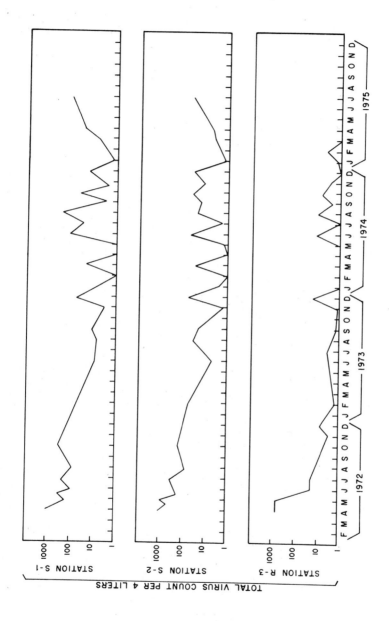

Figure 6. Virus concentrations monitored at Fulton County through 1975.

ures 4 and 5) in 1972 correspond with the presence of cattle in and around the vicinity of Big Creek at S-1 and S-2 which probably influenced the virus levels. Virus data are presented as monthly counts in Figure 6 to show variations.

Total Virus and Indicator Organism Correlations

Stepwise regression analysis was employed to study correlations among levels of viruses, total coliforms, fecal coliforms and fecal streptococci at the S-1, S-2 and R-3 sampling locations. The level of each organism was tabulated for each sample date and site. A $\log_{10}$ transformation of data was used. Data values below the detectable limits of the tests considered (all "less than" values) were eliminated from consideration, since inclusion of these values resulted in a nonnormal data distribution which affected the correlations.

The regression analysis of all of the data (TV, TC, FC and FS—total of 59 samples) established at the 5 percent confidence interval a positive correlation between fecal coliforms and fecal streptococci. The regression coefficient of determination (r^2) was 0.30, indicating that 30 percent of the variation of one organism was reflected in the other.

Regression analyses employing pooled data from S-1 and S-2, and data from R-3 established no correlations among the levels of indicator bacteria and viruses.

Figures 7 and 8 illustrate the $\log_{10}$ geometric mean of two other reservoirs which were sampled for indicator organisms only, from 1972 through 1975. Reservoir R-12-1 (Figure 7) is located within and receives some runoff from a sludge application area; R-10-7 (Figure 8) is in an area which receives no sludge. As can be seen one cannot distinguish between these two reservoirs in terms of indicator organism population. These data support the conclusion that the sludge application areas have no significant impact on the indicator organism populations of reservoirs within sludge application areas. This is to be expected since the sludge application areas at the Fulton County site are constructed to retain runoff until the indicator organism populations are well below those levels which might seriously jeopardize water quality.

The geometric means of fecal coliforms in samples from wells within the Fulton County site are presented in Table 2. Three wells (W 16, 17 and 18) are control wells outside of and not influenced by sludge application activity. Three other wells (W 11, 14 and 15) are test wells inside the sludge application area and presumably influenced by runoff and soil percolation from the holding lagoons. It can be seen from Table 2 that the control wells and the test wells are virtually identical in terms of fecal coliform concentrations from 1971–1975. These data indicate that the site preparation is adequate to control runoff from the sludge application areas, and that translocation of fecal coliforms and presumably other microorganisms by percolation through the soil strata is not occurring.

Soil test sites are shown in Table 3. There were two control fields (17 and 6) to which sludge has never been applied and four fields to which quantities were applied. The bacterial parameters estimated in these two sets of fields were TC, FC, FS, TPC, *Psuedomonas aeruginosa, Salmonella* species and *Staphylococcus aureus.* Table 3 demonstrates that, for the period of this sampling, the microbial populations are little different in soils to which sludge has been applied and in control soils. Specific pathogenic species (*Staphylococcus aureus* and *Salmonella*) were not recovered from either control or test fields. This could be a function of sample size, but is not unexpected in light of the results with barged and lagooned sludge (Table 1).

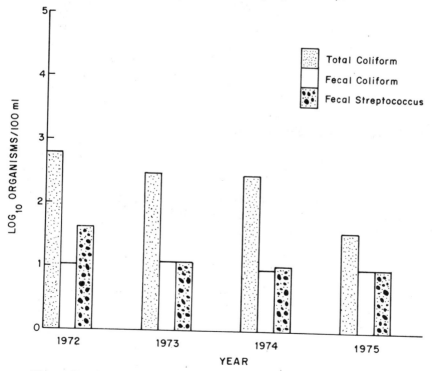

Figure 7.   Geometric mean indicator organism counts 1972–1975, reservoir site R-12-1, Fulton County, Illinois.

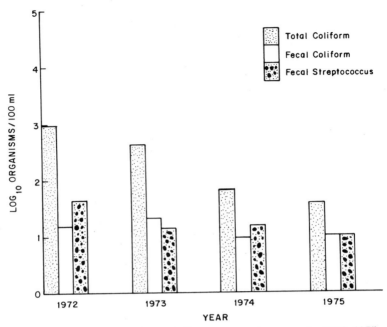

Figure 8.   Geometric mean indicator organism counts 1972–1975, reservoir site R-10-7, Fulton County, Illinois.

ANIMAL PARASITE STUDIES

Fitzgerald and Jolly (15) studied the possibility of transmission of parasitic organisms from digested sludges sprinkled on Sudax forage to cows and calves. Ninety-one pregnant Hereford-Angus-Charolais cows were released on Sudax pasture fertilized with liquid digested sludge from the MSDGC. A herd of 19 preg-

Table 2.   Fecal coliform geometric means for control and test wells, Fulton County, Illinois (1971–1975).

		1971	*1972*	*1973*	*1974*	*1975*
Control wells	W 16	NS	NS	NS	1.8	<2
FC/100 ml	W 17	<4	0	2	0	0
	W 18	<4	0	2	0	0
Test wells	W 11	<4	0	2	0	<2
FC/100 ml	W 14	<4	0	2	0	0
	W 15	<4	0	2	0	<4

Table 3. Microbial populations in control and test fields, 1975 (counts per dry gram of soil).

	TC	FC	FS	TPC	Pseudo-monas aeruginosa	Staphylo-coccus aureus	Salmonella
Control Fields							
17	7.1×10^4	8.9×10^0	4.4×10^2	2.0×10^{10}	4.1×10^3	NR[a]	NR
6	2.0×10^3	6.5×10^2	1.2×10^3	7.1×10^9	6.5×10^2	NR	NR
Test fields							
3	1.0×10^4	7.3×10^2	1.6×10^4	2.0×10^9	1.5×10^3	NR	NR
9	9.0×10^3	6.0×10^3	1.5×10^4	1.1×10^{10}	2.0×10^3	NR	NR
20	1.1×10^4	5.2×10^2	3.3×10^4	1.2×10^{10}	1.6×10^3	NR	NR
21	2.3×10^2	2.2×10^1	5.4×10^2	2.7×10^9	8.7×10^3	NR	NR

[a] NR = not recovered.

nant cows was pastured on Sudax in an area not receiving digested sludge as controls.

Random fecal samples were examined at monthly or shorter intervals to determine the species and relative abundance of parasites of each herd. Comparisons were made of the parasite species present, egg and oocyst discharge quantity and visual clinical signs between the two groups of animals to determine the effects of sludge on animal health.

Fecal specimens (614) from the test animals and 255 specimens from the control herd were examined for ova and parasites over a 14-month period. No significant differences were discernible between the two groups during the course of this study. Nematode egg discharge and coccidian oocyst discharge are plotted in Figures 9 and 10, respectively. Generally, the test and control herds followed a similar pattern for nematode discharge, while the discharge of coccidian oocysts was actually higher in the control herd (Figure 10).

CONCLUSIONS

There is much indirect evidence that disease transmission is not a significant problem with municipal sludge application sites.

The four years of data presented in this paper support this point of view. These data indicate that a relatively large, adequately monitored sludge application site presents no significant risk to surface or ground water quality in terms of indicator

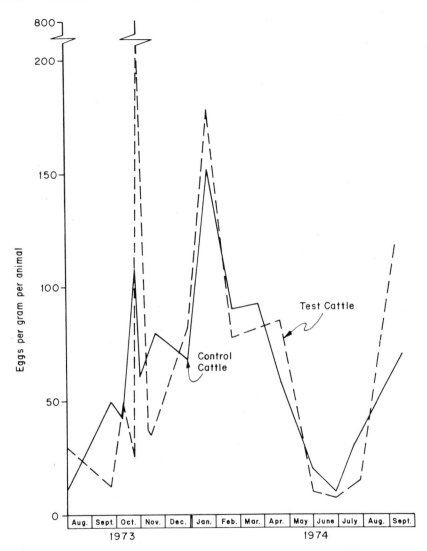

Figure 9. Sample data ova per gram of feces per animal based on random or individual monthly field samples (15).

organisms or viruses. Animals exposed to forage grown with sludge as fertilizer cannot be distinguished from control animals with respect to discharge of nematode eggs or coccidian oocysts. Soils exposed to sludge show little difference in indicator orga-

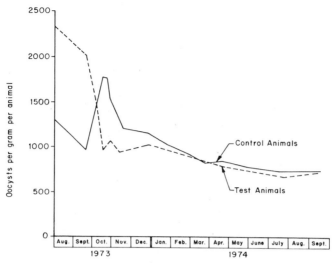

Figure 10. Cumulative total coccidian oocysts per gram of feces per animal based on random or individual monthly field samples (15).

nisms or pathogens from areas not exposed to sludge. Examinations of barged and lagooned sludge indicate low numbers of indicator organisms and *Salmonella*, which demonstrates the effectiveness of anaerobic digestion and subsequent storage as reliable methods for reducing bacterial and pathogen levels in sludge.

The use of liquid-digested sludge as fertilizer and soil amendment is in a totally different category than using raw or treated wastewater effluents as irrigation water. In either case the extent of the possible health hazards relates to the type of soil receiving the sludge or wastewater, the prior treatment of sludge or wastewater and site preparation.

REFERENCES

1. Dean, R. B. "Disposal and reuse of sludge and sewage: what are the options," in *Proc. Conference on Land Disposal of Municipal Effluents and Sludges*, Rutgers University, New Jersey (March 1973).
2. Fuhrman, R. E. Memorandum, "Disposal of sewage sludge," Municipal Waste Water Systems Division, EPA, Washington, D.C. (March 1973).
3. Blakeslee, P. A. "Monitoring considerations for municipal wastewater effluent and sludge application to the land," In *Proc. Joint Conference on Recycling Municipal Sludges and Effluents on Land*, National Association of State Universities and Land Grant Colleges, Washington, D.C. (1973).
4. Benarde, "Land disposal and sewage effluent: appraisal of health ef-

fects of pathogenic organisms," *J. Amer. Water Works Assoc.* 65(6): 432–440 (1973).

5. Braude, G. L., C. F. Jelinek and P. Corneliussen. "FDA's overview of the potential health hazards associated with the land application of municipal wastewater sludges," In *Proc. 1975 National Conference on Municipal Sludge Management and Disposal,* ORD:USEPA, August 18–20, Anaheim, California (1975).

6. Foster, D. H. and R. S. Engelbrecht. "Microbial hazards in disposing of wastewater on soil," in *Recycling Treated Municipal Wastewater and Sludge through Forest and Cropland,* W. E. Sopper and L. T. Kardos, Eds., Pennsylvania State University Press, University Park and London (1973).

7. *Standard Methods for the Examination of Water and Wastewater,* 13th Edition, APHA, AWWA, WPCF (1971).

8. Kenner, A., and H. P. Clarke. "Detection and enumeration of *Salmonella* and Pseudomonas aeruginosa," *J. Water Pollution Control Fed.* 49(9):2163–2171 (1974).

9. Wallis, C. and J. L. Melnick. "Concentration of viruses on aluminum and calcium salts," *Amer. J. Epidemiol.* 85(3):459–468 (1967).

10. Wallis, C., S. Grinstein, J. L. Melnick and J. E. Fields. "Concentration of viruses from sewage and excreta on insoluble polyelectrolytes," *Applied Microbiol.* 18(6):1007–1014 (1969).

11. Lue-Hing, C., D. Zenz and J. Bertucci. "Report on inactivation of viruses in anaerobically digesting sludge," Metropolitan Sanitary District of Greater Chicago, Research and Development Dept. (May 1973).

12. Bertucci, J., C. Lue-Hing, D. Zenz and S. J. Sedita. "Studies on the inactivation of four enteric picorna viruses in anaerobically digesting sludge," Metropolitan Sanitary District of Greater Chicago, Research and Development Dept., Report No. 74-19 (August 1974).

13. Bertucci, J. J., C. Lue-Hing, D. R. Zenz, and S. J. Sedita. "Studies on the inactivation rates of five viruses during anaerobic sludge digestion," Metropolitan Sanitary District of Greater Chicago, Research and Development Dept., Report No. 75-21 (September 1975).

14. Meyer, R. C., F. C. Hinds, H. R. Isaacson and T. D. Hinesly. "Porcine enterovirus survival and anaerobic sludge digestion," presented at the International Symposium of Livestock Wastes, Columbus, Ohio (April 22, 1971).

15. Fitzgerald, P. R. and W. R. Jolley. "The use of sewage sludge in pasture reclamation: parasitology, nutrition and the occurrence of metals and polychlorinated biphenyls," unpublished report to the Metropolitan Sanitary District of Greater Chicago from the College of Veterinary Medicine, University of Illinois, Urbana, Illinois (1974).

SECTION IV

CASE HISTORIES
OF LAND APPLICATION
OF SEWAGE

22

THE REMOVAL OF NUTRIENTS AND TRACE METALS BY SPRAY IRRIGATION AND IN A SAND FILTER BED

B. H. Ketchum and R. F. Vaccaro
Woods Hole Oceanographic Institution
Woods Hole, Massachusetts

INTRODUCTION

The summer of 1976 will mark the third growing season of the Cape Cod Waste Water Renovation and Retrieval System. This pilot program on Cape Cod, Massachusetts, is being jointly sponsored by the U.S. Environmental Protection Agency and by the Division of Water Pollution Control of the Commonwealth of Massachusetts. The field experiments are conducted at Otis Air Force Base, Cape Cod, which is a military installation located 75 miles south of Boston.

In comparison with other spray irrigation programs currently being conducted at Muskegon, Michigan, and at Penn State University, the Otis program is a small but intensive study of spray irrigation of something less than 10 acres. The Muskegon operation in contrast encompasses some 17,000 acres, while at Penn State the area under spray irrigation is currently about 70 acres with an anticipated expansion to 500 acres.

The objective is to evaluate alternative means of wastewater treatment and to develop design criteria for the recharge of water of potable quality to the ground water reservoir of the coastal outwash plains of Cape Cod and the Islands. The installation features a spray irrigation system wherein animal forage crops are used to remove nutrients and trace metals from secondary effluent provided by the sewage treatment plants at Otis. Additional attention is also being given to the existing sand filter beds which, over the past 33 years, have provided the traditional means of disposing of secondary effluent from the treatment plant at Otis.

The population of Cape Cod continues to increase at a rate of about 4 percent per year and a doubling of the current population is predicted by the year 2000. The ground water is the only significant source of potable water, either as wells or ponds, which has led to an increasing awareness of surface and subsurface water pollution and the fear that the increasing population will ultimately lead to a shortage of potable water.

GEOLOGY AND HYDROLOGY

Otis Air Force Base is located near the intersection of the four towns of Bourne, Sandwich, Mashpee and Falmouth. The site includes a mass of unconsolidated gravel, sand and silt deposited by glacial outwash which is underlain by an impermeable consolidated bedrock roughly 260 ft below the surface. In a north-south direction, the bedrock remains level at an average depth of 170 ft below sea level; however, east of the site, it dips to a depth of 400 ft below sea level (1).

Precipitation and its ultimate storage in ground water is the sole source of fresh water on Cape Cod. An annual rainfall of about 2.14 million gallons per day per square mile, following transpiration losses, provides an estimated 1.07 million gallons per day per square mile to form a water table which reaches a maximum height of 60 ft above sea level north of Otis. Ground water levels then subside in all directions toward the ocean. Monitoring wells installed within the project area reveal ground water elevations from 53.8 to 49.7 ft above sea level, in accordance with the north-south hydraulic gradient previously described by Meade and Vaccaro (2).

Base-line soil samples taken from the pilot irrigation plots show a surface layer of loam underlain by roughly a meter of windblown Enfield sandy loam. Quite frequently, deposits of silt-clay, visually and mineralogically distinct from the surrounding sandy loam, occur in layers or clumps immediatey beneath the surface. Below 2-m depth, the topsoil grades from fine sandy loam to medium sand. Deep geological cores indicate a substrate of medium to coarse sand with occasional cobblestone layers to the vicinity of bedrock 258 ft below the surface.

The sewage treatment plant at Otis Air Force Base has been in operation since 1942 and processes wastewater generated at the base. The primary treatment facility of the plant consists of a comminutor with a by-pass bar screen, a Parshall flume, a grease-skimming snd flocculation tank and two Imhoff tanks. Two trickling filters of 3-ft depth and two settling tanks or clarifiers comprise the secondary treatment facility. The plant effluent is typically discharged onto sand filter beds for final disposal to the underlying ground water. Sludge removed from the Imhoff and settling tanks is flushed to sand drying beds and periodically removed for burial.

The plant was designed to accommodate an average summer flow of 3.7 MGD and to serve a population of about 37,000. Its load has averaged substantially less than 1 MGD for the past 10 years due to reduction of personnel and activities at the base. Presently, the plant treats slightly less than 50,000 GPD and is principally serving the stable but skeleton population which occupies base housing. Because the waste load is much less than the design capacity, the treated effluent is actually recycled in order to maintain minimum volumes. Only one Imhoff tank, one trickling filter and one secondary tank are currently being used at any one time.

During the history of the plant over 6 billion gallons of water have been discharged onto the sand filter bed system. The area of application which was served by gravity for sand filtration totals 17 acres and is divided into 22 filter beds. At full operation, 3 MGD, the rate of application was 150,000 gallons per acre per day. Presently, only two filter beds are used at any one time due to the reduced flow. Effluent is delivered to the filter beds by gravity mains and distributed by large wooden sluiceways. Cores taken from the beds show a layer roughly 1 ft thick of graded sand underlain by natural deposits of Enfield sandy loam and some silt deposits within the first few feet. At greater depths, 18 to 20 ft of medium sand and cobblestone occur down to groundwater level.

SYSTEM DESIGN AND INSTALLATION

To evaluate a spray irrigation-cropping program for wastewater management, the pilot facility was designed to provide year-round irrigation to test the water quality effects of a holding basin, to test prechlorination before irrigation and to compare the performance of fixed versus rotary rigs for wastewater distribution. The supply of wastewater comes from the Otis Air Force Base sewage treatment plant, and enters a holding basin or lagoon having a capacity of 300,000 gallons of secondary effluent, which represents but a small diversion of the total plant output. Residence, or holding time, in the lagoon is about 3 weeks.

The remainder of the system includes a pump house and control station, a 2200-ft force main, the irrigation equipment and irrigated fields. The main pump has a capacity of 160 gal/min and is started with a secondary priming pump. Safety valves are installed to prevent high pressures and to provide automatic shutoff if pressure drops suddenly. A semi-automatic gas chlorinator is used for chlorination. The effluent is pumped through a 4-in. Johns-Manville PVC force main to the irrigation site.

The relationship between the original Otis sewage treatment installation and the facilities added for experimental purposes is shown in Figure 1. The figure shows two irrigation sites A and B which have been seeded with forage crops. Irrigation site A has been planted in reed canarygrass and employs fixed deflection head sprinklers. Within site A three subplots provide rates of application of 1, 2 and 3 in. per week of effluent over three equal areas of 0.225 acres. Site B contains a 150-ft rotary irrigator anchored on a center pivot. Six deflection heads 8 ft apart mounted downward and of decreasing opening diameter toward the center deliver at the rate of 2 in. of effluent per week to each of 4 subplots, each encompassing 0.445 acres. Both site A and site B include appropriate control areas which are nonirrigated and whose monitoring provides a basis for background comparison. Each subplot is equipped with banks of lysimeters which are used to sample interstitial ground water from depths of 0.5, 1, 2, 3 and 4 ft. A total of nine wells has been installed throughout both areas which are oriented according to groundwater flow so that groundwater monitoring can be provided throughout the experimental area.

THE MONITORING PROGRAM

A monitoring program provides information on short- and long-term trends in terms of nitrogen, phosphorus and trace metals after various degrees of treatment. All wells are constructed of PVC plastic which allows sampling for ground water nutrients, metals and pesticides and many other organic substances without undue contamination. Measurements at the sand filter beds show ground water levels at 21 ft below surface and from 48 to 55 ft below surface on the experimental irrigation fields. The locations of nine ground water sampling wells used to monitor sites A and B are numbered in Figure 1.

Monitoring within the sand filter beds is also provided by a series of suction lysimeters along with seven observation wells

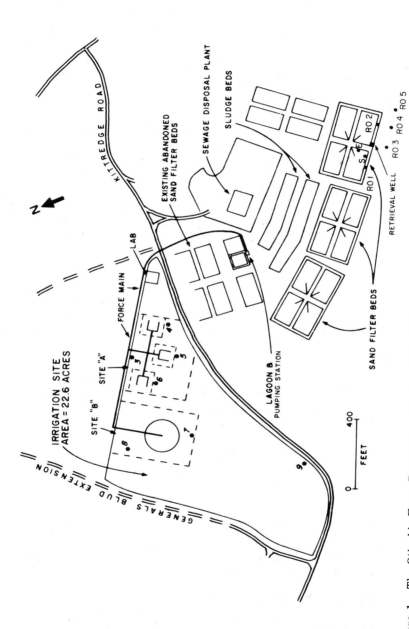

Figure 1. The Otis Air Force Base sewage treatment plant, lagoon, pumping station and irrigated areas. Location of wells for ground water sampling given by numbers 1–9.

to determine the extent of wastewater renovation and the quality of the recharged water.

Reed canarygrass is the crop at all site A locations, while at site B smooth brome, timothy, a mixture of timothy and alfalfa and reed canarygrass are grown. The control area at site A is planted with reed canarygrass while control areas of site B have been seeded with timothy.

Table 1. Wastewater characteristics of Otis treatment plant secondary effluent (concentrations as mg/l; ppm).

Constituent	Typical Secondary Treatment Effluent	Otis Secondary Effluent (mean; St. Dev.)	
Nitrogen (as N)	20		
Total inorganic	–	17.90	± 3.25
Nitrate	–	10.18	± 5.17
Nitrite	–	0.29	± 0.15
Ammonium	–	7.43	± 5.46
Phosphorus (as P)			
Total dissolved	10	8.49	± 0.83
Orthophosphate	–	6.93	± 0.78
Other Elements [a]			
Cadmium	0.01–0.03	0.00024	± 0.0001
Calcium	24	10.77	± 0.70
Chloride	45	26.78	± 2.39
Chromium ($Cr^{+6} + Cr^{+3}$)	0.02–0.14	<0.01	
Copper	0.07–0.14	0.050	± 0.020
Iron	0.10–4.3	0.502	± 0.07
Lead	0.01–0.03	0.00054	± 0.0001
Magnesium	17	3.79	± 0.17
Manganese	0.02	0.021	± 0.016
Mercury (Hg^{+2})	0.01	<0.00027	± 0.0001
Potassium	14	8.94	± 1.09
Sodium	50	37.91	± 6.34
Zinc	0.20–0.44	0.047	± 0.025

[a] Typical values from Pound and Crites (4) or from Driver et al. (3).

Table 1 shows typical concentrations of some chemical constituents of Otis secondary effluent. Comparison with accepted averages for secondary effluents as prepared by Driver et al. (3) and by Pound and Crites (4) is provided.

SAMPLING AND ANALYTICAL TECHNIQUES

All water samples are collected in 2-liter polyethylene bottles, a 200-ml aliquot of each sample being filtered (0.45 μ porosity) in the field using a Swinnex-Millipore filtration unit. The filtered aliquot is refrigerated and used for chemical analyses of plant nutrients, i.e., PO_4-P, NH_4-N, NO_2-N, and NO_3-N.

Well water samples are obtained with a vacuum-pressure apparatus modified from the design of Wood (5) for deep lysimeter sampling. With this equipment samples of water from depths greater than 25 ft below the ground surface are obtainable.

Lysimeters are used to collect interstitial soil water by applying vacuum to permeable ceramic cups (pore size ca. 1.0 μ) buried below ground level. During dry periods, particularly in the nonirrigated control plots, it is not always possible to acquire an adequate volume of water. Mature crops are cut, field-dried, and baled according to accepted agricultural practices. Desiccated subsamples are macerated in a blender and stored pending chemical analyses for nitrogen and phosphorus content along with other selected anions and cations. Soil samples are removed from trench sidewalls exposed by use of a back hoe. This technique has proven much more satisfactory for our purposes than the more commonly used coring devices.

In general, the chemcial analytical methods used are those described in the EPA "Manual of Methods for Chemical Analysis of Water and Wastes" (6). Where unusual analytical sensitivity is required, i.e., on certain groundwater samples, more sensitive methods are employed as referenced below.

Inorganic plant nutrients, PO_4-P, NO_2-N, NO_3-N and NH_4-N, are all determined colorimetrically with a Beckman DU spectrophotometer. For phosphate, the single reagent method described by Murphy and Riley (7) is used. Nitrite is measured in acid solution using sulfanilamide as recommended by Strickland and Parsons (8). Nitrate is analyzed by the brucine method described in the EPA manual cited above. Ammonia is determined with phenolhypochlorite reagent as described by Solorzano (9).

Elementary analyses for carbon, hydrogen and nitrogen in solid materials are measured in a Perkin-Elmer 240 elementary analyzer. Total elementary phosphorus, in labile form, is measured after ignition of hay and soil samples at 600°C for 2 hr followed by an acid persulfate digestion at 15 lb steam pressure (240°C) for 30 min. The inorganic phosphate released by the above treatments is then measured by the Murphy and Riley (7) method.

Cations of trace metals were originally analyzed by chelation-extraction using ammonium pyrrolidine dithiocarbamate (APDC) and methyl isobutyl ketone (MIBK) as per Brewer, Spencer and Smith (10). Currently, these elements are being determined more efficiently via heated graphite atomization (HGA) wherein a small amount of water or digestate is injected into an electrically heated graphite furnace connected to an atomic absorption spectrophotometer. Major cations are analyzed using flame atomic absorption with direct aspiration of known standards as recommended by the EPA (6).

CHANGES IN NITROGEN AND PHOSPHORUS DURING TREATMENT

The removal of phosphorus from secondary effluent increases with the stage of treatment. The phosphorus concentrations (as orthophosphate-P) in the secondary effluent before lagooning ranged from about 8 to 10 (mean = 9.25) ppm. The phosphate concentrations were higher during winter than in summer (Figure 2). Following chlorination and distribution over the

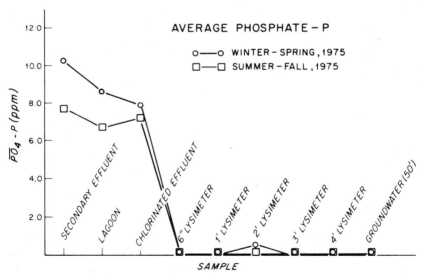

Figure 2.   Winter and summer phosphate-phosphorus concentrations (ppm) for secondary effluent after lagooning, chlorination, irrigation and percolation through the ground. Site A-west; nominal rate of application 3 in./wk.

irrigation field the phosphorus concentration was reduced to less than 0.5 ppm in the interstitial water recovered from 6 in. below the surface. At successively greater depths, no significant changes in phosphorus were observed down to the depth of the groundwater sampling at 50 ft.

Nitrogen is more versatile than phosphorus chemically and is affected by a variety of biological processes. Gains or losses of nitrogen to or from the system are regulated by the relative importance of nitrogen fixation, ammonification, nitrification and denitrification. Thus, much greater uncertainty exists regarding nitrogen distributions as compared with phosphorus. Phosphorus rather than nitrogen has been selected as the more appropriate element for mass balance analysis.

Total inorganic nitrogen concentrations in Otis secondary effluent before lagooning range from about 12.5–18.5 (mean = 11.6) ppm and show a comparable reduction pattern to that of phosphorus following successive stages of treatment (Figures 3 and 4). During the winter-spring period, ammonia is the most abundant form of inorganic nitrogen in secondary effluent although appreciable amounts of oxidized nitrogen ($NO_2^- + NO_3^-$) also occur. During the summer-fall months oxidized forms of nitrogen are more abundant than ammonia. In the interstitial soil water, the ammonia is almost completely absorbed in the first foot of soil, but the oxidized forms of nitrogen penetrate at reduced concentrations to the ground water. The data on inorganic nitrogen distribution suggest that soil nitrification is maintained at all seasons of the year.

APPLICATION OF NITROGEN AND PHOSPHORUS TO THE AGRICULTURAL PLOTS

The measured amounts of irrigation water and the amounts of phosphorus and nitrogen applied on sites A and B during the period July 1974 through November 1975 are shown in Figure 5. Due to inevitable shutdowns associated with such necessities as liming, mowing and drying, the annual average amount of irrigation actually achieved is less than that predicted from the rates of application. The amounts of phosphorus applied on site A ranged between 1.5 to 3.8 g per sq ft (*ca.* 160–380 kg per ha) depending upon the rate of irrigation. All quadrants of site B received 3.5 g per sq ft (*ca.* 377 kg per ha). The range of nitrogen applied at site A was between 1.9 to 4.7 g per sq ft (*ca.* 204–505 kg per ha) and the amount applied to the site B plots was 4.3 g per sq ft (*ca.* 463 kg per ha).

The efficiency of the spray irrigation system as a sewage treatment process depends upon the amounts of nitrogen and phosphorus removed from the secondary effluent by the crops and by accumulation in soil so as to limit the amount of these elements reaching the underlying ground water. Thus, we have compared the combined amounts of phosphorus and nitrogen

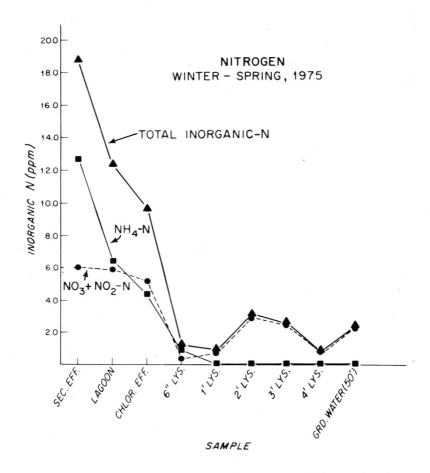

Figure 3. Winter-spring ammonia, oxidized and total inorganic nitrogen concentrations (ppm) for secondary effluent after lagooning, chlorination, irrigation and percolation through the ground. Site A-west; nominal rate of application 3 in./wk.

harvested with crops and accumulated in the soil with the amounts supplied in the irrigation water.

SIZE AND CHEMICAL COMPOSITION OF CROPS

The total amounts of hay harvested from site A between September 1974 through November 1975 (four cuttings) along with the phosphorus-nitrogen contents of the roots and harvested crops

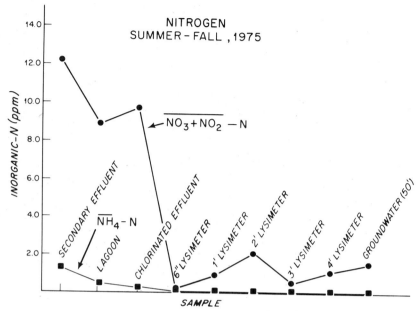

Figure 4. Summer-fall ammonia, oxidized and total inorganic nitrogen concentrations (ppm) for secondary effluent after lagooning, chlorination, irrigation and percolation through the ground. Site A-west; nominal rate of application 3 in./wk.

are shown in Figure 6. In general, the tonnage of hay harvested increases with an increase in the amount of irrigation water applied. The phosphorus content of the crop also appears to vary directly with the amount of irrigation. There appeared to be a reversal in terms of the nitrogen removal since the crop on the south field (2 in. per week) accumulated less nitrogen than that on the east field where the irrigation was 1 in. per week. However, the application of 3 in. per week on the west field yielded the highest amount of crop nitrogen. During the time of our observations, approximately one-half of the phosphorus and nitrogen which entered into site A plant tissue was retained in the developing root systems.

The harvested amounts of hay for four cuttings from site B over the same 17-month period are shown in Figure 7. These yields are significantly greater than those recorded for the 20 in. per week application on site A, the highest recorded yield (1.6 tons) being obtained from the timothy-alfalfa subplot. Other yields (smooth brome, timothy and reed canarygrass) were relatively consistent within the range 1.2 to 1.3 tons. The phospho-

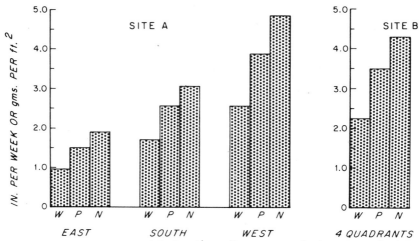

Figure 5. Actual rates of irrigation and amounts of nitrogen and phosphorus applied on sites A and B from July 1974 to November 1975. W = water volume (in./wk); P = Phosphate P (g/ft²); N = total inorganic nitrogen (g/ft²).

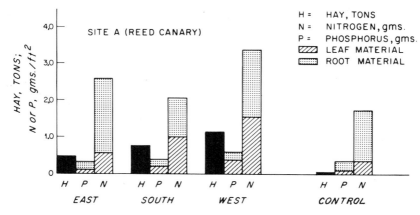

Figure 6. Site A total hay harvested (tons) (September 1974–November 1975) and the phosphorus and nitrogen content of leaf and root tissue expressed in g/ft² of subplots. Nominal irrigation rates: east—1 in./wk; south—2 in./wk; west—3 in./wk (site A plots are 0.225 acres each).

rus content of these crops remained relatively stable between 0.5 and 1 g per sq ft (54–108 kg per ha). Nitrogen variations were considerably larger and ranged between 3 and 6 g per sq ft

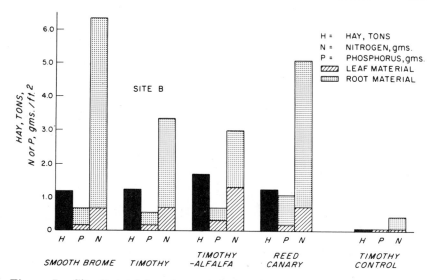

Figure 7.   Site B total hay harvested (tons) (September 1974–November 1975) and the phosphorus and nitrogen content of leaf and root tissue expressed in g/ft² of subplots. Nominal irrigation rate = 2 in./wk (site B plots are 0.445 acres each).

(325–650 kg per ha) with the smooth brome providing the greatest, and timothy-alfalfa showing the least nitrogen content. On site B more than half of the plant nitrogen and phosphorus was associated with root structure.

We have also studied the accumulation of other cations in the agricultural crops. The cation content of reed canarygrass in three consecutive harvests from site A west, where the actual irrigation rate was 2.59 in. per week, is shown in Table 2. Within this period, significant increases in calcium, magnesium, copper, chromium and nickel were observed. The data also indicate that there was a tendency to reduce the amount of sodium and potassium with time.

NITROGEN:PHOSPHORUS
RELATIONS IN SOIL AND CROPS

Prior to October 1973, the agricultural plots were all woodlands that were cleared and planted to our hay crops. In the fall of 1975, after the development of a good root system, soil profiles to a depth of 4 ft were taken on each plot, and the upper foot separated into soil and root samples. Analyses for the amounts of phosphorus and nitrogen accumulated within the

Table 2.   Cation content, reed canarygrass—site A, west; irrigation = 2.59 in. per week.

| | Harvest Date and Concentration (mg/kg) | | |
	September 1974	June 1975	November 1975
Cadmium	0.32	0.22	0.61
Calcium	1,750	1,129	9,489
Chromium	1.8	17	44
Copper	6.8	6.7	15
Iron	451	216	651
Lead	4.5	1.5	5.4
Magnesium	1,700	2,406	8,076
Manganese	62	24	84
Nickel	1.8	7.7	25.2
Potassium	22,500	15,319	9,826
Sodium	556	265	385
Zinc	35	23	47

upper foot of soil are given in Table 3. At each of the subplots, irrigation has resulted in an increase in phosphorus content above that recorded for the appropriate control soil. Two site B soils, however, the timothy-alfalfa and smooth brome plots, differed by less than one standard deviation from the mean of all or from the mean of the controls. Regarding nitrogen, the situation is not so clear-cut. Except for the timothy plot, the nitrogen content of all soils was within one standard deviation of the mean, and the apparent differences are not significant. The timothy soil value may be in error, but was rechecked and found not to be an analytical error. An "unusual" soil sample at this location may have been obtained. Nitrogen accumulation in the upper foot of the soil appears to have been obscured by crop requirements.

It is well known that sewage is not a well-balanced fertilizer for crop needs, with phosphorus being present in excess amounts relative to nitrogen as compared to the normal composition of both terrestrial and aquatic plants. The phosphorus: nitrogen ratios in leaf, root and soil samples collected from each of the agricultural plots were compared. The data, summarized in Table 4, show that root material of forage grasses consistently contains less phosphorus relative to nitrogen (P:N = 0.09–0.21) than do the leaf structures (P:N = 0.18–0.40). For entire plants the ratio ranged from 0.13 to 0.22. By comparison the phosphorus: nitrogen ratio in irrigation water is about 0.80, a large excess of phosphorus based on crop requirements. Thus, crop limitation in terms of nutrient supply is more likely to be related to a nitrogen rather than to a phosphorus deficiency. The phosphorus:nitrogen ratios in the soils averaged 0.47 for the irrigated plots.

Table 3. Nitrogen-phosphorus content of Otis soils (gm/ft^2 to a depth of 1 ft).

	Phosphorus	Nitrogen
Site A (Reed Canary)		
East	10.0	21.7
West	10.7	19.2
South	11.3	19.8
Control	8.7	25.6
Site B		
Smooth brome	8.2	21.5
Timothy	10.4	7.0
Timothy-alfalfa	7.1	17.3
Reed canary	9.5	20.8
Control	6.8	17.8

A comparison between the average P:N ratio of the plants (0.16) and that of the effluent used for irrigation (0.80) suggests that an excess of 0.64 of phosphorus for each gram of nitro-

Table 4. Phosphorus:nitrogen ratios (Otis spray irrigation system).[a]

	Leaf	Roots	Entire Plant (P:N by Wt)	Soil
Crops				
Site A (Reed Canary)				
East	0.18	0.11	0.14	0.46
South	0.22	0.15	0.18	0.56
West	0.40	0.11	0.17	0.57
Control	0.33	0.13	0.19	0.34
Site B				
Smooth brome	0.23	0.09	0.13	0.38
Timothy	0.23	0.14	0.18	1.48 [b]
Timothy-alfalfa	0.24	0.21	0.22	0.41
Reed canary	0.23	0.21	0.22	0.46
Control, timothy	0.25	0.14	0.18	0.47
All Irrigated Plots, Mean Σ P/ Σ N	0.22	0.11	0.16	0.47

[a] The P:N ratio for the chlorinated lagooned effluent, derived from the mean of analyses made throughout the period of irrigation, was 0.80. This differs from the ratio derived from data in Table 1 defining secondary effluent characteristics, presumably because of processes taking place in the lagoon.

[b] The high ratio results from spurious low N in the soil sample. Soil data from this plot were not used in computing the average ratio.

gen assimilated by the plants is being provided. This corresponds to about four times the estimated crop requirement for phosphorus. Compared to the P:N ratio of the soils the excess is 0.33 of phosphorus for each gram of nitrogen. If all of the applied phosphorus and none of the applied nitrogen were retained by the soils, the change in the P:N ratio in the soils would not be significant compared to the variability of our analyses.

CROP EFFICIENCIES IN TERMS OF NITROGEN AND PHOSPHORUS REMOVAL

Comparisons of the amounts of nitrogen and phosphorus taken up by crops versus the quantities supplied in irrigation water have been used to assess the overall efficiency of nitrogen and phosphorus removal. The data in Table 5 show that nitrogen up-

Table 5.   Nitrogen exchanges—Otis irrigation plots.

| | Nitrogen in Crops | | | | |
	N Applied (g/ft^2)	Leaf[a]	Roots[b] (g/ft^2)	Total	Percent N in Crops
Site A (Reed Canary)					
East	1.82	0.566	2.01	2.58	142
South	3.17	1.00	1.06	2.06	65
West	4.83	1.55	1.85	3.40	70
Control	0.10	0.380	1.86	2.24	
Site B					
Smooth brome	4.34	0.684	5.67	6.35	146
Timothy	4.34	0.704	2.65	3.35	78
Timothy-alfalfa	4.34	1.31	1.69	3.00	69
Reed canary	4.34	0.726	4.38	5.11	118
Control	0.10	0.440	0.450		

[a] Total, four harvests.

[b] Established over 17-month period—integrations to a depth of 1 ft.

take by the various grasses consistently exceeded 65 percent of the total amount of nitrogen provided. In three instances crop-nitrogen actually exceeded supply by a significant amount (percentage values > 100) thereby reinforcing the possibility of a nitrogen withdrawal from the soil reserve or the recruitment of new nitrogen via nitrogen fixation.

Unlike nitrogen, crop removal efficiencies for phosphorus (Table 6) were consistently less than 31 percent, re-emphasizing the likelihood of an excess phosphorus supply in terms of plant needs. This observation is consistent with our previous description of phosphorus accumulation in the soil. The adsorptive capacity of the soil column above groundwater level will determine the amount of phosphorus which can be retained without addition to the ground water.

Table 6. Phosphorus exchanges—Otis irrigation plots.

		Phosphorus in Crops			
	P Applied (g/ft^2)	Leaf[a]	Roots[b] (g/ft^2)	Total	Percent P in Crops
Site A (Reed Canary)					
East	1.46	0.102	0.225	0.327	28.0
South	2.61	0.216	0.160	0.376	14.5
West	3.85	0.389	0.208	0.597	15.5
Control	0.09	0.110	0.248	0.358	
Site B					
Smooth brome	3.49	0.154	0.504	0.658	18.9
Timothy	3.49	0.164	0.374	0.538	15.7
Timothy-alfalfa	3.49	0.314	0.355	0.669	19.2
Reed canary	3.49	0.168	0.914	1.082	31.1
Control, timothy	0.09	0.003	0.062	0.065	

[a] Total, four harvests.

[b] Established over 17-month period—integrations to a depth of 1 ft.

PHOSPHORUS MASS BALANCE

More complete information on the phosphorus exchanges is shown in Table 7 which, in addition to crop removals, includes an assessment of the phosphorus accumulation in soils above that of the appropriate control. At site A the combined amount of phosphorus in crops and soil consistently amounted to more than 80 percent of the total amount of phosphorus supplied. Comparable results for site B are more variable, especially the probably aberrant value, 27 percent recovery, recorded for the timothy-alfalfa plot. Excepting this entry, the remaining data for site B indicate that between 59 and 121 percent of the phosphorus supplied has either entered the crops or has accumulated in the soil. The excess phosphorus accumulation in the soil is 3.2 times the amount found in the total plants (crop and roots) substantiating the fourfold excess of P derived from the P:N ratios.

TRACE METAL ACCUMULATIONS
ON A SAND FILTER BED

The original design of the Otis sewage treatment plant provided sand filter beds for the terminal disposal of secondary effluent. Over the past 33 years more than 6 billion gallons of effluent have been discharged into the 17 acres reserved for this purpose. The latter area consists of 22 separate beds which measure 200 ft in length and 100 ft in width as shown in Figure 1.

Monitoring capability is provided by seven wells located at strategic points downstream with respect to ground water flow of one sand filter bed as shown in the lower right of Figure 1. A high-capacity retrieval well which is capable of pumping ground water from a depth of about 60 ft (19 m) at a rate of 300 gal/min also has been installed. These wells have been used to measure rates of ground water drawdown, recovery and aquifer hydraulic conductivity. The same facilities have also been used to conduct experimental observations on the dissipation of various constituents of secondary effluent in the absence of an agricultural crop.

The ability of sand filter beds to remove trace metals has been evaluated by sampling the soil at various depths. The heavy metals tend to accumulate within the upper few centimeters of sand. The concentration in the effluent and the 33-year supply of copper, zinc, cadmium, lead and chromium to a typical sand filter bed is shown in Table 8. At Otis, wet sludge is separated and deposited on sludge beds for drying and ultimate disposal by burial. The filter beds have been exposed to only a small fraction of the total heavy metal load produced by the treatment plant since the major concentration of these heavy metals is found in the dried sludge material as compared with the secondary effluent (Table 8). The sludge beds have not been studied as a possible source of ground water contamination.

In terms of supply, the order of abundance for heavy metal additions to the sand filter bed was Cu > Zn > Cr = Pb > Cd. For comparison, the distribution of these same metals within the upper 52 cm of soil surface is shown in Table 9. Here the marked tendency for heavy metals to accumulate within the upper 6 cm of sand is clearly demonstrated and the order of abundance was Cu > Zn > Pb > Cr > Cd. The surface enrichment factor, derived by comparison with a control soil, varied from 20- to 360-fold, the resulting order being Cu > Cd > Pb = Zn > Cr.

By comparing the integrated amounts of these metals distributed within the upper 52 cm with the heavy metal supply over

Table 7.   Phosphorus mass balance—Otis irrigated plots.

| | P Applied (g/ft^2) | Phosphorus in | | | Recovery Percent |
		Crops	Soil [a] (g/ft^2)	Crops & Soil	
Site A (Reed Canary)					
East	1.46	0.327	1.30	1.63	115
South	2.61	0.376	2.00	2.38	92
West	3.85	0.597	2.60	3.20	83
Site B					
Smooth brome	3.49	0.658	1.38	2.04	59
Timothy	3.49	0.538	3.66	4.21	121
Timothy-alfalfa	3.49	0.669	0.28	0.95	27
Reed canary	3.49	1.082	2.37	3.45	99

[a] Phosphorus content in the upper foot of soil. The amount in control soil was 8.7 g/ft^2 for Site A and 6.8 g/ft^2 for Site B. The values listed are the amounts present in excess of these controls.

the past 33 years it is possible to assign removal efficiencies for each of the five metals examined. As recorded in Table 9, our data indicate that 86 percent of the Cu, 49 percent of the Zn, 113 percent of the Cd, 128 percent of the Pb and > 59 percent of the Cr supplied are bound in the surface soil. Thus of the metals examined only Zn and possibly Cr would appear to have a reasonable chance of becoming entrained in the ground water some 640 cm (21 ft) below the surface.

Table 8.   Heavy metal supply to experimental sand filter bed.

| | Metal Concentration | | | | |
	Cu	Zn	Cd	Pb	Cr [a]
Sample Source					
Secondary Effluent			(ppb)		
Soluble	80	55	0.24	0.50	<10
Particulate	32	12	0.58	9.30	0.6
Dry Otis Sludge	2,799	246	(ppm) 13	358	96
Total 33-yr Metal Loading			(g/ft^2)		
Soluble	4.45	3.12	0.014	0.028	<0.570
Particulate	1.82	0.67	0.033	0.530	0.034
Total	6.27	3.79	0.047	0.558	<0.604

[a] Soluble chromium concentration was less than the sensitivity of the analytical method.

Table 9. Heavy metal build-up in a sand filter bed. Metal concentrations corrected for appropriate control analysis.

Depth (cm)	Metal Concentration (mg/kg)				
	Cu	Zn	Cd	Pb	Cr
Surf	679	320	10.5	166	82
0–4	236	67	1.7	28	4.7
4–6	272	60	1.8	26	10.0
14–16	7.2	5.5	0.11	0.00	0.10
24–26	6.1	9.3	0.25	0.00	2.30
29–31	2.4	1.3	0.09	0.00	0.00
44–46	2.5	3.9	0.07	0.80	0.90
50–52	0.7	0.0	0.07	0.00	0.00
Control Soil, Surf	1.9	5.8	0.07	2.70	4.30
Surface Enrichment Factor	360	56	150	60	20
Integrated Metal Content (g/ft^2 to 52 cm) [a]	5.36	1.87	0.053	0.718	0.353
Percent Retention Content/Loading	86	49	113	128	>59

[a] Assumes soil density of 1.85, i.e., 86.9 kg under a ft^2 to a depth of 52 cm.

CONCLUSIONS

Use of secondary effluent for irrigation of hay fields serves as a tertiary treatment which can supply water of potable quality to the ground water reservoir. Commonly, agricultural wastewater applications are located in semi-arid regions where water conservation rather than water quality improvement is the more important consideration. It appears that three desirable goals, i.e., water conservation, crop irrigation and fertilization, and water quality improvement may be simultaneously achieved. If so, domestic wastewaters could be elevated to the level of a valuable national resource.

On Cape Cod, Massachusetts, the geology and hydrology are ideally suited for spray irrigation-cropping and a considerable amount of undeveloped acreage could be committed to this purpose. Furthermore the rapidly expanding population of Cape Cod is entirely dependent upon ground water for its water supply. Thus, there is developing an extreme reluctance to deplete the ground water reserve by the construction of additional marine outfalls. Our activities at Otis Air Force Base over the past two years were planned to assess the advantages provided by adding forage crop irrigation to the traditional method of effluent disposal by direct soil infiltration via sand filter beds or dispersed septic tanks and cesspools.

The major accumulation of heavy metals on Otis sand filter beds after 33 years of service occurs within the uppermost 6 cm of sand. The extent of enrichment for Cd, Cr, Cu, Pb and Zn ranges from 20 to 360-fold. Excepting Zn and possibly Cr, more than 85 percent of the applied amounts of each of the above metals resides within the upper 52 cm of sand.

The crops grown on plots irrigated at more than 1 in./wk have provided 3.5 to 5 tons per acre per year. Moreover, these crops accumulate nitrogen, phosphorus and the cations calcium, magnesium, copper, chromium and nickel within the plant tissues. There is excess phosphate-phosphorus in the effluent, but like the heavy metals bound on the sand filter beds, this excess phosphorus becomes bound within the upper foot of soil. Significant phosphorus or heavy metal leaching into the ground water is not anticipated within the foreseeable future since more than 50 ft of soil above ground water level remains available for adsorption. Excessive nitrate-nitrogen concentrations (50 ppm) in the wells monitoring the sand filter bed (11) have been found, but the crops grown on the irrigation plots provide adequate control of this contaminant and only trace amounts of nitrate (about 2 ppm) in the wells monitoring these irrigated plots have been found.

ACKNOWLEDGMENTS

The authors gratefully acknowledge the assistance from Woods Hole Oceanographic Institution colleagues who helped acquire the information presented in this study. Particular thanks are due to Messrs. Paul Bowker, Nathaniel Corwin and Mark Dennett for sampling and chemical analytical measurements and to Ms. Patricia Pykosz for secretarial assistance and manuscript preparation. This paper is Woods Hole Oceanographic Institution Contribution No. 3802.

REFERENCES

1. Kerfoot, W. B. and B. H. Ketchum. "Cape Cod waste water renovation and retrieval system, a study of water treatment and conservation," Technical Report, W.H.O.I. 74–13, Woods Hole Oceanographic Institution, Woods Hole, Massachusetts (1974).
2. Meade, R. and R. F. Vaccaro. "Sewage disposal in Falmouth, Massachusetts. III. Predicted effects of inland disposal and sea outfall on ground water," *J. Boston Society of Civil Engineers* 58(4):278–297 (1971).
3. Driver, C. H., B. F. Hrutfiord, D. E. Spyridakis, E. B. Welch and D. D. Woolridge. "Assessment of effectiveness and effects of land meth-

odologies of wastewater management," Technical Report, Contract No. DACW 73–73–C–0041, University of Washington, Seattle, Washington (1972).

4. Pound, C. E. and R. W. Crites. "Wastewater treatment and reuse by land application," Environmental Protection Technology Series EPA–660/2–73–006a, Office of Research and Development, U.S. EPA, Washington, D.C. (1973).

5. Wood, E. D., F. A. Armstrong and F. A. Richards. "Determination of nitrate in sea water by cadmium-copper reduction to nitrite," *J. Mar. Biol. Ass. U. K.* 47:23–31 (1967).

6. EPA Standard Methods. "Methods for chemical analysis of water and wastes," Environmental Protection Agency, NERC Analytical Control Laboratory, Cincinnati, Ohio (1974).

7. Murphy, J. and J. P. Riley. "A modified single solution method for the determination of phosphate in natural waters," *Anal. Chim. Acta.* 26:31–36 (1962).

8. Strickland, J. D. H. and T. R. Parsons. "A practical handbook for sea water analyses," Fisheries Res. Bd. Canada, Bulletin 167; 311 pp. (1968).

9. Solorzano, L. "Determination of ammonia in natural waters by the phenolhypochlorite method," *Limnol. Oceanog.* 14:799–801 (1969).

10. Brewer, P. G., D. W. Spencer and C. L. Smith. "Determination of trace metals in sea water by atomic absorption spectrophotometry," Special Technical Report 443, American Society for Testing Materials, pp. 70–77 (1969).

11. Kerfoot, W. B., B. H. Ketchum, P. Kallio, P. Bowker, A. Mann and C. Scolieri. "Cape Cod waste water renovation and retrieval system, a study of water treatment and conservation, first year of operation," Technical Report, W.H.O.I. 75–32, Woods Hole Oceanographic Institution, Woods Hole, Massachusetts (1975).

23

EVALUATION OF AN INFILTRATION-PERCOLATION SYSTEM FOR FINAL TREATMENT OF PRIMARY SEWAGE EFFLUENT IN A NEW ENGLAND ENVIRONMENT

M. B. Satterwhite
Department of the Army, Corps of Engineers,
  Waltham, Massachusetts
G. L. Stewart
University of Massachusetts, Amherst, Massachusetts

INTRODUCTION

Treatment of municipal wastewater is a major concern of many communities faced with present water quality legislation and realities of implementing conventional wastewater treatment facilities. Many land-oriented wastewater treatment systems have operated for years and have proven to be an effective alternative for wastewater treatment. With proper care and management, the infiltration-percolation systems can provide a less expensive, more reliable wastewater treatment.

Although infiltration-percolation systems have been employed to treat varying qualities of sewage effluent in New England for many years, little or no data exist to describe their effectiveness in renovating wastewater effluent or the longevity of these systems.

During 1973, investigations were conducted at Fort Devens, Massachusetts, to determine the effectiveness of an infiltration-percolation system to renovate unchlorinated primary sewage effluent. The system had been in operation for more than 30 years.

DESCRIPTION AND PRESENT OPERATION
OF THE TREATMENT FACILITY

Fort Devens is a U.S. Army military installation located about 52 km northwest of Boston and 35 km northeast of Worcester, Massachusetts. The number of persons residing or working on the post has fluctuated with the mission of the installation. In 1973, the daytime population was about 15,000 of which 10,400 were permanent residents.

The present sewage treatment facility, constructed in 1942, has provided continuous service to the installation (Figure 1). Sewage from the cantonment and housing area receives the equivalent of primary treatment using three Imhoff tanks. Wastewater retention time in the Imhoff tanks is estimated to be 6 hr based on a design flow volume of 11,355 m³/d (3.0 MGD). The actual retention time varies, however, in relation to the influent flow volume and the depth of sludge accumulation in the tanks. Kitchen grease, fats and various oils are removed from the sewage in grease traps enroute to the treatment site.

Final treatment of the unchlorinated primary sewage effluent is achieved by discharging the effluent to 22 infiltration-percolation basins which average about 0.32 ha (0.79 ac) in size. The basins were constructed on a large, oval-shaped, steep-sided kame composed of unconsolidated stratified sand and gravel which rises approximately 21 m (70 ft) above the floodplain of the Nashua River. The treatment basins are rotated by inundating three basins for a 2-day application period followed by a 14-day "rest" or "dry-up" period. During holidays, weekends, or when high effluent flows are anticipated, four treatment basins may be used.

The basins are operated throughout the year. Ice and snow often accumulate on the basins during the winter months; however, the effluent is sufficiently warm (8 to 12° C) to melt the ice. Although the infiltration-percolation rates are reduced somewhat during colder periods, the applied effluent continues to infiltrate and percolate through the gravelly sand and sandy gravel strata.

During the summer the treatment basins have a good stand of naturally occurring annual grasses: fall panicum (*Panicum dicotomiflorium* Michx.), barnyard grass [*Echinochloa crusgalli* (L.) Beauv.], and *Carex* sp.

The treatment basins require periodic excavation to a depth of 0.3 m (1 ft) with replacement using bankrun sand and gravel. The most recent "cleaning" operation was completed in October

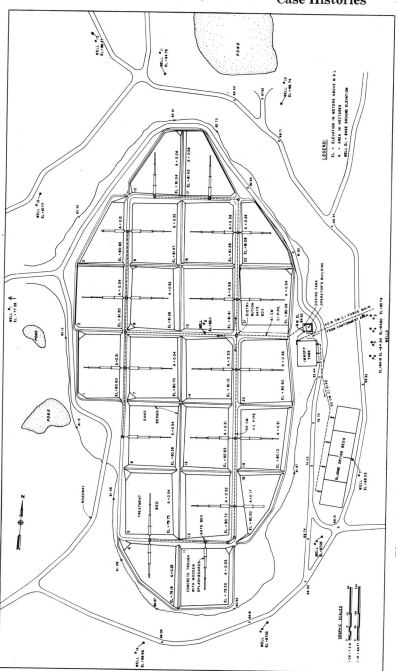

Figure 1.   Sewage treatment facility layout plan, Fort Devens, Massachusetts.

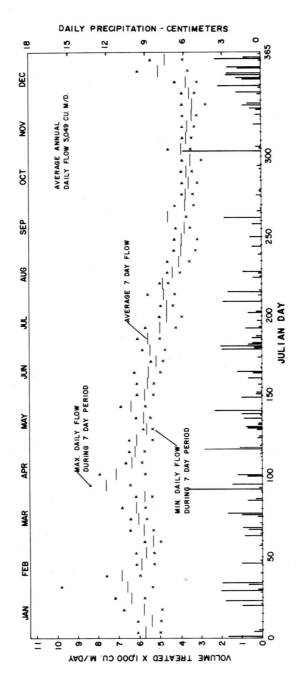

Figure 2. Maximum, minimum and average 7-day wastewater flows and precipitation for the Fort Devens sewage treatment facility (1973).

1968, at which time 18 of the 22 basins were excavated to 0.45 to 1.22 m (1.5 to 4.0 ft). Excavation below the specific 0.3 m (1 ft) was necessary to remove a "tarlike" layer which was about 0.45 m (1.5 ft) thick.

Normal operation and maintenance of the treatment facility is carried out by two full-time employees. Operation of the facility has been directed toward maintaining the hydraulic capacity of the treatment basins.

METHODS AND MATERIALS

Ground water quality beneath the application site and the surrounding area was monitored by collecting and analyzing liquid samples from 14 observation wells (Figure 1). Each well screen was positioned so that the upper portion of the water table could be sampled. All wells were installed by wash drilling, using tap water to avoid possible contamination. Following installation, each well was pumped for 30 min to remove any fine sediment.

Ground water samples were taken biweekly from each observation well. Samples were placed directly into sterile 1-liter (0.26-gal) labeled beakers using a hand-operated suction pump. A refrigerated proportional sampler cooled to 5° C was used to collect an 18-liter composite sample of the primary effluent for the 24-hr period preceding each sampling event. After thoroughly mixing the composited effluent sample, a liter subsample was drawn off for laboratory analysis. Ground water and effluent samples were analyzed for the following parameters according to procedures outlined in *Standard Methods* (1): pH (glass electrode method); 5-day biochemical oxygen demand (BOD_5) (Winkler determination); chemical oxygen demand (COD); organic Kjeldahl nitrogen (potentiometric titration); ammonia nitrogen (NH_4-N) (Nesslerization); nitrite-nitrogen (NO_2-N) (colorimetric]; total and orthophosphates (PO_4-P) (stannous chloride method); and total coliform bacteria (membrane filter technique). Nitrate-nitrogen (NO_3-N) (2) and chlorides (Cl) (3) were determined utilizing specific ion electrodes.

RESULTS AND DISCUSSION

The volume of unchlorinated primary sewage effluent applied to the 22 treatment basins varied between 2,676 and 9,841 m³/d (0.7 to 2.6 MGD) with an average daily volume of 5,049 m³/d (1.3 MGD) (Figure 2). Following the 16-day operation cycle, the volume of effluent applied to the basins was 27 m/yr (90 ft/yr).

Ground water flow in the vicinity of the treatment basins was toward the Nashua River. Elevation difference in the water tables of the various wells showed a ground water gradient of about 1 percent toward the river or adjacent surface waters. The dynamic ground water mound directly beneath the application site had ground water gradients of 1.1 to 1.6 percent from the site.

Results of the chemical and bacteriological analyses of the primary effluent and loadings of various wastewater constituents to the treatment basins are summarized in Table 1. Because negatively charged chloride ions are not readily fixed or ab-

Table 1. Chemical and bacteriological characteristics of primary effluent and annual additions to treatment basins (1973).

Parameter[a]	Effluent		Total Additions[b] (kg/ha-yr) [c]
	Range	Mean	
pH (standard units)	6.2–8.0	—	—
BOD_5	30–185	112	30,750
COD	110–450	192	52,720
Total nitrogen	19–78	47	12,910
Organic nitrogen	11.5–32.8	23.4	6,425
NH_4-N	6.2–42	21.4	5,880
NO_3-N	0.4–2.8	1.3	360
NO_2-N	0.002–0.06	0.02	5.0
Total PO_4-P	6–16	11	3,020
Ortho PO_4-P	3–15	9	2,470
Chloride	56–330	150	41,190
Total coliform bacteria x 10^6/100 ml	18–53	32	—

[a] mg/1 unless otherwise indicated.

[b] Based on an effluent flow volume of 5,059 m^3/d.

[c] kg/ha-yr x 0.8929 = lb/ac-yr.

sorbed by soil, chloride levels in the well samples were used to monitor the impacts of the percolate on ground water quality. Soil analysis at the site had shown that chlorides were carried through permeable horizons by percolating water (4). Chloride additions to the basins totaled about 41,000 kg/ha. However, chloride levels in the substrate were less than 200 kg/ha-m (15 ppm) and those in the primary effluent varied between 56 and 330 mg/l with a mean value of 150 mg/l.

Table 2. Chemical and bacteriological characteristics of ground water in selected observation wells (average values).

Parameter [a]	Well									
	2	3	7	8	9	10	11	12	13	14
pH (standard units)	6.8	6.3	6.4	6.6	6.5	6.1	6.3	6.6	6.2	6.5
BOD_5	12	2.5	2.1	2.0	1.4	0.9	0.8	1.2	1.0	0.9
COD	26	19	22	15	8	10	9	10	13	11
Total nitrogen	14.5	19.5	28.0	19.8	10.4	20.3	12.1	3.7	1.9	9.7
Organic nitrogen	8.3	2.3	3.4	4.2	3.7	1.2	1.0	0.8	1.2	1.5
NH_4-N	5.3	1.3	4.7	4.5	3.2	0.5	1.0	0.3	0.3	0.4
NO_3-N	0.9	15.6	19.5	10.7	3.5	18.6	10.1	2.6	0.4	7.8
NO_2-N	0.03	0.3	0.4	0.4	0.02	0.02	0.02	0.01	0.01	0.02
Total PO_4-P	5.9	0.9	0.8	0.8	1.4	1.3	1.9	0.6	0.7	1.1
Ortho PO_4-P	5.6	0.2	0.1	0.1	0.4	0.1	0.2	0.1	0.1	0.1
Chloride	85	230	257	220	144	257	221	40	15	162
Total coliform bacteria (#/100 ml)	3900	210	110	158	230	620	130	120	370	120

[a] mg/1 unless otherwise indicated.

Chloride levels reported (5) for native ground water (15 to 20 mg/1) were comparable to chloride levels observed in the background well, number 13 (Table 2). Chloride levels in ground-water samples obtained from the other observation wells were substantially higher than the background levels found in well number 13 (6 to 20 mg/1).

Ground water chloride levels indicate the relative influence of the percolate upon the ground water quality and permit the classification of each well into one of three arbitrary groups: native ground water, well number 13 with a mean chloride level of less than 30 mg/1; moderately impacted wells (2 and 12) containing mean chloride levels of 30 to 100 mg/1, and strongly impacted wells (3, 7, 8, 9, 10 and 11) containing mean chloride levels of more than 100 mg/1. Grouping the wells according to the relative influence of the percolate revealed that those wells with the highest mean chloride levels were located north and east of the treatment site. These data supported the ground water gradient data which concluded that ground water flow away from the treatment site was toward the north and east.

The COD and the BOD_5 of primary effluent averaged approximately 192 and 112 mg/1, respectively. Additions of COD to the treatment basins averaged about 145 kg/ha-d while BOD_5

averaged approximately 85 kg/ha-d. Although additions of organic matter to the basin were substantial, these materials were readily removed in the treatment basins. COD values averaged less than 15 mg/l in most wells, although wells 3, 7 and 8 had COD values ranging from 20 to 70 mg/l. These high values were recorded when the water table was close to the soil surface, but as depth to ground water increased, the COD levels decreased to less than 20 mg/l. This suggested that some of the COD observed in the ground water resulted from the eluviation of natural organics.

BOD$_5$ levels in most observation wells were less than 2 mg/l, while BOD$_5$ in wells 3, 7 and 8 average 2.0 to 2.5 mg/l. The COD and BOD$_5$ levels observed in most wells averaged approximately 5 percent and 2 percent respectively of the mean COD and BOD$_5$ levels observed in the primary sewage effluent.

Total effluent nitrogen varied annually from 20 to 70 mg/l and averaged 47 mg/l. Organic nitrogen and NH$_4$-N were found in about equal amounts, 23.4 and 21.4 mg/l, respectively, while NO$_3$-N and NO$_2$-N were present in small amounts, 1.3 and 0.1 mg/l, respectively. Total nitrogen additions to the treatment basins were about 13,000 kg/ha during 1973. Total nitrogen in the effluent appeared inversely related to wastewater flow. During low flow periods of later summer through early fall, high nitrogen concentrations were observed. During late winter through spring, when high flows were recorded, nitrogen levels were somewhat lower (Figure 3). Seasonal variations in nitrogen levels and effluent flow volume may reflect the infiltration of ground water into sewer lines and the direct flow of surface runoff into manholes that occasionally become inundated.

Comparing mean total nitrogen levels observed in ground water with the mean effluent concentration revealed that ground water levels were substantially less (Figure 4). Nitrogen levels in observation wells 3, 8, 10 and 11, which were most affected by percolate from the treatment basins, contained about 10 to 20 mg/l total nitrogen. These levels were 21 to 43 percent of the effluent nitrogen levels. The mean total nitrogen in well 7 averaged 28 mg/l, which was believed to show the additional effect of the leachate from the adjacent sludge dewatering beds on ground water quality.

Organic nitrogen in the effluent ranged from 11 to 33 mg/l while organic nitrogen levels in the ground water (wells 3, 7, 8 and 9) were only 2 to 3 mg/l greater than the native ground water levels of 1 to 2 mg/l. The effluent NH$_4$-N varied from 6 to 42 mg/l during the study period. During this same period, NH$_4$-N levels in wells 2, 7 and 8 were about 5 mg/l; wells 3, 9, 10, 11 and

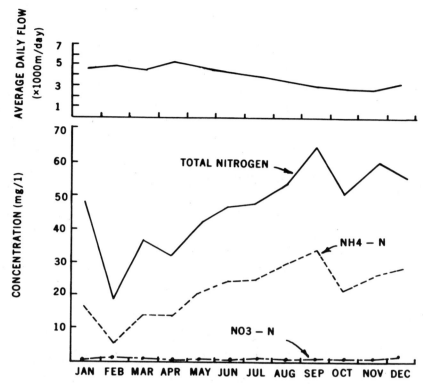

Figure 3. Total nitrogen, ammonium-N and nitrate-N in primary sewage effluent (1973).

12 had less than 4 mg/l NH$_4$-N. Native ground water (well 13) contained about 0.3 mg/l NH$_4$-N.

Nitrate-N comprised the major portion of total ground water nitrogen in those wells influenced by the percolate from the application area and sludge beds. Nitrate-N levels in wells 3, 7, 8, 10 and 11 varied from 10 to 20 mg/l throughout the year, although seasonal fluctuations were observed. The highest nitrate levels were observed in ground water samples during March through May and the lowest levels during October and November (Figure 5). Peaks in the nitrate curves probably reflected the warmer temperatures which enhanced microbial activity and the nitrification processes.

Nitrite-N was found at levels less than 0.05 mg/l in most observation wells, although concentrations of 0.5 to 1.0 mg/l were occasionally observed in wells 3, 7 and 8.

Changes in the concentration of the various forms of nitrogen

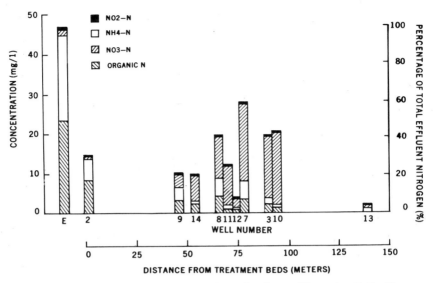

Figure 4. Organic-nitrogen, ammonium-N, nitrate-N, and nitrite-N concentration in primary effluent and ground water (1973).

in the effluent and the subsequent nitrogen levels observed in the ground water provide direct evidence that nitrification of effluent organic-nitrogen and NH_4-N had occurred. Effluent nitrogen was primarily organic-nitrogen and NH_4-N with small amounts of NO_3-N and NO_2-N, while ground water contains predominantly NO_3-N with lesser amounts of organic-nitrogen, NH_4-N and NO_2-N.

In view of the substantially lower total nitrogen levels in the ground water beneath the treatment site and the surrounding area, several mechanisms accounting for nitrogen reduction were considered. Biological assimilation of the nitrogen could account for limited nitrogen removal only. Nitrogen taken up by grasses and other surface macrophytes growing on the treatment basins accounted for about 0.4 percent of the annual nitrogen inputs. Another 1 percent of the total nitrogen additions may be assimilated by bacteria and other microorganisms in the treatment basins. Because the surface plants and microbes were not removed annually, any nitrogen assimilated would eventually be released through microbial degradation.

Fixation of nitrogen as relatively stable organic matter in the soil could remove some nitrogen. Total nitrogen levels in soil samples receiving primary effluent for 5 years were only 0.06 to 0.4 mg/g of soil greater than background levels (4). As-

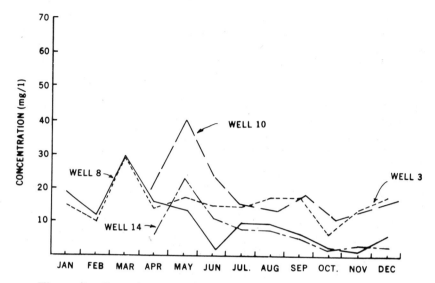

Figure 5. Ground water Nitrate-N levels in selected wells (1973).

suming primary effluent quality during 1973 was indicative of
the effluent quality for the previous 5-year period, the amount
of nitrogen removed by this mechanism was calculated to be 83
to 134 kg/ha-yr or about 1 percent of the total annual nitrogen
additions.

NH_4-N sorption in the soil could account for substantial nitrogen removal (6). NH_4-N retained in the soil was calculated to
be about 0.034 mg NH_4-N per gram of soil. The concentrations
of calcium and magnesium were approximately 20.5 and
4.84 mg/l, respectively, and cation exchange capacity in the surface 1 m of soil averaged 1.22 meq/100 grams soil for three
treatment basins (4). In these conditions, 1 m of the substrate
in three treatment basins could remove the NH_4-N in 1.47 m^3
(388 gal) of primary effluent containing 21.4 mg/l NH_4-N.
This would be about the amount of NH_4-N applied to three treatment basins during 3 days of effluent application at a flow
volume of 5,061 m^3. However, this removal mechanism requires
that the exchange sites be made available before the next application. Lance (7,8) observed in laboratory column studies that
ammonia retained on the soil exchange complex was nitrified
during the drying period where aerobic conditions existed. If
the NH_4-N absorption capacity of this soil was not rejuvenated,
then subsequent effluent applications could result in ammonia
being carried in the percolate to the ground water. The ammo-

nium retained on exchange sites and nitrified during the ensuing recovery period could create high nitrate concentrations in retained soil water. During the next inundation period, NO_3-N would be leached through the sand and gravel material, thus increasing ground water nitrate levels.

The NO_3-N levels observed here in the ground water were indicative that ammonification of nitrogenous organics and nitrification of ammonium nitrogen had occurred. In ground water samples, organic nitrogen and ammonia were 17 and 20 percent, respectively, of effluent levels, while NO_3-N levels increased 8 to 15 times.

It is uncertain whether the lower level of total nitrogen in the ground water had resulted from denitrification. Nitrogen loss through denitrification appears to be the most likely mechanism.

Because large amounts of organic matter were applied to the treatment basins in the primary effluent and by the vegetation growing on the basins, sufficient carbon was probably available for denitrification processes. St. Amant and McCarty (9) determined that 1.3 mg of carbon is required for the reduction of 1.0 mg of nitrogen to nitrogen gas. Total organic carbon available in the primary effluent was calculated to 60 mg/l, which gives a C:N ratio of 1.29:1 for the primary sewage effluent. This ratio is probably higher for effluent in the treatment basins because additional organic carbon is added annually by plants growing on the treatment beds.

Although substantial differences were observed in the total nitrogen levels of the primary effluent and ground water, occurrence of high nitrate levels in ground water indicates that optimum conditions do not exist at the site by which to facilitate complete denitrification. These results suggest the 2-day inundation of treatment basins practiced at the Fort Devens facility was adequate for the nitrification process, but was of insufficient duration to achieve the level of denitrification desired. Increasing the inundation period could result in additional nitrogen reduction by the denitrification processes; however, the magnitude of any further nitrogen reduction by denitrification requires further investigation.

Some reduction in total nitrogen could have been achieved through mixing of percolate with native ground water. In view of the relatively short flow distance from the observation wells to the treatment basins plus ground water chloride levels compared to effluent concentration, it would appear that dilution was not a major consideration in the reduction of total nitrogen levels.

Phosphorus in the primary effluent varied between 6 to 16 mg/l with an average value of 11 mg/l. Annual total phosphorus input to the treatment beds was about 3,400 kg/ha PO_4-P of which about 80 percent was orthophosphorus. Although substantial quantities of phosphorus were applied to the treatment basins, concentrations observed in the ground water were quite low. Total phosphorus directly beneath the application area (well 2) was about 6 mg/l, but levels decreased to less than 2 mg/l in wells located 45 to 90 m from the application area. Phosphorus levels in peripheral wells were approximately 11 percent of the levels observed in the primary sewage effluent.

Total coliform bacteria in the effluent ranged 18 x 10⁶ to 53 x 10⁶ per 100 ml of effluent with a mean value of approximately 32 x 10⁶/100 ml. Bacterial analysis of the ground water samples revealed total coliform bacteria in well 2 was about 4,000/100 ml but were less than 200/100 ml in wells 3, 7, 8, 9, 11 and 14. Because well casings had not been sealed, there was good probability total coliform bacterial levels reported for peripheral wells were indigenous coliform bacteria and did not originate from the primary effluent.

SUMMARY

Investigations conducted at Fort Devens, Massachusetts, wastewater treatment facility during 1973 ascertain the effectiveness of rapid infiltration to renovate unchlorinated primary effluent. Pretreatment of the 5,049 m³/day domestic sewage is accomplished using Imhoff tanks. Final treatment of unchlorinated effluent is achieved within 22 sand treatment basins. Effluent loading to the beds, following a 2-day inundation and a 14-day recovery operational cycle, was about 27.4 m during 1973.

Ground water COD and BOD_5 varied about 10 to 20 mg/l and 1.0 to 2.5 mg/l, respectively, which were about 10 and 2 percent, respectively, of the effluent COD and BOD_5 levels. Organic and ammoniacal nitrogen were generally less than 5 mg/l in ground water adjacent to the disposal site, while effluent concentrations were 11 to 33 mg/l and 6.2 to 42 mg/l, respectively. Nitrate-N in observation wells varied from 1 to 2 mg/l during 1973; however, several wells impacted by percolate from the treatment basins had NO_3-N levels of 10 to 20 mg/l. Comparing mean total effluent nitrogen with ground water concentrations from the observation wells over the study year showed ground water contained only 21 to 43 percent of nitrogen levels found in the effluent. This and the chloride data suggest that some dilution occurs because of ground water flowing into the area that does not originate at the treatment basins. Although ground water samples were taken near the water table, the extent of ground water dilution cannot be determined without a detailed hydrologic investigation. Of the 10 to 20 mg/l total nitrogen found in various observation wells, the greatest proportion was as NO_3-N, which was often greater than levels set for drinking water standards. Concentrations of the various forms of nitrogen in the effluent percolate suggest that organic and ammoniacal nitrogen was transformed to NO_3-N, a substantial portion of which was presumed to be denitrified.

Total PO_4-P in ground water was about 2 mg/l of which 0.2 mg/l was ortho PO_4-P. Total phosphorus in the ground water was only about 10 percent of that found in the primary effluent, which suggests that the treatment basin sediments were effectively removing phosphorus from the applied effluent.

Infrequent determinations for fecal coliform bacteria in the ground water samples proved negative although substantial numbers were present in the primary effluent. Mean total coliform bacteria levels in the ground water ranged from 100 to 200/100 ml while coliform in the primary sewage effluent averaged 32 x 10^6/100 ml.

CONCLUSIONS

The infiltration-percolation system serving Fort Devens, Massachusetts, is treating unchlorinated primary sewage effluent to a quality comparable to that achieved by conventional tertiary sewage treatment facilities. The treatment basins which have been operated for more than 30 years were found to greatly reduce the levels of COD, BOD_5, organic-nitrogen, ammonia-nitrogen, phosphorus, and total coliform bacteria in the applied effluent. With the exception of ammonia and nitrate, ground water quality in the area peripheral to the application site was within drinking water standards for most water quality parameters determined. Because these forms of nitrogen are readily affected by biological processes, the ammonia and nitrate levels in the ground water may be reduced with some modifications in the present operation of the treatment basins.

ACKNOWLEDGMENTS

The study was sponsored by Cold Regions Research and Engineering Laboratory as part of RDTE Project 4A062112A891, Task 05, Environmental Quality for Army Construction, and was conducted with the cooperation of the Massachusetts Agricultural Experiment Station, University of Massachusetts, Amherst.

REFERENCES

1. American Public Health Association, AWWA, WPCA. *Standard Methods for Examination of Water and Wastewater*, 13th Ed. APHA, New York (1971).
2. Langmuir, O. and R. L. Jacobson. "Specific ion electrode determination of nitrate in some freshwater and sewage effluent," *Environ. Sci. Technol.* 4(10):835 (1970).
3. Orian Research. Instruction Manual: Chloride ION Activity Electrode, Model 92–17, Orian Research, Inc., Cambridge, Massachusetts (1967).
4. Satterwhite, M. B., G. L. Stewart, B. J. Condike and E. Vlach. Rapid Infiltration of Primary Sewage Effluent at Ft. Devens, Massachusetts, USACRREL Special Report, in progress.
5. Motts, U. S. and M. Sains. "The occurrence and characteristics of groundwater contamination in Massachusetts," Water Resources Research Center Publication No. 7, Amherst, Massachusetts (1969).
6. Lance, J. C. "Nitrogen removal by soil mechanisms," *J. Water Pollution Control Fed.* 44:1353–1361 (1972).
7. Lance, J. C. and F. D. Whisler. "Nitrogen balance in soil columns intermittently flooded with secondary sewage effluent," *J. Environ. Quality* 1(2):180–185 (1972).
8. Lance, J. C. and F. D. Whisler. "Nitrogen removal during land filtration of sewage water," *Proc. Intern. Conf. on Land for Waste Management*, Ottawa, Canada (October 1–3, 1973).
9. St. Amant, P. P. and P. L. McCarty. "Treatment of high nitrate waters," *J. Am. Water Works Assoc.* 61:659–662 (1969).

24

EFFLUENT REUSE
IN LUBBOCK

D. M. Wells and R. M. Sweazy
Department of Civil Engineering
Water Resources Center, Texas Tech University, Lubbock, Texas
F. Gray
Farmer, Lubbock, Texas
C. C. Jaynes
Department of Plant and Soil Science,
  Texas Tech University, Lubbock, Texas
W. F. Bennett
College of Agricultural Sciences,
  Texas Tech University, Lubbock, Texas

EFFLUENT REUSE IN LUBBOCK

All of Lubbock's treated wastewater has been reused for some purpose since 1925, when reuse for irrigation was started. The city entered into a contract with Frank Gray and his associates for reuse for irrigation in 1937. All the effluent produced by the city from 1925 until 1964 was, therefore, used for irrigation by Gray and his predecessors. In 1964, the city constructed a small new treatment plant (the northwest plant) near the Texas Tech University campus, and Texas Tech entered into an agreement with the city to take the effluent from that treatment plant for irrigation. In 1971, Southwestern Public Service Company built a new power plant in the vicinity of the old (southeast) sewage treatment plant, and entered into a contract with Gray and the city that allows it to take up to 5 million gallons of effluent per day for use as cooling water and boiler makeup water. The power company then entered into a contract with another local farmer providing for reuse of its blowdown water, after dilution with effluent to specified TDS levels, for irrigation. The entire operation can be termed highly successful, not only from the standpoint of benefits to agricultural production, but also from the standpoint of saving the City of Lubbock large sums of money that would necessarily have been spent on tertiary treatment if its wastewater had been discharged to the normally dry stream that traverses the city.

Bids have been taken, and a contract will be awarded shortly for construction of 27 wells on Gray's farm and for about 14 miles of pipeline and associated pumping equipment for the purpose of recovering percolated effluent from beneath Gray's farm. This recovered wastewater will be reused again for watering parks, cemeteries and recreational areas, for supplying cooling water to a small power plant owned by the city, and for maintaining the water level in a series of small recreational lakes that are nearing completion in the city.

Local Setting for the Projects

Lubbock is located in the semi-arid high plains of West Texas at an elevation of about 3300 ft. Annual precipitation averages about 18 in. The entire area is a flat, almost featureless plain with a general slope of about 10 ft per mile to the southeast. There is, for practical purposes, no well-developed system of surface drainage networks. Rather, virtually all the runoff that occurs from the land drains to shallow depressions—called playa lakes locally—that occur at an average density of about one per square mile throughout the area. Most of the lakes are ephemeral in nature, containing water for only a few days or weeks following very heavy rains. Some of them, however, contain water throughout the year.

The Ogallala aquifer underlies thousands of square miles of land in the high plains of West Texas and New Mexico, and extends northward into the Dakotas. It is the source of practically all water used for all purposes on the high plains. About 95 percent of the water usage in the area is for irrigated agriculture, and most of the rest is used for municipal purposes.

For practical purposes, there is no recharge to the Ogallala aquifer. Thus, the economy of the region is based heavily on mining of fossil water stored in the Ogallala, which is being depleted at a rate of 1 to 3 ft/yr.

Until 1968, Lubbock obtained all its water from the Ogallala.

In that year Lubbock and 12 other cities and towns in the area started receiving parts of their supplies from Lake Meredith, which is located on the Canadian River about 160 miles north of Lubbock. Since 1968, Lubbock has received roughly two-thirds of all its water from Lake Meredith. The remainder continues to be supplied by the Ogallala.

Prior to 1925, the city discharged its treated wastewater to the North Fork of the Double Mountain Fork of the Brazos River, a small stream that has cut a canyon 45 to 75 ft deep through the city in geologic times. This stream is normally dry, having had zero total flow for 97 of the 120 months it was gauged in the period from 1939 to 1949. Two or three privately owned recreational lakes existed in the canyon several miles below the sewage treatment plant in 1925, and the existence of these lakes plus poor quality in the normally dry water course prompted the city to look for alternative means of disposal.

Gray and his associates originally contracted with the city for the use of about 1 to 1.5 MGD of effluent on 200 acres of land, 100 acres of which was leased from the city and 100 acres owned by Dr. Fred Standefer. The contract, which has since been extended four times, calls for the city to deliver all its water to Gray continuously, and for him to dispose of it in a manner acceptable to the state and local health departments. Among other things, he is not permitted to allow any of it to run off the surface of his land.

Since 1937, the city has experienced rapid and continuous growth, so that it now produces a total of about 16.5 MGD of treated effluent, about 15 million of which is treated at the southeast plant. To avoid being inundated with sewage, Gray has had to expand his farming operations apace with the growth of the city. He now controls about 3,000 acres of land, and he supplies water on a periodic basis to about 2,000 acres of his neighbors' land. He has bench or field leveled about 1,600 acres, has installed about 40 miles of pipeline and has installed six center pivot sprinkler systems. Each sprinkler unit covers about one quarter section of land on some of the rougher portion of his farm. In addition, at this time, he is furnishing an average of somewhat less than five million gallons of water per day to Southwestern Public Service Company.

The small plant near the Tech campus now produces about 1.5 MGD, which Texas Tech utilizes for irrigation of about 1,200 acres of farm land. At this time, Texas Tech has available about 1,600 ac-ft of water per year to use on 1,200 acres of land, and Gray has about 11,000 ac-ft per year to use on 3,000 to 5,000 acres, as he sees fit. The primary difference in the two operations at this time, and the difference was much more pronounced until very recently, is that Gray has historically been plagued by a surplus of water, an average of 5 ac-ft per acre per year, and that Texas Tech has not generally had nearly enough to supply the water needs of the crops it was producing. Gray sometimes applied as much as 12 to 15 feet per year on parts of the land. He was, however, using only forages on these acres. Grasses, of course, are high users of water.

Characteristics of the Wastes

The average quantities of sewage treated by Lubbock's sewage treatment plants for the last 15 years are shown in Table 1.

Table 1.   Lubbock sewage treatment plant flow rates (1960–1975).

Year	Population	Average Sewage Flow Rates in MGD			gal/day/capita
		Northwest Plant	Southeast Plant	Total	
1960	128,691	0	9.96	9.96	77
1965	145,900	0.17	12.64	12.81	88
1970	149,101	0.67	13.46	14.13	95
1975	165,000 [a]	1.5 [a]	15.0 [a]	16.5 [a]	100 [a]

(Source: Freese, Nichols and Endress, 1971).

[a] Estimated from city sources.

It is unfortunate that different people have been analyzing the effluent from the two sewage treatment plants in Lubbock for different purposes over the years they have been in operation. Environmental engineers have been concerned primarily with the concentrations of BOD, COD, suspended solids, and similar constituents of normal concern to sanitary engineers, while agronomists, who are utilizing the water at Texas Tech, have been concerned with the concentrations of total salts, sodium, calcium, potassium, nitrate, and similar constituents that are of primary concern to agronomists. Nevertheless, some of the interests overlap, and some representative analyses of effluents are presented in Table 2.

As noted earlier, the primary difference, and it is a significant difference, between the Texas Tech operation and Frank Gray's operation is that, in general, Gray has had a surplus of wastewater and has been continuously faced with the problem of how to dispose of a significant quantity of unwanted water.

Table 2. Representative concentrations of pollutants in effluent from City of Lubbock sewage treatment plants and concentrations in local well water.

Constituent	Effluent from Northwest Treatment Plant (Effluent to Tech)	Southeast Treatment Plant (Effluent to Gray)	Well Water at Texas Tech
TDS	2318	1200	1140
Sodium	687	–	126
Calcium	77	–	99
Magnesium	47	–	114
Potassium	31	–	18
Nitrate	37	2.5	43
Chloride	653	318	228
pH	7.1	7.4	7.7
SAR	16.2	–	2.06
SSP	79%	–	27.1%

Texas Tech, on the other hand, has never yet had enough water to supply the needs of the crops grown, and has been faced with the problem of how to allocate a scarce supply of water to promote the maximum benefit from it. Therefore, the two operations have not been parallel, even though both were using effluent from the City of Lubbock. A further difference is the fact that most of the effluent that Texas Tech receives is derived from the Texas Tech campus, and is characterized by a very high concentration of total dissolved solids. This high concentration of TDS is thought to be related to the fact that the campus heating and cooling plant utilizes large quantities of demineralized water, and that the dissolved minerals that were originally contained in all of this demineralized and distilled water are discharged to the sewer that leads to the northwest treatment plant. Thus, the concentration of total dissolved solids in the influent to the northwest treatment plant is approximately double the concentration in the influent of the southeast treatment plant, and the concentration in the effluent is also approximately double that in the effluent from the southeast plant. Of further significance is the fact that effluent from the northwest treatment plant has a very high concentration of sodium and a correspondingly high sodium absorption ratio of about 16, as compared with a ratio of about 6 in the effluent from the southeast treatment plant.

In general, the concentration of BOD in the effluent from the southeast treatment plant has been very high. Average concentration in samples collected by researchers at Texas Tech have been about 75 mg/l over the past ten years. BOD concentrations in the effluent from the northwest treatment plant have generally been very low. This plant, because it has usually been loaded at a rate much lower than its designed capacity, has generally produced an effluent with a BOD of less than 10 mg/l. Thus, the effluent that has been available to Frank Gray over the years has usually contained a high concentration of organic matter and a relatively low concentration of dissolved inorganic salts, while the effluent that Texas Tech has received from the northwest treatment plant has had a high concentration of dissolved salts and a relatively low concentration of organic matter.

Original and Current Application Rates

Gray Farm

Until about 1974, Gray's primary problem was that he was not able to buy or lease land fast enough to keep pace with the growth of the city and the increase in sewage production. He was therefore faced with the problem of disposing of an average of about five ac-ft of water each year for each acre of land under his control. Since most crops grown in the area require only about 12 to 18 in. of supplemental water for maximum production each year, he was frequently forced to maintain 15 or 20 percent of his land as a disposal (infiltration) operation. On land used for this purpose, he allowed weeds or any other vegetation to grow as rapidly as possible, and he also planted a significant amount of his land to forage crops such as alfalfa and grasses, as mentioned earlier. This practice, while it might have been abhorred by his neighbors, undoubtedly resulted in improvement of his land through the production of large tonnages of green vegetation that could be plowed under. It also resulted in the recharge of large quantities of water to the underlying formation so that, at this time, the water table under all of Gray's land is within 10 or 15 ft of the land surface. Had the geological formation not had the natural subsurface drainage it has, he would not have been able to have continued to use so much water per acre.

In general, soil characteristics in the South Plains of West Texas are such that nitrogen is almost the only fertilizer needed by most crops. The high concentration of calcium in the soil has resulted in a correspondingly high concentration of phosphorus, and potassium is also available in amounts needed for most crops. Consequently, as a result of the availability of sewage with its fairly significant concentration of nitrogen, Gray has not used very much commercial fertilizer on his farm since 1937. The only exception to this statement is that, as a result of soil tests he had made a few years ago, he added some small quantities of trace elements such as molybdenum to most of his land. He is a very progressive farmer who practices crop rotation in addition to having an abundant supply of sewage available, and his farm consistently outyields other irrigated farms in the area (Table 3). Crop quality, as well as yield, is better under irrigation.

Table 3.   Usual yields of crops grown most commonly in the Lubbock area.

Crop	Dry Land	Irrigation with Water from Ogallala Formation (with Fertilizer)	Irrigation with Sewage Effluent without Chemical Fertilizers
Grain sorghum (lb)	800–1000	4000–5000	6500
Wheat (bushels)	10–12	30–40	80
Lint cotton (lb)	150–225	600–800	1250

(Source: Gray, 1968).

Gray's operation can be summarized by stating that it has been successful for almost 40 years without any significant amount of scientific study. No one has been concerned, historically, with the build-up of toxic pollutants, with loss of productivity resulting from excessive sodium or boron, or with any other aspect of the operation. Gray, who was a student at Texas Tech at the start of his operation, has simply assumed from the beginning that sewage was good irrigation water—and for that matter, good stock watering water—and has built a very successful operation without worrying about any adverse impacts that might result. He has shown by accident, more than by design, that this is indeed the case. The geological formation, again, has been a big factor.

In general, it can be stated that his experience indicates that, in a semi-arid area, at least, irrigation with sewage improves the soil, improves the yield of crops, eliminates or lessens the need for fertilizer, and does not appear to have any adverse effects on the soil. To the contrary, soils that were originally very poor and composed primarily of caliche have been improved to a significant degree by his practices. Almost 40 years of experience in utilizing effluent for irrigation with no apparent detrimental results would indicate that, at least in semi-arid regions, municipal effluent can be used with no reservations for the irrigation of ordinary field crops, and for the watering of livestock.

Texas Tech Farm

While Gray contracted with the city for use of municipal effluent for one purpose—the production of crops—Texas Tech entered into a similar contract for two purposes. Texas Tech wanted to assure itself of a supply of water, but, more importantly, researchers at Texas Tech wanted to determine the effects of municipal effluent on soil properties, yield and quality of crops such as Midland Bermuda grass, cotton, grain sorghum and soybeans. Investigations of these effects were started in 1964 and were concluded in 1975.

As has already been mentioned, one primary difference between the Texas Tech investigations and the Frank Gray operation was that, while Gray has generally had a surplus of water and has been faced with a problem of disposal of excess water, Texas Tech has generally not had enough effluent to supply all the water required by the crops being grown, without regard to nutrient requirements. Nevertheless, researchers in the Department of Plant and Soil Science at Texas Tech have gained a great deal of useful information on the effects of effluent on both crop yields and soil characteristics in the 10-year duration of the experiment.

The research was conducted on an Amarillo fine sandy loam soil located on the Texas Tech University agronomy farm. The soil was nearly level and had excellent infiltration properties. The experimental design used was a split plot arrangement in a randomized complete block with eight replications. The data collected were subjected to analysis of variance.

For each crop, the field plan provided eight strips, each 800 ft long. Bermuda grass borders were 13.33 ft wide, and cotton, grain sorghum and soybean borders were 26.66 ft in width.

The crops grown were cotton, grain sorghum, soybeans and Midland Bermuda grass. Cultural practices other than fertilizer rates were those in common use. Two sources of water—well water and treated municipal effluent—were utilized. Fertilizer treatments on the row crops consisted of 0-0-0, 80-0-0, 0-80-8, and 80-80-0. The Bermuda grass fertilizer treatments were 0 and 400 pounds of nitrogen. Fertilizer treatments were applied with each source of water. An estimated 30 pounds of nitrogen as nitrates and 38 pounds of potassium per acre per year were contained in the municipal effluent. The effect of effluent on average yields of Bermuda grass is shown in Table 4.

Table 4. The effect of sewage effluent on the average yields of Midland Bermuda grass.

	1966–1972 *(Pounds of Dry Matter Per Acre)*	
	Fertilizer P/A	
	No Nitrogen	*400 lb Nitrogen*
Well Water		
1966	3,865	11,268
1967	660	3,909
1968	514	9,484
1969	1,224	10,981
1970	380	5,214
1971	857	6,002
1972	1,299	10,491
	8,799	57,349
Average	1,257.0	8,192.7
Sewage Effluent		
1966	6,336	13,705
1967	983	5,402
1968	1,817	12,081
1969	2,972	13,474
1970	493	6,834
1971	1,665	7,764
1972	3,972	12,328
	18,238	71,588
Average	2,605.4	10,226.9

Yields from the row crops were determined in the fall at harvest. Bermuda was sampled each month during the growing season. The green forage was dried and weighed, and the yield per acre was calculated. Protein content and chemical composition were determined for Bermuda grass. Well water and the municipal effluent were sampled periodically. Soil sampling was conducted in 1971 and again in 1975.

Effect on Ground and Surface Water Characteristics

Neither the Gray operation nor the Texas Tech operation have had any effect upon surface water characteristics other than the positive benefit to surface water quality that results in not discharging municipal effluent to a normally dry water course, and the negative benefit that results from allowing the normally dry water course to remain dry by not discharging municipal effluent to it. Both operations, however, have had beneficial and detrimental impacts on ground water quality. The beneficial impacts are that, in both cases, the depleted aquifer underlying all or parts of the irrigated land and/or the storage lagoons associated with the irrigation operations have been recharged to some extent. Under Gray's farm, the aquifer is, for practical purposes, fully recharged. Under the Texas Tech farm, except for the area surrounding the storage lagoon, recharge has been fairly insignificant but the water table has not declined to the same extent that it has in surrounding irrigated areas.

The major impacts on ground water quality have been very large increases in the concentration of nitrates, the concentration of total dissolved solids, and the hardness in the ground water. If the ground water is to continue to be used primarily for irrigation, none of these impacts is, in itself, of major significance to the water quality. Since, however, the city of Lubbock is shortly to begin drilling wells on the Gray farm to recover water for utilization as irrigation water, as power plant cooling water, and as makeup water for the Canyon Lakes, the high concentrations of both nitrates and hardness achieve a much higher degree of significance. Representative concentrations of various pollutants found in well water from the Gray farm, from a well adjacent to the sewage storage pond on the Tech farm, and from a well on the Tech farm that is located some distance from the sewage storage pond are shown in Table 5.

As noted, the recharge of treated municipal effluent has a significant impact on the concentrations of nitrates, total dissolved solids, and hardness in the ground water under the land irrigated for a long period of time with treated municipal effluent. However, none of these impacts are of particularly serious detrimental value to the further use of the water for irrigation, for which about 95 percent of all ground water in the area is used.

Effect on Crops and Soil Conditions

As indicated previously, Gray has been concerned primarily with crop production in his operation. He has been concerned with the impact of effluent application on soil characteristics only to the extent that such application not decrease the productivity of the soil. His own observation, which is borne out by crop yields but not by long-term analyses of soil, is that the land to which sewage has been applied longest has benefited the most from its application. That is, the 200 acres on which he started his operation was very poor soil. It is now very good soil that yields excellent crops. Since Gray is a very progressive and intelligent farmer who keeps good records of crop yields, one is drawn to the irrefutable conclusion that irrigation with sewage has made an improvement in the characteristics of the soil to which it has been applied, and has not resulted in a build-up of any materials that are toxic to the plants grown in the area.

Table 5.   Representative analyses of well water.

Constituent	Gray Well	Tech Well (Near Sewage Pond)	Tech Well (Remote from Sewage Pond)
		Concentrations in mg/l	
TDS	1,700	2,300	1,140
Ca	125	37	99
Mg	79	–	114
Na	297	–	127
K	18	–	18
O-PO$_4$	0.03	0.04	0.16
HCO$_3$	456	–	335
SO$_4$	256	–	329
Cl	405	–	228
NO$_3^-$	118	23	43
Alky	374	–	–
Hardness	815	1,350	–
pH [a]	7.9	7.3	7.7

[a] Negative log of hydrogen ion concentration.

Researchers at Texas Tech have been more specifically concerned with changes in soil characteristics, and have monitored soil characteristics along with crop production and quality since the project was started at Texas Tech (Table 6). Also, the quantity of water available at Texas Tech, as previously mentioned, has been barely adequate or inadequate to supply the water needs of the crops grown, and has not been sufficient to supply any very significant amount of fertilizer to any of the crops grown.

The effect of effluent on crops grown on the Tech farm can be summarized as follows:

Table 6. Chemical content of dry Midland Bermuda grass forage as
affected by water source.

| Element | Water Source | |
	Sewage	Well
Phosphorus (%)	0.250	0.189**
Potassium (%)	1.750	1.490**
Calcium (%)	0.454	0.438ns
Magnesium (%)	0.228	0.218ns
Sulfur (%)	0.380	0.321*
Iron (ppm)	334.000	332.000ns
Manganese (ppm)	78.600	77.200ns
Zinc (ppm)	34.900	34.000**
Copper (ppm)	8.040	7.660ns
Boron (ppm)	10.600	11.500*

Note: Numbers followed by ** and * are significantly different from the number immediately preceding them to the left at the 1% and 5% probability levels, respectively. Numbers followed by ns are not significantly different from the preceding number.

Bermuda Grass

Dry matter yields were increased significantly each year of the experiment (nine for Bermuda grass) by the application of effluent, but neither effluent nor well water produced economical yields without the use of nitrogen fertilizer. Thus, the greatest value of effluent was in the water rather than in nutrients which it contained.

Effluent had no detrimental effect on the chemical composition of dry matter produced from Bermuda grass. The feeding value was not impaired. A sod type of grass such as Bermuda might be a good place to dispose of effluent if other uses could not be found. Bermuda grass might respond to much larger quantities of water than were used, and it conceivably might produce reasonably high yields without fertilizer. In this experiment 400 pounds of nitrogen should have produced greater yields. It is believed that it would have produced more had more water been applied. A good irrigated pasture requires at least twice the amount of water needed by cotton or grain sorghum.

Cotton, Grain Sorghum, and Soybeans

Effluent did not affect these crops to any significant degree. In some years the yields were greater than those produced from well water, in other years there were no appreciable differences, and a decrease in yield was noted occasionally. Quality was not measured in these crops. It may be that differences among years were due entirely to seasonal factors. In two of the years, effluent was in short supply when the crops needed water. No doubt this had greater influence on yield than any other factor.

The results of soil analyses made in 1971 and 1975 can be summarized as follows:

1. Effluent caused a significant increase in P_2O_5 under both cotton and Bermuda grass, with the increase under cotton being most pronounced in the surface two inches and the increase under the Bermuda being most pronounced in the surface six inches (Figures 1 and 2). Under cotton, the P_2O_5 tended to stay at the surface, but it tended to move down under Bermuda grass. Some of the P_2O_5 may have been changed into the organic form under Bermuda and in this form it would move more easily. Further increases in phosphate were noted in 1975.

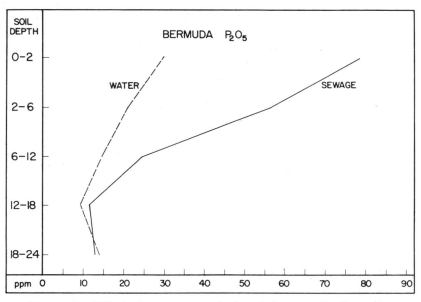

Figure 1.   Effect of sewage on soil phosphate concentration under Bermuda grass.

2. Soluble salt concentration was significantly higher under both cotton and Bermuda grass that were irrigated with effluent (Figures 3 and 4). This increase was evident throughout

the soil profile, with differences often being greater at lower depths. The soluble salts remained at higher levels at the surface under Bermuda grass than under cotton. Soluble salt and sodium levels continued to increase in 1975. Some plots are beginning to approach a level at which production would be expected to be reduced slightly. However, the high calcium content in these soils will continue to counteract the detrimental effect of sodium to a certain extent.

Levels at which soluble salts become detrimental depend upon the soil and crop. Bermuda grass and cotton are both salt-tolerant crops. Certain crops (vegetables) have a low salt tolerance. The soluble salt levels are not now critical but with time they may become detrimental.

3. Nitrogen levels in the soil continue to fluctuate depending upon crops and rainfall. There were no significant effects on potassium, calcium or magnesium concentrations.

People Interactions and Problems That Have Resulted

No significant problems have arisen as a result of either operation. In general, Gray's neighbors look with favor upon his operation simply because it has resulted in a significant rise in

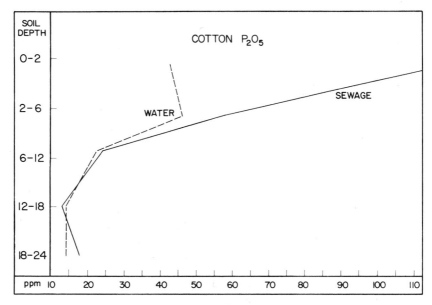

Figure 2. Effect of sewage on soil phosphate concentration under cotton.

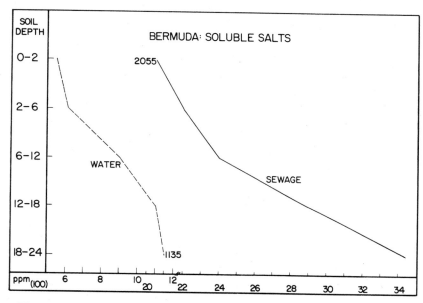

Figure 3.   Effect of sewage on soil soluble salts concentration under Bermuda grass.

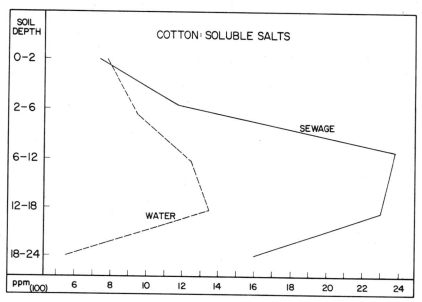

Figure 4.   Effect of sewage on soil soluble salts concentration under cotton.

their water tables. Thus, any detrimental impacts are far outweighed by the benefits of having their aquifer recharged, albeit inadvertently, by Gray. The Texas Tech operation is almost surrounded by developed sections of Lubbock, yet no complaints have been reported. In fact, part of the area irrigated lies between Methodist Hospital and the Tech Medical School–Lubbock County Teaching Hospital, and no steps have been taken to discontinue the operation. It may well be that most people in the surrounding homes are not aware of the fact that Tech uses sewage for irrigation, and that complaints would arise if they were aware of it. Any odors associated with the operation are likely blamed on the animal feedlots that are presently located in the middle of the campus, and complaints may arise when these feedlots are moved within the next year or so.

Relative Costs

No specific cost data are available to form a basis of comparison of the cost of irrigating with sewage and with well water. However, it can be stated without qualification that the energy cost is much less for utilizing effluent. The pumping head on wells in the area is about 160 feet, and the yield of most wells is relatively small—about 300 gpm. The only added cost associated with utilizing sewage is the necessity to provide reservoirs for storing the water when it is not needed. The amount of reservoir storage required seems to be on the order of two or three months production of effluent to provide maximum flexibility in utilizing the effluent efficiently. The alternative to having an adequate storage capacity is to utilize some of the available land as a disposal area rather than as a crop-producing area. In some years in the past, when the city was growing at a faster rate than Gray's farm was expanding, he has been forced to leave up to 20 or 25 percent of his land idle and utilize it as a disposal area. This practice, obviously, would be a significant cost item if it were required in all cases. Again, had the geological formation had different characteristics, he could not have accommodated the amounts of water used per acre.

25

LAND TREATMENT OF WASTEWATER: CASE STUDIES OF EXISTING DISPOSAL SYSTEMS AT QUINCY, WASHINGTON AND MANTECA, CALIFORNIA

R. P. Murrmann
USDA-ARS, Starkville, Mississippi
I. K. Iskandar
U.S. Army Cold Regions Research and Engineering Laboratory,
Hanover, New Hampshire

INTRODUCTION

Treatment of wastewater has become one of the major environmental concerns of many nations in the world. Sewage treatment plants have long been the most popular method for wastewater renovation. However, the discharge of primary or even secondary treated domestic effluents into open water has resulted in serious deterioration in surface water quality. This is primarily because wastewater nitrogen and phosphorus enhance eutrophication of open water (1). Accumulation of potentially toxic substances in sediments and aquatic life also has been shown to be related to wastewater discharge (2,3).

As a result of more restricted standards on the quality of wastewater discharged to ground or surface waters (Federal Water Pollution Control Act Amendments of 1972) the efficiency of treating wastewater using conventional sewage treatment plants is being questioned. Attention has focused recently on more effective alternatives such as advanced waste treatment and treatment of wastewater by application on land.

The concept of application of wastewater to land is not new. Wastewater and sludges have been used for agricultural purposes in Europe since at least the sixteenth century (4). The present emphasis on wastewater application on land, however, is for treatment purposes, as contrasted to disposal on land or agricultural use. In this respect the term "land treatment" implies controlled management rather than uncontrolled application or "land disposal" (5). With few exceptions (6–8) very little information is known about the long-term effects of wastewater application on land. Because land treatment of wastewater effluents on a wide area basis, like any other new engineering or scientific concept, poses many uncertainties, evaluation of the success or failure of small-scale, existing land disposal sites is expected to provide information necessary for selecting and designing new land treatment systems. For this reason evaluation of the performance and effects of existing land disposal facilities has been a part of the Wastewater Management Program at the U.S. Army Cold Regions Research & Engineering Laboratory (CRREL) since 1973.

The objective of this chapter is to discuss the findings obtained for slow infiltration disposal type sites located at Manteca, California, and Quincy, Washington.

EXPERIMENTAL PROCEDURES

Sampling

The major objectives of evaluating existing land disposal facilities are to determine current renovative performance as well as long-term changes that have taken place in the soil system. Since initial performance and site conditions are not known, it is necessary to deduce the changes that have resulted from wastewater application by comparison of the disposal sites to a control site that has never received wastewater but has the same soil type.

At both Manteca and Quincy, soil cores to 160 cm were taken from wastewater application and control sites. The control fields were under agriculture production. Composite samples from four cores from each field were taken for chemical analysis. Samples were composited at 15-cm intervals to 60 cm and, thereafter, at 30-cm intervals. At Manteca, samples were taken from both 2-yr and 11-yr application fields, and at Quincy from 17-yr and 20-yr application fields. Water samples were collected from the pretreatment facilities, lagoons, drainage ditches, and drainage tiles at each site. Soil solution samples were collected using suction lysimeters installed at 80- and 160-cm depths in the soil of both the disposal and control fields. Ground water samples were also collected from existing wells at each site.

Three sampling trips were made to both locations during 1974 to take into account seasonal changes in water quality parameters. Sampling trips to Manteca were made in June, September, and November, and to Quincy in May, August, and November. During each site visit, water analysis was conducted at each sampling station over a three-day interval.

Soil and Water Analysis

Soil samples were air dried (25°C), thoroughly mixed and sieved, and a < 2-mm fraction was taken for chemical analysis. Soil pH (1:1wt/vol), soluble salts (measured as specific conductance), cation exchange capacity, exchangeable Ca, Mg, Na, K, organic carbon, organic nitrogen, free iron oxides, organic-P, total-P (acid digestion), exchangeable-P, and soluble-P (1.2 soil to water) were determined according to Black (9). Two forms of heavy metals were determined—total and acid-extractable (plant-available). Details of the analytical procedures and instruments used for heavy metal analysis are presented in Iskandar (10).

Water samples were analyzed in the field for NH_4-N, NO_3-N, pH and ortho-P using a Hach Model DR/2 spectrophotometer according to Hach Chemical Co. (11). Total heavy metals (Ni, Zn, Cu, Cr and Cd) were determined in the laboratory. For heavy metal analysis, water samples were acidified to pH < 1 with concentrated HNO_3. The analyses were performed directly using a Perkin-Elmer Model 304 atomic absorption Spectrophotometer with a graphite furnace (HGA 70).

SITE CHARACTERISTICS

The factors involved in selecting land treatment over other wastewater treatment alternatives are complicated and not fully understood. However, knowledge of site characteristics is among the important information required to make a determination. Site characteristics include many diverse factors such as local ground water hydrology and quality, climate, soil types, wastewater quality and volume produced, land pattern, land use and projected population growth in the area. This type of information available for the two study sites is summarized below.

Manteca

The Manteca wastewater treatment plant, which serves a population of 16,000, is located approximately 150 km (80 miles) east of San Francisco. Manteca has a mild climate: based on a

10-yr data record, the mean maximum temperature (July) is 34.5°C (94.2°F) and the minimum (Jan.) is 2.6°C (36.7°F). The overall average temperature is 16.2°C (61.1°F). The mean number of days with temperatures below 0°C is 23. Average annual precipitation is approximately 30 cm (11.1 in.). The design flow of the sewage treatment plant is 9.5 x 10⁶ l/day (2.5 MGD). In 1973, the maximum and minimum flows were 5.84 x 10⁶ l/day (Sept.) and 3.30 x 10⁶ l/day (Jan.), with an overall average flow of 4.7 x 10⁶ l/day. Wastewater has been given secondary treatment (activated sludge) prior to application on land since 1971. The wastewater is stored, when necessary, in a holding pond (Figure 1). Sludge from the secondary treatment plant is cycled into the holding pond where settlement occurs. Probably a considerable amount of the sludge is applied to the land during irrigation. Solids skimmed from the top of the treatment plant reservoirs are collected and buried in an area directly east of the main treatment plant facility.

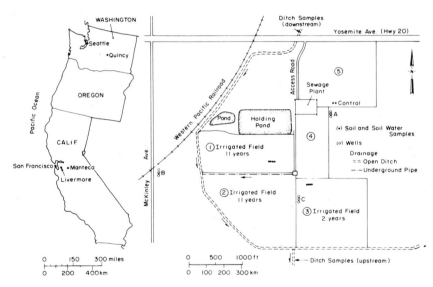

Figure 1. Wastewater disposal site and sampling locations at Manteca, California.

The disposal site itself consists of four different areas (Figure 1), roughly 64.8 ha (160 acres) in total area. Field 1 received wastewater for 11 years, and it has not been graded for irrigation purposes since that time. Field 2 also received wastewater for 11 years, but it was graded in 1971. Field 3 has been in operation only since 1971. Field 4 was established in 1971 but has received wastewater on only an intermittent basis for about one year. Each of the major fields is subdivided into smaller cells in which water can be impounded for irrigation purposes. Normally, each cell is flooded until water is impounded on the surface, using a regular rotation scheme. Each individual cell receives an average of approximately 1.8 cm/ha/wk (1.8 in./A/wk) of undisinfected wastewater at a frequency of approximately one day per week. Due to the mild climate, there were no problems associated with wintertime operation. A control site that had never received wastewater was selected (Figure 1) for comparison during this study.

The principal vegetation on the site is ryegrass, which has never been harvested or removed; this should be expected to adversely affect nitrogen renovation. The soil type in the disposal and control areas is classified as very deep ($>$ 150 cm) Tujunga loamy coarse sand. Particle size analysis indicates about 20 percent clay (<5 μm) and 80 percent silt and sand (>5 μm). The parent material is alluvial soil of mixed origin. The soil at the disposal areas is the same as in the control field, but it is classified as less well-drained due to impact of wastewater irrigation. The water table in the 11-yr disposal field fluctuated from 45 to 90 cm and that in the 2-yr field from 65 to 90 cm, while the water table in the control field was 90 to 150 cm below the soil surface.

Most of the wastewater applied at the site infiltrates to the ground water table, although during flooding operations in Fields 2 and 3, water may run overland since a small collection ditch has been constructed from which the water can be pumped back to the holding pond. Ground water flow appears to be in a southwesterly direction towards a river in the area, but possibly a substantial portion of the water is intercepted by a major drainage ditch that bounds the north, west, and southern sides of the disposal area (Figure 1).

There are no published data on the quality of water infiltrating through the soil. However, three sets of monitoring wells (A, B and C, Figure 1) were installed when the disposal site was constructed. At each well site, individual wells were driven to depths of 2, 3 and 6 m. The plant operator has occasionally collected water samples for nitrogen analysis from the wells and from upstream and downstream locations of the drainage ditch. He has been unable to detect any adverse impact due to operation of the site.

Quincy

Quincy (population 3200) is located about 407 km (220 miles) east of Seattle, Washington, in an arid region that depends completely on irrigation for crop water supply. The main industry in the area is vegetable processing. The city has separate wastewater treatment facilities for industrial and municipal waste products. Therefore, industrial wastes are not mixed with the municipal effluent at the location under study. The 10-year average annual precipitation is 20 cm (8 in.). The maximum, minimum, and mean temperatures based on a 10-year record, are 31.9°C, 89.5°F (July), -9.4°C, 15.0°F (Feb.) and 9.6°C, 49.3°F, respectively. There is a meteorological station maintained by the treatment plant operator at the site.

The domestic waste treatment facility was established in 1954 with a design flow of 1.9 x 10⁶ liters/day (0.5 MGD). However, the capacity of the system has been overloaded for some time, the 1973 load being approximately 2.6 x 10⁶ liters/day (0.7 MGD) in the fall and about 3.8 x 10⁶ liters/day (1.0 MGD) in the summer. There are no flow records available for the operation of the facility for the period 1954 through 1968. From 1968, there are detailed records summarizing the average monthly and yearly flows.

Raw effluent passes from the city collection lines to a connection device (grinder), aeration tank and then through a lift station to a sedimentation tank clarifier (Figure 2). Water from the clarifier is stored in three holding ponds, two of which are in series. Sludge from the clarifier goes through a digester to drying beds, after which the dried sludge is apparently deposited in a local dump. Some consideration has been given to application of the dried sludge to the disposal site land, but apparently this idea was never implemented. After preliminary treatment, the wastewater is directed to one of the two holding pond areas. From a holding pond, the water moves through head ditches where it is fed by gravity flow into five fields that are under agricultural production by a tenant farmer. The fields are about 2 ha (5 acres) in size with a total of 11 ha (27 acres) available for irrigation.

In early 1954, when the treatment facility was constructed, there was no sludge digester and only one pond, referred to as Pond 1 in Figure 2. In 1957, Ponds 2 and 3 were added. The sludge digestion system was constructed in 1960. According to the history of the site, field C has been irrigated since 1954, and fields A, B, D and E have been irrigated since 1957. Crops grown on the disposal site were primarily corn and wheat. The farmer

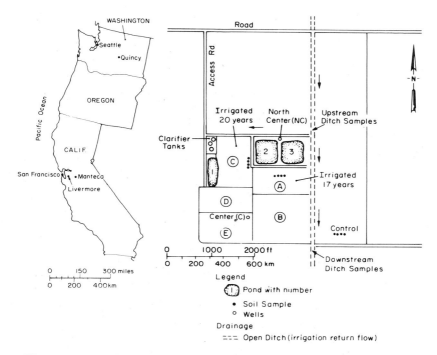

Figure 2. Wastewater disposal site and sampling locations at Quincy, Washington.

attempts to irrigate the crops during the summer on an as-needed basis, with application of excess water to a fallow field held in reserve for this purpose; however, it is necessary to add more water than required to irrigate the crop.

The ditch distribution system is in poor condition, and as a consequence, the different disposal fields have not necessarily received equal amounts of water. In winter months, the fields are vacant with no cover crops. During this period water over-flows from the lagoons through the irrigation ditches primarily to Fields A, B and E (Figure 2). No previous data are available on the quality of ground water directly under the disposal sites.

According to soil survey information, the soil is classified as a well-drained Warden very fine sandy loam (2 percent slope). It has developed from wind-deposited material and reworked lake sediments. The soil texture is very fine sandy loam to a depth of about 45 cm (18 in.), with calcareous silt loam from 45 to 150 cm (60 in.). Water erosion hazard is slight or non-existent and wind erosion hazard is moderate. Ground water in the area is commonly found at depths of from 1.5 to 4.5 m. At the disposal site, water is frequently ponded on the surface during winter months, but the water table is generally at a deeper level, indicating an overloading of the soil in terms of its infiltration capacity.

The ground water was reported to flow in a southeasterly direction and should be intercepted by the large, irrigation return-flow collection ditch. However, seepage of the ground water to the collection ditch is seldom observed, except when water is ponded in the field during the winter. Recognizing overloading conditions at the disposal site, the city of Quincy has acquired additional land for increasing the size of the disposal area when funds are available for its development.

RESULTS AND DISCUSSION

Manteca

Table 1 summarizes the seasonal variation in pH, NO_3-N, NH_4-N and ortho-P in water samples from the treatment plant effluent, the soil solution at two depths, and the monitoring wells. There were no significant differences in pH, which ranged from 7.1 to 8.3 among samples, due to location or sampling period. Concentrations of NH_4-N in samples of the secondary effluent and from the storage lagoon were rather low—in a range from 8.8 to 12.7 and from 7.6 to 11.3 mg NH_4-N/l, respectively. Soil solution from the treated and control sites contained < 1 mg NH_4-N/l. A lower concentration of NH_4-N in soil solution compared to that in applied effluent is always expected, due to the transformation of NH_4 to NO_3 in surface soil (nitrification).

Average concentrations of NO_3-N in soil solution with depth in the disposal areas were much higher than those at the control site (Table 1). Soil solution at 0.8- and 1.6-m depths from the 11-yr field during the three sampling trips contained from 1.0 to 7.0 mg/l NO_3-N. For the 2-yr field, a concentration as high as 19.7 mg NO_3-N/l was found during the summer (June 1974) at 0.8 m. The concentrations of NO_3-N at this depth were 8.0 and 6.4 mg NO_3-N/l in September and November 1974, respectively. At 1.6 m, soil solutions collected from the 2-year field in June, September and November, 1974 contained 2.6, 16.4 and 12.2 mg NO_3-N/l. One explanation for this variation is that nitrogen, accumulated as NH_4 during colder winter months leaches from the surface soil during the following months. The reason for the higher NO_3-N found in leachate from the 2-yr field as opposed to the 11-yr field is probably the mineralization of organic-N during early years of operation. Previous

Table 1. Average values for water quality at wastewater disposal site, Manteca, California.

Sample[a]	pH			NO$_3$ (ppm N)			NH$_4$ (ppm N)			Ortho-P (ppm P)		
	6/74	9/74	11/74	6/74	9/74	11/74	6/74	9/74	11/74	6/74	9/74	11/74
Secondary treatment	7.6	7.5	7.9	1.4	1.9	1.2	12.7	12.2	8.8	4.1	6.1	3.8
Storage lagoon	8.3	7.3	7.9	<0.1	2.0	1.2	7.6	11.3	9.6	4.2	12.0	6.1
Drainage ditch downstream	8.1	7.7	8.1	1.3	4.5	3.8	0.8	0.5	0.5	0.5	0.3	1.3
Wells A - northeast	7.8	7.4	8.1	0.5	0.7	1.0	1.3	1.0	0.8	0.4	0.6	0.6
Wells C - center east	7.7	7.3	7.9	2.6	2.5	2.5	0.4	0.3	0.4	0.8	1.8	1.6
Wells B - center west	7.9	7.4	7.9	11.0	11.6	9.6	0.4	0.3	0.4	0.3	0.6	0.7
Control field - 0.8 m	7.6	ns[b]	ns	4.9	ns	ns	0.4	ns	ns	<0.1	ns	ns
1.6 m	7.7	7.9	8.1	0.4	1.1	0.7	0.2	0.6	0.2	0.7	0.8	0.8
2-yr field - 0.8 m	7.6	7.3	7.8	19.7	8.0	6.4	1.0	0.7	0.4	7.3	9.2	10.0
- 1.6 m	7.6	7.2	7.7	2.6	16.4	12.2	0.4	0.8	0.2	9.0	9.1	8.0
11-yr field - 0.8 m	7.6	7.2	7.8	5.0	6.9	1.0	0.8	0.8	0.4	8.6	8.3	8.6
11-yr field - 1.6 m	7.7	7.1	7.9	2.7	7.0	3.8	0.6	0.8	0.4	9.9	9.8	7.7

[a] Sampling location at Manteca, California.
[b] ns indicates insufficient sample for analysis.

investigators (10,12) have also reported on both the higher rate of nitrogen mineralization in the early years of land treatment operation and on the increased NO_3-N in soil solution in early summer. Lower soil organic-N and organic-C values at the 2-yr field, compared to the control and to the 11-yr fields, also indicate that a large amount of organic-N was mineralized during the first two years of disposal operation.

Comparison of water quality data from Wells A and C (Table 1) indicates that two years of wastewater application has had little effect on ground water quality. Average values for NO_3-N in ground water from the treated area (Well C, Figure 1) were 2.6, 2.5, 2.5 mg NO_3-N/l in June, September and November, 1974, respectively. Ground water from the control site (Well A) collected in June, September and November, 1974, contained < 1 mg NO_3-N/l. The appearance of higher NO_3-N (9.6-11.6 mg NO_3-N/l) in water from Well B was presumably due to impact of the disposal site although, even in this case, the NO_3-N values are only slightly higher than the proposed standard of 10 mg/l.

Soil solutions from the disposal sites contained much higher concentrations of PO_4-P, compared to those from the control site. A value as high as 9.9 mg PO_4-P/l was found in soil solutions from the 11-yr field at 1.6 m depth while < 1 mg PO_4-P/l was found at the control site. The removal of PO_4-P by soils in both the 2-yr and the 11-yr sites is apparently not very efficient. The concentration of PO_4-P in soil solution from the two sites at 1.6 m was similar to, or sometimes even higher than, the current value in the secondary treatment and the storage lagoon wastewater (Table 1). It is important to note that contamination of ground water by phosphorus does not seem to be significant, in spite of the high phosphorus concentration in soil solution. The reason for high ortho-P in soil solution from treated plots is not understood. It would appear from both soil solution and soil analysis data (Table 2) that phosphorus has leached downward in the soil profiles. However, this is generally contrary to the findings of others, which state that phosphorus had not leached significantly after 7 years of wastewater application (13).

The normal ground water PO_4-P values are probably due to the large volume of soil available for phosphate retention as well as dilution effects. While the possibility of sampling error due to channeling of wastewater directly into the suction lysimeters cannot be entirely ruled out, this seems extremely unlikely since concurrent contamination by NH_4 should also have occurred. While these data are not conclusive, they should serve

Table 2. Soil chemical data for wastewater disposal site, Manteca, California.

Depth Interval (cm)	Conductivity, mmhos/cm			Free Iron Oxides, %			Total-P, ppm			Extractable-P, ppm			Organic-P, ppm			Soluble-P, ppm		
	Control	2-yr	11-yr	Control	2-yr	11-yr	Control	2-yr	11-yr	Control	2-yr	11-yr	Control	2-yr	11-yr	Control	2-yr	11-yr
000015	0.71	0.35	0.26	0.69	0.67	0.56	506	480	699	40	55	117	56	64	94	0.49	1.20	1.29
015030	0.36	0.20	0.31	0.69	0.62	0.55	459	455	512	39	58	85	28	25	53	0.41	0.70	0.27
030045	0.21	0.23	0.28	0.65	0.63	0.53	439	441	515	44	52	69	19	19	26	0.56	0.68	0.22
045060	0.15	0.20	0.16	0.63	0.63	0.55	424	404	450	34	41	70	16	13	29	0.38	0.54	0.24
060090	0.12	0.20	0.29	0.55	0.59	0.49	363	424	476	29	40	38	14	15	21	0.16	0.42	0.38
090120	0.14	0.20	0.24	0.55	0.55	0.46	340	419	435	28	38	36	14	12	15	0.19	0.09	0.27
120150	0.20	0.62	0.24	0.52	0.46	0.38	378	386	409	29	30	37	13	11	13	0.23	0.13	0.20

Depth Interval (cm)	Total-Hg, ppb			Extractable-Hg, ppb			Total-Cd, ppb			Extractable-Cd, ppb			Total-Cu, ppm			Extractable-Cu, ppm		
	Control	2-yr	11-yr	Control	2-yr	11-yr	Control	2-yr	11-yr	Control	2-yr	11-yr	Control	2-yr	11-yr	Control	2-yr	11-yr
000015	928	928	1056	34	35	56	148	128	232	68	21	188	16.8	15.4	23.7	2.3	1.2	4.6
015030	640	832	958	138	37	30	107	136	182	42	13	109	16.4	23.4	19.8	2.2	1.6	2.1
030045	1488	712	945	66	70	32	99	82	56	41	24	65	14.5	22.5	22.1	1.8	6.1	1.7
045060	992	848	1241	36	62	44	58	77	78	42	20	45	13.7	20.1	16.4	1.4	1.6	1.5
060090	960	1216	1147	222	30	64	49	36	75	20	11	42	13.0	19.3	15.5	1.6	1.8	1.2
090120	971	1360	763	366	24	22	98	32	79	66	10	38	12.9	19.6	18.3	1.2	1.6	1.3
120150	944	1376	875	44	32	35	61	78	64	47	5	40	20.4	20.2	16.2	1.1	1.8	1.6

Depth Interval (cm)	Total-Zn, ppm			Extractable-Zn, ppm			Total-Ni, ppm			Extractable-Ni, ppm			Total-Cr, ppm			Extractable-Cr, ppm		
	Control	2-yr	11-yr	Control	2-yr	11-yr	Control	2-yr	11-yr	Control	2-yr	11-yr	Control	2-yr	11-yr	Control	2-yr	11-yr
000015	39	38	71	13.8	10.2	40.6	7.2	7.4	8.8	1.8	1.7	8.2	43	71	48	4.06	3.71	14.50
015030	35	40	48	7.7	10.4	13.0	6.9	8.5	11.7	1.9	1.6	2.6	44	43	45	3.97	3.80	4.45
030045	35	39	46	7.0	12.6	13.4	5.7	6.2	10.1	1.9	1.6	2.1	43	51	47	3.94	3.62	4.02
045060	31	95	40	11.8	6.4	5.6	6.4	8.6	10.8	1.8	2.4	1.9	41	52	49	3.88	3.74	3.81
060090	32	52	36	4.2	7.6	4.5	6.7	8.4	11.9	1.8	2.2	1.9	42	54	38	3.99	3.65	3.53
090120	35	53	36	3.2	4.3	8.4	6.9	9.4	10.6	1.6	2.5	2.0	43	47	39	3.67	3.63	3.58
120150	38	48	38	9.6	5.9	6.2	7.1	9.8	9.8	1.7	2.3	1.8	44	47	38	3.55	3.97	3.96

to earmark a potential problem requiring additional research. It is of interest that even the background PO_4-P values are several orders of magnitude higher than the 0.03 mg P/l level reported by Sawyer (14) as causing eutrophication in fresh water lakes.

Comparison of the soil chemical characteristics of wastewater treated and untreated fields is shown in Table 2. These data are important for evaluating the long-term impact of wastewater application. Total-P in surface soil from the 11-yr field was 699 $\mu g/g$, compared to 506 $\mu g/g$ in the control. This relatively small increase is another indication that phosphate applied in wastewater was not retained under conditions prevailing at this location. Khin and Leeper (15) and Johnson et al. (8) reported several orders of magnitude higher phosphorus concentrations in wastewater-irrigated soils compared to unirrigated. From 3 to 11 percent of total P was present as organic-P and a similar percentage as extractable-P, while soluble-P was at a level of only 0.2 to 1.3 $\mu g/g$.

The increased CEC (Figure 3) of surface soil from the control and disposal fields is closely related to the higher content of organic matter at the surface. The vertical distribution of CEC at the three sites is similar to that of both organic-C and organic-N (Figure 3). The highest CEC (8.8 meq/100 g) was found in surface soil of the 11-yr field, followed by the control site (5.1 meq/100 g). The reason for low CEC and organic-C and organic-N in soils from the 2-yr field, compared to the control, is probably the mineralization of organic matter during the early years of wastewater application on land, as previously discussed. Johnson et al. (8) obtained an empirical relationship between organic-C and CEC (r = 0.91). They reported an increase of 1.4 meq for each 1 percent increase in organic-C. Of the exchangeable cations (Figure 4), calcium was the most predominant, followed by magnesium, then sodium. Exchangeable calcium was approximately 6 meq/100g in surface soils from the 11-yr field. Soils from the 2-yr field contained less total exchangeable calcium than the 11-yr field, due to the lower CEC in the former. The increase in exchangeable sodium in wastewater disposal sites, compared to the control, is due to the relatively high salt content of the applied effluent. In contrast, the normal irrigation water applied to the control site is of unusually low salt content. The exchangeable sodium in the disposal site soil, however, did not exceed 15 percent of the base saturation and no alkalinity problems were observed.

The data presented in Table 2 show that application of wastewater effluent did not change in a consistent manner the distri-

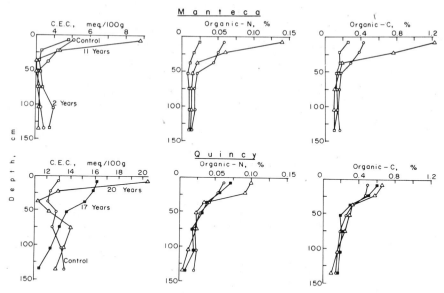

Figure 3.   Distribution of CEC, organic-N and organic-C with depth in soils from wastewater disposal and control sites at Manteca and Quincy.

bution of heavy metals in soils. The increase of total heavy metals in soil at the 11-yr site was most noticeable in the distribution of nickel, zinc, copper and chromium. There was a slight accumulation of heavy metals, but in general the ranges were within those reported by Bowen (16) for most soils. This is not surprising since the domestic wastewater applied contains low concentrations of heavy metals (data not presented). Analysis for acid-extractable heavy metals (plant-available) showed larger differences between the 11-yr disposal area and the control (Table 2). The increase in extractable heavy metals in the surface soils of the 11-yr site paralleled the increase in organic matter. The presence of high amounts of hydrous iron and aluminum oxides in surface soils may contribute to the surface accumulation of heavy metals. The data in Table 2 indicate that extractable chromium, zinc, cadmium and nickel concentrations in the 11-yr field are higher than the typical values reported in the literature (8, 16-18). Although the increase in plant-available heavy metals in the wastewater irrigated soils was not very high, this is an indication that plant uptake should be monitored to evaluate the potential problem of introducing heavy metals at toxic levels into the food chain. As shown in Figure 5, the surface

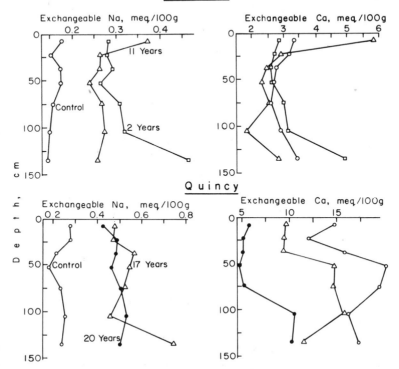

Figure 4. Distribution of exchangeable sodium and exchangeable calcium with depth in soils at Manteca and Quincy.

soil at the disposal sites increased in pH almost 1 unit (from about 6 to 7). This increasing trend was noted throughout the profile and is probably related to the increase in exchangeable sodium discussed above.

Quincy

The results for Quincy are qualitatively similar to those discussed for Manteca. However, the wastewater at Quincy (Table 3) contained much higher concentrations of nitrogen compared to Manteca, and the system was stressed with a higher rate of wastewater application. As is normally expected, most of the nitrogen in the pretreated effluent was in the NH_4 form. As high as 25.6 mg NH_4-N/l was found in samples from the primary clarifier. Water samples from the storage lagoon contained 12.2, 7.2 and 13.6 mg NH_4-N/l in May, August, and November 1974, re-

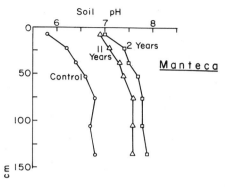

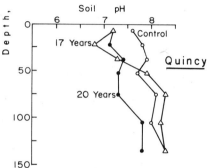

Figure 5.   Distribution of soil pH with depth at Manteca and Quincy.

spectively. The data collected from the monitoring wells and the drainage ditch did not indicate contamination of ground water by any form of nitrogen. Nitrate-N was < 4 mg/l on all sampling dates. Soil solution, however, contained higher levels of NO_3-N, especially with the samples collected in May 1974. As high as 71.2 mg NO_3-N/l average concentration was found in soil solution from the 17-yr field at the 0.8-m depth. Again, this is probably due to leaching of stored NH_4-N after conversion to NO_3 in the surface soil in spring.

Similar findings were reported by Iskandar *et al.* (19). They reported concentrations as high as 120 mg NO_3-N/l in the leachate at 150 cm from a sandy loam soil treated with domestic wastewater of similar nitrogen content. The increased NO_3 concentration in early summer also was observed in the 20-yr field at Quincy (Table 3) but to a lesser extent than that in the 17-yr field. The reason for this difference is not clear, but may be due to a differential rate of application or to passage of the nitrate peak prior to the first sampling date (May 1974). There were slight increases in organic-N (Figure 3) and organic carbon

Table 3. Average values for water quality at wastewater disposal site, Quincy, Washington.

Sample[a]	pH			NO$_3$ (ppm N)			NH$_4$ (ppm N)			Ortho-P (ppm P)		
	5/74	8/74	11/74	5/74	8/74	11/74	5/74	8/74	11/74	5/74	8/74	11/74
Primary clarifier	7.3	7.1	7.3	1.9	2.2	1.7	21.9	11.3	25.6	4.1	2.1	3.1
Storage lagoons	8.3	8.4	7.4	0.8	0.8	2.8	12.2	7.2	13.6	3.9	2.7	5.3
Drainage ditch	7.6	7.7	7.6	1.2	1.1	3.7	0.3	0.3	0.2	0.1	0.2	0.1
Tile to drainage ditch	7.6	7.5	7.6	2.8	2.5	3.3	0.1	<0.1	<0.1	0.2	0.1	0.1
well - north center	7.2	7.2	7.2	0.8	0.9	0.3	0.4	0.9	0.8	0.3	0.2	0.2
well - center	7.1	6.9	7.0	1.2	2.5	1.4	1.7	0.3	0.2	1.6	4.3	4.6
Control field - 0.8 m	7.2	7.2	ns[b]	2.7	1.2	ns	0.2	0.2	ns	0.2	0.2	ns
- 1.6 m	7.2	7.2	ns	0.7	1.4	ns	0.3	0.3	ns	0.2	0.2	ns
17-yr field - 0.8 m	6.9	6.8	6.9	71.2	13.8	3.7	0.7	0.5	<0.1	1.0	1.1	1.7
1.6 m	6.8	6.9	6.9	7.9	3.6	1.2	1.2	0.6	0.2	2.8	2.2	2.9
20-yr field - 0.8 m	6.7	6.7	6.8	1.5	2.7	17.9	0.5	0.2	0.2	1.1	1.5	1.2
1.6 m	7.1	7.1	6.9	19.2	4.0	4.7	0.4	1.5	0.1	0.6	1.2	0.5

[a] Sample location at Quincy, Washington.
[b] Insufficient sample for analysis.

in the 20-yr soils, compared to the control and the 17-yr fields. This is reflected in a corresponding increase in CEC in the surface soil of the 20-yr disposal field.

The data in Table 4 show that total-P increased from about 900 ppm in the topsoil of the control site to about 1100 and 1300 ppm in the 17- and 20-yr disposal fields, respectively. There was an increasing trend in total-P throughout the profile in the disposal fields. Soil extractable-P was four- to five-fold higher than the control. Soil solution (Table 3) at 1.6 m contained 2.2-2.9 and 0.5-1.1 mg PO_4-P/l in the 17- and 20-yr fields respectively. These data again indicate that phosphate may not be retained by soil as efficiently as commonly expected.

The vertical distribution of the exchangeable cations (Figure 4) showed some interesting trends. Application of wastewater for 17 and 20 years, respectively, decreased the exchangeable calcium from about 15 meq/100g to 10 and 5 meq/100g in the top 45 cm. In contrast to calcium, exchangeable magnesium and exchangeable sodium increased after irrigation. Exchangeable magnesium increased from 1 meq/100g to 3-4 meq/100g in the top 45 cm of the irrigated soils .The increase in exchangeable sodium in both the 17- and 20-yr fields was more than two-fold. However, exchangeable sodium did not appear to create a serious soil alkalinity problem. Also, the soil salinity as measured by specific conductance was not affected by wastewater application. The specific conductance of the 1:2 (soil:water) extract was less than 0.8 mmhos/cm in both the irrigated and the control soils, a value much less than the 4 mmhos/cm reported for saline soils (20).

Of the heavy metals studied, only total cadmium increased in the irrigated soils as compared to the control (Table 4). There were no consistent trends in the distribution of other heavy metals in the soils from the control and the treated sites.

Soil pH decreased in the top 45 cm of the soil from approximately 8 in the control site to about 7 in both the 17- and 20-yr disposal sites (Figure 5). This drop in soil pH is possibly due to nitrification as explained by Alexander (21). It is possible that soil pH would have decreased even further if the applied wastewater had contained lower concentrations of sodium, which probably tend to counter the nitrification effect.

Comparison of Sites

Facilities at both Manteca and Quincy are designed and operated as disposal sites rather than as managed facilities for treatment of wastewater. They were not intended to produce water quality that meets the goals of PL 92-500 (1972). The pretreat-

Table 4. Soil chemical data for wastewater disposal site, Quincy, Washington.

Conductivity, mmhos/cm

Depth Interval (cm)	Control-I	17-yr	20-yr
000015	0.35	0.25	0.21
015030	0.35	0.78	0.27
030045	0.26	0.31	0.15
045060	0.40	0.29	0.19
060090	0.50	0.34	0.36
090120	0.81	0.36	0.51
120150	0.41	0.40	0.69

Organic-P, ppm

Depth Interval (cm)	Control-I	17-yr	20-yr
000015	53	61	83
015030	44	48	72
030045	42	38	40
045060	33	22	31
060090	36	26	34
090120	38	20	22
120150	21	7	14

Total-Cd, ppb

Depth Interval (cm)	Control-I	17-yr
000015	263	381
015030	179	272
030045	173	344
045060	212	369
060090	159	222
090120	122	197
120150	166	204

Free Iron Oxides, %

Depth Interval (cm)	Control-I	17-yr	20-yr
000015	0.97	1.32	1.10
015030	0.84	0.98	0.93
030045	0.93	0.82	0.76
045060	0.94	0.84	0.79
060090	0.86	0.79	0.79
090120	0.84	0.76	0.81
120150	0.59	0.65	0.69

Soluble-P, ppm

Depth Interval (cm)	Control-I	17-yr	20-yr
000015	1.26	4.39	6.76
015030	1.23	2.44	4.47
030045	0.37	2.18	4.44
045060	0.24	1.16	0.64
060090	0.13	1.03	0.44
090120	0.14	0.56	0.29
120150	0.18	0.63	0.32

Extractable-Cd, ppb

Depth Interval (cm)	Control-I	17-yr	20-yr
000015	214	159	183
015030	186	131	160
030045	62	92	122
045060	26	107	101
060090	32	143	64
090120	60	113	66
120150	45	94	71

Total-P, ppm

Depth Interval (cm)	Control-I	17-yr	20-yr
000015	883	1121	1274
015030	902	1100	1155
030045	850	1072	969
045060	965	1041	1002
060090	951	1029	951
090120	960	1011	1005
120150	894	1001	982

Total-Hg, ppm

Depth Interval (cm)	Control-I	17-yr	20-yr
000015	753	886	496
015030	622	1021	832
030045	433	466	1296
045060	848	576	1152
060090	878	692	1232
090120	1239	684	1200
120150	752	752	1424

Total Cu, ppm

Depth Interval (cm)	Control-I	17-yr	20-yr
000015	35.4	35.4	28.8
015030	36.2	37.1	32.4
030045	35.0	34.0	30.8
045060	36.2	40.5	28.6
060090	35.8	36.8	35.1
090120	34.1	33.5	41.3
120150	38.2	32.7	33.7

Extractable-P, ppm

Depth Interval (cm)	Control-I	17-yr	20-yr
000015	37	106	169
015030	33	97	184
030045	20	58	111
045060	4	49	123
060090	3	45	90
090120	4	21	52
120150	7	24	51

Extractable-Hg, ppb

Depth Interval (cm)	Control-I	17-yr	20-yr
000015	34	51	50
015030	56	238	96
030045	440	90	247
045060	394	82	104
060090	204	80	51
090120	88	88	66
120150	86	53	122

Extractable-Cu, ppm

Depth Interval (cm)	Control-I	17-yr	20-yr
000015	5.0	5.4	4.3
015030	5.1	5.8	4.6
030045	0.4	5.9	3.5
045060	0.4	4.3	4.3
060090	0.4	0.9	4.2
090120	0.3	0.6	1.7
120150	0.4	0.7	0.6

	Total-Zn, ppm			Extractable-Zn, ppm			Total-Ni, ppm			Extractable-Ni, ppm		
000015	86	83	85	21.4	14.1	15.0	31.2	29.2	28.0	3.33	3.07	0.95
015030	88	84	81	25.2	11.8	15.5	30.4	26.6	26.4	3.03	2.96	1.02
030045	85	72	74	1.0	4.8	12.3	34.4	28.7	27.2	0.55	2.67	1.09
045060	75	78	73	0.9	3.2	8.1	34.9	34.3	26.5	0.60	3.00	1.03
060090	69	75	76	0.8	2.8	6.4	33.2	31.7	29.3	0.58	2.11	1.00
090120	66	73	78	0.8	1.6	2.5	34.3	32.2	36.0	0.73	0.98	0.59
120150	73	80	78	0.9	2.8	1.4	33.8	29.4	35.7	0.99	1.03	0.55

	Total-Cr, ppm			Extractable-Cr, ppm		
000015	46	39	40	4.16	3.55	1.73
015030	47	41	38	3.92	3.77	1.34
030045	44	39	35	1.44	3.46	0.95
045060	45	38	34	1.25	3.32	0.83
060090	43	37	37	1.28	2.54	0.81
090120	41	36	36	1.29	0.65	0.74
120150	43	38	38	1.24	0.87	0.61

ment facilities at Quincy were designed in 1954 for 1.9 x 10^6 l/day (0.5 MGD), but at the present time approximately 2.6 x 10^6 liters/day in the fall and 3.8 x 10^6 liters/day (1 MGD) in the summer are treated. The area of land designated for this site is inadequate, although steps have since been taken to enlarge the area. Wastewater is being applied at rates greater than 15 cm (6 in.) per week, as compared to < 5 cm (< 2 in.) per week in Manteca.

One of the interesting similarities between Quincy and Manteca is the presence of a period of higher NO_3-N concentration in soil solution in early summer, presumably the consequence of NH_4 accumulation in winter months. This effect was more prevalent at Quincy, which has a colder winter, a finer textured soil with higher CEC and water holding capacity, and a higher N-loading rate compared to Manteca. Data from controlled slow-infiltration test cells in a similar cold climate (19) have shown similar periods of high NO_3 concentrations in early summer. With the exception of the NO_3 peak in early summer, percolate water of suitable quality (about 10 mg NO_3-N/l or less) was produced at both disposal locations.

At both disposal sites there was an indication of movement of applied phosphate to a depth of at least 1.6 m (5 ft) in the profile; however, no increase in phosphorus was found in ground water. Therefore, while this is an indication of a potential problem area, current phosphate renovation seems satisfactory.

Soil pH is another parameter of importance in land treatment. It is usually anticipated that soil pH will decrease as a result of nitrification-induced acidity (21). This was true at Quincy in the upper noncalcareous zone, but not at Manteca where, in fact, soil pH increased from 5.8 to 7.0 in the top 15 cm. This increase in soil pH is probably due to the high sodium concentration (exchangeable sodium increased substantially) in applied wastewater with relatively low amounts of NH_4. Analysis of the soils for heavy metals at each site showed no indication of accumulation or unusual levels. No attempts were made to study viruses and pathogens or vegetative responses to heavy metals at either disposal site.

There have been no local complaints on the operation of the disposal sites. Problems are apparently of a technical and operational nature. With proper site management at Manteca and Quincy, and expansion of the pretreatment facilities and land area at Quincy, it is expected that the quality of water produced could meet anticipated effluent water quality standards.

ACKNOWLEDGMENTS

The authors wish to thank Mr. Ronald Roduner, Plant Operator, Sanitary Treatment Plant, Quincy, Washington, and Mr. Manuel Oliveira, Sanitation Plant Operator, Wastewater Quality Control Facilities, Manteca, California, for both their willingness to cooperate and the assistance they rendered during the course of this study. We also want to acknowledge the technical participation of Mr. Bruce Brockett, Physical Science Technician, USACRREL, who was largely responsible for collection of field samples and data collection and reduction, and to Mr. D. C. Leggett for technical review. The study was conducted as part of the Civil Works Research and Investigation Project under the work unit "Evaluation of Existing Facilities for Wastewater Land Treatment" (CWIS 31280) and under the Military Project 4A762720A896 "Land Treatment of Wastewater."

REFERENCES

1. Lee, G. F. "Eutrophication," Occasional Paper No. 2, Eutrophication Information Program, Water Resources Center, University of Wisconsin, Madison, Wisconsin (1970).
2. Syers, J. K., I. K. Iskandar, and D. R. Keeney. "Distribution and background levels of mercury in sediment cores from selected Wisconsin lakes, *Water, Air, Soil Poll.* 2:105–118 (1973).
3. Irukayama, K. "The pollution of Minamata Bay and Minamata disease," *Adv. Water Pollution Research Proc.* Int. Conf. 3rd, 3:153–180 (1967).
4. Evans, J. O. in: "Research needs—land disposal of municipal sewage wastes," *Recycling Municipal Wastewater and Sludge Through Forest and Cropland*, Pennsylvania State University Press (1973).
5. Reed, S., H. McKim, R. Sletten and A. Palazzo. "Pretreatment requirements for land application of wastewaters," paper presented at ASCE 2nd National Conference on Environmental Engineering Research, Development and Design, Gainsville, Florida (July 1975).
6. EPA. "Survey of Facilities Using Land Application of Wastewater," American Public Works Association, prepared for EPA, PB 227–351 (1973).
7. EPA. "Land treatment of municipal wastewater effluents, case histories," EPA Technology transfer seminar publication (1976).
8. Johnson, R. D., R. L. Jones, T. D. Hinesly and D. J. David. "Selected chemical characteristics of soils, forages, and drainage water from the sewage farm serving Melbourne, Australia," Report to U.S. Army, Corps of Engineers (1974).
9. Black, C. A., Ed. *Methods of Soil Analysis.* Agronomy Series No. 9, American Society of Agronomy, Madison, Wisconsin (1965).
10. Iskandar, I. K., "Urban waste as a source of heavy metals in land treatment," *Proc. International Conf. on Heavy Metals in the Environment*, October 27–31, 1975, Toronto, Ontario, Canada (1976).
11. Hach Chemical Co. *Handbook for Water Analysis.* Ames, Iowa (1973).

12. Reinhorn, T., and Y. Aunimelech. "Nitrogen release associated with the decrease in soil organic matter in newly cultivated soils," *J. Environ. Quality* 3:118–121 (1974).

13. Kardos, L. T., and W. E. Sopper. "Renovation of municipal wastewater through disposal by spray irrigation," In: *Recycling Municipal Wastewater and Sludge Through Forest and Cropland*, Pennsylvania State University Press (1973).

14. Sawyer, C. N. "Basic concepts of eutrophication," *J. Water Pollution Control Fed.* 38:737–744 (1966).

15. Khin, A., and G. W. Leeper. "Modifications in Chang and Jackson's procedure for fractionating soil phosphorus," *Agrochimica* 4:246–254 (1960).

16. Bowen, H. J. *Trace Elements in Biochemistry*. Academic Press, New York (1966).

17. Buckman, H. O., and N. C. Brady. *The Nature and Properties of Soils*, 7th Edition, The MacMillan Co., New York (1969).

18. Leeper, G. W. "Reactions of heavy metals with soils with special regard to their application in sewage wastes," Prepared for Department of the Army, Corps of Engineers, Contract No. DAW73–C–0026 (1972).

19. Iskandar, I. K., R. S. Sletten, D. C. Leggett and T. F. Jenkins. "Wastewater renovation by a prototype slow infiltration land treatment system," CRREL Report (in print), USACRREL, Hanover, New Hampshire (1976).

20. Richards, L. A., Ed. "Diagnosis and improvement of saline and alkali soils," Agriculture Handbook No. 60, USDA, U.S. Government Printing Office (1954).

21. Alexander, M. "Nitrification," In: *Soil Nitrogen*, W. V. Bartholomew and F. E. Clark, Eds. Agronomy Monographs No. 10, American Society of Agronomy, Madison, Wisconsin (1965).

26

PRELIMINARY EVALUATION OF 88 YEARS RAPID INFILTRATION OF RAW MUNICIPAL SEWAGE AT CALUMET, MICHIGAN

C. R. Baillod, R. G. Waters
Department of Civil Engineering,
  Michigan Technological University, Houghton, Michigan
I. K. Iskandar and A. Uiga
U.S. Army Cold Regions Research and Engineering Laboratory,
  Hanover, New Hampshire

INTRODUCTION

Land treatment systems for wastewater have been classified according to hydraulic loading as slow infiltration and rapid infiltration. In a typical slow infiltration system, the wastewater is applied at loading rates ranging from 5 to 15 cm/week. At these relatively low loading rates, a significant amount (perhaps 20 to 80 percent) of the nutrients applied are taken up by vegetation. Additional physical, chemical and biological reactions are generally effective in accomplishing a high degree of treatment as the wastewater percolates through the soil matrix. A significant amount of research has been conducted on slow infiltration-spray irrigation systems (1,2) and many have recently been constructed (3). However, in rapid infiltration systems, the hydraulic loading rates are typically an order of magnitude larger than those employed in slow infiltration and range from 10 to 250 cm/week. Consequently, the nutrient uptake by vegetation is less significant, and the transformations occurring as the wastewater percolates through the deeper soil layers must be relied upon to accomplish the desired treatment. Major advantages of rapid infiltration are that much less land is needed and significant ground water recharge may be accomplished.

Since a high rate of infiltration is necessary, the wastewater is usually applied to infiltration beds, which may or may not be vegetated, by ponding or flooding. Research (4-7) conducted at the U.S. Water Conservation Laboratory at Phoenix, Arizona has provided a better understanding of the factors influencing the operation of rapid infiltration systems.

One point which still requires further study, particularly for rapid infiltration, concerns the long-term influence of land treatment on the soil-ground water environs. To investigate this, studies are being conducted on existing land treatment systems that have been in continuous operation over long periods of time. This chapter is concerned with one such rapid infiltration system which has served the Village of Calumet, Michigan for the past 88 years. This installation may be the oldest land treatment site currently in operation in the U.S.

BACKGROUND INFORMATION ON THE SITE

Calumet is located on the Keweenaw Peninsula in the Upper Peninsula of Michigan as shown in Figure 1. The location of the wastewater disposal site is indicated in Figure 2. For this area, the mean monthly temperature ranges from 15° F (-9.5° C) in January to 66° F (19° C) in July with an annual average temperature of about 40° F (4.5° C). The average date of the first fall temperature of 32° F (0° C) or colder is September 21, and

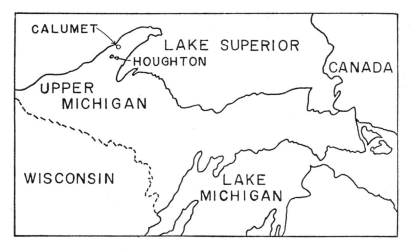

Figure 1.   General location map.

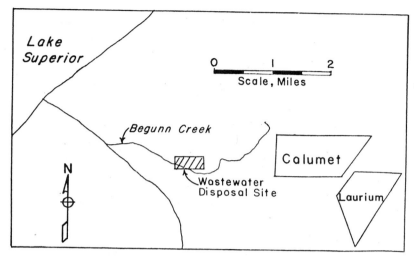

Figure 2.   Site location map.

the last day in the spring is May 28. Average annual precipitation is about 32 in. (81 cm) with snowfall averaging about 180 in. (457 cm). Monthly evaporation potentials applicable to this area have been determined by Wiitala et al. (8).

The surface topography of the area is largely controlled by the bedrock formations, the most prominent of which is the Copper Range. This formation, consisting of Keweenawan lava flow and conglomerates, forms the central highland of the Keweenaw Peninsula rising to elevations of more than 600 ft (183 m) above Lake Superior. The wastewater disposal site is underlain by the Copper Harbor Conglomerate which is composed of light red to brown cemented, nonporous boulder conglomerates and minor amounts of pebble conglomerate and feldspar-rich sandstone (9).

Most of the Keweenaw Peninsula bedrock is covered by salmon-colored till (unstratified glacial drift deposited directly from the ice with little water sorting—a heterogenous mixture of clay, sand, gravel and boulders) varying from a few inches to more than 300 ft (91 m) in thickness. The overburden deposit underlying the wastewater disposal site has been characterized as a loose-textured pink till containing beds of lacustrine sand and poorly sorted gravel, frequently overlain or washed over by thin lake clay or sand (9).

Figure 3 shows the local topography and drainage in the immediate vicinity of the site along with the measured depths of overburden (total thickness of material overlying the bedrock)

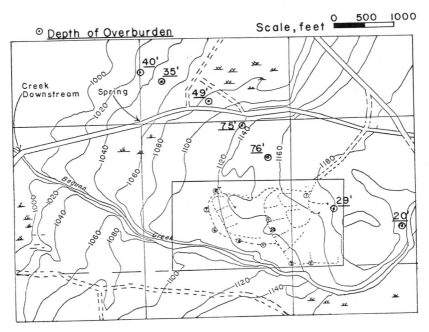

Figure 3. Topography and depths of overburden in vicinity of wastewater disposal site.

at several locations. These depths were determined from the logs of several exploratory diamond drill holes made by the Calumet and Hecla Mining Company during the 1920s. The general bedrock surface dips to the northwest toward Lake Superior, with the dip roughly paralleling the line of drill holes shown in Figure 3. It is possible that, directly under the disposal site, the bedrock surface may dip more toward the west as the location of Begunn Creek may be partially determined by bedrock topography. Because of this possibility, it is difficult to estimate the thickness of overburden directly under the site. Based on the information indicated in Figure 3, it is estimated that the overburden thickness directly under the site could range from 30 ft (9.1 m) to 70 ft (21 m). The soil of the site is classified as typic haplorthod—sandy, mixed, frigid—and is similar to the Kalkaska Soil Series.

In the early 1840s, just a few years prior to the discovery of gold in California, the Keweenaw Peninsula of Michigan became the site of America's first mining boom. Subsequently, this area became one of the world's greatest copper mining districts. Nothing resembling the native copper deposits of the Keweenaw Peninsula is known anywhere else in the world (9).

The Village of Calumet was initially settled during the 1860s and became incorporated in 1875. Throughout most of its history, a paternal relationship existed between the village and its major employer, the Calumet and Hecla Consolidated Copper Company and its predecessor companies. Most of the land and many of the residences were owned by the mining company. It is not surprising, therefore, that the company provided many municipal services including water supply and sewerage. During the 1880s, the mining company built a separate sanitary sewerage system for the village, and in 1887 constructed a "sewage farm" on a 100-acre site 2 miles (3.2 km) west of the village. Initially, this system served roughly 3000 people; by 1916 a report submitted to the Michigan Department of Public Health indicated that the system served 14,000 people. Subsequently, over the 1920 to 1950 period, the local mining industry became depressed, and the population declined drastically. Between 1950 to 1970, the population served by the system remained relatively constant at about 4800 people.

From its inception in 1887 until 1961, the system was owned and operated by the Calumet and Hecla Company. The system operated without problems and was presumably well-maintained as, during much of this period, the mining company assigned a full-time employee to live at, manage and maintain the "sewage farm." There is no evidence of any agricultural crops being harvested from the site during this period. From 1961 to 1971, the system was owned and operated by the Northern Michigan Water Company, and in 1972 ownership was transferred to the neighboring village of Laurium in order that federal grant funds could be expended to improve the site.

During the summer of 1972, the land treatment system was reconstructed and modified to accept the wastewater from an additional 3300 people located in the neighboring Village of Laurium. This modification consisted of building additional dikes, infiltration areas and distribution trenches. Presently, the system is operated by the Northern Michigan Water Company under contract with the Village of Laurium.

DESCRIPTION OF ORIGINAL AND RECONSTRUCTED TREATMENT SITE

Original Site

As pointed out earlier, the population served by the site was increased from 4800 to 8100 during the fall of 1972. A topographic map of the site prior to the 1972 modifications is shown in Figure 4. The sewage entered at the northeast corner of the

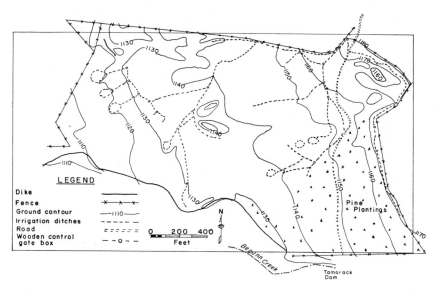

Figure 4. Topographic map of wastewater disposal site prior to 1972 reconstruction.

site and was routed to the various areas within the site by a system of open trenches and manually operated gates. In this condition, the site apparently functioned as a combination of overland flow and rapid infiltration. There were about ten fairly well defined infiltration beds occupying about two acres. In addition, there was an undeterminable area over which overland flow probably occurred. This overland flow area existed in the vicinity of the percolation pond inlets as well as the terminal ends of a few distribution trenches.

During the several years prior to reconstruction of the system, the 4800 residents tributary to the system used an average of about 42 gpcd (159 lpcs) or 0.2 MGD (757 m³/day) of water. However, because of the age and condition of the collection system and because part of the collection system consisted of combined sewers, the average daily flow to the disposal site was approximately 0.5 MGD (1892 m³/day). Although no wastewater strengths were measured during the period, the sewage would have been rather weak.

A flow of 0.5 MGD (1892 m³/day) discharged to the 2 acres

(0.8 ha) of percolation pond areas would have resulted in a loading rate of 9 in./day (23 cm/day) which is indicative of a rapid infiltration system. The actual hydraulic loading was less because of the volume absorbed during overland flow. The operating procedure had been to flood one infiltration area each day on a seven-to-ten cycle, *i.e.*, flood on day one, rest for days two to seven and flood again on day eight. Since the infiltration rate had been higher than the loading rate, the system operated without any surface effluent.

Reconstructed Site

Figure 5 shows a topographic map of the reconstructed site as it presently exists. The modifications made during the summer of 1972 are evident from a comparison of Figures 4 and 5. The infiltration pond areas were increased considerably to a total area of roughly 12 acres (4.85 ha) by building additional dikes. However, the area which was formerly available for overland flow was considerably reduced. Presently, extensive overland flow can be effected only in the pine plantings located in the southwest corner of the site. A series of observation wells with depths ranging from 10 to 30 ft (3 to 9 m) was installed around the site perimeter and adjacent to two of the ponds.

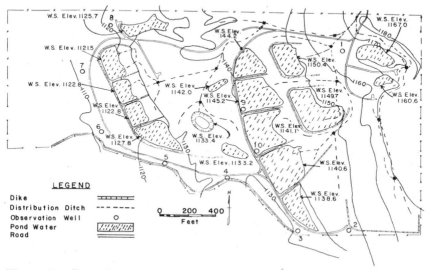

Figure 5. Ground surface topography of reconstructed wastewater disposal site showing ponded water elevations at beginning of study (5/30/75).

At this time, an additional 3300 people using 42 gpcd (159 lpcd) or 0.14 MGD (530 m³/day) of water were added to the sewage system. This increased the total population tributary to the system to 8100 using 42 gpcd (159 lpcd) or 0.34 MGD (1287 m³/day) of water. Figure 6 shows the seasonal flow variations

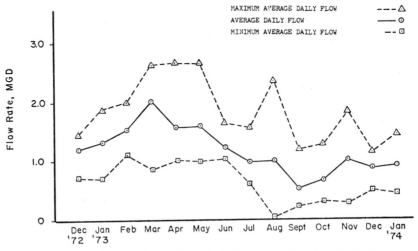

Figure 6. Seasonal variations in influent wastewater flow rate.

measured at the site during 1973. It is apparent that very large quantities of infiltration/inflow are entering the collection system. The average annual flow rate of 1.2 MGD (4542 m³/day) is 3.5 times the average water usage and amounts to 148 gpcd (560 lpcd). The infiltration/inflow problem is less severe during the summer and fall months but is magnified considerably during the spring snowmelt period.

The average loading rate on the 12 acres (4.85 ha) of percolation area amounts to 3.68 in./day (9.35 cm/day) or 25.7 in./week (65.3 cm/week). This loading rate is within the range indicated for rapid infiltration systems.

SAMPLING AND ANALYTICAL PROCEDURES

The ideal approach to determining the long-term influence of wastewater application on the soil-ground water environs would be to compare data obtained prior to wastewater application with data obtained from the same site after a long period at operation. However, in the case of old systems, such as the one studied in this chapter, the prior data are usually not available. For these systems the research approach must be based on comparisons of observations made on "control" and "treated" areas at the same point in time. A potential difficulty arises here in that one can never be completely certain that observed differences are due entirely to the wastewater application.

The general approach to assessing the effects on shallow ground water quality involved comparison of water quality data obtained from an observation well which was interpreted to be receiving ground water relatively unaffected by the treatment site with water quality data from wells interpreted to be receiving percolating wastewater. Likewise, the approach employed to assess the effects on soil chemistry involved comparisons between an infiltration area and nearby control areas which had never received wastewater.

Eight of the ten observation wells, as well as Begunn Creek upstream and downstream of the site and the spring located 1700 ft (518 m) northwest of the site, were sampled during June, July and August of 1975. At the initial sampling, at least two well volumes of liquid were withdrawn and discarded prior to collecting a sample for analysis. The water samples were withdrawn from the wells by means of a portable vacuum pump powered by a gasoline engine and were collected in a standard vacuum flask. Subsequently, the samples were transferred to acid-washed bottles and transported to the laboratory for analysis. A total of four composite samples of influent wastewater was collected on days which reflected both wet and dry weather flows. Twenty-four individual grab samples of influent were collected at hourly intervals by means of a portable sampler. These individual samples were then composited according to flow to make the 24-hr composite samples which were subsequently analyzed.

An aliquot of each water sample was filtered, acidified and stored at 4° C for subsequent analysis for metallic ions. An aliquot of the filtrate was analyzed for chloride, orthophosphate, nitrate and nitrite. An aliquot of the unfiltered sample was analyzed for pH, alkalinity, suspended solids, total phosphorus, ammonium, chemical oxygen demand (COD), total organic carbon (TOC), and organic nitrogen. The analyses for pH, alkalinity, suspended solids, and forms of nitrogen were performed as soon as possible after sample collection, generally within 24 hr (pH, alkalinity and suspended solids were performed within 4 hr). The remaining parameters were determined on samples suitably preserved and either refrigerated at 4° C or frozen.

Analyses for metallic ions were performed by atomic absorption spectrophotometry according to methods outlined by EPA (10). A Perkin Elmer Model 360 atomic absorption unit equipped with a graphite furnace was employed for the measurement of copper, lead, nickel, chromium and cadmium. All other metallic ions were measured using a Perkin Elmer Model 305-B flame atomic absorption unit. Analyses for substances other than metallic ions followed the methods given in *Standard Methods* (11). Total organic carbon was determined by means of a Beckman Model 915-A Carbon Analyzer.

A composite sample from at least three soil cores from each site was taken for chemical analysis. Soil samples were taken each 2-12 in. as specified, air dried at 25° C, thoroughly mixed, sieved, and the minus 2-mm fraction was employed for chemical analysis. Soil pH (1:1 wt/vol); soluble salts (measured at specific conductance); cation exchange capacity; exchangeable Ca, Mg, Na and K; organic carbon; organic nitrogen; free iron oxides; organic-P; total-P (acid digestion); extractable P; and soluble-P (1:2 soil to water) were determined according to Black (12). Total-P was also determined by fusion with Na_2CO_3 according to Jackson (13). Total and extractable heavy metals were extracted according to Iskandar and Keeney (14) and subsequently determined using a Perkin Elmer Model 304 atomic absorption unit.

RESULTS AND DISCUSSIONS

Influent Wastewater

A summary of the suspended solids, chemical oxygen demand, and 5-day biochemical oxygen demand data obtained on four influent composite samples is shown in Table 1. Unfortunately, meaningful average flow measurements were not recorded on

Table 1.   Influent wastewater characteristics.

Sampling Date		6/30/75	7/30/75	8/25/75	8/29/75	"Normal Raw Sewage"	Avg.
Flow condition	Infiltration	+++	++	+	+		
	Inflow	–	–	–	+		
Suspended solids (mg/1)		50	56	36	70	240	53
COD (mg/1)		69	144	472	223	500	228
BOD (mg/1)		20	61	(180)[a]	(85)[a]	200	87

[a] Estimated from COD.

these days because of mechanical problems. However, the flow data for 1973, illustrated in Figure 6, and the precipitation data recorded during the 1975 study period were used to roughly estimate the flow on the sampling dates. These rough estimates were in agreement with visual observations and spot flow readings made on the sampling days and indicated that clear water tributary to the collection system was predominantly infiltration and that the infiltration component was greatly increased during the spring and early summer as well as during the period following substantial precipitation. This observation has been applied to indicate the relative contributions of ground water infiltration and surface inflow to the wastewater flow on the various sampling dates shown in Table 1. This table indicates that even during periods of relatively low infiltration/inflow, the influent wastewater is somewhat weaker than normal raw municipal sewage. On an average annual basis, it is estimated that the influent concentration would be less than half that of normal municipal sewage.

Ground Elevations

At each sampling time, ground water elevations in the observation wells were recorded. In addition, water levels in the percolation ponds were determined at the beginning of the study and again at the conclusion of the study. Figure 7 shows the

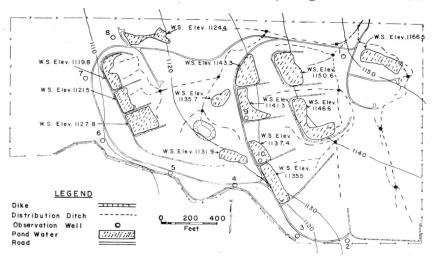

Figure 7. Topographic map of ground water table contours and ponded water areas and elevation at the end of the study (August 25–29, 1975).

ground water contours along with the ponded water areas and elevations at the end of the study. All elevations are with respect to sea level equal to zero and Lake Superior water surface equal to approximately 602. The data indicate that the ground water gradient slopes to the southwest. Furthermore, comparison of the ponded water elevations and outlines given in Figures 5 and 7 shows that a marked decrease of ponded water inventory occurred over the summer.

Impact of Sewage Infiltration on Ground Water Quality and Soil Chemistry

Figure 8 depicts the areal distribution of the average chloride concentrations observed in the samples from the influent, observation wells, creek and spring. Interpretation of these data is aided by consideration of the site surface topography shown in Figures 3 and 5 and the ground water contours shown in Figure 7. The latter figure indicates that the percolating wastewater generally travels toward the west-southwest. Along the northern boundary of the site, the ground water movement is more westerly, whereas along the eastern boundary the apparent ground water movement is toward the southwest. The areal distribution of chloride concentration agrees well with the apparent direction of ground water movement. The data of Well 2 in par-

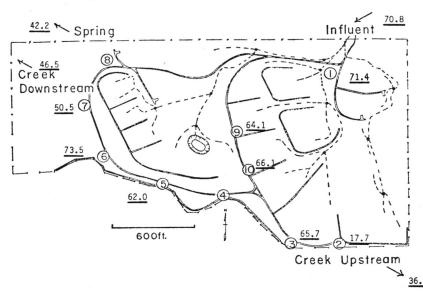

Figure 8.   Average chloride concentrations (mg/l).

ticular are noteworthy in that the concentrations of nearly all substances were considerably lower than in the other wells. Prior to late August 1975, sewage had not been fed to the overland flow-pine planting area at the southeast corner of the site for 10 years and only infrequently (if at all) before then. Because of this and because of the pattern of ground water movement, it appears that Well 2 can be considered as a "control" or indicator of "background" water quality levels. Thus, by comparing the Well 2 water quality data with those of the influent, other wells and spring, it is possible to make a tentative assessment of the impact of the wastewater percolate on ground water quality. The water quality data observed in the spring are of particular value in that they indicate the potential for movement over longer distances.

Table 2 summarizes the areal distribution of average concentrations for the various parameters measured. The values ob-

Table 2.   Average values of water quality parameters.

Parameter	"Control" Well 2	Influent	Interior Wells 1,9,10	Perimeter Wells 3-7	Spring	Creek Upstream	Creek Downstream
Na^+	0.7	3.4	2.3	2.5	1.3	1.3	1.8
K^+	19	64.0	61	65	34	33	45
Mg^{++}	5	13.0	12.0	10.0	12	5.0	8
Ca^{++}	14	57.0	45.0	32.0	64	19	38
Alkalinity (as $CaCO_3$)	51	155	195	131	169	65	113
Cl^-	18	71	67	63	42	36	46
Org-N	0.09	0.8	0.07	0.11	0.00	0.33	0.13
NH_4^+-N	1.5	24.0	6.0	3.0	0.00	0.01	0.02
NO_2^--N	0.1	0.01	0.3	0.3	0.3	0.04	0.22
NO_3^--N	0.5	0.2	0.4	2.0	3.4	0.8	0.7
T.K.N.	2.1	24.2	6.7	5.3	3.7	0.8	0.9
Ortho-P	0.03	1.7	0.05	0.02	0.00	0.01	0.00
Total P	0.05	3.5	0.4	0.1	0.03	0.07	0.07
COD	9	228	58	33	14	35	29
TOC	1	44	16	15	2	12	11
Soluble TOC	1	22	11	10	2	11	9
Cu	0.018	0.043	0.022	0.010	0.002	0.013	0.017
Pb	0.002	0.005	0.005	0.004	0.000	0.001	0.001
Ni	0.003	0.005	0.006	0.006	0.003	0.004	0.002
Cd	0.001	0.026	0.022	0.015	0.012	0.011	0.022
Cr	0.010	0.027	0.021	0.020	0.009	0.003	0.003

All Concentrations Expressed as mg/l

served in the interior observation wells 1, 9 and 10 were averaged as were the values observed in the perimeter observation wells 3, 5, 6 and 7. Ground water concentrations of sodium and potassium followed the same pattern with Well 2 being the lowest, the spring concentration being on the order twice that of Well 2, and the remaining wells generally showing values slightly less than the influent. The pattern for calcium and magnesium was different from that of sodium and potassium in that the concentrations observed in the spring were comparable to the influent and some of the intermediate wells. This could be due to the low soil pH (5.5-6.1) which is not suitable for calcium and magnesium carbonate formation. However, soil chemistry data given later in Figure 12 showed that exchangeable calcium, magnesium and organics all were substantially greater in the soil treated with wastewater than in the control soils. Although the treated soil showed higher values of exchangeable sodium than did the control soils, all values were relatively low [Figure 12(b)]. These properties of the treated soil (high values of exchangeable calcium, magnesium and organics, low values of exchangeable sodium) are conducive to the formation of soil aggregates which, in turn, help to maintain soil permeability. Thus, favorable soil chemistry appears to be one factor responsible for 88 years of successful rapid infiltration at this site.

Figures 9 and 10 show the distribution of ammonium and ni-

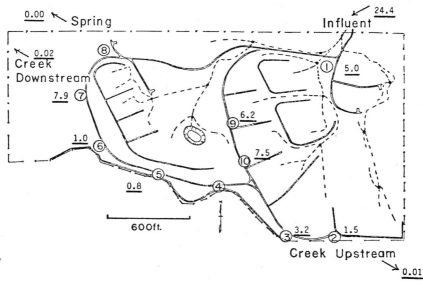

Figure 9.   Average ammonia nitrogen (NH$_3$$^-$) concentrations (mg/l).

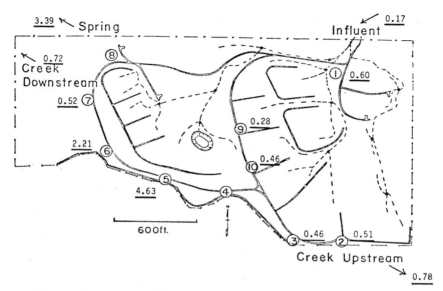

Figure 10. Average nitrate nitrogen (NO_3^-) concentrations (mg/l).

trate nitrogen concentrations in ground water. Appreciable amounts of ammonium were found in some of the perimeter wells indicating incomplete nitrification or that the soil in the site vicinity may be nearing saturation with absorbed ammonia. However, the oxidized nitrogen concentrations observed at the spring suggest that nitrification is occurring in the ground water as it percolates. Nitrate levels comparable to the 3 to 4 ppm observed at the spring in this study have been reported for this sampling location in the past. These concentrations are much greater than those normally found in the base flow of streams in the area. Thus, there is an effect of wastewater disposal on nitrate-N concentrations in the ground water.

The total phosphorus data shown in Figure 11 indicate very effective absorption of phosphorus. Soil analysis showed that total, extractable and soluble phosphorus increased in the treated area compared to the control sites [Figures 12 (i) to 12 (l)]. Total phosphorus increased to almost 4000 μg/g. It may be that the top 12 in. (30.5 cm) of soil are near saturation with respect to phosphorus. Treatment with wastewater apparently caused the soluble phosphorus to increase from 1-2 μg/g to 10-15 μg/g in the top 12 in. (30.5 cm). The soluble phosphorus profile [Figure 12 (k)] suggests that, at depths greater than 20 in. (51 cm), the soluble phosphorus content of the treated soil approached that of the controls. This may indicate that this soil is not yet saturated with phosphate. Leaching of phosphorus is apparently minimal as water samples collected from observation wells contained very low (generally <0.1 mg/l) concentrations of total phosphorus.

The TOC and COD data presented in Table 2 indicate that the percolating wastewater does have some effect on ground water quality at the perimeter wells. The TOC and soluble organic carbon data indicate that the higher levels of organic material were associated with particulate matter. The COD concentrations of 9 ppm and 14 ppm observed at Well 2 and the spring respectively were probably indicative of background levels. COD values of this magnitude may be due to reduced inorganics.

The general distribution pattern for heavy metals was similar to that shown for the sodium and potassium. However, the concentrations were considerably lower. Also, the spring concentrations were equal to or lower than the Well 2 tentative background levels. Soil analysis (Table 3) indicated that heavy metals in the applied wastewater were retained in the top 12 in. (30 cm). Concentrations of copper were as high as 78 μg/g. Soil analysis performed on the top organic layer which had accumulated on a different infiltration bed showed copper contents up to 800 μg/g. Concentrations of zinc, cadmium, chromium and nitrogen were as high as 360, 0.7, 25 and 14 μg/g, respectively. With the exception of copper and zinc, these values are similar to those reported by Bowen (15) for typical soils.

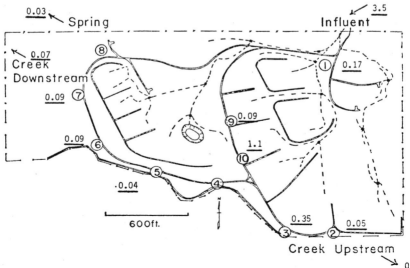

Figure 11.   Average total phosphate concentrations (mg/l).

Water samples from selected wells were examined for the presence and approximate concentrations of total and fecal coliform bacteria. The results are listed in Table 4. The July sample data seem plausible and indicate coliform concentrations that might be reached in a normal surface water. It seems possible that these quantities of organisms could have moved through the soil. However, the extremely high coliform concentrations indicated for the October samples are more difficult to interpret. It may be that an "aftergrowth" of organisms occurred within the observation wells. Water samples taken from the spring generally showed no coliforms. A few spring samples showed apparent total coliform concentrations of 5-13 per 100 ml, but it is likely that these results were due to improper sample collection.

Two samples of grassy vegetation were collected by carefully cutting the vegetation from a 1 sq ft area. One vegetation sample was obtained adjacent to Well 9 and the other was obtained from the northeast corner of the site, an area relatively remote from sewage contact. Each vegetation sample was washed with sterile distilled water, and coliform concentrations were determined on the wash water. The results showed 2.5×10^3 coliforms per gram of dry vegetation for the site adjacent to Well 9 and 7.8×10^3 coliforms per gram of dry vegetation for the site.

In general, the water quality measurements made on samples from the observation wells indicate that the shallow ground water in the proximity of the site does not meet commonly accepted drinking water standards. Concentrations of ammonia and organics were higher than those normally accepted. Although EPA has not set a limit for ammonia in drinking water, the World Health Organization European Drinking Water Standards set a limit of 0.5 mg/l NH_4 (16). Although the concentrations of coliform bacteria were variable, all samples showed values much higher than acceptable for drinking water.

It appears that the controlling ground water quality parameters are bacteriological, forms of nitrogen, and organics, and that additional data are needed.

CONCLUSIONS

Based on the results of this preliminary study, the following conclusions were drawn concerning the Calumet-Laurium wastewater treatment site.

The influent raw sewage contains a large amount of infiltration/inflow and is characteristically dilute. On an average annual basis, the influent volume of raw sewage is 3.5 times the volume of water used by the tributary population.

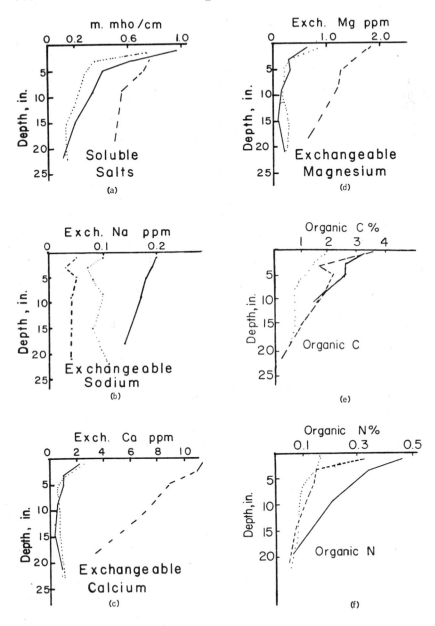

(a) Soluble Salts

(b) Exchangeable Sodium

(c) Exchangeable Calcium

(d) Exchangeable Magnesium

(e) Organic C

(f) Organic N

Key: Forested Control – – – –
        Grassy Control   . . . . .
        Infiltration Bed _____
            (Treated)

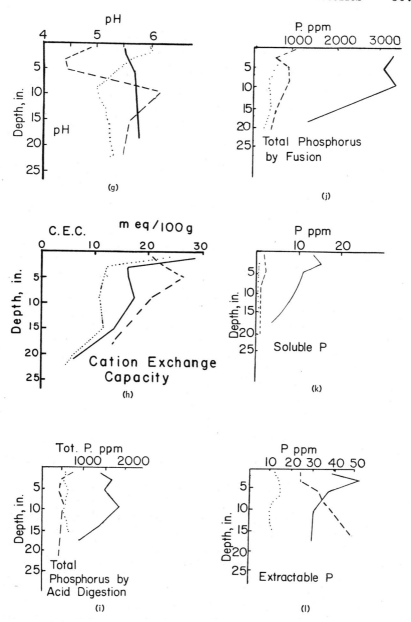

Figure 12. Vertical distribution of soil chemical parameters.

Table 3. Total heavy metals in soils.

Field	Depth (in.)	Concentration (µg/g)				
		Cd	Zn	Cu	Ni	Cr
Forested control	0–2	0.2	54	25	10	10
	2–4	0.2	51	10	7	10
	4–6	0.8	34	10	10	38
	6–12	0.5	16	15	21	34
	12–18	0.4	61	10	13	10
	18–24	0.2	13	16	12	6
Grassy control	0–2	–	38	14	11	4
	2–4	–	40	8	6	5
	4–7	–	–	7	5	6
	7–10	–	13	4	6	10
	10–20	–	35	7	7	10
Infiltration bed (treated)	0–2	0.7	120	78	14	10
	2–4	0.5	360	63	10	14
	4–6	0.5	110	57	7	25
	6–12	–	104	46	14	18
	12–24	0.9	210	19	4	10

Table 4. Bacteriological observations.

Sample	Observation	Well 1	Well 2	Well 9
July	Total coliform per 100 ml	10–100	10–100	100–1000
July	Fecal coliform per 100 ml	1–10	1–10	1–10
October	Total coliform per 100 ml	1000	2×10^5	1.2×10^5

The average annual hydraulic loading rate on the 12 acres (4.85 ha) of percolation area amounts to 3.68 in./day (9.35 cm/day). This is typical of rapid infiltration systems. The maximal monthly and maximal daily hydraulic loading rates occur during the spring snowmelt and are approximately 170 and 220 percent, respectively, of the average. The seasonal flow patterns indicate that ground water infiltration into the collection system is the major contributor to the high maximal monthly flow. The high seasonal flows persist for a period of several months.

Water quality measurements on shallow ground water samples withdrawn from observation wells in the immediate vicinity within 300 ft of the percolation ponds indicated that, in general, the shallow ground water did not meet current drinking water standards. Concentrations of ammonia, organics and coliform bacteria were higher than those normally accepted for drinking water. Water quality measurements made on a spring, 1700 ft (518 m) down slope from the site showed some apparent effects of the percolating wastewater. However, except for a few questionable bacteriological samples, the spring water met accepted drinking water standards.

Concentrations of phosphorus and heavy metals in ground water were low and did not indicate any potential problems. However wastewater application caused soil soluble phosphorus to increase substantially in the top 12 in. (30 cm) which may indicate that this layer is becoming saturated with phosphorus.

ACKNOWLEDGMENTS

This study was partially supported by the U.S. Army Cold Regions Research and Engineering Laboratory under contract DACA89-75-1569 as part of the work unit, "Evaluation of Existing Facilities for Wastewater Land Treatment." Support was also received from the Department of Civil Engineering at Michigan Technological University. The cooperation at the Northern Michigan Water Company is gratefully acknowledged.

REFERENCES

1. Kardos, L. T. et al. "Renovation of secondary effluent for reuse as a water resource," U.S. EPA Publication, EPA-660/2-74-016 (February 1974).

2. Reed, S. et al. "Wastewater management by disposal on the land," U.S. Army Cold Regions Research and Engineering Laboratory, Special Report 171 (1972).

3. Malhotra, S. K. and E. A. Myers. "Design, operation and monitoring of municipal land irrigation systems in Michigan," paper presented at the 1974 WPCF Conference, Denver, Colorado (1974).

4. Bouwer, H., R. C. Rice and E. D. Escarcega. "High rate land treatment I: Infiltration and hydraulic aspects of the Flushing Meadows project," J. Water Pollution Control Fed. 46:834–843 (1974).

5. Bouwer, H., J. C. Lance and M. S. Riggs. "High rate land treatment II: Water quality and economic aspects of the Flushing Meadows project," J. Water Pollution Control Fed. 46:845–859 (1974).

6. Lance, J. C. and F. D. Whisler. "Nitrogen balance in soil columns intermittently flooded with secondary sewage effluent," J. Environ. Quality 1:180–186 (1972).

7. Rice, R. C. "Soil clogging during infiltration at secondary effluent," J. Water Pollution Control Fed. 46:708–716 (1974).

8. Wiitala, S. W., T. G. Newport and E. L. Skinner. "Water resources of the Marquette Iron Range area, Michigan," U.S. Geological Survey Water Supply Paper 1482 (1967).

9. Doonan, C. J., G. E. Hendrickson and J. R. Byerlay. "Ground water and geology of Keweenaw Peninsula, Michigan," State of Michigan Department of Natural Resources, Geological Survey Division, Water Investi-

gation 10, Lansing, Michigan (1970).

10. *Methods for the Chemical Analysis of Water and Wastes 1971*, U.S. EPA, Publication 16020–07/71 (1971).

11. *Standard Methods for the Examination of Water and Wastewater*, 13th Edition, APHA, WPCF, AWWA, New York (1971).

12. Black, C. A., Ed. *Methods of Soil Analysis Parts 1 and 2*, American Society of Agronomy, Madison, Wisconsin (1965).

13. Jackson, M. L. *Soil Chemical Analysis—Advanced Course*, published by author, Madison, Wisconsin (1969).

14. Iskandar, I. K. and D. R. Keeney. "Concentration of heavy metals in sediment cores from Wisconsin lakes," *Environ. Science and Technol.* 8:165–175 (1974).

15. Bowen, H. J. M. *Trace Elements in Biochemistry*, Academic Press, New York (1966), 241 pp.

16. McKee, J. E. and H. W. Wolf. "Water Quality Criteria," 2nd edition, California State Water Resources Control Board, Publication No. 3-A (1963).

27

WASTEWATER REUSE AT
LIVERMORE, CALIFORNIA

A. Uiga, I. K. Iskandar and H. L. McKim
U.S. Army Cold Regions Research and
  Engineering Laboratory, Hanover, New Hampshire

INTRODUCTION

The application of wastewater to land has historically been used as a means of disposal. In addition, some regions of the U.S. have reused wastewater since the 1880s for irrigation purposes. Present practices use land application for multiple purposes including treatment, disposal and reuse. Typically, the application of wastewater has been used on golf course and farmland areas and in other water-deficient areas. Wastewater reuse has also been encouraged as a method of water conservation.

A degree of uncertainty exists when wastewater reuse is considered. The magnitude of the uncertainty is dependent upon numerous factors, including wastewater characteristics, management objectives, site operations and specific site data. Data are required from operational sites to provide design information for use in future land application systems. Some questions arise concerning the fate of applied nitrogen, phosphorus, trace metals, dissolved solids, and possibly pathogenic organisms. The questions of whether land application systems can perform successfully over continued periods of operation and whether land application is safe and environmentally compatible are also of principal concern. Data to answer the above questions are generally not available.

The U.S. Army Cold Regions Research and Engineering Laboratory selected Livermore, California, as an existing site for evaluation. The primary objectives of the study were to compare multiple reuse practices for treated wastewater effluent, and to evaluate short-term (annual) and long-term (8 years) effects of the wastewater reuse options. The Livermore study and others (2–4) are a component of an overall USACRREL research program on land treatment of wastewater.

SITE DESCRIPTION

The City of Livermore is located about 80 km (50 mi) east of San Francisco along Interstate 580 (Figure 1). The area is growing as a result of its proximity to the San Francisco Bay area. Local industry includes the Lawrence Livermore Radiation Laboratory and agronomic interests, notably farming and wineries. The 1974 population was 48,360 with an approximate growth rate of 8 percent since 1960 (5). The average per capita water consumption in the area tributary to the wastewater treatment plant was 490 liters/day (130 gal per day).

The climate is Mediterranean, with hot, dry summers and mild,

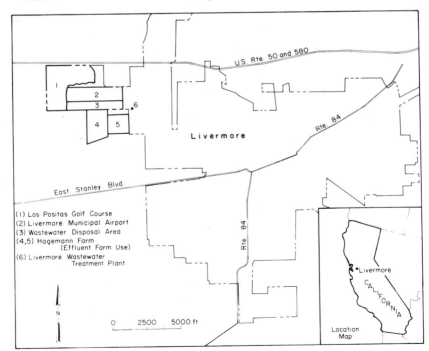

Figure 1.   Location map—Livermore, California.

wet winters. The mean annual temperature is 14.9°C (58.9°F) with a mean daily minimum of 1.9°C (35.4°F) in January and a mean daily maximum of 32.2°C (89.9°F) in July. The mean number of days with temperature $\geqslant$ 32.2°C (90°F) is 59, occurring from May through October. The 30-yr recorded average growing season is 261 days. Annual rainfall averaged 37.3 cm (14.7 in.) from 1951 to 1960, occurring mainly from September through May with less than 1 percent occurring during June, July and August. Records at the treatment site show a total evaporation of 205 cm (81 in.) in 1970 (6).

Water sources in the Livermore area contain high total dissolved solids, high chlorides, and occasional high nitrate concentrations. The total dissolved solids (TDS) of the ground water varied according to location from 500 to over 1000 mg/l due to recharge of sodium chloride water from Altamont Creek; subsurface flow from marine sediments; and, to a limited extent, stream discharge of wastewater treatment plants. Potable water meets a 400 mg/l TDS limit by a blending of higher quality transported water and lower quality local water. Irrigation practices use a high TDS water, whether it is a reclaimed effluent, local ground water, or surface water.

The wastewater treatment system at Livermore has continually expanded to meet the needs of an increasing population. Before 1927 all effluent from the rural community was treated in a septic tank and disposed on a sewage farm. Primary treatment (by Imhoff tanks) was used from 1927 to 1957 with effluent disposal on percolation beds. In 1957, the treatment plant was relocated on the edge of town in a farm area. The design provided secondary treatment to a domestic wastewater flow of 0.11 m³/sec (2.5 MGD). Effluent was discharged to an intermittently flowing drainage ditch.

In 1967 the plant was upgraded to provide an increased degree of treatment for an average dry weather flow of 0.22 m³/sec (5 MGD). The present plant includes preliminary treatment, trickling filters, activated-sludge secondary treatment, and pre- and postchlorination facilities. Since 1967, part of the effluent has been used on a seasonal basis to irrigate a municipal golf course, an airport, and an adjacent farmland. Effluent limitations (Regional Water Quality Control Standards) have been 20 mg/l BOD_5, 20 mg/l suspended solids, 1 mg/l grease, and 5 Most Probable Number (MPN) per 100 ml coliforms during the study period (7).

The raw wastewater can be classified as medium strength in BOD_5 (216 mg/l) and suspended solids (233 mg/l), but high in total dissolved solids (TDS 723 mg/l) and chlorides (163 mg/l)

(8). The treated wastewater effluent is low in BOD$_5$ (8.5 mg/l) and suspended solids (17.5 mg/l), but higher in chlorides (180 mg/l) and TDS (788 mg/l) than the raw wastewater. Other wastewater parameters of concern in land application were total nitrogen (24.9 mg/l), nitrate nitrogen (22.8 mg/l), potassium (11.4 mg/l), total phosphate phosphorus (15.0 mg/l), and boron (1.2 mg/l) (9). The average electrical conductivity was 1110 micromhos/cm, and the sodium adsorption ratio (SAR) varied from 3.7 to 7.5 with an adjusted SAR from 4.9 to 11.6 (10). The wastewater effluent was classified as a high-saline and low-to-medium-sodium hazard water according to Richards (11). Some trace elements were analyzed in the treatment plant effluent, but the concentrations of arsenic, cadmium, chromium, lead and selenium were less than the maximum allowable levels for the use of continuous irrigation (9, 12).

The wastewater effluents from the Livermore Wastewater Treatment Plant can be diverted to four locations: (a) Arroyo Las Positas by discharge through outfall; (b) a golf course by spray irrigation; (c) an adjacent farm by flood irrigation; and (d) airport grounds by spray irrigation. Two operational constraints are imposed for wastewater reuse by the State of California: zero runoff and the bacterial quality (5 MPN/100 ml) for reuse must be met (6).

The soils of the studied area are classified as Yolo-Pleasanton and Clearlake-Sunnyvale Association (13). The Yolo loam and Sycamore silt loam have a moderate permeability rate and few restrictions that limit their use and are ideally suited to agricultural use. Several soils that do have limitations because of either a compact subsoil or gravelly texture are the Livermore gravelly loam, Sycamore silt loam over clay, and Yolo gravelly loam. The Sunnyvale clay loam composes most of the soils found on the golf course. It occurs on recent floodplains and has moderate limitations due to excess water at 1.5 m (5 ft) of the soil surface. The Diablo clay has severe limitations and would require the proper selection of plants, special conservation practices or both in order to be used in a land treatment.

Golf Course

Since 1967 wastewater has been sprayed on the golf course from a series of four interconnected storage lakes. The nightly application of about 1.9 x 10^6 liters (500,000 gal) (0.94 cm or 0.37 in.) is sprayed upon half of the 40.5 wetted hectares (100 acres) of turfgrass. The wastewater irrigation contains the following nutrients.

Parameter	Average Value (mg/l)	Weekly[a] Application (per ha)	Seasonal[b] Application (per ha)
Water volume	—	8.2 cm	254 cm
Nitrogen	24.9	8.2 kg	254 kg
Phosphorus	15.0	4.9 kg	152 kg
Potassium	11.9	3.9 kg	121 kg
Total dissolved solids	788	261 kg	8091 kg
Boron	1.2	0.4 kg	12 kg

[a]Based on application every other day.
[b]Based on 31-week application season, April to October.

The spray schedule for the Diablo clay, Sunnyvale clay loam, and Sycamore silt loam requires high-frequency, small volume application based on water needs of the turfgrass and daily access to the site. The turfgrass is a mixture of Kentucky Bluegrass, Creeping Red Fescue, Highland Colonial Bentgrass, and Seaside Creeping Bentgrass. The golf course manager sends samples for soil testing and analysis on a regular basis. These tests have been used to determine the need for soil amendments.

Farm Area

Approximately 18.2 ha (45 acres) of an adjacent farm field receives from 61 to 94 cm (24 to 36 in.) of effluent per season (111 to 164 x 10^3 m³ in total application) applied by flood irrigation. The treated wastewater was supplied free of charge to the farm field; however, a contract is presently being negotiated to cover annual costs which are expected to be $6,000 to $7,000 an irrigation season. The seasonal application of effluent provides 152 to 228 kg N/ha, 92 to 138 kg P/ha, and 73 to 109 kg K/ha. The wastewater is approved for reuse by the State of California on the row crops of safflower, sugarbeets, beans, cucumbers and squash.

Airport Area

The 22.3-ha (55-acre) airport area adjacent to the runways receives wastewater from April to October. Previous operations have grown alfalfa; however, present practices grow barley and Sudangrass. The wastewater is applied on a daily basis for 2 to 4 weeks at a rate of approximately 1.3 cm/day (2.6 x 10^6 1/22.3 ha) until crop growth is complete. The soil is allowed to dry which permits equipment access and harvest. The cycle is repeated for 2 to 3 cuttings per season. In 1975 a revenue of $17,000 was obtained for the municipality-run operation (6). Soil samples were not collected at this area.

EXPERIMENTAL PROCEDURES

The purpose of this study was to document changes that occurred in the soils based on data obtained from current site investigation and site history. Information about wastewater application to the site was more difficult to obtain because of a general lack of records, changes in management practice, and variations in the quantity and composition of the wastewater. However, a comparison was made between the treated wastewater soil and a "control" site that had not received wastewater, had the same soil type, was adjacent to the treated area(s), and had a fairly well-known history.

At Livermore, soil cores were taken at 6-in. intervals to a depth of 90 cm (36 in.) from five sites (Figure 2). The treatment sites included a clay (GC1) and a clay loam (GC2) site on the golf course, a clay control area, a gravelly loam farm field (AR1) irrigated with effluent and a loam control area. Samples were not taken from the airport area. A composite sample from at least three cores was taken from a 15- or 30-cm (6- to 12-in.) interval as specified. All samples were collected during July 1975.

The soil samples were air dried (25°C), thoroughly mixed and passed through a < 2-mm sieve. Soil pH (1:1 wt/vol), soluble salts (measured as specific conductance), cation exchange capacity, exchangeable Ca, Mg, Na, K, organic carbon, organic nitrogen, free iron oxides, organic-P, total-P (acid digestion), exchangeable-P, and soluble-P (1:2 soil to water) were determined according to Black (14). Heavy metals were extracted and subsequently determined by atomic adsorption spectrophotometry (15).

RESULTS

The exchangeable sodium percentage (ESP) of the surface soil (GC1) from the golf course was 11.4 compared with a control value of 1.6 (Figure 3). The soil of the second treated site (GC2) also showed an increase in ESP to 8.5 even though it received less wastewater due to the location of the site at the edge of a fairway. The ESP values of the treated sites decreased sharply with depth, and at 74 cm (29 in.) values for the treated sites were less than the control. The increase in ESP is important because soil alkalinity may occur at values approaching 15 (11). The maximum ESP observed in the treated soils was less than the equilibrium value for the adjusted sodium adsorption ratio (SAR) of the applied wastewater (10,11). The ESP for the agricultural control and effluent reuse site (AR1) did not show the same large increase as the treated golf course soil (GC1 or

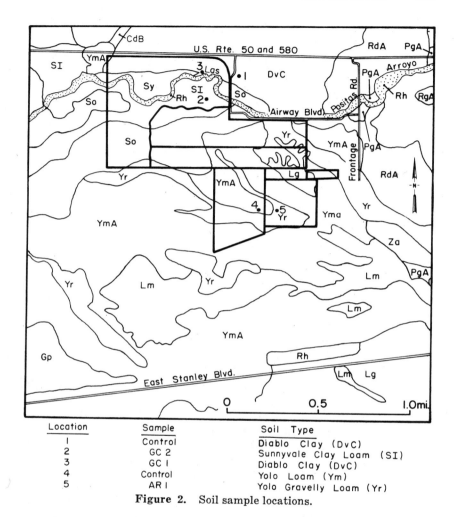

Location	Sample	Soil Type
1	Control	Diablo Clay (DvC)
2	GC 2	Sunnyvale Clay Loam (SI)
3	GC I	Diablo Clay (DvC)
4	Control	Yolo Loam (Ym)
5	AR I	Yolo Gravelly Loam (Yr)

Figure 2.   Soil sample locations.

GC2) (Figure 3). The slight increase from a control of 2.5 to
3.5 in the surface soil of AR1 was probably due to better soil
drainage, larger application volumes, and less frequent applications. Exchangeable sodium percentage decreased from 8.5 (March
1975) to 3.5 (July 1975) as a result of the addition of gypsum
to the land (16).

The soluble salt content of the soil samples was generally low
for all tested sites except the treated site (GC1) on the golf
course (Figure 4). This treatment site received the largest quantity of wastewater and showed an increase to 2.5 μmhos/cm as

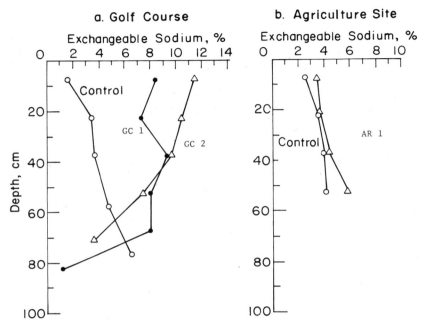

Figure 3.   Exchangeable sodium percentages for golf course and agricultural sites.

compared with values of $\leqslant 0.6$ μmhos/cm at the other sites. The increase for the GC1 treatment site agrees with the (observed) increase in ESP, indicating the dominant salt is sodium.

Total phosphorus in the soil increased above control values for both the treated sites (GC1 and GC2) in a manner consistent with observed application quantities and as observed by other investigators (3, 15). One of the treated sites (GC1) showed an increase of total phosphorus to 2300 mg/l as compared to a control value of 400 mg/l. Surface accumulation occurred in both of the treated sites (GC1 and GC2) (Figure 5). The GC1 values at a depth of 74 cm (29 in.) were larger than the control values (1500 mg/l as compared with 300 mg/l, respectively). For GC2, similar phosphorus values were obtained at below 29 in. for the treated and the control sites.

The treated soil (AR1) at the agriculture site had greater P values than the control in the surface horizon, but with depth the reverse was found (Figure 4). The increase of 350 mg/l on the agriculture site was considerably less than that observed for the golf course site (GC1 and GC2).

The soluble phosphorus concentration increased to 30 ppm

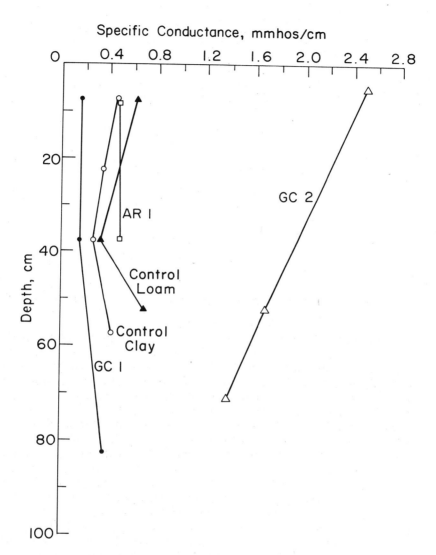

Figure 4.   Soluble salts for golf course and agricultural sites.

in the upper 30 cm (12 in.) at the golf course treated sites (GC1 and GC2) (Table 1). The values below 30 cm (1 to 3 ppm) were similar to the control values (3 ppm). The agricultural reuse site (AR1) showed an increase in soluble phosphorus from 5 to 30 ppm in the upper 15 cm (6 in.). An increase was noted at

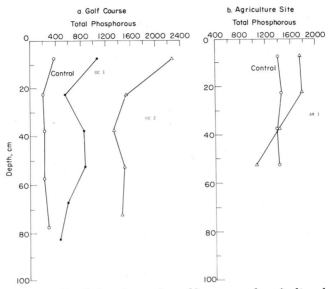

Figure 5.   Total phosphorus for golf course and agricultural sites.

60 cm (24 in.) for the treated site, AR1 (12 ppm compared with a control value of 3 ppm).

The pH of the soil at the golf course sites was different from that of the control (Figure 6). One treated site, GC2, showed an increase in soil pH of 0.2 to 0.6 pH units (from 6.4 to 7.0) over the control, while the other treated site, GC1, showed a decrease of 0.7 to 1.4 pH units (from 6.6 to 5.4). The agriculture treated and control sites showed only minor changes.

The organic carbon contents in the soil profile of the treated golf course sites, GC1 and GC2, were greater than the values observed in the control (Figure 7). The treated and control sites showed a maximum organic carbon content in the surface horizons (upper 10 cm); however, the treated sites (GC1 and GC2) showed an increase with depth from 20 cm downward. The golf course control site showed the opposite trend. The agriculture treated and control sites showed similar trends with the soil organic carbon decreasing with depth. The agriculture control showed larger values than those from the treated site (AR1) except for the surface soil horizon.

An increase of free iron oxides was observed for the treated site (GC1) compared with the control site (Figure 8). The treated site (GC1) showed an accumulation of free iron oxides from 40 to 60 cm, but the values at the 70- and 80-cm depths were

Table 1. Selected soil chemical data.

	Depth (cm)	Free Iron Oxides (%)	Total P Acid (ppm)	Organic P (ppm)	Exch P (ppm)	Soluble P (ppm)	Organic N (%)	C:N
			Golf Course Spray Area					
Control	0–15	0.16	400	175	20	3	0.09	9.6
	15–30	0.16	400	127	18	3	0.09	5.2
	30–45	0.24	350	100	18	3	0.08	5.3
	45–70	0.15	350	125	21	7	0.07	5.0
	70–86	0.16	450	25	21	3	0.07	3.1
Treated, GC2	0–15	0.24	1150	425	37	40	0.17	6.6
(clay loam)	15–30	NSS [a]	950	275	23	19	0.10	5.8
	30–45	0.24	800	175	21	3	0.11	5.6
	45–60	0.24	700	225	21	1	0.09	6.9
	60–84	0.24	650	275	53	1	0.16	4.3
Treated, GC1	0–15	0.47	700	175	92	33	0.20	5.4
(clay)	15–30	0.46	400	150	63	14	0.12	3.4
	30–45	0.64	350	NSS	54	8	0.09	5.2
	45–60	0.71	450	NSS	53	2	0.12	4.7
	60–75	0.32	400	NSS	53	1	0.15	4.2
	75–90	0.08	350	175	32	3	0.18	5.3
			Agriculture Flood Reuse					
Control	0–15	0.32	900	250	24	5	0.12	5.6
	15–30	0.40	750	200	26	5	0.12	5.9
	30–45	0.32	900	225	20	2	0.11	5.5
	45–60	0.33	750	275	20	3	0.09	7.0
Treated, AR1	0–15	0.23	900	300	33	31	0.25	3.0
	15–30	NSS	950	475	25	26	0.12	4.8
	30–45	0.24	900	100	22	20	0.04	11.8
	45–60	NSS	700	75	17	12	0.10	4.6

[a] NSS = not sufficient sample.

similar to those of the control site and the other treated site (GC2).

The total concentrations of heavy metals in soils from the golf course and agricultural site are summarized in Tables 1 and 2. There was no apparent trend for enrichment of heavy metals, probably because the wastewater is strictly municipal with no industrial discharge. All the values of heavy metals that were observed are within the ranges normally encountered for agricultural soils (18).

DISCUSSION

The observed changes in soil conditions in regard to salinity, cation balance, phosphorus and organic carbon can be explained

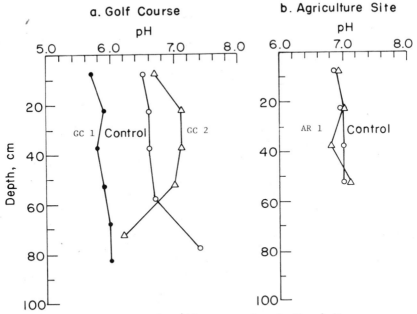

Figure 6.   pH for golf course and agricultural sites.

by existing agronomic practices. The change in soil pH cannot be fully explained on the basis of available data. A possible explanation may be related to organic matter residue; however, no conclusion is stated in this paper.

One of the principal concerns in any land treatment system is the fate of the applied nitrogen. The behavior of nitrogen is largely unexplained for any of the Livermore reuse sites. The golf course operates with essentially little percolation during the dry summer months. This was apparent from the lack of success in obtaining water samples at 0.3-m (1-ft) and 0.6-m (2-ft) depths using suction lysimeters. The influence on the ground water, which varies from 9.8 m (32 ft) to 23 m (75 ft), was not determined. However, the soil organic nitrogen of the treated sites increased in the surface 25 cm and also in the 50- to 80-c zone (Table 1). Similar distribution patterns occurred for organic nitrogen and organic carbon in the soil profiles.

The problems of soil salinity are normally encountered in semi-arid and arid regions where water needs exceed available rainfall. The salinity problems that result from the application of waste-water effluent to the land have been reported (19) as well as methods to compute the magnitude of the problem for variable

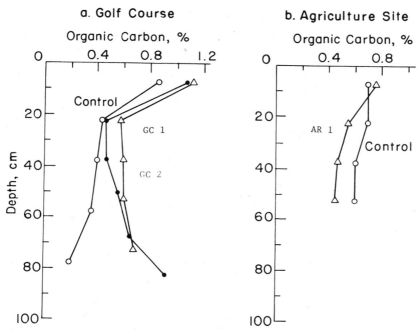

Figure 7.   Organic carbon for golf course and agricultural sites.

water quality and management practices (20). A decrease in soil salinity and an increase in crop productivity can be achieved by less frequent application and larger application volumes (21). Salinity problems occur on the golf course soils due to contradictory site requirements. It is not possible to allow daily access to a dry site for recreation and still maintain the required leaching volumes to remove the salt build-up resulting from reuse of the high-TDS water. The low permeability of the clayey soil results in restricted percolation. The remainder of the golf course, farm area, and airport area do not have the same problems because site access is not critical and soil permeability is not restricted.

An increase in total phosphorus concentration for the treated golf course sites (GC1 and GC2) indicated a phosphorus build-up throughout the soil profile. The concern for the treated site (GC2) arises because the values at the 75-cm depth are considerably larger than those of the control. Additional sampling would be necessary below a 75-cm depth to evaluate phosphorus build-up with depth. The fairly uniform values from 40 to 75 cm may indicate that background total P levels were higher for the clay

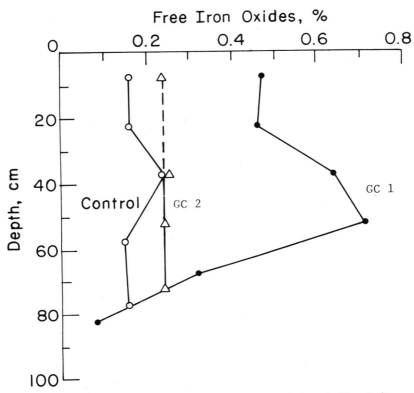

Figure 8.   Free iron oxides for golf course and agricultural sites.

loam soil (GC2) than for the clay golf course control. If this is the case, the increased total P values at the 70-cm depth may not be indicative of a phosphorus build-up but may reflect *in situ* phosphorus conditions.

The soluble phosphorus content of the soil profile may be related to the phosphorus content in the percolate. Below 30 cm, the golf course sites and control site exhibited similar values. The agricultural site showed increased values of soil soluble P with depth (to 60 cm).

Trace metals have not been a problem because the treated effluent levels are below recommended maximum irrigation limits (Table 3). The soil analyses show concentrations typical of agricultural soils for all the test sites (Table 2).

The increase in organic carbon for the treated golf course sites can be attributed to the turfgrass cutting procedure. The cuttings remain on the site for decomposition under high tempera-

Table 2. Total heavy metals in soils.

Location	Depth (cm)	Cr	Cu	Zn (μg/g)	Mn	Ni
Golf course control	0–7.5	45	14	29	550	42
	7.5–15	48	16	27	640	42
	15–30	47	15	28	640	42
	30–45	53	19	28	700	52
	45–70	56	20	28	690	52
	70–85	49	22	37	630	52
Treated, GC2	0–5	62	46	69	770	96
(clay loam)	5–10	61	44	66	760	103
	10–15	64	44	65	770	102
	15–30	60	44	64	760	104
	30–45	76	45	67	820	113
	45–60	62	47	66	790	110
	60–85	58	43	77	640	83
Treated, GC1	0–5	51	28	40	600	42
(clay)	5–10	54	20	28	950	42
	10–15	55	22	27	610	42
	15–30	51	19	27	870	42
	30–45	45	19	31	630	42
	45–60	50	21	35	770	57
	60–75	51	19	29	660	47
	75–90	55	21	29	780	47
Agriculture control	0–15	103	40	45	850	130
	15–30	105	42	48	940	135
	30–45	108	46	50	970	140
	45–60	109	44	44	1090	135
Agriculture treated, AR1	0–15	78	57	77	840	143

Table 3. Trace element analysis—Livermore effluent (City of Livermore, 1968–1974).

	1968	1969	1970	1971[a]	1972[a]	1973[a]	1974[a]	Irrigation[b] Limit
				(values expressed as mg/l)				
Arsenic	<0.01	0.001	0.0015	0.0012	0.002	0.001	0.002	0.1
Cadmium	<0.01	0.0015	<0.01	<0.01	0.002	0.001	0.001	0.01
Hexavalent Chromium	<0.0014	0.015	<0.01	<0.01	0.008	0.002	0.006	0.1
Lead	<0.03	0.035	0.07	0.06	0.043	0.022	0.007	5.0
Selenium	0.0035	0.0055	<0.001	0.0023	0.003	0.005	0.001	0.02

[a] Three-month preserved composite.
[b] National Academy of Sciences.

tures and moisture conditions which would result in high organic carbon in the surface soil.

The change in soil pH is not consistent and cannot be explained on the basis of the available data. A decrease in soil pH as a result of nitrification has been observed for sites where the wastewater nitrogen is applied in the ammonium form (22). However, the Livermore effluent is being applied in the nitrate form. An increase in exchangeable sodium ions should not show as both an increase and a decrease in soil pH (14). Although it is believed that microbial denitrification can raise pH (23), insufficient data are available to verify or deny this hypothesis. The conditions favorable for denitrification to occur at the site include organic carbon throughout the soil profile, warm air temperatures, moist soil conditions from frequent wastewater application, and nitrogen applied in the nitrate form.

Occasional problems of plant toxicity from boron occurred on the golf course. The high concentrations (1.2 mg/l) and insufficient leaching of the clay soil were probably responsible for this effect, although verification of this hypothesis was not undertaken.

In general, the use of a high salinity water for irrigation at the Livermore sites cannot be avoided. The treated wastewater effluent had a slightly higher TDS than the existing ground water. The wastewater effluent that is not applied to the golf course, airport area, and farmland is being discharged to Arroyo Las Positas for recharge of the Niles Cones ground water aquifer. Salinity problems occurred on the clay soil portion of the golf course as a result of conflicting requirements for salt leaching and recreational access.

Alternative Stream Discharge

The disposal of wastewater by discharge to surface water instead of application to the land was investigated by evaluating the seasonal changes in water quality in the Arroyo Las Positas stream. The Arroyo Las Positas (Figure 2) is an intermittently flowing stream with observed total dissolved solid (TDS) concentrations above the Livermore treatment plant outfall varying from 200 to 1700 mg/l with an average TDS of 623 mg/l. Wastewater discharge to surface water increased total dissolved solids by 200 to 300 mg/l throughout most of the year as shown in Figure 9. The exception occurred from January to March when the TDS of the stream increased sharply (as high as 1700 mg/l) due to rainfall-induced overflowing of salt marshes in the tributary foothills. During this period, the wastewater dil-

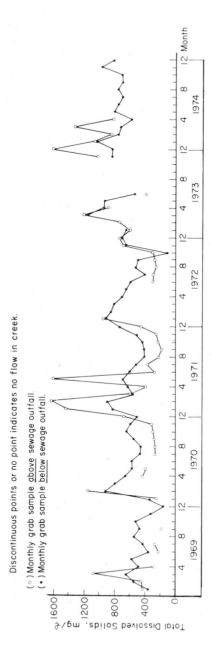

Figure 9. Total Dissolved solids—effects of wastewater discharge to Arroyo Las Positas.

uted the TDS. The overall effect was a slight increase (from 622 to 646 mg/l) in TDS when yearly comparisons were made. The wastewater effluent provided the only stream flow during frequent periods of high water demand and low rainfall.

A second environmental effect results from the requirement imposed by the Regional Water Quality Board that the effluent should have a median number of coliform bacteria not to exceed 5 MPN per 100 ml. This requirement is being satisfied by disinfection of the effluent with chlorine to a residual concentration of 8.5 to 17.5 mg/l (Table 4). The applied chlorine dosage

Table 4.   Chlorine usage at Livermore, California.

Year	Daily Flow (m^3/sec)	Annual Flow (m^3)	Annual Chlorine Use (metric tons)	Calculated Dose (mg/l)	Chlorine Residual (mg/l)
1969	0.16	4.9×10^6	105	21.1	8.5
1970	0.14	4.5×10^6	127	28.1	8.8
1971	0.15	4.7×10^6	169	36.4	13.9
1972	0.17	5.4×10^6	182	33.8	15.9
1973	0.16	5.1×10^6	193	37.8	16.7
1974	0.18	5.7×10^6	221	39.1	17.5

to achieve this residual concentration has varied from 21.1 mg/l to 39.1 mg/l. While the dosage is extremely effective in disinfection, as illustrated by the coliform count and no incidents of effluent-related disease, the discharge of large quantities of chlorine has been shown to be harmful to the receiving water and crop growth (12,24). A frog and fish kill was observed in the summer of 1974 for the last two holding lakes on the golf course. The exact cause of the kill was not determined but the high level of chlorination can be considered a contributing factor. Present plans at the treatment plant include installation of dechlorination facilities for the effluent.

CONCLUSIONS

The seasonal reuse of treated wastewater effluent has continued successfully for 8 years on a farmland, an airport irrigation area, and most portions of a golf course. The salt build-up on the soil and increase in exchangeable sodium percentage with reuse on the clay soil require salinity control management and soil testing for successful operation. Salinity problems occur on the clay soil portions of the golf course. Poor soil infiltration and high salinity effluent create contradictory requirements of daily site access and insufficient application volume for leaching. A golf course location under similar conditions, *i.e.,* clay soil, high-TDS water and climate, is not recommended.

The accumulation of wastewater components (organic carbon, nitrogen, phosphorus and heavy metals), except salts in the soil do not represent a potential problem which would prevent continued effluent reuse. Salinity problems are unavoidable, since the wastewater, local groundwater, and surface water all contain high total dissolved solids.

SUMMARY

Wastewater reuse occurs at Livermore, California by application of treated effluent to a golf course, to a farmland, to an airport area and to a stream. Salinity problems occurred on the clay soils of the golf course because requirements for daily site access and wastewater application were contradictory. The effluent was successfully reused at the agriculture site and disposal area. The outfall discharge increased the total dissolved solids of the receiving water and discharged large quantities of chlorine. Soil chemical analysis showed that exchangeable sodium percentage, total phosphorus, soluble phosphorus, pH, and organic carbon were changed but not critically by effluent reuse. The changes, except in pH, could be explained by existing agronomic techniques for irrigation in a semi-arid climate.

ACKNOWLEDGMENTS

The authors would like to acknowledge the assistance given by the following in the preparation of this chapter: Mr. Daniel Lee and Mr. William Loftin of the City of Livermore, California, for their unpublished data; Mr. August Hagemann for his comments on effluent reuse; Mr. Sherwood C. Reed for technical review of the paper; Mr. Thomas F. Jenkins for technical review of the paper and assistance in collecting soil samples; and Mr. Robert S. Sletten for helpful comments during the report writing. This project was funded by the Office of the Chief of Engineers, under the Civil Works Research Unit 31280, *Evaluation of Existing Facilities for Wastewater Land Treatment.*

REFERENCES

1. Sullivan, R. H., M. M. Cohn and S. B. Baxter. "Survey of facilities using land application of wastewater," Office of Water Programs, U.S. Environmental Protection Agency, Washington, D.C. (1973).
2. Baillod, C. R., R. G. Waters, I. K. Iskandar and A. Uiga. "Preliminary evaluation of 88 years rapid infiltration of raw municipal sewage at Calumet, Michigan," presented at the 8th Annual Waste Management Conference, Rochester, N.Y. (April 28–30, 1976).
3. Murrmann, R. P. and I. K. Iskandar. "Land disposal of wastewater: case studies of existing systems at Quincy, Washington, and Manteca, California," presented at the 8th Annual Cornell Waste Management Conference, Rochester, N.Y. (April 28–30, 1976).
4. Satterwhite, M. B. and G. L. Stewart. "Evaluation of an infiltration percolation system for final treatment of primary sewage in a New England environment," presented at the 8th Annual Cornell Waste Management Conference, Rochester, N.Y. (April 28–30, 1976).
5. Lee, D. Personal communication, Director of Public Works, City of Livermore (1975).
6. Loftin, W. Personal communication, Sewage Treatment Plant Superintendent, City of Livermore (1976).
7. U.S. Environmental Protection Agency. "Upgrading existing wastewater treatment plants—case histories," Technology Transfer (1974).
8. Metcalf and Eddy, Inc. *Wastewater Engineering,* McGraw-Hill Book Company, New York (1972).
9. City of Livermore. "Monthly average of weekly 24-hour composite samples," unpublished data (1969–1974).
10. University of California. "Water quality—guidelines for interpretation of water quality for agriculture," January 15, 1975, University of California, Committee of Consultants (1975).
11. Richards, L. A., Ed. "Diagnosis and improvement of saline and alkali soils," Agriculture Handbook No. 60, U.S. Department of Agriculture (1954).
12. National Academy of Sciences—National Academy of Engineering. "Water quality criteria," U.S. Government Printing Office (1972).
13. Anon. Soil Survey—Alameda Area, California, Soil Conservation Service, U.S. Department of Agriculture (1966).
14. Black, C. A., Ed. *Methods of Soil Analysis,* Agronomy Series No. 9, American Society of Agronomy, Madison, Wisconsin (1965).
15. Iskandar, I. K. "Urban waste as a source of heavy metals in land treatment," *Proc. International Conference on Heavy Metals in the Environment,* October 27–31, 1975, Toronto, Ontario, Canada (1976).
16. Hagemann, A. Personal communication, cooperating farmer, Livermore, California (1975).
17. Johnson, R. D., R. L. Jones, T. D. Hinesly and D. J. David. "Selected chemical characteristics of soils, forages, and drainage water from the sewage farm serving Melbourne, Australia," Report to the U.S. Army Corps of Engineers (1974).
18. Bowen, H. J. *Trace Elements in Biochemistry,* Academic Press, New York (1966).
19. Peterson, H. B. "Salt buildup from sewage effluent irrigation," *Proc. Municipal Sewage Effluent for Irrigation,* Louisiana Polytechnic Institute (1968).
20. Oster, J. D. and J. D. Rhoades. "Calculated drainage water compositions and salt burdens resulting from irrigation with river waters in the western United States." *J. Environ. Quality* 4(1), (1975).

21. Noel, J., *et al.* "Effect of water quality and irrigation frequency on farm income in the Imperial Valley," *California Agriculture* (November 1975).
22. Iskandar, I. K., R. S. Sletten, D. C. Leggett and T. F. Jenkins. "Wastewater renovation by a prototype slow infiltration land treatment system," CRREL Report (in preparation), USACRREL, Hanover, New Hampshire (1976).
23. Alexander, M. *Soil Microbiology*, John Wiley & Sons, New York (1961).
24. Palazzo, A. J., R. S. Sletten and H. L. McKim. "The effects of wastewater disinfection on crop yields and tissue analysis of four forage grasses," *Agronomy Abstracts*, American Society of Agronomy (1975).

28

FEASIBILITY STUDY OF
LAND TREATMENT OF WASTEWATER
AT A SUBARCTIC
ALASKAN LOCATION

R. S. Sletten and A. Uiga
U.S. Army Cold Regions Research and
Engineering Laboratory, Hanover, New Hampshire

INTRODUCTION

Many municipalities and military facilities in arctic and sub-arctic Alaska depend on lagoons for secondary treatment of waste-water. The U.S. Environmental Protection Agency (EPA) has established secondary treatment standards for which compliance will be difficult for the existing systems. Limited effluent data suggest that achieving the 30 mg/l suspended solids and BOD standards is a year-round problem but is especially acute in the summer when extended periods of daylight often lead to high algal concentrations.

The feasibility of land treatment to meet the 1977 secondary treatment standards was investigated at Eielson Air Force Base (AFB) near Fairbanks, Alaska. Three test plots approximately 25 ft square were sprayed with aerated lagoon effluent at the rate of 2, 4 and 6 in. per week. The undisinfected lagoon effluent was taken from the first cell of a two-cell aerated lagoon to in-sure worst case conditions. Weekly water quality samples were taken from the applied lagoon effluent, from catchment lysim-eters at a depth of 6 in. below the surface of the test plots, and from well points approximately 4.5 ft deep. Weather and soil data were also collected.

Data collected during two summers of effluent application (1974 and 1975) indicated that the use of land treatment to achieve secondary treatment criteria appears feasible at Eielson AFB.

Wastewater treatment in arctic and subarctic Alaska has always been a difficult undertaking. In addition to the long periods of below freezing air temperatures, many of the municipalities and military installations in northern and interior Alaska face the added difficulties of a remote location, lack of skilled personnel, and limited financial resources for routine operation and maintenance of a treatment system. Recently, however, the normal difficulties associated with wastewater treatment and disposal in Alaska have been magnified by the requirement to meet EPA-defined secondary treatment standards by July 1, 1977. Secondary treatment has been defined as meeting the following criteria (1) :

1. *Biochemical oxygen demand, 5-day (BOD$_5$).* The arithmetic mean of the values for effluent samples collected in a period of 30 consecutive days shall not exceed 30 mg/l.
2. *Suspended solids (SS).* The arithmetic mean of the value for effluent samples collected in a period of 30 consecutive days shall not exceed 30 mg/l.
3. *Fecal coliform bacteria (FC).* The geometric mean of the value for effluent samples collected in a period of 30 consecutive days shall not exceed 200 per 100 ml.

It is reasonable to question whether or not the uniform application of these standards is the most efficient and realistic way to protect both public health and the environment. At this point, however, there is every indication that the standards will have to be met, although perhaps at a later date than originally scheduled (2). Although treatment plant monitoring data are extremely sparse, indications are that the EPA regulations require a better quality effluent that is normally attained from most treatment systems presently in use in Alaska. Thus it is necessary to investigate the feasibility of methods to upgrade the present facilities to meet EPA standards.

One possible upgrading technique is land application of wastewater. Even though wastewater has historically been applied to the land in other areas of the country (3), present requirements for engineered systems required a further assessment of past practices (4–6). Previous Alaskan practices have relied on the simplest possible disposal methods, moving gradually from raw sewage discharge to simple biological systems, generally lagoons where suitable land is available and extended aeration package plants where it is not. Faced with stringent new effluent standards and the necessity to meet them in an efficient, cost-effective manner, land application may very well prove to be a reasonable treatment technique in certain areas of Alaska.

This chapter will discuss an investigation of the feasibility of land application as a treatment process to meet EPA secondary effluent standards at Eielson AFB, during the summers of 1974 and 1975. This work was performed under the sponsorship of the Military Construction Directorate, Office, Chief of Engineers, Army Corps of Engineers.

FAIRBANKS AREA CLIMATE AND SOILS

Land treatment of wastewater is dependent to a large degree on the suitability of the climate and soils at the prospective treatment site. The following overview of soil and climatic conditions in the Fairbanks area is taken from a 1963 Soil Survey by Rieger *et al.* (7). Eielson AFB, located approximately 30 miles south of Fairbanks, is included in the Fairbanks area (Figure 1), and the following information is considered to be applicable to Eielson.

Climate

The Fairbanks area, located in the Interior Basin of Alaska, has extreme seasonal variations in temperature. Nearly all of the extreme temperatures, ranging from a high of 99°F to a record of 66°F below zero, have been recorded in the Interior Basin. The dates for freeze-free periods shown in Table 1 are a fairly dependable guide for design of a land application system. The data most applicable to Eielson is that at the Fairbanks Airport which shows a potential application period of 3.5 to 4 months. During the months of June, July and August, possible sunshine averages slightly more than 19 hours per day.

Most of the rain in this area falls during the growing season, and the amount may vary greatly within short distances. The frequency and intensity of showers tend to increase as the summer season progresses. The average monthly precipitation is less than 0.25 in. in April but increases to slightly more than 2.25 in. in August. Snowfall averages between 50 and 60 in. per year. Snow is usually on the ground from mid-October to mid-April. The maximum depth of snow on the ground during the winter averages between 25 and 30 in.

Soils

The soils of the Fairbanks area can be classified into two geographical groups, either upland soils or alluvial soils, as shown in Figure 2 (7).

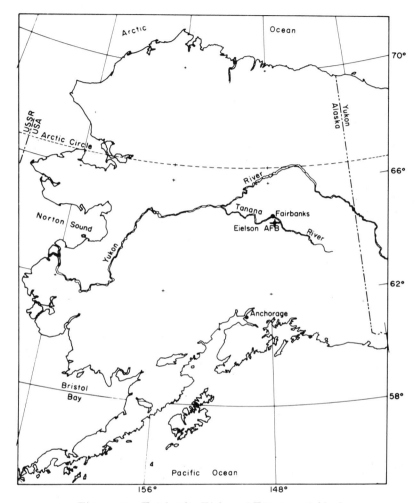

Figure 1.   Fairbanks-Eielson AFB area of Alaska.

Since development in the Fairbanks area (including Eielson AFB) has been predominantly in the alluvial plain, a general description of the soils will be restricted to the alluvial plain. These soils have developed in sandy or silty, water-deposited material. Those near the course of the main streams are generally sand, and permafrost in them is deep or absent. Those away from the streams are silty and have permafrost closer to the surface. Gravel underlies all the silty alluvial soils at a depth ranging from less than 1 ft to more than 50 ft.

Table 1. Average dates, for beginning and end of season, on which temperature is equal to or above the temperature indicated (7).

Temperature	University Experiment Station, College [a]		U.S. Weather Bureau Airport Station, Fairbanks [b]	
	Terminal Dates	Number of Days	Terminal Dates	Number of Days
32° F	May 29–Aug 24	88	May 19–Sep 2	106
28° F	May 18–Sep 9	114	May 7–Sep 13	129
24° F	May 9–Sep 20	134	Apr 29–Sep 24	148
20° F	Apr 26–Oct 1	158	Apr 22–Oct 1	162
16° F	Apr 24–Oct 6	165	Apr 19–Oct 7	171

[a] Data period, 1931 through 1960.

[b] Data period, 1934 through 1960.

The well-drained, sandy Salchaket soils border the principal rivers in the area and are the most extensive soils of the alluvial plains. The silty, imperfectly drained soils occupy large areas between the streams and the uplands. The availability of the alluvial soil in the Fairbanks area as determined from a detailed survey is given in Table 2 (7).

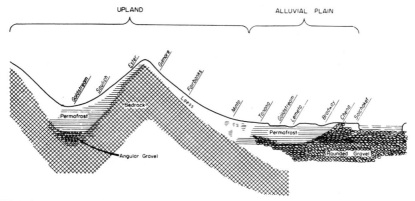

Figure 2. Relationship of soil series, underlying material, and permafrost (7).

The most favorable soil series for a land treatment system in the alluvial plains is the Salchaket series. The Salchaket series consists of nearly level, well-drained soils that have developed in recently deposited water-laid material along the Tanana and Chena Rivers. These soils are dominantly sandy but commonly

Table 2.   Availability of alluvial soil in Fairbanks area (7).

Soil Type	Restriction	Detailed Acres	Percent [a] of Detailed Area
Salchaket	May be droughty	54,737	25.3
Tanana	First year wetness	16,525	7.6
Goldstream	May need drainage	37,623	17.4
Bradway	May need drainage	12,182	5.6
		121,067	55.9

[a] Total area of detailed survey is 216,471 acres.

contain layers of silty material. They are underlain by thick deposits of coarse sand and gravel.

PRESENT EIELSON AFB WASTEWATER TREATMENT SYSTEM

The wastewater treatment system at Eielson AFB consists of a two-cell aerated lagoon following previously existing conventional primary treatment with sludge digestion. The system is operated and maintained by competent, well-trained civilian and military personnel. System characteristics, summarized in Table 3, show an average flow rate of approximately 1 MGD and retention time of 16 days. The design effluent BOD of 40 mg/l represents approximately an 85 percent reduction in the design influent BOD of 245 mg/l. Figure 3 represents the results of

Table 3.   Characteristics of the aerated lagoon located at Eielson AFB, Alaska.

Characteristic	Predicted
Volume (million gallons)	16.56
Liquid Depth (ft)	9.00 (winter)
	10.00 (summer)
Aeration Rate (ft^3/min)	1,017
Average Flow Rate (gal/day)	1,035,630
Retention Time (days)	16
Influent BOD_5 (mg/l)	245
Effluent BOD_5 (mg/l)	40
Temperature (^oC)	
Maximum	20
Minimum	0.5
Ice Thickness (ft)	1

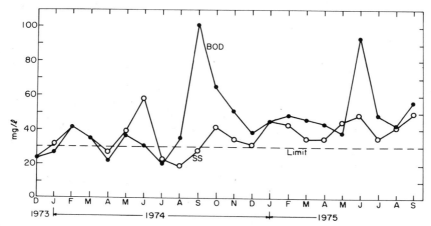

Figure 3. Effluent BOD and SS at the Eielson AFB aerate lagoon.

22 months of effluent sampling and shows that neither BOD nor SS meet EPA secondary effluent criteria. The average effluent BOD for the period of data is about 45 mg/l, only slightly above the design limit of 40 mg/l. Examination of the data and a trend analysis indicated that the effluent quality was deteriorating over the period of data. This suggested that operational changes alone would be unlikely to improve the effluent to secondary criteria and that some upgrading process was in order. Although not clear from the data, one possible cause of the high SS may be summertime algae growth and/or inadequate settling time in the winter.

DESCRIPTION OF PROJECT

A test area consisting of three test plots was located adjacent to the aerated lagoons (Figure 4). Although this area was known to have been disturbed several years earlier during construction of the lagoons, native vegetation similar to that in surrounding undisturbed areas had been reestablished at the time of this study and the test plots were thought to be representative of locally available land. Effluent from the first cell of the two-cell aerated lagoon system was applied to the test plots. This was done to provide worst case conditions regarding algae and solids to evaluate the performance of the receiving land at the airbase. The lagoon effluent was delivered by a submersible pump through ball valves and meters to each individual test plot, and then through distribution nozzles onto the test plot. The nozzles were adjusted

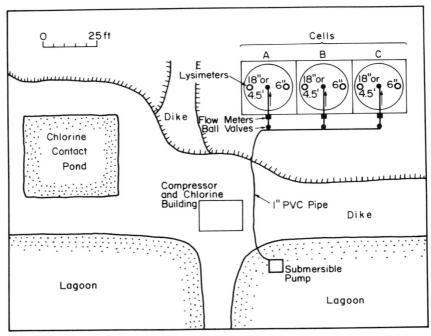

Figure 4.   General plan of land application pilot study at Eielson AFB.

to spray wastewater to a 20-ft diameter area in the center of each 25-ft square plot. Each plot was loaded individually on separate days.

Water quality samples were taken weekly from the applied wastewater and from lysimeters and well points located in each test plot. During the first application season (1974), lysimeters were located 6 and 18 in. below the ground surface in each test plot. Due to problems in collecting an adequate sample volume in the 18-in. lysimeters, they were replaced the second year (1975) with well points approximately 4 ft below the ground surface. The first year, samples were analyzed for TOC, SS and fecal coliform. The second year, the TOC determination was replaced by one for BOD. All analyses were conducted immediately following sample collection using the procedures in *Standard Methods* (8).

The primary purpose of this project was to demonstrate the feasibility of the land treatment process to meet the 1977 criteria. Should it be desired to use land treatment to meet more stringent standards, a considerable amount of additional information would be needed regarding vegetative responses to wastewater application and the chemical and biological interactions of nitrogen and phosphorus in the soil profile.

TEST PLOT SOIL CONDITIONS
AND HYDRAULIC APPLICATION RATES

The general characteristics of the soil in the chosen test area were evaluated by examination of an existing SCS Soil Survey (7), conducting sieve analyses, and a field verification of estimated permeability during the first year of the experiment. The results of the sieve analyses are given in Table 4 for all

Table 4.   Results of test plot soil analyses at Eielson AFB.

Test Plot	Depth (in.)	Sieve Analyses [a]			Classification [c]
		Gravel (%)	Sand (%)	Silt and Clay (%)	
A	0–6	12	44 [b]	44	Gravelly loam
A	6–12	2	36	62	Silt loam
B	0–6	Sample lost			
B	6–12	13	27	60	Gravelly silt loam
C	0–6	4	25	71	Silt loam
C	6–12	4	28	68	Silt loam

Percolation tests: Plot A: 4 in./hr

C: 7.5 in./hr

[a] Minimum effective sieve size, 0.075 mm.

[b] Soil survey maps indicate most small particles can be assumed to be silt rather than clay.

[c] SCS classification.

three test plots. These tests, coupled with the soil classification map for the area showed that the soil can be classified as a silt loam or very fine sand loam of the Salchaket series. For this reason, a slow infiltration mode of land application was chosen as opposed to rapid infiltration (very permeable soils), or overland flow (essentially impermeable soils) (4).

Experimental hydraulic loading rates were chosen to cover the range generally considered to be slow infiltration, that is 2 to 6 in. per week. In addition, consideration was given to operational practice in that the entire weekly hydraulic loading was applied on one day of the week. Design and actual hydraulic loading rates are given in Table 5 for the three test plots over

Table 5. Design and actual hydraulic loading rates (in./wk) for 1974 and
1975 application seasons.

Loading	Plot A	Plot B	Plot C
Design 1974–75	2	4	6
Actual 1974 [a]	3.2	5.1	7.2
Actual 1975 [b]	3.1	4.7	6.2

[a] 8-week application period.
[b] 9-week application period.

both seasons' applications. The discrepancy between design and
actual loading is attributed to the automatic timers used to control the application rates, which were not completely reliable
and required close monitoring to ensure meeting design loadings.
Also, personnel were not always available to provide the necessary monitoring of the timers, pumps and valves. Weather conditions were monitored during both application seasons. Parameters measured included precipitation, temperature, wind speed
and direction, and relative humidity. No extreme or unusual
values were observed.

1974 TEST RESULTS

The results of the first year's weekly applications (August
and September, 1974) are presented in Table 6. Test plot effluents are from the 6-in. level only since an adequate sample volume
could not be obtained at the 18-in. level. During the first year,
it was necessary to substitute TOC analyses for BOD at the
6-in. level for the same reason. Since TOC includes BOD as one
part, a TOC reading of 30 mg/l or less indicates that the BOD
standard has been met.

The tabulated data indicate that both the BOD and SS criteria
were met during the 1974 application period. The SS never exceeded the EPA criterion during the application period, and
the mean TOC was less than 30 mg/l for all three test cells indicating that the EPA BOD criterion was also met. Although the
maximum TOC values exceeded 30 mg/l, experience elsewhere indicates that it is unlikely the corresponding BOD values would
have been greater than 30 mg/l (9). The SS and TOC results do
not appear to be correlated with wastewater application rates.

The EPA fecal coliform standard was not met at the 6-in. level.
Reasons for this are unknown, but it is thought that a greater
depth of soil is required for effective filtering of coliform orga-

Table 6. Results of first year wastewater application to Eielson test plots.

Constituents	Applied Wastewater	Test Plot Percolates (6-in. depth)		
		A	B	C
Total Organic Carbon (mg/l as C)				
Mean	53.6	21.5	24.6	23.0
Maximum	90	47	34	36
Minimum	32	13	10	13
Number	5	6	5	6
Suspended Solids (mg/l)				
Mean	64.6	6.0	3.6	13.0
Maximum	128	10	8	28
Minimum	42	3	2	5
Number	7	5	8	3
Fecal Coliforms (per 100 ml)				
Mean	–	377	10504	–
Maximum	–	880	36000	–
Minimum	–	40	210	–
Number	0	3	5	0

nisms. Since adequate samples were not obtained below the 6-in. level, the question of whether or not a greater depth of soil could remove coliform organisms was not answered during the first year.

1975 TEST RESULTS

During the second year of wastewater application, the 18-in.-deep lysimeters were removed and wellpoints were installed. The wellpoints were positioned so that they just penetrated the surface of the ground water table approximately 4.5 ft below the soil surface. The lysimeters at the 6-in. depth were also dug up, inspected, cleaned and reinstalled. The primary purpose of the second year application was to determine whether or not the fecal coliform criteria could be met and also to confirm the first year BOD and SS data.

Results of the second year water quality analyses for both shallow and deep sampling points are shown in Table 7. Examination of the data indicates that mean BOD was well within the 30 mg/l criterion thus confirming the TOC data collected during the first year. The maximum BOD values of 50 and 89 mg/l for test plot B appear to be an anomaly in the data since they both occurred on the same sampling day and no other samples

Table 7. Results of second year water quality analyses for both shallow and deep sampling points.

Test Plot	Second Year Data 9-Week Application Period	BOD (mg/l) 6 in.	BOD (mg/l) 4.5 ft	SS (mg/l) 6 in.	SS (mg/l) 4.5 ft	FC (P/100 ml) 6 in.	FC (P/100 ml) 4.5 ft
Applied Wastewater	Mean	97		44		3.4×10^5	
	Max	240		59		$>10^6$	
	Min	17		21		1.1×10^4	
	Number	9		9		9	
	Mean	3	5	21	1553	803	407
A	Max	8	17	36	5934	2400	2800
	Min	0	0	11	11	0	0
	Number	7	8	6	8	7	7
	Mean	11	21	11	9912	20,350	7
B	Max	50	89	29	>10000	102000	16
	Min	2	2	4	8	0	0
	Number	9	7	9	7	9	7
	Mean	5	4	18	1539	7656	6
C	Max	18	19	53	9088	36,000	16
	Min	0	0	7	51	16	0
	Number	9	9	8	9	9	9

collected during the application period were greater than 11 at 6 in. or 37 at 4.5 ft. Sampling error is suspected as the cause for the high values, but this cannot be confirmed.

The mean SS was also well within the 30 mg/l criterion at the 6-in. level although somewhat higher than the 1974 values. During both 1974 and 1975, test plot B showed the best performance, but no special significance is attached to this fact. The native ground water was undoubtedly disturbed during installation of the wellpoints, so high SS readings were expected at the 4.5-ft. level. One possible reason the wellpoint samples remained high in SS is that the wellpoints were pumped only once each week during collection of the sample, and the pumping action apparently resuspended the very fine silty material which appeared in the samples. The SS data taken at the 6-in. level generally confirm the 1974 results and indicate the ability of these test plots to meet the EPA SS criterion after passage through 6 in. of soil.

The fecal coliform data for 1975 are very similar to those of 1974 at the 6-in. level in that there was some reduction but the values greatly exceeded the EPA criterion. At 4.5 ft, however, test plots B and C show almost complete removal. Test plot A had one value in excess of 200/100 ml which made the average exceed EPA criteria. All other values were less than 20, however, and the data as a whole for all test plots show effective removal of fecal coliform after passage through 4.5 ft of soil.

The summary of data from both years presented in Figure 5 graphically illustrates the ability of the land application system

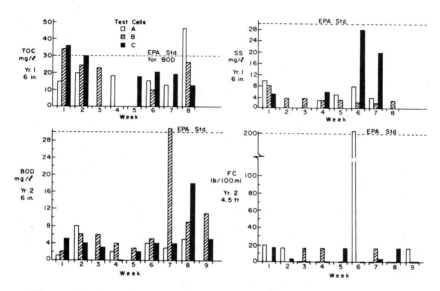

Figure 5.   Summary of results of two years of level application pilot testing at Eielson AFB.

to meet EPA secondary effluent criteria. TOC values are generally less than 30 mg/l, representing even lower BOD values. This is confirmed by the second year BOD data where all values except one are well below EPA criteria. All SS values meet criteria at the 6-in. level during 1974. Average SS during 1975 was slightly higher for all three test plots. All fecal coliform values except one met criteria at the 4.5-ft level. The data taken collectively support conclusions of other studies which have found that BOD and SS are effectively removed in the upper layers of the soil profile, and coliform organisms require additional depth for removal (9, 10).

DISCUSSION AND ANALYSIS

It was expected at the beginning of this study that land application could effectively polish lagoon effluent to a level that would meet EPA secondary treatment criteria. What was not known was whether land application could be adapted to the unique climatic and soil conditions of a subarctic Alaskan location such as Eielson AFB. Although the application periods during the two summers of pilot testing in this study were quite short (8 weeks in 1974 and 9 weeks in 1975), the results must be considered encouraging. Land application of wastewater to meet EPA secondary effluent criteria appears to be quite promising in some sections of Alaska. For the test plots used in this study, there appeared to be no correlation between hydraulic application rate and product water quality. Since the secondary treatment criteria were satisfied with all application rates tested, it would be most economical to design a system for the maximum application rate of 6 in. per week. However, additional field verification of soil permeability would be needed prior to a full-scale design. No evidence of soil clogging at the surface was noted during the application periods, and it is believed that the one day per week application schedule used in this study provides an adequate resting period for the soil between loadings.

Although it has been suggested that a maximum hydraulic loading rate may be used in order to minimize land area required, it should be kept in mind that the results presented here apply to a land application system to meet 1977 criteria. No attempt was made during this study to characterize native vegetation responses to wastewater application other than visual observation. Based on the results of other studies, however, it is felt that nitrogen would be a controlling parameter in a land treatment system to meet more stringent water quality standards. For example, rather than hydraulic loading, the vegetative responses would be a key element in nitrogen removal (9). Although some data were collected on nitrogen forms in the applied wastewater and percolate, they were too sparse to evaluate. A much more rigorous and comprehensive demonstration project would be required to demonstrate in Alaska the feasibility of land treatment to meet higher quality standards.

SUMMARY AND CONCLUSIONS

During two summers of pilot testing at Eielson AFB, Alaska, slow infiltration land application of wastewater was shown to be able to meet EPA secondary effluent criteria. Wastewater application by spray irrigation on three 25-ft-square test plots reduced BOD and SS levels to less than 30 mg/l after passage through 6 in. of soil and fecal coliform bacteria to less than 200 per 100 ml after passing through 4.5 ft of soil at application rates ranging from 3 to 7 in. per week. From these results it is concluded that slow infiltration at a maximum hydraulic loading rate of approximately 6 in. per week could be used to meet secondary effluent criteria at Eielson AFB and in other areas of Alaska with similar soil conditions. Evaluation of land treatment to meet higher water quality criteria would require a more rigorous testing program, particularly with respect to vegetative responses to wastewater application and nitrogen transformations in the soil.

ACKNOWLEDGMENTS

The authors acknowledge the efforts of the following people for their contributions to this research: Mr. T. D. Buzzell, project concept and design and technical review of this paper; Mr. S. C. Reed, helpful suggestions and comments throughout the project and technical review of paper; Mr. J. O'Neill and Dr. D. W. Smith, data collection and analysis; and Mr. A. F. Gidney, experimental facility construction and operation.

REFERENCES

1. U.S. Environmental Protection Agency. Title 40—Protection of Environment, Chapter 1, Subchapter D—Water Programs, Part 133—Secondary Treatment Information, *Federal Register* 38(159):22298 (17 Aug 1973).
2. Reynolds, W. F. "Public works report from Washington," *Public Works* 106(11):14 (November 1975).
3. U.S. EPA. "Survey of facilities using land application of wastewater," EPA 430/9–73–006 (July 1973).
4. Reed, S. C. *et al.* "Wastewater Management by Disposal on the Land" *Special Report 171,* Cold Regions Research and Engineering Laboratory, Hanover, New Hampshire (May 1972).
5. U.S. EPA. "Wastewater Treatment and Reuse by Land Application, Vol. II," EPA–660/2–73–006 b, Washington, D.C. (August 1973).
6. U.S. EPA. "Design seminar for land treatment of municipal wastewater effluents," prepared for U.S. EPA Technology Transfer Program, Washington, D.C. (1975).
7. Rieger, S., J. A. Dement and D. Sanders. "Soil survey of Fairbanks area, Alaska," USDA and Alaska Agricultural Experiment Station, Series 1959, No. 25 (1963).
8. APHA, AWWA and WPFC. *Standard Methods for the Examination of Water and Wastewater,* 13th edition (1971).
9. Iskandar, I. K. *et al.* "Wastewater renovation by a prototype slow infiltration land treatment system," CRREL Report, U.S. Army Cold Regions Research and Engineering Laboratory, Hanover, New Hampshire (March 1976, in press).
10. U.S. Environmental Protection Agency. "Conference on recycling treated municipal wastewater through forest and cropland," EPA 66/2–74–003 (March 1974).

SECTION V

APPLICATION OF SLUDGES TO LAND

29

CHICAGO PRAIRIE PLAN—
A REPORT ON EIGHT YEARS
OF MUNICIPAL SEWAGE SLUDGE
UTILIZATION

C. Lue-Hing, B. T. Lyman, J. R. Peterson and J. G. Gschwind
The Metropolitan Sanitary District of Greater Chicago
Chicago, Illinois

INTRODUCTION

The Metropolitan Sanitary District of Greater Chicago (District) serves the city of Chicago and 120 adjacent suburban communities covering an area of approximately 900 square miles. The district's sewered population of 5.5 million people represents about 70 percent of the Illinois total sewered population. The commercial and industrial population equivalent of 4.5 million brings the total population equivalent served by the district to 10.0 million people. From this residential and industrial complex, the district collects and treats 1.4 billion gallons of sewage every day.

The solids produced from the treatment of the district's sewage for ultimate disposal amounts to approximately 800 dry tons/day.

Prior to 1971, processed sewage solids were either stored in lagoons in liquid form, were heat-dried, and sold in bulk as a dry fertilizer, or were air-dried (Imhoff–digested sludges only), and stored on district property. It is not difficult to understand that in a large urbanized center such as Chicago, storage lagoon space was not limitless. Indeed, additional storage lagoon space is no longer available in or around Chicago. For the three disposal modes mentioned, the problems were multiple and include: for lagooning—no net reduction in volume, no more available land, and occasional complaints of nuisance; for air drying—the problems were no net reduction and no more available land; for heat drying—the problems were a freeze on natural gas allocation and a cost increase for air pollution control and natural gas combined which was almost exponential.

To say that the district was confronted with a solids disposal problem would be the understatement of the sludge age. Faced with this veritable insurmountable and ever increasing problem, the governing board of the district in 1967 charged the staff with developing recommendations for a long-term sludge disposal scheme which would incorporate the following criteria:

1. Eliminate air and water pollution problems caused by other methods of disposal.
2. Conserve and utilize all resource materials for beneficial purposes.
3. Be economical, or at least, be no more costly than other methods.
4. Be permanent, *i.e.*, complete the natural cycle into perpetuity.

Following several years of surveys, feasibility studies, laboratory experimentation and field demonstrations within Metropolitan Chicago and elsewhere in the state, the Board of Commissioners, on the recommendation of staff, established the policy that properly processed municipal sludge was a valuable natural resource and must be utilized. With this policy, the Prairie Plan was born, and implementation began in 1970 with the initial purchase of 7100 acres (28 km²) of partly strip-mined lands in Fulton County, Illinois, approximately 190 miles southeast of Chicago (1–4).

SOIL ENRICHMENT DEMONSTRATION PROJECTS

Many soil enrichment demonstration projects provided valuable scientific and technical information which led ultimately to the decision to launch the Prairie Plan. Those projects conducted within Metropolitan Chicago and Cook County are shown in Figure 1. The more obvious benefits of these projects were technical and scientific. However, their sociological and psychological benefits cannot be underestimated. Their location and management were selected such that they could, and did, serve as an effective barometer of public acceptance. Observations of these aspects will be discussed later.

A brief technical description of some of these projects is presented.

Northwestern University

The problem was to establish a permanent vegetative cover on a former 5-acre sand dune landfill which was to be the site of Northwestern's Norris Cultural Center on the shores of Lake

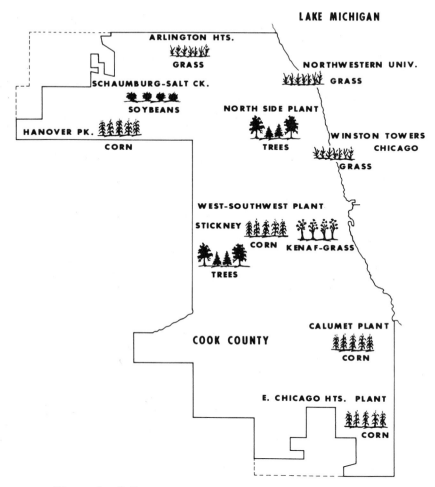

Figure 1.   Soil enrichment projects in Cook County, Illinois.

Michigan in Evanston, Illinois (5). The site is bounded by Lake Michigan and a constructed lagoon. The dune area was divided into two sites, A and B, for sludge application. In 1968, site A received 90 tons/acre of sludge solids (201 Mt/ha), directly from the Calumet digester; and in 1969, site B received 150 tons of sludge solids/acre (336 Mt/ha) from a storage lagoon. The sand surface was topped by 9-18 in. of clay for stabilization. The soil changes in the A horizon of site A the year following digested sludge application are shown in Table 1.

Table 1. Soil characteristics of A horizon at Northwestern University the year after the application of 201 metric tons/ha of digested sludge (Unpublished Data, MSDGC, 1971).

		Sludge Applied (metric tons/ha)	
		0	201
Cation exchange capacity	meq/100 g	24.6	21.9
Bulk density	g/cm^3	1.07	1.05
Field moisture capacity, 1/3 bar	%	28.8	33.7
Organic matter	%	1.9	3.2
pH 1:1 soil/water ratio		7.6	7.2
Soluble salts	mmhos/cm	0.85	0.37
Exchangeable cations	μg/g		
K	"	115	156
Ca	"	3,980	3,043
Mg	"	451	718
Na	"	181	137
Total N	"	1,000	1,600
NO$_3$-N	"	72	128
P-Bray 1 extractable [a]	"	8	58
Cl	"	29	14
B-H$_2$O extractable	"	1.63	1.57
Metals 0.1 N HCl extractable			
Zn	μg/g	1.5	47
Fe	"	108	145
Cu	"	14	14
Mn	"	16	30
Ni	"	137	113
Pb	"	144	95
Cr	"	25	35
Cd	"	18	27

[a] Bray and Kurtz (6).

Although the treated soil was still in a very dynamic state, organic matter, available P, NO$_3$ and total N contents were considerably higher (5). Extensive monitoring of test wells, surface runoff, the lagoon, and Lake Michigan has shown no detrimental effects. This site now supports a lush growth of grasses and evergreens.

Calumet Sanitary Landfill

In 1968, a 135-acre former landfill site, located on district property, east of the Calumet Wastewater Reclamation Plant, was subdivided into various sized plots to receive liquid digested sludge. The landfill, which was formerly part of Lake Calumet, contained fly ash, glass and other debris over a silty clay loam.

It was anticipated that a high treatment with sludge would suf-ficiently reclaim the tract for possible crop use. In 1969, the site received 201 tons/acre of digested sewage sludge (452 Mt/ha). Surface soil samples (0–15 cm) were collected from the site be-fore, and after, 3 months' fertilization with sludge at 12 tons/ acre (25 Mt/ha), the soil contents of available P (6) and ex-changeable K were increased, while the 0.1 N HCL extractable Zn and Cu, the exchangeable Na and Ca, and soluble salts de-creased. Winter wheat ("Balboa") was seeded in September 1969 after an additional 72 tons/acre (161 Mt/ha) of Calumet di-gested sludge. The wheat grew vigorously, was a deep green in color, showed no nutrient deficiency symptoms, and produced 41 bu/ac (47 hcl/ha).

Corn and wheat have been planted every year for the past six years and high loadings of digested sludge have been applied.

There was no crop in 1975. Water samples are taken monthly from three piezometers in the fill area, approximately 10 ft deep and a monitoring well 130 ft deep. The 1975 results for ground water analyses are summarized in Table 2 (7).

These results showed that leachate water samples from the three piezometers were highly mineralized and contained high concentrations of NH_3-N, N-Kjeldahl, Zn, Pb, Hg and fecal coli-form. The 130-ft-deep well samples water from the first natural aquifer (water-bearing stratum) underlying the landfill site. Table 2 shows that this water also is highly mineralized but es-sentially a good quality water. There is no evidence in the deep well of the contamination found in the piezometer waters.

Soil analyses have shown that liquid fertilizer beneficially in-creases the water holding capacity, cation exchange capacity, and plant available phosphorus.

Hanover Park Experimental Corn Plots

The most extensive demonstration study undertaken within Metropolitan Chicago is the district's Hanover Park experimental farm.

In 1968, a 7-acre field at the Hanover Park Water Reclama-tion Plant was selected for development of experimental corn plots for sludge fertilization. In order to grade plots for furrow irrigation, the topsoil was stripped, the subsoil graded to desired slope, and the topsoil was replaced. The soil is a disturbed Drum-mer silty clay loam with poor natural drainage. The original ex-perimental design was a randomized block with five replications having grades of 0.5, 1.5 and 2.5 percent confounded in the repli-cations (Figure 2).

Table 2. Mean of 12 monthly chemical analyses and the geometric mean of fecal coliform density for the three piezometers and well at the Calumet landfill site for 1975.

		Piezometers [a]			Well [b]
		1	2	3	
pH (range)		6.5–8.4	7.9–9.1	7.4–8.5	8.5–9.1
Total P	mg/1	0.65	0.85	5.23	0.49
CL⁻	"	243	525	272	47
SO_4=	"	777	1115	907	54
N-Kjeldahl	"	2.9	7.9	284	1.1
N-NH_3	"	1.6	6.7	240	0.4
N-NO_2+NO_3	"	0.97	6.6	2.95	0.06
Alk as $CaCO_3$	"	302	833	1006	426
E.C.	umhos/cm	1633	4241	3731	2106
K	mg/1	23	56	79	2
Na	"	62	910	221	166
Ca	"	261	34	138	7
Mg	"	98	241	101	8
Zn	"	29	4.6	14.8	0.0
Cd	"	0.01	0.01	0.01	0.00
Cu	"	0.25	0.04	0.07	0.01
Cr	"	0.01	0.00	0.01	0.00
Ni	"	0.13	0.02	0.03	0.00
Mn	"	0.47	0.11	0.53	0.04
Pb	"	0.61	0.14	0.28	0.03
Fe	"	81.5	3.7	57.8	1.1
Al	"	3.8	1	0	2
Hg	μg/1	1.5	0.17	0.2	0.0
Fecal Coliform per 100 ml		<15	<7	<94	<4

[a] Piezometers are located within the original landfill site and are 10 ft deep.

[b] Well is 130 ft deep and located in the uppermost natural aquifer.

Four monitoring wells approximately 20 ft in depth were placed at the corners of the field of plots for sampling ground water leachate. A single tile drain bisects the field of plots draining from south to north. This tile was overlaid with gravel.

Methods

The monitoring wells and field tile drain were sampled weekly. These were analyzed according to *Standard Methods for the Examination of Water and Wastewater*, 13th edition (8).

Soil samples were taken in spring and were composites of 20 cores, 0–6 in. deep. The samples were air-dried and then analyzed for pH (9) and electrical conductivity (10); 1/3 bar moisture content (11); organic carbon content (12); cation exchange capacity

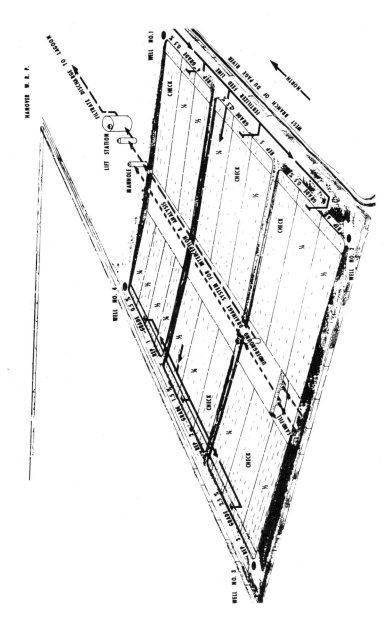

Figure 2.   Liquid-digested sludge utilization on land (Hanover W.R.P. research site).

and exchangeable K, Na, Ca, Mg (13) ; 0.1 N HCl extractable Zn, Cd, Cu, Cr, Fe, Ni, Pb, Mn and Al (1:10 soil to 0.1 N HCl solution) ; and available P (14).

Corn grain analysis for trace metal content began with the 1973 corn crop. Grain subsamples were taken during corn yield determinations. Grain was dry-ashed at 450°C, taken up in 0.1 N HCl, and analyzed by atomic absorption spectroscopy.

Each spring the plots were plowed and harrowed. Ridges and furrows were formed parallel to field slope. Sludge was applied from gated irrigation pipes at the upper end of the plots and allowed to flow down the furrows.

The plots have been fertilized with sludge every year beginning in 1968. Sludge application has not been replicated according to the original design. Most of the sludge applied came directly from heated anaerobic digesters at the Hanover Park Sewage Plant. On a few instances, lagooned digested sludge was also applied. The average analysis for digester sludge applied each year is presented in Table 3.

Effect of Sludge Application on Soil

Analyses of the soil sampled in 1970, 1973 and 1975 are presented in Table 4. The control plot (0 in./wk) showed a depletion of some major nutrients and organic carbon over time. No nitrogen or phosphorus fertilizer has been applied to the control plots. Potash fertilizer was applied to all plots in 1968 and 1972–1975. There was no influence on soil pH from the sludge application.

The 1975 soil analysis showed that Zn, Cd, Fe, available P, $NO_2 + NO_3 - N$, organic carbon, and ⅓ bar water-holding capacity all increased rapidly with the increase in the levels of application of sludge. The maximum values of these constituents lie outside the experimental ranges used. The level of Ni and cation exchange capacity (CEC) exhibited a parabolic trend indicating that if higher levels of sludge application were to be used that the Ni and CEC level would not increase any further. The soil concentration of Cu, Pb and Cr increased linearly with increased sludge application.

Effect of Sludge Application on Corn

Metal uptake in the grain was determined in the 1973 and 1975 crops. These data are presented in Table 5.

The only significant increases in metal content in 1973 were K and Ca. These increases occurred with the ¼-in. per week sludge application rate and the reason for these increases is not apparent. The 1975 grain crop had a significant increase in Zn

Table 3. Average annual chemical content of the liquid-digested sludge applied to the Hanover Park experimental corn plots. These plots received liquid sludge from 1968 through 1975.

	Year					
	1969	*1970*	*1971*	*1972*	*1973*	*1975*
pH	7.2–8.1	6.8–8.4	7.1–7.9	6.8–7.7	7.1–7.6	4.7–7.6
Total Solids (%)	3.92	4.41	3.13	5.74	3.40	2.24
Vol. Solids (% of TS)	41.1	40.8	49.9	32.5	46.6	52.2
				% dry basis		
Total P	3.45	1.60	2.72	1.06	1.42	1.86
Total N	5.6	5.39	4.29	2.63	4.61	6.42
NH_3-N	3.7	4.04	2.38	1.52	2.44	2.84
Cl	–	0.12	–	0.28	0.77	1.14
Alk as $CaCO_3$	8.20	1.80	7.83	4.31	6.37	11.7
Fe	1.8	3.3	1.55	1.89	1.72	1.64
Zn	0.07	0.069	0.068	0.080	0.072	0.089
Cu	0.045	0.054	0.087	0.056	0.070	0.11
Ni	0.005	0.049	0.042	0.017	0.008	0.01
Mn	0.07	–	–	0.02	0.051	0.049
K	0.68	–	–	0.23	0.26	0.44
Na	0.3	–	–	0.29	0.50	0.77
Mg	1.64	–	–	1.19	1.22	1.22
Ca	5.05	–	–	2.56	2.89	3.01
Pb	0.08	0.051	0.017	0.011	0.017	0.011
Cr	0.005	0.015	0.037	0.021	0.022	0.056
Cd	0.005	0.012	0.010	0.002	0.004	0.00089
Al	–	3.21	1.73	1.57	1.05	4.0

content with the addition of sludge and a highly significant increase in Cd, Cu and Fe content. The Fe increase was linear in behavior for the sludge application rates used, while the Cu response was curvilinear indicating a maximum might be near.

The Cd concentration of the grain grown on the ¼- and ½-in. per acre treatments were 0.091 and 0.096 $\mu g/g$, respectively.

The survey conducted by Pietz, Peterson, Lue-Hing and Welch (15) on 15 corn fields located throughout Illinois, where all fields were fertilized with inorganic fertilizer, yielded a mean Cd concentration of 0.037 $\mu g/g$ and a range of 0.007 to 0.294 $\mu g/g$ in the grain. This survey reported the following means and ranges in $\mu g/g$ for Zn, Cu and Fe: 19.6 (13.0–32.6); 2.62 (1.32–6.31); and 20.3 (10.4–84.0), respectively. The concentrations of Zn, Cu, Ca and Fe at the Hanover Park farm were all within the range reported by Pietz *et al.* (15).

The corn yields are presented in Table 6. In 1975, the yield on the ¼-in. per week application exceeded the ½-in. per week application. In previous years the ½-in. per week application usually gave higher yields.

Table 4.   Chemical and physical properties of a disturbed Drummer silty clay loam collected in 1970, 1973 and 1975 from the Hanover Park experimental corn plots. These plots received liquid sludge from 1968 through 1975.

Parameter	1970			1973			1975		
	Weekly Sludge Application (in.)								
	0	1/4	1/2	0	1/4	1/2	0	1/4	1/2
pH	7.9	7.7	7.7	7.3	7.1	7.1	7.8	7.5	7.6
E.C. (mmhos/cm)	N.A.[a]	N.A.	N.A.	0.41	0.51	0.61	0.48	0.98	0.90
CEC (meq/100 g)	33.4	34.0	32.2	33.9	33.7	33.3	31.4	34.5	32.3
1/3 bar H_2O (%)	38.0	38.0	36.8	32.2	33.1	32.2	32.0	34.7	33.8
C-organic (%)	3.08	3.21	3.15	2.61	3.37	3.33	2.66	3.35	3.01
P-available (μg/g)	N.A.	N.A.	N.A.	9.8	51.5	79.8	31.3	102.9	113.9
Exchangeable									
K (μg/g)	138	158	153	126	149	126	106	122	117
Na (μg/g)	95	131	117	25	30	44	22	84	78
Ca (μg/g)	5440	5660	5530	5140	6300	5650	4300	4680	4590
Mg (μg/g)	1520	1600	1600	1660	1700	1690	1060	1100	1060
0.1 N HCl extractable									
Fe (μg/g)	813	827	835	301	339	346	247	409	412
Mn (μg/g)	253	281	290	198	200	195	211	222	235
Zn (μg/g)	30.5	39.0	37.5	24.4	23.4	27.6	22.9	50.0	51.7
Cd (μg/g)	0.7	0.7	0.5	0.68	0.58	0.61	0.39	0.49	0.50
Cu (μg/g)	18.3	66.6	33.8	6.72	13.3	15.8	9.3	23.2	30.2
Cr (μg/g)	1.7	2.30	5.2	0.93	1.26	1.47	0.42	1.02	1.46
Ni (μg/g)	11.3	10.9	10.9	8.60	9.11	8.93	6.78	8.15	7.39
Pb (μg/g)	16.2	11.8	13.3	8.25	9.21	9.68	9.1	12.0	13.9
Al (μg/g)	2010	2750	2680	1700	1770	1820	1206	1315	1344

[a] Not applicable.

Effect of Sludge Application on Ground Water

Monitoring wells in each corner of the field plots (Figure 2) were sampled weekly since 1969. These results were averaged annually for ground water effects (the average of four wells) and are presented in Table 7.

Most of these constituents have been somewhat cyclic with no apparent net effects for eight years of sludge application to the field. Nitrite and nitrate-nitrogen concentrations did increase from 1968 to 1970 (0.07–0.34 mg N/l), but since 1970 the concentration has held from 0.20 to 0.34 mg N/l.

Summary—Hanover Park

After eight years of digested sludge application which has totaled 110 dry tons per acre, or about 14 T/acre/yr, on the high

Table 5. Chemical concentration of 1973 and 1975 corn grain grown at the Hanover Park experimental corn plots which have received liquid sludge from 1968 through 1975.

Treatment	Sludge Applied to Date	Zn	Cd	Cu	Cr	Fe	Ni	Pb	K	Na	Ca	Mg	Mn
In./wk	T/A					($\mu g/g$ dry basis)							
1973													
0	0	25.0	<0.5	1.74	0.77	39.2	<0.5	0.81	3030[b]	26.2	57.9[b]	1330	5.07
1/4	52	25.8	<0.5	2.01	0.76	34.4	<0.5	0.83	3600	27.2	70.6	1340	5.51
1/2	79	23.8	<0.5	1.70	0.71	36.9	<0.5	0.79	3158	32.1	55.0	1240	5.65
1975													
0	0	18.2[a]	0.067[b]	1.81[b]	0.076	23.6[b]	0.63	0.54	3260	9.71	61.4	1070	5.70
1/4	85	25.8	0.091	2.32	0.058	29.8	0.43	0.53	3490	2.78	57.8	1140	6.52
1/2	110	25.8	0.096	2.29	0.067	30.3	0.48	0.51	3380	3.39	54.2	1160	6.52

[a]Values in a column are significant at the 0.05 level of probability for that year.

[b]Values in a column are significant at the 0.01 level of probability for that year.

Table 6.   Corn yields for the Hanover Park experimental corn plots which have received liquid sludge from 1968–1975.

Weekly Application Rate	Year[a]						
	1968	1969	1970	1971	1972	1973	1975
in.	Corn Yield (bu/acre @ 15.5% H_2O						
0	32.9	40.8	24.0	54.8	11.0	28.9	49.0
1/4	62.8	138.6	46.0	84.7	68.8	77.8	102.0
1/2	63.8	141.7	46.9	96.7	78.7	98.7	82.4

[a] The corn crop in 1974 was lost due to a very early frost.

treatment plots, and 85 dry T/acre or about 11 T/acre/yr, on the ¼-in. per week treatment, increases in most of the chemical constituents in the soil as well as the CEC and organic carbon content have occurred. The corn yields have responded favorably to sludge additions. The corn has shown increases in the Zn, Cd, Cu and Fe concentrations with the addition of sludge. These concentrations, however, are well within the Illinois mean values reported by Pietz *et al.* (15).

The ground water quality at this site has remained constant since 1970.

PUBLIC ACCEPTANCE

Since public reaction to sludge utilization may be one of acceptance, indifference, curiosity, objection, rejection or downright hostility, it was important that the district's demonstration projects provide an assessment of these reactions within the Chicago area. In general, it has been our experience that the least amount of adverse reactions are encountered, if:

1. The sludges used are generated in the general vicinity in which the utilization site is located.
2. There are obvious tangible benefits, be they aesthetic or financial, which will accrue to the local community.
3. There is a strong focus of local support from public officials, environmental groups and influential local residents.
4. The operating agency has undertaken an effective program of informing the public in order to gain public support.

Since sludge utilization is essentially an agricultural undertaking, it is obvious that many large urbanized areas may experience some difficulty in distributing or utilizing large quantities of sludge locally. This problem, however, is not insurmountable, and can be resolved quite satisfactorily through an effective program

Table 7.   Mean annual chemical and fecal coliform content of leachate collected weekly from four monitoring wells at the Hanover Park experimental corn plots. These plots have received liquid sludge from 1968 through 1975.

Constituent ($\mu g/l$ except as noted)	MDL^a	Year						
		1968	1969	1970	1971	1972	1973	1975
Total P	0.01	0.25	0.21	0.24	0.50	0.15	0.53	0.18
Cl^-	1.0	10	13.3	11.6	18.0	27.3	23.8	22.5
$SO_4^=$	1.0	246	156	167	136	119	87	187
N-Kjeldahl	1.0	2.2	0.80	0.5'	0.58	0.9	0.72	0.82
$N-NH_3$	0.1	0.78	0.43	0.4	0.38	0.31	0.30	0.35
$N-NO_2+NO_3$	0.01	0.07	0.11	0.34	0.33	0.24	0.34	0.20
Alk as $CaCO_3$	1.0	355	337	323	317	268	302	256
E.C. umhos/cm		928	729	695	649	669	634	
K	1.0	4.6	1.87	1.21	0.90	1.30	1.13	1.25
Na	1.0	14.7	9.47	10.8	13.6	7.9	7.84	11.2
Ca	1.0	177	113	91.7	88.0	81.5	80.1	85.8
Mg	1.0	83.3	65.2	62.7	66.3	65.0	61.8	62.5
Zn	0.1	14.3	2.23	0.55	0.64	0.20	0.12	0.22
Cd	0.1	0	0	0.01	0	0.01	0	0
Cu	0.01	0.53	0.04	0.07	0.08	0.09	0.09	0.16
Cr	0.02	0.1	0	0.01	0.01	0.01	0.00	0
Ni	0.1	0.01	0	0.06	0.03	0.04	0.02	0
Mn	0.1	0.47	0.22	0.21	0.21	0.23	0.19	0.28
Pb	0.03	0	0	0.24	0.05	0.06	0.05	0.04
Fe	0.1	35.7	10.60	15.0	19.1	14.1	15.0	22.2
FC^b per 100 ml	2		11	11	11	6	1	1

^{a} MDL = Minimum detection limit of laboratory.

^{b} Geometric mean.

of public information and the cultivation of local interest. The following short summary of the district's Nu-Earth program illustrates one approach to resolving this problem.

Nu-Earth

Beginning in the early 1930s, the district has been stockpiling air-dried Imhoff digested sludge at a storage site on Lawndale Avenue adjacent to the West-Southwest (world's largest) sewage treatment plant in Stickney, Illinois. Although this material has always been made available to local residents on a come-and-get-it-yourself basis, the quantities actually removed were miniscule compared to production; and consequently, the "parking lot" became full in 1973. The possible solutions considered at that time were:

1. promote and expand local distribution,
2. remove by contract scavengers,
3. get into the landfill business, and
4. convert the storage site into a high-rise "sludge mountain" with retaining walls up to 200 ft high.

After much internal debate and review option 1 was selected, and contacts were made with the various news media. This option became increasingly attractive when contract scavengers quoted prices in excess of $9.00/cu yd for disposal.

On Sunday, February 24, 1974, the "Home-Garden" section of the *Chicago Tribune* carried a half-page feature story on the benefits of organic fertilizer, showing a detail map of how to reach the Lawndale Avenue air-dried Imhoff solids pick-up site. With the arrival of warmer weather in March, the *Tribune* article sparked an unprecedented rush of gardeners to the Lawndale site. The district's flower display at the McCormick Place Chicago Flower and Garden Show during the last week of March further stimulated interest in organic fertilizer. In April, the lines of cars at the Lawndale site were filmed by two major TV channels. To better accommodate the public, three new pick-up points were announced on April 24th: one at the North Side Sewage Treatment Works; another at the Calumet Sewage Treatment Works; and the third, on Lombard Avenue at the West-Southwest Sewage Treatment Works.

On June 6, 1974, the Board of Trustees adopted a new name for its air-dried fertilizer. The name "Nu-Earth" was selected as being most descriptive of the composted, partially dewatered material in the Harlem Avenue Storage site. This name has since been registered.

On June 27th, the district commenced free deliveries of Nu-Earth to any resident of Cook County calling in a minimum order of one truckload. By September, requests for Nu-Earth outstripped the district's capacity to load and deliver. By December 1974, the backlog of orders had been substantially reduced as a result of seasonal slack and completion of an all-weather haulage road system at the Lawndale storage site.

The new distribution program was effective in producing for the first time a net reduction in volume of Nu-Earth in storage at the Harlem Avenue storage site. The average cost of the free delivery program was $3.56/cu yd.

Nu-Earth distributed in 1974 amounted to 93,336 cu yd, which is comprised of 80,232 cu yd via contract deliveries and 17,104 cu yd via noncontract deliveries. Imhoff sludge input to the Harlem Avenue storage site was estimated at 84,889 cu yd at 25 percent solids. This sludge will dewater ultimately to 34,370 cu yd of Nu-Earth at 60 percent solids and 10,000,857 gallons of runoff to a MSD sewer. Nu-Earth distributed during 1974 was approximately 175 percent more than the effective 1974 input of 34,370 cu yd.

Operations at the Harlem Avenue storage site consisted of three pieces of equipment, the direct labor of two hoisting engineers and one maintenance laborer A, and direct supervisory and planning costs of $32,467. The low-pressure tractor bulldozer was purchased about the first of the year for $49,000 and was operated by a district hoisting engineer on road building and grading and sometimes for towing Nu-Earth delivery trucks stuck in the mud. The crane and the services of an operating engineer were rented for $55/hr and used mainly for stockpiling material and loading trucks. The crane can load a 24-cu yd truck in 6 minutes. The busiest crane loading day totaled 70 trucks. The crane became available through a rental contract on a full-time basis in late September. Purchase orders were used for rental of the crane prior to the contract. The end loader was purchased about mid-year for $32,500 and was operated by a district hoisting engineer first for loading and material stockpiling, and then for road building and grading. The maintenance laborer A was stationed at the storage site for dispatching the Nu-Earth contract delivery trucks.

About eighty 20-cu yd truckloads of material for road building were used at the Harlem Avenue storage site. A general contractor working in Cicero provided 40 truckloads of broken concrete free of charge, including the loading into MSD trucks.

Dewatering at the Harlem Avenue site was accomplished by pumping surface runoff to a MSD sewer and by draining two 20-ft sump holes, dug with the crane.

Cost of Nu-Earth Operations

Since the open literature seldom offers real cost figures relating to sludge handling and disposal, an approximate cost analysis of the 1974 Nu-Earth program is given below. However, certain miscellaneous costs—such as fuel for the equipment, rail car transportation (for the small district-owned and -operated inplant railroad system), and charges from departments, other than maintenance and operations—are not included. The cost analysis is presented in Table 8.

The costing data and experience for 1974 clearly show that the Nu-Earth program has been successful not only in terms of economics but also in terms of public acceptance and public participation. In 1974, approximately 97,000 cu yd were distributed; in 1975, 213,000 cu yd were distributed with the 1976 distribution estimated to be 300,000 cu yd. In addition, the general inflation in operating costs has not affected the Nu-Earth distribution program. On the contrary, there has been a small but steady decrease in the cost of operating this program since its inception in 1974.

The material is currently being used as a soil conditioner fertilizer for house gardens, golf courses, landfill cover, sod farming, and general landscape restoration.

THE PRAIRIE PLAN

Outside the district, the Prairie Plan is symbolized by the Fulton County project. However, this plan includes all the district's sludge utilization projects. The Prairie Plan is really a conceptual mechanism for safely recycling sewage sludge fertilizer into the natural environment. Since the original purchase in 1970 of 7,100 acres (28 km^2), the Fulton County site has expanded to 15,528 acres (61.25 km^2) which are currently distributed by usage as shown in Table 9. It should be noted that the 5,480 acres (2246 ha) of recycle fields represent only 35 percent of the total district holdings and emphasize the district's commitment to maintain this site as a multiple-use facility, including lakes, picnic areas and a nature preserve for the Canada geese.

Monitoring and Controls

Prior to applying any sludge on the property, the district met with the Illinois EPA (IEPA), and the Fulton County Health Department (FCHD) to determine what program for monitoring and control should be instituted. Once established, the monitoring program was incorporated into all operating permits issued by both agencies. An important requirement of the monitoring program is the establishment of baseline, pre-existing conditions and characteristics of soil (surface and subsurface), surface and ground water resources. The project always has been, and is now, subject to all the water, air and land pollution control rules and regulations of the state and all applicable ordinances of the local Board of Health, and zoning requirements of the Fulton County Planning Commission, which reviews all land-use plans for approval.

The IEPA permit monitoring requirements include chemical, physical and biological analyses of sludges shipped, used and in storage on the site; surface and ground water quality, including

Table 8. Cost analysis of Nu-Earth operations for 1974.

I. Estimated Input to Harlem Avenue Storage Site

Volume in 1973:	87,324 cu yd via 3,318 railcars
Volume in 1974:	84,889 cu yd via 3,394 railcars
Average input solids content:	25%
Input dry tonnage:	18,329 dry tons
Average Nu-Earth solids content:	60%
Volume occupied by input solids assuming they concentrate from 25% to 60%:	35,370 cu yd

II. Nu-Earth Distributed in 1974

Delivery contract 74–53:	23,616 cu yd
Delivery contract 74–61:	30,768 cu yd
Delivery contract 74–68:	25,848 cu yd
Total for delivery contracts	80,232 cu yd
Railcar deliveries to N.S. pickup site:	308 cu yd
Railcar deliveries to Lawndale pickup site:	9,576 cu yd
Lindahl truck deliveries to Cal. pickup site:	1,464 cu yd
Lindahl truck deliveries to N.S. pickup site:	1,464 cu yd
Cal. MSD truck to Cal. pickup site:	500 cu yd
WSW MSD truck to Lombard pickup site:	3,792 cu yd
Total for non-contract deliveries	17,104 cu yd
TOTAL FOR ALL DELIVERIES	97,336 cu yd

III. Amortization of End Loader

Procurement date:	7/1/74
Price:	$32,500
Lifetime:	10 yr
Cost of bond funds for MSD:	5%
Annual cost:	$4,200 (figure rounded off)

IV. Amortization of Low-Pressure Tractor Bulldozer

Procurement date:	1/1/74
Price:	$49,000
Lifetime:	10 yr
Cost of bond funds for MSD:	5%
Annual cost:	$6,350 (figure rounded off)

V. Crane Rental Cost (Including Operator & Oiler)

Size:	2-cu yd drag line basket
Two purchase orders:	$4,888
Contract 74–59 period in 1974:	September 30 to December 31
Contract 74–59 expenditure:	504 hr x $55/hr = $27,720
Total 1974 expenditure:	$32,608.00

Table 8.   (Continued)

VI. Direct Supervisory and Planning Cost

Position	Annual Salary	Annual Time	Cost
C.E. IV	$26,172	1/4	$ 6,543
C.E. I	13,848	1/4	3,462
C.E. II	18,671	1/2	9,335
E.T. IV	11,906	1/2	5,953
Hstg. Engr. Foreman	22,464	1/4	5,616
			$24,956
	Overhead factor		x 1.301
	Total direct labor cost		$32,467

VII. Cost of Nu-Earth Deliveries Under Contract 74-53

Contractor:	A & J Cartage
Nu-Earth delivered:	23,616 cu yd
Period:	June 27 to September 27
Delivery cost (five zones):	$48,144.00
Waiting time 74.5 hrs @ $24/hr:	$ 1,788.00
Total contract cost	$49,932.00
Unit cost	$2.11/cu yd

VIII. Cost of Nu-Earth Deliveries Under Contract 74-61

Contractor:	Riemer Brothers
Nu-Earth delivered:	30,768 cu yd
Period:	September 23 to November 22
Delivery cost (five zones):	$90,329.12
Waiting time 290.3 hr @ $29/hr:	$ 8,418.70
Total contract cost	$98,747.82
Unit cost	$3.20/cu yd

IX. Cost of Nu-Earth Deliveries Under Contract 74-68

Contractor:	A & J Cartage
Nu-Earth delivered:	25,848 cu yd
Period:	November 25 to December 31
Delivery cost (seven zones):	$61,593.50
Waiting time 194.3 hr @ $25/hr:	$ 5,037.50
Total contract cost	$66,631.00
Unit cost	$2.58/cu yd

X. Direct Labor Cost

MLA for dispatching (6 mo)	$ 5,818.80
Hoisting Engr. for tractor bulldozer (1 yr)	21,424.00
Hoisting Engr. for end loader (6 mo)	10,712.00
	$37,954.80
Overhead factor	x 1.301
Total direct labor cost	$49,379.19

Table 8.   (Continued)

XI. Summation of Costs

A. *Costs Not Attributable Directly to Delivery Contracts*

1974 amortization of end loader:	$ 4,200.00
Annual amortization of low-pressure tractor bulldozer:	6,350.00
Crane rental cost:	32,608.00
Direct supervisory and planning cost:	32,467.00
Direct Labor Cost:	49,379.19
Lindahl truck deliveries (3 P.O.):	6,000.00
Sub-total	$131,004.19

B. *Delivery Contract Costs*

Contract 74–53:	$ 49,932.00
Contract 74–61:	98,747.82
Contract 74–68:	66,631.00
Sub-total	$215,310.82
Total of all costs	$346,315.01

Total Nu-Earth Output	97,336 cu yd
Unit Cost for Total 1974 Nu-Earth Program	$3.56/cu yd

Table 9.   Ultimate land use for Prairie Plan.[a]

Total Site	15,528.5 acres
Not in Present Program	1,702.1 acres
Net Available for Development	13,826.4 acres
Land Use	
Recycle fields	5,480.0 acres
Holding basins	275.0 acres
Roads	200.0 acres
Administrative center, farm buildings and feed lots	30.0 acres
Fulton County conservation area	440.0 acres
Lakes (excluding conservation area)	975.0 acres[b]
Existing floodplain areas; stream valleys and other steeply sloping or forested areas, not considered economical to grade; and small irregularly-shaped fields inconveniently located for distribution–combined with lakes.[b] These constitute multiple-use areas.	6,426.5 acres
Net Available for Development	13,826.5 acres

[a] As of November 1975.

[b] Total lakes on all parcels equal approximately 1,200 acres.

indicator organisms and viruses from streams, lakes, wells and field runoff basins; ambient air ammonia content measured at the sludge storage basins; sludge application rates per application field; and quality of aquatic life in lakes, which drain the project site. Monitoring reports are submitted monthly for 25 parameters from samples taken from each of 25 wells, 11 streams, 10 lakes, 58 retention basins and 1 spring, both to the IEPA and FCHD. In addition, the FCHD has established ordinances for sludge quality per barge shipment, air quality on and off the project site, conducts its own independent monitoring program, and maintains a 24-hr hot line for receiving and responding to all citizen complaints and or comments. The district reimburses the FCHD for all costs related to monitoring of the project.

The sludges used on this project are produced at the West-Southwest Sewage Treatment Works, where stabilization is by high-rate anaerobic digestion for 14 days minimum at a temperature of 37°C. The mean values and range of the principal constituents of sludges applied from May 5 to October 30, 1975 are shown in Table 10. The year-to-date application quantities for fields 1

Table 10. Mean values and range of the principal constituents of the liquid fertilizer applied to Fulton County fields from May 5 to October 30, 1975 and the mean content per dry ton. Results are based on 26 weekly composite samples.

	Mean	*Maximum*	*Minimum*	*Mean*
pH		8.1	7.1	
	– – – – – – –	*(mg/l)*	– – – – – – –	*(lb/dry ton)*
Kjeldahl-N	2661	5640	1750	113
NH_3-N	1164	2100	574	49.3
Total-P	1791	3160	630	75.9
K	184	290	150	7.8
Fe	2229	3529	1300	94.4
Zn	185	253	126	7.8
Cu	86.9	136.0	60.7	3.7
Ni	20	27	14	0.8
Mn	18.1	28.0	6.8	0.8
Na	108	130	90	4.6
Mg	560	750	400	23.7
Ca	1312	1836	810	55.6
Pb	43.8	60.0	34.5	1.9
Cr	180	252	141	7.6
Cd	14.3	18.7	10.1	0.6
Al	612	1020	420	25.9
Hg	0.327	2.720	0.117	0.014
	– – – – – –	*(g/l)*	– – – – – – – –	
Total solids	47	68	31	2000
Total volatile solids	22	41	12	800

to 25 are shown in Table 11 (16). There are currently 48 application fields. The IEPA permitted application rates range from 75 tons/acre/yr (170 Mt/ha/yr) for spoiled lands to 20 tons/acre/yr (45 Mt/ha/yr) for placed lands. In the case of spoiled lands, the rates decrease from 75 to 20 tons/acre/yr over the initial 5 years. A sample data sheet for atmospheric ammonia for the month of August (the worst month for 1975) is shown in Table 12 (17). In order to keep pace with this massive program of monitoring and field quality control, the district maintains on site a 6-person laboratory with support from the main laboratories in Chicago.

Effects on Ground Water Quality

As previously mentioned, there is an extensive network of monitoring wells associated with the project. The major part of this network was installed in 1971 to serve the dual role of determining baseline pre-existing and operational ground water characteristics following extensive sludge applications in 1973.

Table 11.   Fulton County land reclamation project liquid fertilizer application quantities, year to date—1975.

Field No	Volume (cu yd)	Wet Tons (tons)	Wet Tons Per Acre (tons/acre)	Dry Tons (tons)	Dry Tons Per Acre (tons/acre)
1	34857	29350	638	1336	29.0
2	23250	19576	376	955	18.4
3	11992	10097	246	417	10.2
4	75788	63813	797	3065	38.3
5	29911	25185	839	1199	40.0
6	0	0	0	0	0.0
7	66020	55589	505	2457	22.3
8	51283	43180	568	1933	25.4
9	18657	15709	179	588	6.7
10	50888	42847	455	1884	20.1
11	18813	15841	754	739	35.2
12	12262	10325	382	519	19.2
13	14879	12528	348	535	14.9
14	25665	21610	308	929	13.3
15	35370	29782	676	1326	30.1
16	102535	86334	575	3979	26.5
17	46479	39135	447	1825	20.9
18	0	0	xxx	0	0.0
19	9944	8372	220	332	8.8
20	23690	19947	174	1033	9.1
21	11645	9805	251	472	12.1
22	18369	15467	297	662	12.7
23	10467	8813	400	379	17.3
24	0	0	0	0	0.0
25	28473	23974	630	1101	29.0
Sub-totals	721248	607291		27676	

Ground water data from six placed land and six mine-spoil monitoring wells, respectively, were selected for statistical analysis of water quality (18).

Monthly ground water samples from December 1971 to December 1973 revealed that former strip-mining operations for coal up to 1960 in the Pennsylvanian geologic formations had significantly affected the ground water quality in mine-spoil areas. Electrical conductivity (EC), alkalinity and concentrations of Cl^-, $SO_4^=$, Kjeldahl-N, NH_3N, K, Na, Ca, Mg, Zn, Cd, Cr, Ni, Mn, Pb and Fe were significantly higher, at the 0.01 level, in mine-spoil ground waters. In ground waters from both nondisturbed and mined areas the mean concentrations of NH_3-N, Fe, Pb, Mn and Zn exceeded public water supply recommendations. Concen-

Table 12. Atmospheric NH$_3$ concentrations and sampling conditions at the MSDGC liquid fertilizer holding basins in Fulton Couty during August 1975.

Date Sampled	Location of Sampling Basin #	Berm	Starting Time	Total Time (min)	Air Volume Sampled (liters)	Dew Point (°C)	Air Temp (°C)	Wind Direction	Mean Wind Speed (mph)	Atmospheric NH$_3$ (ppm − v/v)
8-1-75	2	SW	1050	90	956.71	27	36	NE	12.6	1.08
8-4-75	2	E	1355	90	1044.08	26	33	W-NW	2.7	1.04
8-6-75	3A	SW	1333	90	1018.63	20.5	26	NE	7.5	1.36
8-8-75	2	N	1012	90	1066.43	19	24.5	S-SE	8	1.51
8-11-75	1	NE	1455	90	1506.42	20.5	22	SW-SE	4	0.27
8-13-75	1	NE	1336	90	986.11	22.5	25	SW	11	1.29
8-15-75	3A	S	1045	90	922.94	20.5	23.5	N	5	1.33
8-18-75	2	W	1352	90	885.52	23	28	SE	7	1.06
8-20-75	2	N	1000	90	881.71	24	28.5	S-SE	14	1.38
8-22-75	1	N	1103	90	492.48	25.5	30	S-SW	6	1.57
8-25-75	1	E	1436	90	846.28	22	27	W-SW	<5	1.11
8-27-75	2	W	1000	90	833.34	23	25	E	<5	0.64
8-29-75	1	N	1347	90	840.74	24	26.5	S-SW	8	1.67

trations of Cd, Cr, Ni, Pb and Zn in mine-spoil waters were approximately two to five times those detected in placed land ground waters. The frequency of trace metal detection in mine-spoil waters was higher than for nondisturbed areas.

Mine-spoil ground waters were generally characterized by a greater number of significant monthly, seasonal and well variations for the 23 chemical characteristics examined. An evaluation of the relationships between ground water pH, NH_3-N, NO_3 + NO_2 − N, and EC, and the same chemical characteristics in deep soil borings, 35–50 ft, using linear regression analysis showed higher linear correlations in nondisturbed lands. Significant linear correlations between ground water and soil pH and EC were obtained in placed land areas and between ground water and soil EC in mine-spoil areas.

Results indicated that the crushing of shale, limestone and sandstone rocks and their redistribution in or near the water table during former mining operations altered the chemical-physical composition of the soil and geologic profiles. This alteration, resulting in an extremely heterogeneous spoil material, was reflected in the chemical quality of mine-spoil ground waters.

Using a typical mine-spoil well, confidence limits were developed to evaluate future ground water quality. The 1971–1973 monitoring data was considered to be baseline. Monitoring data for 1974 were tested and found to be within the established confidence limits. Data for wells routinely monitored during 1975 have been summarized and compared to 1974 data. Table 13 shows the number of wells in 1975 with either a less than or greater than 10 percent change in water composition as compared to 1974. Decreases (>10 percent) were noted for Cl⁻, EC, Ca, Mg, Zn, Cd, Mn and Pb. Increases included Kjeldahl-N, Cu and Hg.

Fulton County Soil Changes

The fields developed from mine spoil by regrading and plowing under existing sod resulted in a decrease in organic carbon content. Liquid sludge application gradually increased the organic carbon content. A larger effect was noted for the levels of 0.1 N HCl extractable metals and exchangeable Ca which increased with sludge application. There was little effect on soil pH.

Fields developed from undisturbed land showed responses similar to mine-spoil fields when sludge was applied. There was a greater increase in organic carbon content, and no decrease was noted after initial development. This was probably due to the prior row crop use and to the minimal regrading required compared to mine-spoil fields.

Table 13. Number of wells at Fulton County, Illinois with less than or greater than 10 percent change in annual mean composition in 1975 compared to 1974 mean annual composition.

Constituent	< 10% Change	≥10% Change	
		Increase	Decrease
	Number of Wells in Each Category		
Total P	7	9	9
Cl	8	3	14
SO$_4$	2	12	11
N–Kjeldahl	5	13	7
N–NH$_3$	6	11	8
N–NO$_2$+NO$_3$	13	6	6
Alk. as CaCO$_3$	15	6	4
E.C.	5	0	20
K	17	2	6
Na	13	4	8
Ca	11	3	11
Mg	13	0	12
Zn	5	7	13
Cd	19	1	5
Cu	8	14	3
Cr	17	7	1
Ni	10	9	6
Mn	5	6	14
Pb	3	0	22
Fe	6	10	9
Al	0	13	12
Hg	5	17	3
Fecal coliform	18	3	4

Fulton County Yields and Crop Quality

The 1975 corn harvest records of sludge-fertilized fields in Fulton County showed a yield range from 25.8 bu/acre to 63.9 bu/acre. In comparison, experimental plots on mine spoil receiving sludge since 1972 had a yield range from 79.0 to 86.3 bu/acre in 1975. These replicated plots are maintained jointly by MSD and the Agronomy Department of the University of Illinois. The yields obtained reflect the closer management and are indicative of the yields possible from sludge-fertilized mine spoil.

The metal concentration of the grain grown on land fertilized with commercial fertilizer and sewage sludge is presented in Table 14. The Zn, Mn and Ni levels have increased while Cu and Cr concentrations have declined with time. The Zn, Fe and Cd concentrations reported for 1975 were within the range reported for corn grain in Illinois (15).

Table 14.   Metal content of corn grain harvested on MSDGC land in Fulton County, Illinois, with and without sludge application for 1972–1974.

	K	Mg (%)	Ca	Zn	Fe	Mn	Cu	Ni	Cr	Pb	Cd
						(μg/g, dry weight)					
1972											
Nonsludged	0.37	0.16	0.035	NA	46	10.9	2.3	NA	NA	NA	NA
Sludged	0.31	0.15	0.025	NA	40	9.3	2.5	NA	NA	NA	NA
1973											
Nonsludged	0.31	0.12	0.057	24	26	7.2	3.3	0.8	0.4	0.3	0.4
Sludged	0.32	0.13	0.07	26	25	8.1	3.4	0.5	0.3	0.3	0.3
1974											
Nonsludged	0.26	0.13	0.005	28	22	8.5	2.3	0.8	0.07	0.1	0.02
Sludged	0.37	0.13	0.008	31	47	9.5	2.7	1.8	1.1	0.3	0.13
1975											
Sludged	0.28	0.19	0.006	27	79	8.6	2.3	1.2	0.4	0.6	0.22

Animal Feed Studies—Cattle

Beginning in September 1972, the district in cooperation with the College of Veterinary Medicine, University of Illinois, and Norris Farms Inc., initiated a cattle-feeding study (19), the objectives of which were:

1. To determine whether there is transmission of biological materials or organisms, or accumulation of heavy metals, chemicals including polychlorinated biphenyls, or other unknown contaminants harmful to cattle on vegetation fertilized with digested sludge.
2. To ascertain the effect of fertilizing new pastures with digested sludge.
3. To ascertain the feasibility of using digested sludge as a fertilizer and water source to aid in establishing new pastures, upgrading marginal pastures, and maintaining existing pastures.

The results of this study to date may be summarized as follows (20):

1. Sludge irrigation apparently has not introduced unusual or additional common bovine pathogens to the pasture environment, nor otherwise increased the incidence of common bovine disease or unthriftiness.
2. The parasite load carried by cattle in the experimental and control groups were essentially the same. Careful examination of the carcasses of control and experimental animals failed to reveal differences in the parasite load. The same kinds of parasites were present in the G.I. tracts of both groups of animals.
3. Digested sludge, under the laboratory conditions specified is generally detrimental to development and survival of *Eimeria bovis*, *Eimeria zuernii* and *Haemonchus contortus*.
4. At this time neither toxic metals nor polychlorinated biphenyls

have accumulated in the soil or vegetation to levels which can be termed significant by comparison with those in control pastures.

5. The quality of sudax grass grown on sludge-amended strip-mine soil is nutritionally equal or superior to that grown on the bottom-land pasture used, as determined by laboratory analysis.

6. Acceptance of the sludge-irrigated sudax grass by the cattle is apparently equal to that of nonsludge-irrigated sudax.

Animal Feeding Studies—Swine

In a companion project using swine, the district is cooperating with the College of Veterinary Medicine and Agricultural Experiment Station, University of Illinois, the U.S. EPA, and the Illinois Institute for Environmental Quality as a co-sponsor (21). A summary of this work to date is as follows:

> Swine (two groups) fed a ration containing 79 percent corn from sewage sludge-fertilized plots for 8 weeks may have had a modest performance advantage over swine fed corn from plots heavily fertilized with commercial nitrogen. Sludge-fertilized corn feeding-related deviations from population normals for electroencephalograms, electrocardiograms, clinical chemistry or histopathology were not observed; however, swine fed sludge-fertilized corn differed significantly from the control swine for certain physiologic and biological characteristics, indicating possible interference with glucose metabolism and microsomal mixed-function oxidase activity. Control swine had mild microcytic anemia associated with parasitism, but swine in the principal groups were less adversely affected, probably due to higher iron intake. Corn heavily fertilized with sewage sludge from an urban-industrial complex can be safely fed to swine, based on these results, but it might be advantageous to feed on a short-term or intermittent schedule.

Residues of heavy metals in the animal products are currently being examined. Tables 15 and 16 show, respectively, analyses of corn grain used in feed and body weight gains of experimental swine (22). This work is being conducted by the University of Illinois and is continuing.

Public Participation and Information

The district is acutely aware of the need for public participation and support of all sludge utilization projects, and especially projects like the Prairie Plan which are large, remotely located from the area of sludge generation, and thus have the potential for accentuating urban rural differences. In response to this awareness, the district established in 1971 an advisory group—the Fulton County Steering Committee—comprised of elected public officials,

Table 15.   Elemental analysis of corn incorporated in gilt feed (22).

Element (ppm exeept as noted)	Control Corn	Sludge-Fertilized Corn	Significance[a]
N (%)	1.74	1.98	ND
P	3,038 ±159	3,293 ±62	5%
Zn	26.8 ± 2.0	51.9 ± 1.0	1
Cd	0.10 ± 0.01	0.56 ± 0.02	1
Ni	0.66 ± 0.06	1.28 ± 0.06	1
Cu	1.99 ± 0.06	2.22 ± 0.10	1
Mn	7.40 ± 0.12	6.98 ± 0.09	1
Fe	14.23 ± 1.37	18.45 ± 0.53	1
Mg	1,238 ± 70	1,258 ±15	NS
Ca	23.48 ± 2.35	26.65 ± 0.41	5
K	4,678 ±281	4,787 ±75	NS
Na	7.09 ± 2.52	15.10 ± 0.76	1

[a]By analysis of variance. Data are reported as mean (four replicate samples) ±SD.
ND = not determined; NS = not significant.

Table 16.   Body weight gains in gilts fed sludge-fertilized corn (22).

Day	Control Group (n=3)	Group 2 (Fed Sludge-Fertilized Corn for 14 days) (n=3)	Group 3 (Fed Sludge-Fertilized Corn for 56 days) (n=3)
Before Start of Experimental Feeding			
		Initial Weight (kg)	
–16	23.3 ± 4.2	23.5 ± 2.7	23.5 ± 3.6
		Cumulative Weight Gain (%)	
–9	11.4 ±11.5	26.7 ± 4.1	24.5 ± 3.0
–3	40.0 ± 4.5	41.5 ± 6.7	40.8 ± 3.8
After Start of Experimental Feeding			
4	65.4± 5.9	68.3 ± 12.5	63.6 ± 8.0
11	91.4± 11.5	87.7 ± 12.2	89.5 ± 9.7
18	115.2± 14.7	106.8 ± 14.6	114.9 ± 17.7
25	112.7± 13.6	111.9 ± 18.0	115.4 ± 17.0
32	150.9± 16.1	150.6 ± 29.0	152.9 ± 20.1
48	195.0± 18.4	199.6 ± 43.4	203.2 ± 20.9
56	204.1± 19.1	206.2 ± 44.4	220.5 ± 24.7

[a] Data expressed are mean ±SD; n = number of gilts.

representatives of organized civic groups, unattached private citizens, and district staff personnel. The public officials who sit on this committee include District Board members, President of the Planning Committee of the Fulton County Board, Chairman of the Fulton County Planning Commission, Fulton County Planning Administrator, local mayors, a representative of the Governor's office and the Chairman of the Fulton County Soil and Management District.

Meetings of this committee are held monthly, are public and open to all interested persons including the news media, and provide a forum for any and all parties interested in the planning, operation or performance of the project. Every land-use plan related to the project is first submitted to the committee for review and comments; as a result, the district has made substantial changes to some land-use plans prior to seeking formal approval from the Fulton County Planning Commission. In addition, any citizen may request that an item be placed on the agenda for discussion.

It is important to emphasize here that there probably is no mechanism including extensive public participation which will eliminate complaints. However, it is most important that public officials, planners, designers and managers associated with projects such as the Prairie Plan be provided with an effective barometer of public sentiment. Experience shows that the Fulton County Steering Committee serves this function very well. Experience also shows that to be effective, a public participation and information program must be vigorous and literally remain in perpetual motion.

Socio-Economic Effects

The most obvious local economic impact of the Prairie Plan is the creation of new jobs for local residents. The project has created approximately 120 skilled and unskilled contract jobs for area workers, who average 6 to 8 months of employment yearly. During 1975, the district paid approximately $1.0 million for contract employees and $300,000 to full-time staff personnel.

Contributions of the project to local public finance is substantial and is increasing each year. In 1971, the district paid a total of $43,126 in real estate taxes. By 1974, this had increased to $102,667, plus $43,893 in personal property taxes or a total of $146,560 which represents a substantial boost to local public finances. Real estate taxes paid in 1974 amounted to 1.4 percent of total revenue, and 3.5 percent of personal property taxes, respectively.

REFERENCES

1. Kudrna, F. L. "The Prairie Plan," *J. Urban Planning and Development Division, ASCE*, 99(UP2):205–215, Proc. Paper 9971 (September 1973).
2. Hinesly, T. D., R. L. Jones and B. Sosewitz. "Use of waste treatment

plant solids for mined land reclamation," *Mining Congress J.* (September 1972).

3. Kudrna, F. L., and G. T. Kelly. "Implementing the Chicago Prairie Plan," In: *Recycling Treated Municipal Wastewater and Sludge through Forest and Cropland*, Sopper, Ed., Penn State Univ. Press (1973).

4. Dalton, F. E. and R. S. Murphy. "Land reclamation—the natural cycle," presented at the 45th Annual Conf. of the Water Pollution Control Federation, Atlanta, Georgia (October 1972).

5. Peterson, J. R., T. M. McCalla and G. E. Smith. "Human and animal wastes as fertilizers," In: *Fertilizer Technology and Use*, R. A. Olson, et al., Eds., Soil Science Society of America, Inc., Madison, Wisc., pp. 557–596 (1971).

6. Bray, R. H. and L. T. Kurtz. "Determination of total, organic, and available forms of phosphorus in soils," *Soil Sci.* 59:39–45 (1945).

7. The Metropolitan Sanitary District of Greater Chicago, Department of Research and Development. "1975 Annual Report."

8. American Public Health Association. *Standard Methods for the Examination of Water and Wastewater*, 13th edition, APHA, New York (1971).

9. Peech, M. "Hydrogen-ion activity," In: *Methods of Soil Analysis*, C. A. Black, Ed., Am. Soc. Agron., Madison, Wisc., pp. 914–926 (1965).

10. Bower, C. A. and L. V. Wilcox. "Soluble salts by electrical conductivity," Chapter 62, In: Methods of Soil Analysis, C. A. Black, ed., Am. Soc. Agron. Madison, Wisc., pp. 933–951 (1965).

11. Richards, L. A. "Physical condition of water in soil," In: *Methods of Soil Analysis*, C. A. Black, Ed., Am. Soc. Agron., Madison, Wisc. pp. 128–152 (1965).

12. Allison, L. E. "Organic carbon," In: *Methods of Soil Analysis*, C. A. Black, Ed., Am. Soc. Agron., Madison, Wisc., pp. 1367–1378 (1965).

13. Chapman, H. D. "Cation exchange capacity," In: *Methods of Soil Analysis*, C. A. Black, Ed., Am. Soc. Agron., Madison, Wisc. pp. 891–901 (1965).

14. Olsen, S. R. and L. A. Dean. "Phosphorus," In: *Methods of Soil Analysis*, C. A. Black, Ed., Am. Soc. Agron., Madison, Wisc., pp. 1035–1049 (1965).

15. Pietz, R. I., J. R. Peterson, C. Lue-Hing, and L. F. Welch. "Variability in the concentration of 12 elements in corn grain," unpublished data, Dept. of Research and Development, The Metropolitan Sanitary District of Greater Chicago (March 1976).

16. The Metropolitan Sanitary District of Greater Chicago, Department of Research and Development. Report No. 76-1, "Environmental protection system report for Fulton County, Illinois," (November 1975).

17. The Metropolitan Sanitary District of Greater Chicago, Department of Research and Development. Report No. 75-25, "Environmental protection system report for Fulton County, Illinois," (August 1975).

18. Pietz, R. I., J. R. Peterson and C. Lue-Hing. "Groundwater quality at a strip-mine reclamation area in west central Illinois," presented before the Second Research and Applied Technology Symposium on Mined-Land Reclamation, Coal and the Environment Technical Conference, Louisville, Kentucky (October 22–24, 1974).

19. Fitzgerald, P. R., College of Veterinary Medicine, University of Illinois, Urbana, Illinois. "The use of sewage sludge in pasture reclamation: parasitology, nutrition and the occurrence of metals and polychlorinated biphenyls," a proposal submitted to the Metropolitan Sanitary District of Greater Chicago (May 1973).

20. Fitzgerald, P. R. and W. R. Jolley. "The use of sewage sludge in pasture reclamation: parasitology, nutrition and the occurrence of metals and polychlorinated biphenyls," College of Veterinary Medicine, University of Illinois, Urbana, Illinois (1974).

21. Hinesly, T. D., O. C. Braids, R. I. Dick, R. L. Jones and J. A. E. Molina. "Agricultural benefits and environmental changes resulting from the use of digested sludge on field crops," The Metropolitan Sanitary District of Greater Chicago and USEPA Grant No. DO1–UI–00080 (1974).

22. Hansen, L. G., J. L. Dorner, C. S. Byerley, R. P. Tarara and T. D. Hinesly. "Effects of sewage sludge-fertilized corn fed to growing swine," *Am. J. Vet. Res.* 37(6):711–714.

30

THE AGRICULTURAL USE
OF MUNICIPAL SLUDGE

D. A. Ardern

Lee Division of the Thames Water Authority, Old Harlow,
Essex, England

INTRODUCTION

Prior to the passage of the 1973 Water Act, functions now dealt with by the ten water authorities in England and Wales were dealt with by:

> 161 local authority water departments and water companies
> 1,300 sewerage and sewage disposal organizations
> 368 land drainage boards

The new water authorities are all-purpose units based on one or more river basins handling all aspects of the water cycle with the exception (at the moment) of navigation and the collection of the contents of cesspools, septic tanks and liquid industrial wastes. Thus they are concerned with water from the time it falls to the ground through land drainage, river regulation and flood protection to extraction for water supply or irrigation whether from river or underground source and thence through reservoirs, water treatment works and supply mains. After use by households or industries they are again responsible for controlling trade discharges and managing the whole sewerage and sewage treatment process back to the discharge into the river or underground strata. The protection of aquifers, the development of water-based leisure activities including fisheries, sailing, and bird watching are also in their terms of reference. The controlled disposal of sludge arising from sewage treatment works is one of the many facets of their work.

The Thames Water Authority, commonly known as Thames Water services the largest population of any of the new authorities. It has an arduous task because of the high population densities, low rainfall and important manufacturing, commercial and recreational areas under its authority. Some statistics of Thames Water are:

Area	5,058 sq mi
Population served	12 millions
Employees	12,000
Budget 1975/76	£250 million
Rateable value served	£2,500 million
Rivers maintained	2,500 miles
Water mains	26,000 miles
Main sewers	31,000 miles
Water storage capacity	54,000 million gallons
Water supplied	710 million gals per day
Sewage treated (D.W.F.)	944 million Imperial gals per day
Water treatment works	60
Sewage treatment works	450
Sludge production (dry solids)	314,500 Imperial tons per annum

Thames water is organized into nine divisions as follows:

Single purpose functions

Metropolitan water divisions	London water supply
Metropolitan public health	London sewerage and sewage treatment
Thames conservatory division	River Thames and tributaries

Multipurpose functions

Lea	Land drainage, tidal defenses, water supply, sewerage and sewage disposal
Chiltern Cotswold Lambourn Southern Wales	Water supply, sewerage and sewage disposal.

Thames Water is said to be the biggest organization in the world to deal with all aspects of the water cycle and has recently set up an international consultancy service to make its expertise available to other countries. It already has been consulted by Middle East interests.

SINCE REORGANIZATION

In the two years since reorganization very significant improvements have been made in all aspects of the water cycle, and nowhere is this more apparent than in sewage treatment. Many outdated and often grossly overloaded sewage works have been shut down and the flow to them diverted to larger, more modern and more effective plants. This has made significant improvements in river conditions despite the two dry years since reorganization, which has greatly reduced the natural flow in the rivers and the dilution of effluents. At the same time there has been a considerable reduction in the number of men employed, which has been accompanied by an improvement in the pay and conditions of the remainder of the labor force.

Soon after Thames Water was set up, an analysis of the various sludge treatment and disposal processes operating in the 450 Sewage Treatment Works was instituted and revealed over 40 different methods in use. This is in part a reflection of the varied outlooks of the precedent authorities but is also an indication of the many processes available and the effect of economic, geographical and environmental considerations.

At reorganization the majority of the sewage works taken over were full to brimming over with sludge, and the first few months were a desperate battle to deal with the excess. Extensive tankering, mostly by contractors, was used to move this surplus sludge to works that had spare capacity. The construction of a large number of temporary earth sludge lagoons occurred. These actions plus the commissioning of substantial new treatment works has stabilized the situation and overcome the worst of the problems.

The simplicity and reliability experienced with the tankering arrangements was influential in the development of present trends, which are to reduce the number of works at which sludge is treated while retaining the works for sewage treatment only. Sludge is then transported to the larger works for proper treatment. Most of the smaller works originally used simple drying beds, manually lifted, either with or without cold digestion. The sludge was disposed of mostly to allotments after stock piling. This was chancy and labor-intensive.

There were also a number of small and rather sophisticated processes such as centrifuges, disc filters, filter presses, belt presses and a Zimpro installation. Some of these were portable, particularly the centrifuges and disc filters, and were transported from works to works. These mobile plants were high in labor requirement both in operation and maintenance. Many of the chassis were fairly new and so were stripped of their equipment, which was, where suitable, installed permanently at a suitable works and the chassis converted into tankers.

SLUDGE DISPOSAL—GENERAL

Some 75 percent of the sludge produced in the Thames Water

area is digested. The rest is mainly from small works. Most of the sewage from the Greater London area is carried by major trunk sewers alongside the Thames to treatment works north and south of the river some ten miles below Tower Bridge. Sludge from these works, after digestion, is loaded into specially built vessels for disposal at sea 50 miles from the works.

This has been going on since 1887 and despite extensive monitoring no adverse effects have been detected in the condition of the estuary. Indeed, every year there is an improvement in the fish life in the tidal Thames due to better sewage treatment.

The sludge disposal fleet, which is part of the Metropolitan Public Health Division, now comprises five ships. These carry up to 2,300 tons of sludge, have a top speed when loaded of 12.5 knots, and run continuously around the clock including weekends and all public holidays except Christmas and Boxing Day. They are fitted with sophisticated navigation and communication equipment as they have to operate in some of the most congested waters in the world with a very bad weather record.

The outer areas of London beyond the catchment area of these main sewers have a number of substantial works and as there is little agricultural land in their immediate vicinity, the dominant method of sludge treatment is air drying on beds, or in lagoons, followed by cartage of the dry digested sludge to agricultural land.

In the areas outside the London conurbation there are a number of large works that have useful areas of agricultural land in their immediate vicinity, and these normally use air drying of digested sludge on large remotely controlled mechanically lifted beds or distribute liquid digested sludge onto agricultural land.

There are also a number of works that press the raw primary sludge after treatment with lime and copperas, aluminum chlorohydrate or a polyelectrolyte. Pressed cake conditioned with lime and copperas is used on arable land. That containing aluminum chlorohydrate or polyelectrolyte is usually used as top covering for completed landfill sites or tipped to waste. Some surplus activated sludge is spread directly on land without further treatment.

There are several hundred small works in remote rural areas where, in the past, air drying of undigested sludge with hand lifting from beds or lagoons was carried out. This still continues but to a lesser extent.

At present there is no uniform method of disposal within Thames Water and it seems probable that this will always be the case. Each works' sludge will be dealt with by the means that best suits the type of sludge, the analysis of it and the geographical, agricultural and environmental constraints subject to overall policy guidelines.

LEA DIVISION PRACTICE

In the Lea Division there is a well-tried system of disposal of most sludge after heated digestion and thickening in lagoons and secondary digesters. This system uses small tankers with capacity between 1,000 and 1,500 Imp. gallons on four-wheel drive chassis. This system was started 18 years ago with three old Air Force fuel tankers that were suitably modified. With these, experience was gained in the acceptability of the sludge to farmers and the problems that arise when using this method. After this trial, specifications were drawn up for suitable tankers that were quickly filled from an overhead steel tank and discharged onto the land through two "ski-jump" type spreaders. These spreaders are remotely controlled by the tanker driver who does not need to leave his cab to start spreading.

There is a large amount of permanent grassland near most of the larger sewage works in the Division, some used for grazing either by strip or paddock systems, the rest for hay or silage. The high content of nitrogen in the sludge makes it a very suitable fertilizer for this crop. The normal method of application is to spread once per year at rates between 5,000 and 10,000 gallons per acre, with the most usual rate being about 7,000. This is equivalent to less than two tons of solid material to the acre. The demand for sludge by farmers is so great that it is impossible to meet, particularly in summer.

A difficult problem arises therefore in allocating sludge to farms. Currently under consideration is a charging policy since no charge is made at present. The sludge and its distribution on the fields would continue to be without charge but if a farmer wanted delivery at a specific time he would buy priority, *i.e.*, by booking and paying say 50p a thousand gallons he would be guaranteed delivery within a week. This scheme would introduce complications. Experience in other Divisions has shown that the demand at critical times continues to exceed the supply and the priority question is not resolved. Payment makes the service a commercial matter and items such as damage to fields, gates left open, mud on farm roads, argumentative drivers, and thin sludge can cause friction and require more staff attention.

At the moment the principles applied are:

1. Supply farmers nearer the works rather than those further away. Normally this means delivery within a 4-mile radius.
2. Give priority in summer to those who allowed their lands to be used in winter (sometimes at the price of rutted fields).

The practice employed in the light of experience is for individual drivers to be allotted to individual farmers who give them detailed instructions on the area to be sprayed. Under this system a friendly relationship is established and a flexible distribution pattern can be used. For instance if the ground is frozen early in the morning, the tanker can go onto an area that could not otherwise be reached; similarly overnight rain might mean the tanker would have to switch from one field to a drier one.

Much of the permanent grass is on shallow light soils on the top of gravel, and panning of the surface through consolidation under the wheels of a tanker is not generally a problem. There is some evidence that the low pressure tires act like agricultural rollers in encouraging tillering of the grass and so develop the best pattern of growth of the grass.

In an attempt to ameliorate the worst of the summer demand problem, the Division has built at its own expense on farmers' land, after a thorough investigation and clearance of the site by the geology and pollution prevention departments, a number of unlined earthen banked lagoons holding 0.5–1 million Imp. gallons of sludge adjacent to surfaced farm roads. These lagoons are topped up as required by hired tankers of 4,500-gal capacity. The farmer then draws the sludge as required and applies it to the land either with his own irrigation equipment or an Ehrwig tanker trailer of 2,000-gal capacity fitted with a Molex pump and spray equipment towed behind his own tractor. The tanker trailer is owned by Thames Water and lent to the farmer who operates it with his own men as he requires. Such a scheme has many merits for both parties. As far as Thames Water is concerned additional sludge storage is made available very cheaply and on the farmer's land where any minor smell or fly nuisance is less than many other farming operations. The need for storage on sewage works, where sites are often restricted, is reduced. It also enables economy of scale and flexibility in haulage to be achieved. The farmer has the surety of supply, and the timing and method of operation are entirely under his control.

Although it has been possible to keep the small tankers going for 11 months in each year, this has been partly due to the recent dry weather. In an attempt to diversify we have experimented with the use of a large tanker fitted with a Mole pump and monitor used on ex-R.A.F. airfields in the area. The tanker could remain on the concrete runways and perimeter track and blow the sludge up to 60 yds from the tanker, thus covering nearly the whole airfield without going on the grass. The small areas not reached by this method could be covered by the small tankers or the main vehicle after discharge of part of his load, provided the ground was not too wet. The airfields are not used for commercial or military flying and the grass will normally be cut for silage or hay.

The general policy of concentrating sludge treatment at a limited number of sewage works mentioned earlier is well advanced in this Division. It appears probable that within a short time sludge treatment will be concentrated at only 11 of the 53 works in the Division and hand-lifting of undigested sludge will be practically abolished. When this has been done, approximately two-thirds of the sludge will be digested and disposed of to agricultural land, the other third being pressed. Of the pressed cake, about one-half will be used on arable land after composting and the rest will be used as cover to landfill sites and other similar uses.

PRACTICE IN OTHER DIVISIONS

Disposal of liquid sludge to agricultural land in other divisions of Thames Water outside the Greater London area differs somewhat from that in Lea Division due to historical and geographical reasons. In the past some of these areas bought large areas of farmland that are run as commercial farms under a farm manager, and the application of the sludge is tailored to the demands of the form of farming that is practiced.

Because these farms and other farmland supplied with sludge are somewhat more distant from the sewage works than in the Lea Division, there has been more of an emphasis on larger tankers that remain on the roads and supply either "daughter tankers" or irrigation equipment of various types.

A promising type of irrigation equipment under development uses a flexible hose that coils on a large diameter horizontal drum and may be used both as a conduit and a tenon to haul a sledge with a monitor on it to the headland of the field. This means that only the light sledge need cross the field, while the reeling drum, tanker and pump remain at the edge of the field. Earlier versions of this used either a self-powered vehicle with the monitor and reeling drum mounted on it, or a winch and a reeling drum mounted on the headland. All types use a monitor with restricted angular distribution so that the vehicle, sledge, cable or pipe are not sprayed with sludge.

There is some indication that on heavy soils farmers prefer dried sludge rather than liquid both because of its effect on soil structure and because compaction of the ground by tankers and irrigation equipment is avoided. This is consistent with the better utilization of tankers than is normally obtained on light soils. On these heavy soils, particularly clays, there is usually enough potash, so the shortage of this element in the sludge (whether dried or liquid) is not important.

PREVENTION OF POLLUTION

Thames Water is, of course, most concerned that the distribution of liquid sludge does not lead to any contamination of the streams and rivers or of the underground water supplies. The sludge storage lagoons on farms mentioned previously are built only after extensive borehole and geological research.

Similarly sludge is spread only on areas of farmland well away from water supply boreholes, streams and rivers, and spreading on river meadows or low lying land would not normally be practiced. In this respect grassland is better than arable, because heavy rain falling on arable land after application of liquid sludge might wash some of it into the rivers. Although insufficient experimental work had been done it appears that the chemicals in the sludge are "loosely bound" and not as liable to wash away as when applied in an inorganic form.

Nitrates are a matter of particular concern to the Lea because the river is a major drinking water source, because of the intensive agriculture in the area and because of the high proportion of the river flow that is derived from sewage works, particularly in dry summers when virtually all the river flow is used for water supply purposes.

No evidence exists that the distribution of liquid sludge contributes significantly to increased nitrate levels in the rivers. This area has an average moisture deficiency in the soil of the order of 7 in. per annum, which tends to reduce pollution of the rivers due to runoff.

PUBLIC RELATIONS

No attempt was made to conduct a "hard sell" for liquid digested sludge. Instead farmers have been associated with the development of the use of liquid sludge from the outset. After preliminary informal chats and small scale trials, some of the farmers agreed to try using the digested sludge on an experimental basis, and Dr. Coker of the Herts College of Agriculture was brought in as an independent adviser in 1958 and performed considerable research work on the project until 1964.

To publicize the work, secure the cooperation of farmers and control the research work, a Committee was set up under the auspices of the Hertfordshire Branch of the National Farmers' Union. It was chaired by the Branch Chairman and serviced by the Branch Secretary and his staff. On the Committee sat prominent local farmers, the Chief Engineer, Work's Chemist and Work's Superintendent of the Middle Lee Regional Drainage Scheme and staff from the West Herts Main Drainage Authority,

LEA ?

with Dr. Coker acting as their independent agricultural adviser. This Committee has been so successful that the example is being followed in other areas.

Ironically, the farmers who helped so much in publicizing the work are now paying the price of the success of the Committee by being charged for the delivery of sludge in some areas and having to take their place in the long queue elsewhere.

No complaints have occurred from the general public, first because the sludge distribution areas have been kept reasonably away from residential land and the sludge after digestion is comparatively odorless, and second because of the extensive lectures, film shows and organized visits to sewage works and open days that have been arranged to publicize to the greatest possible extent the work. As a result and through strict compliance with ADAS 10 (1) relating to the acceptable limits for heavy metals, no trouble has been encountered in finding suitable outlets for sludge from the main works. The scheme is now being extended to digested sludge from the smaller works. The word sludge has been played down and it has been marketed under proprietory names such as Rygonic and Hydig. This has helped the public accept the idea.

DRIED SLUDGE

With regard to dried digested sludge, the demand is not so consistent, largely because the particle size is too large for drilling with normal agricultural implements. It would be practical to put it through a granulator, turning it into a more suitable form, but at the moment the expenditure of both capital and revenue is not thought to be justified. The material is bought by contractors at 25p per cubic meter (loaded) for use as a top soil substitute—particularly for golf courses—or for application to intensive horticulture fields by machines of the lime spreader type. For such a purpose it is carted distances up to 100 miles. The price charged is likely to be raised shortly.

Discussions have been held with the neighboring County Councils as to the possibility of using this material and similar material from lagoons for use as a final covering for refuse tips, but at the moment sufficient topsoil seems to be available from other sources. As the tips are run commercially rather than directly by the Councils, it has not been possible to make definite arrangements.

One farmer, however, has been helped by applying large quantities of lagoon sludge to a badly reinstated gravel pit. The high humus content plus the fertilizers in the sludge have made a significant improvement to the appearance and behavior of the land.

Because there are many gravel pits in the area that are nearly worked out, it is likely that this will produce a continuing large demand. Experiments are being set up to enable a code of practice to be devised.

In the early days dry sludge was bagged and sold to gardeners, but this was found to be very expensive because

1. the cost of the bags was considerable
2. the labor required to fill and handle was great
3. the disorganization of the works operation caused by people calling in to collect a bag or two was unacceptable
4. the majority of the demand was at the weekends when manning levels were low
5. storage of the filled bags was difficult because the bags tended to deteriorate
6. labor did not like acting as sales hands
7. trouble was experienced with giving change and accounting for bags.

Similarly, although a small truck was bought for delivering sludge, it was both uneconomic and unpopular with staff and labor. Therefore bulk sales to commercial outlets have been found preferable.

COSTS

The Working Party on Sewage Disposal (2) estimated the cost of sludge treatment and disposal as 40 percent of the total costs for a 30:20 standard works, using biological treatment without tertiary stage. This is a very difficult subject on which to give absolute costs. It is not clear at what point in the plant expenses should begin to be allocated to sludge treatment. Are sedimentation tanks to be charged to sewage or sludge treatment? How is the cost of treating secondary sludges priced? Gas, power and heat are produced and used in digestion and power generation, but the allocation of costs is open to many variants.

In addition, the dates at which works were constructed, site conditions, loan charges and land costs appropriate make the true costs of any particular methods of treatment difficult to determine. To use figures from existing works to forecast the best method for future works is bound to be chancy. One can only do one's calculations, make inspired guesses about subsequent trends and hope that future generations will not be too unkind in their comments.

Current figures for the cost of sludge treatment and disposal at Rye Meads, the biggest works in the Lea Division, which has a contributing population of 320,000 and a DWF of 18 million Imperial gals per day, work out as follows:

Total annual running costs of sludge treatment and disposal	approx. 100,000 lb/annum
Sludge production (raw dry solids)	8,900 Tonnes per annum
Total running cost per tanker (1,000/1400 gal) including driver	£6,000 per annum
Cost of sludge treatment and disposal per tonne of raw dry solids	£11.22
Cost of tankering digested sludge to land per tonne of raw dry solids	£9.35

Note that these figures exclude all capital charges. It should also be pointed out that the cost of digestion and subsequent thickening in sludge lagoons is not the difference between £11.22 and £9.35, as the former figure is the average for both sludge drying on mechanically lifted beds and for tanker disposal.

R. W. Horner of the Dept. of Public Health Engineering, Greater London Council, produced some interesting cost comparison figures in 1970. Although the actual costs are now very much out of date, the relative positions have not changed. Extracts are given in the Appendix I (3).

A computer-based management information service is being set up that will give more detailed and comparative costs, including those for sludge collection, treatment and disposal. It appears that typical figures for the Lea Division for 1974/75 deal only with the revenue expenditure, *i.e.*, ignore all capital charges and administration, and are based on the cost per ton of raw dried sludge solids as follows:

1. heated digestion followed by mechanically lifted beds, stockpiling the sludge and loading it free onto contractor's vehicle for them to remove (£7)
2. collection of raw sludge from smaller works average radius about ten miles, digestion at larger works and spreading on land in liquid form (£17)
3. pressing, using lime/copperas and carting to arable land after stockpiling (five miles round trip) but not spreading on arable land (£23).

CHARACTERISTICS OF SLUDGE

One trouble with sludge is that it is not possible to guarantee the composition of either liquid sludge or dried sludge because of the varying nature of the sewage received and the treatment processes. In Britain there is stringent legislation on the composition of fertilizers and it is therefore usual to sell sludge as a soil conditioner rather than as a fertilizer.

Typical analysis results for Rye Meads air-dried digested sludge are given in Table 1.

Dried sludge of this type has a better nitrogen, phosphate and

Table 1. Results for Rye Meads air-dried digested sludge.

Composition		Heavy Metals (mg/Kg	of dry matter)
Moisture	50%	Copper	890
Dry matter	50%	Nickel	250
pH	7.2	Zinc	1600
		Cadmium	50
Dry Matter Analysis		Chromium	550
Organic	60%	Lead	300
Nitrogen (as N)	3%	Iron	12000
Phosphate (as P_2O_5)	2%	Silver	25
Potash (as K_2O)	0.5%	Mercury	5
Calcium (as CaO)	5%		
Magnesium (as MgO)	1.5%		

potash analysis than farmyard manure. It also lends itself to mechanical handling and is less smelly. The trace elements in it can be helpful but a watch must be kept on heavy metals, bearing in mind the existing levels in the fields and the pH value, as this governs the toxic effect of the heavy metals.

Liquid sludge is basically the same composition but is 90 percent water as compared with the 50 percent of the dried sludge, and the potash and nitrogen figures are a higher proportion of the total solids. In particular there is more nitrogen in solution as ammonia, which is more readily available but is more likely to be lost in wet weather.

Considerable field trials have been done by Dr. Coker, particularly on the use of liquid digested sludge with rye grass and barley, and by J. H. Williams of the Agricultural Development and Advisory Service. These show that the result of liquid sludge application is broadly similar to the use of chemical fertilizers of similar composition (4–6).

TRADE WASTE CONTROL

For sewage to be capable of being properly treated and to ensure that the sludge resulting from the sewage is suitable for application to agricultural land, it is imperative that suitable limits are chosen and enforced on all industrialists and others discharging trade wastes to the sewers leading to the sewage works. This requires the employment of experienced qualified inspectors who discuss their manufacturing processes with prospective dischargers before any connection to the sewers is made and who advise what form of pretreatment (if any) will be required to meet the prescribed limits. After connection they will, from time to time, inspect the premises from which the waste discharges to point out any potential hazards such as corroded pipes, leaking tanks, suspect ballcocks, and areas where rain might wash materials into the sewers. They have right of entry at all reasonable times and normally achieve results by persuasion rather than threat.

Routine inspection, sampling and testing of the trade wastes are carried out, the frequency depending on the magnitude and importance of the discharge and the reliability of the operators on the premises. Any infringement of the limits and all accidental spillages are followed up and investigated to reduce the risks of a recurrence. Any costs incurred by the Authority as a result of spillages are normally recovered from the industrialist so it is in his interest to ensure that these do not happen.

In the Lea Division with over 50 sewage works, a staff of ten is employed on trade waste control in addition to the laboratory staff who do the analytical work in the five sewage works control laboratories. Many of the sewage works are very small and handle only domestic sewage, so in practice the trade waste control is concentrated on about 300 trade waste dischargers to ten works. Some of these are of the order of 1 million gal/day, while others, such as launderettes, are trivial. The biggest works in the Division has about 150 trade dischargers. Appended to this chapter are the Trade Effluent Limits used in the Division and Notes for the Guidance of Trade Waste Officers.

FUTURE TRENDS

Apart from the concentration of sludge treatment in a lesser number of works, it would appear that in Britain there will be:

1. few new large drying bed installations constructed because of high land consumption, unpredictable performance because of British weather, and high capital costs.
2. a continual emphasis on the use of mechanical drying equipment with or without previous digestion. The present fashion seems to be for fixed as distinct from mobile equipment and for vacuum disc filters and filter belt presses.
3. an emphasis on the development of higher standards of trade waste control.
4. a more commercial outlook to the disposal of sludge. The water authorities will try to reduce their costs by charging a greater part of the costs of sludge treatment and disposal to the recipients.
5. considerable research (engineering, biological, chemical and market) will be devoted to an economic solution for sludge from conurbations, particularly areas without immediate access to sea disposal.

6. increased quality control of the sludge.
7. composting of undigested dried sludge produced by filter presses, vacuum or belt presses to make a product more acceptable to the potential user.
8. the provision of a comprehensive soil testing and advisory service for use with sludge application. This is already practiced by some areas.
9. a program of research on the precise effects of treatment processes on sludge and the interaction of constituents of the sludge with the soil and crop response.
10. emphasis on the need to retain flexibility in disposal methods so that they can be altered to suit research results, legislation, public opinion and economic trends.
11. the production of a more easily transported and distributed dried sludge form that would be more acceptable to farmers. One promising method is to treat the sludge, possibly using heat from digester gas, to produce a soil conditioner in pellet form that can be drilled by farmers and used by horticulturists.

ACKNOWLEDGMENTS

This chapter is presented with the approval of the Director of Operations of Thames Water but is a personal view and not official policy. My thanks are due to my Divisional Manager, Oscar Addyman, and all my colleagues for their help and encouragement.

APPENDIX 1

Process	Method	Costs[a]	Remarks
Conditioning	Digestion		
	Primary	£0.5	Mesophilic (thermophilic not recommended)
	Secondary	£0.6	
	Physical additive	£1.0	Little advantage
	Chemical conditioning	£2–£4	
Dewatering	Vacuum filtration		
	Raw sludge	£10	£10 per ton on digested sludge solids
	Digested sludge	£6.7	Raw sludge
	Rotoplug concentration	£7.0	Mixed sludges
	Centrifuge	£8.0	
	Air drying		
	Deep lagoons	£3.5–£5	Depending on land value
	Mech. dry beds	£8.5	Mixed sludges
	Pressure filtration	£11	Not suitable for digested sludge
	Rotary kiln	More than £9	Depends on selling price
	Band dryer	More than £9	Depends on selling price
	Flash drying		
	Raymond process	£60	Activated sludge
	Middlesex process	£6.5	After previous drying to 60% moisture
	Multiple hearth	£19.5	As rotary kiln
Wet oxidation	Zimmerman	£12	Costs may be reduced with experience
	Zimpro	£19	Costs may be reduced with experience
Incineration	Multiple hearth	£15	Sludge 90% moisture
	Fluidized bed	No figure	Sludge 80% moisture
	Atomized suspension	£20 ?	
	Combustion with refuse	£5	
Disposal to sea	Ship		
	Raw sludge	£5	
	Digested sludge	£4.3	
	Pipeline	Not given	Depends on construction costs
Disposal on land	Liquid 2.5% solids	£8	Rye Meads practice is to use 5% solids
	Semiliquid 12.5%–17.5%	£4	so equiv. cost at that time would be £4

[a] All costs are per ton of dry raw solids.

APPENDIX 2

Thames Water Authority—Lea Division
Trade Effluent Limits

The chemical requirements set out below are for general guidance only and should not be taken as applying to any particular trade effluent. The actual chemical conditions to be embodied in an Agreement, Consent or Direction may show some variation from the list below to take account of the volume of trade effluent to be discharged, local conditions in sewers and any other relevant circumstances.

1. Cooling or condensing water to be excluded from the trade effluent.
2. The following constituents to be eliminated from the trade effluent before it is permitted to enter the sewer.
   a. All sludges arising from pretreatment of the trade effluent before discharge to public sewer.
   b. All waste substance liable to form viscous or solid coatings or deposit on any part of the sewerage system through which the trade effluent is to pass.
   c. All petroleum spirit and other inflammable solvent not completely miscible with water.
   d. All substances of a nature likely to give rise to fumes or odors injurious to men working in the sewers through which the trade effluent is to pass.
   e. All radioactive substances, unless specifically authorized.
   f. Halogen-substituted phenolic compounds.
   g. Thiourea and thiourea derivatives.
   h. All halogenated solvents such as carbon tetrachloride, chloroform, methylene chloride, trichloroethylene, halogenated ethanes and similar solvents unless specifically approved.
3. Concentrations of specific constituents to be limited as follows:
   The pH value of the effluent discharged to be within the range of 6.0 to 10.0.
   The temperature at which the trade effluent is discharged shall not be greater than 110°F (43°C) at the point of entry into the public sewer.

Grease, oil and other liquids immiscible with water	50 mg/l
Copper	5 mg/l
Zinc	10 mg/l
Iron	25 mg/l
Nickel	4 mg/l
Silver	5 mg/l
Cadmium	4 mg/l
Mercury	1 mg/l
Lead	5 mg/l
Beryllium	1 mg/l
Total chromium	5 mg/l
Cyanides and all compounds from which hydrocyanic acid is liberated on acidification (as CN)[a]	15 mg/l
Suspended solids	400 mg/l
Phenols and cresols	20 mg/l
Sulfate (for Rye Meads)[b]	400 mg/l
Chemical oxygen demand (settled sample)	800 mg/l

[a]10 for small sewers and small works.
[b]Generally for division, 1000 is allowed.

APPENDIX 3

Notes for Guidance of Trade Waste Officer

GENERAL PURPOSE AND OBJECTIVES

The Trade Waste Officer should carry out his duties with the following objectives in view:

1. Safeguard personnel (whether working locally, in sewers, or at purification works).
2. Protect sewer fabric and avoidance of physical blockages.
3. Protect sewage works purification and sludge treatment processes against inhibition, overload and physical damage to structures or equipment.
4. Ensure that due to a trade discharge, sludges from purification processes are not made unacceptable for subsequent disposal from an aesthetic, public health or toxic standpoint.
5. Protect the good condition of rivers with particular reference to their flora and fauna and their acceptability as a source of raw water for eventual potable supplies.
6. Ensure that sufficient information is available regarding discharge strengths and volumes for fair and equitable charges to be made on traders for handling their wastes, and ensure that they understand the need for this and how it is calculated.
7. Be available with advice, information and, if necessary, direction in case of accidental spillages of any solid or liquid material.

PURSUANCE OF OBJECTIVES

1. The Trade Effluent Officer is directly responsible, in all matters, to the Group Manager, but should ensure that effective coordination exists between his own group, other groups and headquarters in the person of the Divisional Trade Effluent Officer.
2. Initially a general survey of Trade Premises should be undertaken and records kept, both in his own files and the group's master file of:
   a. all premises visited
   b. volumes, nature and composition of trade effluent when applicable
   c. details of processes, plant and stores on site
   d. pending supply of detailed plans, a sketch of the trade premises and drainage layout, as determined by himself.
      These records should then be updated at regular intervals, as determined by change in circumstances or collection and supply of further information, but always following a further general survey to be carried out yearly.
3. When a Trade Premise is initially inspected, the procedure adopted should be as follows:
   a. Approach the senior member of staff.
   b. He should be made aware of the definition of a Trade Waste and the relevant legislation concerning the control of its discharge and charging.
   c. He should be made aware of the need for control, limitation on constitution, charging clauses, requirements for plans, metering and if necessary, pretreatment and his responsibility to furnish these requirements.

    d. An inspection of works processes should then follow and an evaluation made of the likelihood of a waste being discharged, and if so, its source, volume and constitution.

    e. He should then be made aware of the necessity of completing a Preliminary Notice and supply of drainage plans as indicated. It is often possible to assist him in completing the notice at this time, which can save time and revisits.

    f. Information on the Preliminary Notice should be checked as correct, then discussed with the Group Manager for his determination of the next step to be taken.

    g. If applicable, supply DTEO with information obtained for records and general appraisal and follow up procedure leading to issue of consent/agreement.

    h. Include the investigated premises in the regular inspection, sampling and meter reading rota, all information obtained to be recorded and the Group Manager notified of unusual occurrences.

NOTE: When a Trader submits a Trade Effluent Notice, it is to be immediately passed to the DTEO, as there is a two-month time limit on the subsequent issue of a consent or the reasons for its refusal.

4. Coordination and cooperation should exist between Traders and Group organization and between the Trade Effluent Section and all other Sections of the Group. It is imperative that the TEO should be made aware by traders of any abnormal discharge, and he should then coordinate any action necessary to monitor and control the discharge. It is his responsibility to inform and assist in any way the Group Manager and any other person likely to be involved by the discharge, either in the Group or Divisionally. If this information is not forthcoming from traders but a discharge is observed, he should, as a matter of urgency, inform all sections as above and trace the discharge to its source, taking necessary action to monitor and control it as soon as possible.

5. The Trade Effluent Officer is probably the only representative of the Authority the general public and Traders in particular are aware of, and as such he should present himself as a tidy, efficient and effective member of our staff, willing to cooperate in the Authority's name but equally willing to make demands when this becomes necessary. To this end he should make himself aware of all relevant legislation relating to his position and become his Group's expert on all things appertaining to Trade Effluent Control.

REFERENCES

1. "Permissible levels of toxic metals in sewage used on agricultural land," Ministry of Agriculture, Fisheries and Food (Agricultural Development and Advisory Service) 10.

2. "Taken for granted," Report of the Working Party on Sewage Disposal, H.M.S.O. (1970).

3. Horner, R. W. "Sludge treatment and disposal," Informal discussion Inst. of C.E. (1970).

4. Coker, E. G. "Experiments in East Hertfordshire on the use of liquid-digested sludge as a manure for certain farm crops," *Inst. Sewage Purification* (1965).

5. Coker, E. G. "The value of liquid-digested sewage sludge, (1) : The effect of liquid sewage sludge on growth and composition of grass clover swards in South East England," *J. Agric Ssc.* 67:91 (1966).
6. Williams, J. H. "Use of sewage sludge on agricultural land and the effects of metals on crops," *J. Inst. Water Poll. Cont.* (1975).

31

STATE OF THE ART
IN MUNICIPAL SEWAGE
SLUDGE LANDSPREADING

D. L. Forster
Department of Agricultural Economics
T. J. Logan, R. H. Miller
Department of Agronomy
R. K. White
Department of Agricultural Engineering
    The Ohio State University, Columbus, Ohio

INTRODUCTION

Landspreading of treated sewage sludge has become an increasingly attractive method of sludge disposal. A recent survey conducted in Ohio indicated that 50–60 communities are using landspreading on agricultural land as their primary method of disposal (1). Furthermore, several communities indicated their interest in investigating landspreading as a future method of disposal.

Several events account for this growing interest. First, the Federal Water Pollution Control Act Amendments of 1972 encouraged municipalities to provide for the "recycling of potential sewage pollutants through the production of agriculture . . ." (2). The Act also provides for grant assistance to construct sewage treatment facilities, and it requires that landspreading be considered when applying for these grants.

While this legislation has encouraged landspreading, economic considerations have improved the acceptability of landspreading. Recent price increases of commercial fertilizer have increased the demand for crop nutrient substitutes such as sewage sludge. Also, the relatively high economic and environmental costs of alternative methods of sludge disposal, such as incineration and landfill, have led to increased interest in landspreading.

In Ohio the increased interest has led to several efforts to investigate landspreading. One of these efforts has been an educational effort by the Ohio Cooperative Extension Service to inform landowners, treatment plant operators, community officials, and the public at large of the engineering, agronomic, environmental, economic, and social impacts of sludge disposal. This effort has been a normative study devoted to the recommended practices in landspreading. A parallel study has involved an identification of the current state-of-the-art in municipal landspreading, the results of which are the focus of this chapter.

OBJECTIVES

The purpose of the research was to inventory the current technical, economic, and social aspects of landspreading in Ohio. While the study focused on Ohio, the results probably can be extrapolated to other midwestern areas. More specifically the objectives were to:

1. determine the degree of adoption of landspreading practices that have been recommended in order to achieve safe landspreading programs
2. determine the type of equipment being used for landspreading and the contractual arrangements being used between landowners and communities
3. determine the monetary benefits and costs of landspreading under existing conditions
4. determine the existing attitudes of selected communities toward landspreading.

Two surveys were conducted to provide information to accomplish these objectives. The first was a survey of waste treatment officials in Ohio communities. Assistance for the interview process was provided by the Ohio Farm Bureau Federation (OFBF) and the Ohio Municipal League (OML). A study by Carroll et al. (3) identified 35 communities in Ohio that spread sludge on land, and these communities were personally visited by OFB personnel. In addition, questionnaires were sent by OML to a sample of Ohio communities not included in the list of 35 landspreading communities.

The second survey focused on the final objective, determining the attitudes of landowners in five counties toward landspreading. The five counties, dispersed in their geographic location, size, and agricultural setting, were chosen as potential sites for a future project demonstrating acceptable management practices in landspreading. While the survey gave an indication of the attitudes of these communities toward landspreading, its primary purpose was to test the degree of community resistance toward the demonstration project. The county agricultural agent drew a random sample of landowners in each county. In addition, all landowners who had practiced landspreading recently were interviewed. The interviews were conducted by representatives of OFBF.

RESULTS

Analysis and Monitoring

The first objective was to examine the landspreading practices currently in use. Recommended practices include analysis of the sludge and soil tests prior to application and monitoring of soil, water, and plant tissue after application.

The application rate suitable for a particular disposal site depends upon the chemical characteristics of the sludge and the properties of the soil receiving the sludge. Thus, analysis of the sludge and soil allows the computation of a suitable application rate. With continued applications, the chemical content of the sludge may affect plant growth, accumulate in plant tissue and affect the food chain, and affect water quality. Thus, monitoring of soil, plant tissue and water quality is recommended to ensure environmentally safe applications.*

The survey of Ohio communities using landspreading provided a picture of the extent of analysis and monitoring programs. The majority had at least a minimal analysis program of the sludge being applied to the land. Only 9 communities out of the total 43 surveyed did not have some type of sludge analysis program (Table 1). Furthermore, 23 of the communities analyzed their sludge monthly or more often, with several analyzing the sludge weekly (Table 1).

Generally, the analyses were one of two types. Either the community analyzed only the total and volatile solids contents of the sludge, or they performed costly analyses of the solids, volatile solids, primary nutrients, and heavy metals. Of the 43 surveyed communities that spread sludge on land, 16 communities performed the minimal analysis, while 18 ran the more detailed analyses (Table 2). Those communities using no sludge analysis program or a minimal program are applying sludge without adequate knowledge of the characteristics of the sludge and the suitability of the sludge for soils in their area.

Several of the smaller communities had limited knowledge of the contents of the sludge that they applied. Of the 11 communities with less than 2.0 million gallons per day sewage flow (MGD), 4 communities did not have a sludge analysis program

---

*Soil limitations are discussed in Reference 4.

Table 1.   Sludge analysis programs as a function of treatment capacity in Ohio communities involved in land application.

Capacity (MGD)[a]	Number of Communities in Category	Number of Communities Analyzing Sludge				
		Weekly	Monthly	Quarterly	Semi-Annually or Less Often	Not Testing
<2.0	11	4	0	0	3	4
2.0–3.9	9	4	1	0	2	2
4.0–7.9	10	5	1	0	2	2
8.0–19.9	9	3	2	1	2	1
>20.0	4	2	1	0	1	0
Total	43	18	5	1	10	9

[a] MGD = million gallons per day.

Table 2.   Type of sludge analysis program as a function of treatment capacity in Ohio communities involved in land application.

Capacity (MGD)[a]	Number of Communities in Category	Number of Communities Using Analyses Program[d]		
		No Analyses	Minimal[b] Analyses	Thorough[c] Analyses
<2.0	11	4	5	2
2.0–3.9	9	2	2	5
4.0–7.9	10	2	4	4
8.0–19.9	9	1	4	4
>20.0	4	0	1	3
Total	43	9	16	18

[a] MGD = million gallons per day.

[b] "Minimal" includes analyses for total and volatile solids content of the sludge.

[c] "Thorough" includes analyses for total and volatile solids content, some primary nutrients (nitrogen, phosphorus, potassium), and some heavy metals (cadmium, zinc, copper, nickel, boron, chromium, cobalt, manganese, mercury, molybdenum, lead).

[d] Communities have a choice of building their own laboratories for sludge analysis or using the services of a commercial laboratory. The communities that perform only minimal analysis tended to analyze the sludge in their own laboratories. Approximately two-thirds of these communities used their own laboratories. On the other hand, the communities performing detailed analysis tended to use commercial laboratories for some or all of their testing. Twelve of the eighteen communities with a "thorough" program used commercial laboratories.

(Table 1), and 3 communities analyzed their sludge less than once every six months. The majority of smaller communities (less than 2 MGD) appeared to have rather limited analyses to assist them in applying sludge at safe rates, with only 2 of the 11 communities having a thorough analysis program (Table 2). Generally, the larger communities have had the resources to establish a regular and more thorough sludge analysis program. Of the 32 communities with capacity of 2 MGD or greater, only 5 did not analyze the sludge, and 16 of the 32 communities had a thorough analysis program with some knowledge of the nutrient and heavy metal content of the sludge (Table 2).

The relative lack of knowledge of sludge contents on the part of small communities does not imply that they are putting sludge on at unsafe rates. Of those responding to the question concerning the rate of application, most appeared to be applying sludge to private land at low to moderate rates (less than 10 dry tons per acre per year).

A few Ohio communities were testing soils before application *and* were monitoring soils to which sludge was applied. However, the majority appeared to have little knowledge of the capacity of the soils to utilize the sludge effectively. Only 4 of the 43 communities surveyed were testing the soil prior to application and only 8 were monitoring the soil after application (Table 3).

It is also suggested that "water sources originating near the sludge application area . . . should be periodically sampled and tested for the presence of trace elements and nitrates" (4). Water quality near disposal sites was being monitored by nearly one-third of the communities (Table 3). Furthermore, there was little variation in the proportion of communities monitoring water quality between size categories above 2 MGD.

It is suggested that plant tissue and the harvestable portion of the plant be monitored for heavy metal uptake. Eight of the 43 surveyed communities were monitoring plant tissue (Table 3).

Application Equipment

Equipment used in sludge disposal largely consisted of tank trucks with gravity discharge or tank trucks with pumped discharge (tank trucks include tank wagons pulled by tractors). The irrigation-sprinkler system was not being used by Ohio communities. Furthermore, the irrigation-overland flow system was found in only 4 of the communities interviewed (Table 4). Land-spreading of sludge by box spreader was found where the treatment process resulted in a sludge with relatively high solids content.

The majority of communities spreading sludge by tank truck

Table 3. Soil testing and monitoring at disposal site as a function of treatment capacity in Ohio communities involved in land application.

Capacity (MGD)[a]	Number of Communities in Category[b]	Testing Soil Prior to Application	Number of Communities		
			Monitoring Soil After Sludge Application	Monitoring Water Quality Near Disposal Site After Sludge Application	Monitoring Plant Tissue On Disposal Site After Sludge Application
<2	11	0	0	1	1
2–3.9	9	1	2	4	2
4–7.9	10	2	2	4	2
8–19.9	9	0	3	3	2
>20	4	1	1	2	1
Total	43	4	8	14	8

[a] MGD = million gallons per day

[b] Several communities were involved in more than one of the activities of soil testing, soil monitoring or plant tissue monitoring; therefore, the sum of the number of communities doing alternative testing and monitoring programs does not equal the number of communities in each category.

Table 4.   Method of treated sludge application by size of community as a
function of treatment capacity in Ohio communities involved in
land application.

Capacity (MGD)[a]	Number of Communities in Category[b]	Number of Communities Spreading Sludge By				
		Irrigation Overland Flow	Tank Truck With Gravity Discharge	Tank Truck With Pumped Discharge	Tank Truck With Soil Injection	Spreader Box
<2	11	1	6	4	0	2
2.0–3.9	9	2	2	4	1	2
4.0–7.9	10	0	4	6	0	0
8.0–19.9	9	0	4	4	1	3
>20	4	1	1	0	0	2
Total	43	4	17	18	2	9

[a] MGD = million gallons per day.

[b] Some communities used more than one type of disposal; therefore, the sum of the number of communities using alternative methods of disposal does not equal the number of communities in each size category.

were using tanks with less than 2,000-gal capacity. A few had larger trucks with the potential to cause soil compaction problems if used directly for field application, with sludge transferred to smaller application trucks at the disposal site.

Contractual Agreements

Contracts between landowners and the communities producing the sludge are viewed by many as a solid foundation for a favorable relationship between the producer and the recipient. Only 20 percent of the communities surveyed had written leases. Provisions in these leases included one or more of the following points:

1. "escape" clause for either party
2. restriction to the type of crops grown on the disposal site
3. restrictions on application during growing season and on application when soils are wet
4. restrictions on application rate
5. placement of any liabilities due to odor, runoff, etc. with the farmer.

The last provision would seem unfair to the farmer recipient unless the analysis of the sludge is known by the farmer. Making the farmer solely liable for the effects of the sludge would be equitable only when the treatment plant is able to provide a detailed analysis of the sludge and is assured that the farmer has been informed of the potential hazards of application.

Most of those landspreading communities without written landowner community agreements said that oral agreements existed. The principal advantage of a written contract would be to make sure that both parties understand the agreement prior to applying the sludge. Often oral contracts are entered with the best of intentions, but the landowner and treatment plant operator have differing notions of the rights and obligations of each party.

Economic Considerations

There is continuing pressure on communities to find less expensive and more environmentally favorable sludge disposal methods. Furthermore, farmers often view sludge as a resource that may be a substitute for a portion of their increasingly expensive fertilizer requirements. This section summarizes the costs of landspreading and the economic benefits accruing in the form of a nutrient supply.

The surveyed communities were asked to identify (a) their capital investment in sludge disposal equipment and storage facilities, (b) the average age of these capital investments, and (c) the annual operating costs of sludge disposal. The capital investments were compounded by an annual rate of 5 percent over the lifetime of the investments in order to have all investments in current dollars. The total annual cost for each disposal system was the sum of depreciation, interest, repair and maintenance, labor, insurance, fuel, administrative, and miscellaneous expenses.

The data from the survey indicated that average sludge disposal cost per unit of treated sewage declined as flow increased until the plan was treating approximately 25 MGD. At this level disposal costs were $24.85 per dry ton (Table 5).

The mean plant capacity was 8 MGD. The average cost of sludge disposal at this capacity was approximately $31 per dry ton. For plants smaller than the mean capacity, disposal costs increased rapidly. The plant with a 4 MGD flow, which was the median plant size, had an average sludge disposal cost of $43 per dry ton.

Sludge disposal costs shown in Table 5 are estimates of total costs and average costs. While variations in costs were seen at each plant size, the model used to calculate the costs shown in Table 5 explained 87 percent of the variation in costs between communities. The model is explained briefly in the footnote under Table 5.

The economic benefits of sludge may be approximated by the value of sludge as a substitute for commerical fertilizer. These substitutes are the nitrogen, phosphate and potash required by

Table 5. Estimated annual total cost and average total cost as a function of treatment flow in Ohio communities involved in land application.

Size (MGD flow per day)[a]	Annual Total Cost[b] ($)	Annual Average Total Cost Per MGD Flow ($/MGD)	Average Total Cost Per Dry Ton[c] ($/dry ton)
2	15562	7781	69.70
4	19657	4914	43.40
6	23881	3980	35.10
8	28235	3529	31.10
10	32720	3272	28.90
12	37334	3111	27.50
14	42078	3005	26.50
20	57091	2854	25.20
30	84712	2823	24.90
40	115583	2889	25.50

[a]MGD flow = million gallons per day flowing through treatment plant.

[b]Depreciation is assumed to be 12.5% of investment (current dollars), and interest is assumed to be 8% of midlife value of investment (current dollars). Total cost is sum of depreciation, interest, repair and maintenance, fuel, labor, insurance, administration, and miscellaneous expenses. Ordinary least squares was used to arrive at the following function:

total cost = 11598 + 1949.67 (size) + 16.25 (size)^2 + 16821 (dummy)

The variable "size" equals the flow per day through the plant in million gallons. The variable "dummy" is equal to 1 when the municipality owns disposal land and 0 when no land is owned. All coefficients are statistically significant at the 0.10 level.

[c]The following function was established from the survey –

Dry tons = 0.32 + 0.311 (size)

where "size" equals the flow per day through the plant in million gallons. The coefficients are statistically significant at the 0.01 level. Annual costs per MGD were converted to costs per dry ton by the following formula:

$$\text{cost per dry ton} = \left(\frac{\text{annual cost per MGD}}{365} \right) 0.31$$

crops. However, the variable chemical composition of sludge and fluctuating nutrient prices make the benefits difficult to calculate.

On the average, a dry ton of sludge contains approximately 100 pounds of nitrogen, 100 pounds of phosphate and 5 pounds of potash. Only about 30 percent of the total nitrogen from sludge is immediately available to the plan. This combination of nitrogen, phosphate and potash is not suitable for most crops. For example, typical fertilizer rates for corn are 180, 50 and 60 pounds per acre of nitrogen, phosphate and potash, respectively. If 6 dry tons of sludge per acre are applied, the sludge would furnish 180 pounds of nitrogen, 600 pounds of phosphate and 30 pounds of potash. Thus, phosphorus buildup likely would occur and potassium fertilizer supplements might be required on low potassium soils. Sludge application can be expected to supplement commercial fertilizers, not to replace them. Applications of commercial fertilizers would be required on most soils to achieve an acceptable balance of nutrients.

The value of sludge as a fertilizer may be reduced drastically by excessive concentrations of trace elements. Sludges high in trace elements may present problems to crops and/or human health. Zinc, copper and nickel may be toxic to plants when used at high application levels. Cadmium is also phytotoxic and may be hazardous to the food chain if concentrations are high. Other trace elements, including chromium, mercury, lead, boron, molybdenum, cobalt, and selenium, are usually at low levels in sludge and do not affect plant growth or human health.

Assuming acceptable concentrations of trace elements and usage of sludge as a supplemental source of nutrients, an approximation of the benefits of sewage sludge is its value as a substitute for commercial fertilizer. Table 6 provides estimates of the value of nutrients in sludge under six alternative assumptions for nutrient content and commercial fertilizer price. The nutrient content is based on data from the *Ohio Guide* (4).

Table 6. Value of one ton of dry sewage sludge under alternative levels of nutrient content and commercial fertilizer prices.

	Value of Nutrients in Sludge	
Nutrient Content	High $N = \$0.30/lb, P_2O_5 = \$0.20/lb, K_2O = \$0.11/lb$	Low $N = \$0.20/lb, P_2O_5 = \$0.15/lb, K_2O = \$0.08/lb$
High ($N = 6.4\%$, $P_2O_5 = 8.7\%$ $K_2O = 0.84\%$)[a]	$49	$36
Medium ($N = 5\%$, $P_2O_5 = 5.25\%$ $K_2O = 0.54\%$)	32	23
Low ($N = 3.5\%$, $P_2O_5 = 1.8\%$ $K_2O = 0.24\%$)	15	10

[a]Approximately 30% of total N and 100% of P_2O_5 and K_2O would be available for crops.

The cost of sewage sludge disposal in the cities surveyed is slightly higher than the value of sludge as a substitute for commercial fertilizer. With current fertilizer prices, the right side of Table 6 would estimate sewage sludge values. Using the medium nutrient content for sludge, the benefits would be $23 per dry ton. For the average communities surveyed it costs $31 per dry ton to spread sludge on the land.

The near equality of landspreading benefits and costs makes landspreading extremely attractive when compared with other methods of sludge disposal. Incineration, landfill, and landspreading are the primary disposal alternatives for the treatment plant operator. Incineration or landfill are practiced in many Ohio communities, but their costs are generally higher than landspreading.

Cost data from Ewing and Dick (5) and Burd (6) indicate that incineration costs are two to three times greater than the costs of landspreading of liquid sludge. Thus, we could expect to find incineration costs in the range of $50–$75 per dry ton. Generally, landfilling of sludge has higher costs than landspreading of liquid sludge. The cost of landfilling with dewatered sludge is approximately 67 percent higher than the disposal costs of landspreading of liquid sludge. Neither landfilling nor incineration would offer any benefits to the municipality or landowner, while landspreading would provide benefits equal to $10–$36 per dry ton if applied to crops as a partial substitute for commercial fertilizer.

Landowner Attitudes

Negative aspects of landspreading sewage sludge may exist. There may be sufficient concentrations of heavy metals in the sludge to cause a buildup of the metals in the soil and toxicity to plants; furthermore, some heavy metals may be taken up by the plants and endanger human health. Runoff of pollutants from sludge disposal areas may affect nearby water quality, and odors may be offensive to neighboring residents. In addition, there has been a continuing concern for the survival of pathogenic viruses and human parasites. However, the adverse impacts of heavy metal buildup in soils, plant toxicity, human health dangers, and noxious odors may not be present or may be safely minimized if the landspreading operation is managed properly.

Although these adverse impacts are controllable, community reaction to landspreading may be negative due to the fear of the unknown and/or improper management. Several communities have proposed landspreading as a method of sludge disposal and have met resistance from local landowners and residents.

The results of the survey indicated that the attitudes of landowners depend upon their level of knowledge of landspreading.

Of those landowners interviewed, 17 percent of the respondents had spread sludge on their land, 45 percent had heard of landspreading sludge, while 38 percent had neither heard of nor practiced landspreading (Table 7). The responses revealed a much different perception of sludge between these three groups. The respondents who had spread sludge tended to be favorable to cautious in their attitude toward landspreading on their land (Question 1, Table 7). Those who had heard of sludge were more cautious in their willingness to accept landspreading on their land, while those who had no knowledge of landspreading had an unfavorable attitude toward spreading (Question 1, Table 7).

All respondents were slightly more negative in their attitude toward a large city spreading sludge on their land (Question 2, Table 7) compared to a local city spreading. Thus, all respondents were more objectionable to others' sludge than local sludge.

Some knowledge of landspreading sludge produced marked improvements in attitudes toward landspreading on neighbors' fields (Question 3, Table 7). Those who had some knowledge (those who had landspread or had heard of it) tended to be agreeable or indifferent towards neighbors' acceptance of sludge. Furthermore, those who had some knowledge did not feel that their neighbors would object to landspreading by the respondent (Question 4, Table 7).

All groups of respondents felt that their neighbors were more opposed to landspreading than were they. The answers to the question asking how respondents would feel about landspreading on a neighbor's fields indicated general indifference (Question 3, Table 7). However, the respondents felt that neighbors would tend to be moderately opposed to spreading on the respondent's fields (Question 4, Table 7).

In general those who had landspread recently seemed anxious to be part of a sludge demonstration project (Question 5, Table 7). The group that had heard of landspreading were less anxious but still tended to accept the demonstration project. Even the unknowledgeable group was not strongly opposed to a demonstration site on their land and appeared to be indifferent toward this education effort.

The landspreading benefits of reduced disposal cost and the value of sludge as a source of plant nutrients were recognized by all groups (Question 6, Table 7).

CONCLUSIONS

Landspreading is receiving renewed interest as a method of sludge disposal due to a variety of price, institutional and attitudinal changes. Sewage sludge may be applied to land in an environmentally safe and economical manner; however, excessive application rates may prove injurious to the soil, plant growth, and water quality.

Results from a recently conducted survey point out the extent of current landspreading in Ohio and the current state-of-the-arts in analysis, monitoring, application techniques and costs. About 50–60 Ohio communities currently are spreading sewage sludge on land. While the acreage covered by this quantity of sludge is a small percentage of the total cropland, sludge application may affect large numbers of farmers, community officials and the community at large.

Survey results indicate that most Ohio communities that apply sludge to land have some notion of the quality of the sludge. Larger sewage treatment plants generally have better sludge analysis programs. Several cities practice only occasional sludge analysis, and approximately 20 percent performed no sludge analysis. Soil testing and soil monitoring programs are not being conducted by most communities. Water quality is being monitored by several of those surveyed.

The vast majority of landspreading communities are using tank trucks as their principal method of disposal. Overland flow irrigation is being used by a few communities as an alternative method of landspreading.

Several communities are using contracts with private landowners in order to solidify the relationship within the community; however, a small proportion of the contracts are written and most agreements are in the form of an oral understanding.

Average disposal costs for surveyed landspreading communities decline as treatment plant capacity increases to 25 MGD. At this size level disposal costs average $25 per dry ton. For the mean size of Ohio treatment plants, disposal costs average $31 per dry ton. Benefits received from sludge range from $10–$36 per dry ton when used as a supplement for commercial fertilizer. These net benefits would appear to make it an attractive sludge disposal alternative for many communities.

The small sample size limits the statistical inferences one may draw from this survey of attitudes toward landspreading. However, some general trends in attitudes are noticeable. First, rural residents do not have much knowledge of sludge disposal even in counties where the cities have conducted landspreading in the past. Either a low profile of landspreading in the past or harmonious community relations have made residents unaware of landspreading. Second, opposition to landspreading is the strongest from those who know little about it. An educational program and

Table 7. Summary of landowners' responses from five-county survey of sludge disposal on farmland.

	Those Who Have Spread	Those Who Haven't Spread But Have Heard of Spreading	Those Who Have Not Spread Nor Heard of Spreading
Percentage of Respondents	17	45	38
Question 1: How would you react to spreading on your land?		- Mean Responses -	
1 = enthusiastically	2.50	3.05	3.81[a]
2 = favorably			
3 = cautiously			
4 = unfavorably			
5 = opposed			
Question 2: How would you react to a large city's proposal to spread on your land?	2.85	3.48	4.06[b]
1 = enthusiastically			
2 = favorably			
3 = cautiously			
4 = unfavorably			
5 = opposed			
Question 3: How would you feel about sludge being spread on neighbors' fields?	1.75	2.00	2.41
1 = agreeable			
2 = indifferent			
3 = moderately opposed			
4 = strongly opposed			
Question 4: How would neighbors react to sludge spread on your land?	1.87	2.71	3.12[c]
1 = agreeable			
2 = indifferent			
3 = moderately opposed			
4 = strongly opposed			

Table 7. (Continued)

	Those Who Have Spread	Those Who Haven't Spread But Have Heard of Spreading	Those Who Have Not Spread Nor Heard of Spreading
Question 5: Would you allow your land to be used as a demonstration site?	1.63	2.48	3.24[a]
1 = yes, I would be most happy.			
2 = yes, if rates are safe.			
3 = No, spreading is all right but I don't want a site.			
4 = No, my neighbors would be opposed.			
5 = No, I am opposed to sludge spreading.			
		- top 3 responses -	
Question 6: What are the reasons for landspreading?	1. reduce disposal costs	1. fertilizer substitute	1. fertilizer substitute
	2. fertilizer substitute	2. reduce disposal costs	2. get it away from city
	3. lessen pollution	3. lessen pollution	3. reduce disposal costs

[a]Responses are significantly different between groups at the 0.10 level.

[b]Responses are significantly different between groups at the 0.05 level.

[c]Responses are significantly different between groups at the 0.01 level.

exposure in the local communications media would raise the willingness of residents to accept landspreading. Third, sludge from large, distant cities is viewed more negatively than local sludge; the sludge of another is viewed with some disdain. Fourth, there is little opposition to spreading on neighbors' fields. However, landowners tend to view neighbors' attitudes as less supportive of landspreading than their own. Finally, most respondents, regardless of their knowledge of landspreading, correctly identify the principal benefits of sludge—providing plant nutrients and a low cost method of disposal.

ACKNOWLEDGMENTS

This study is part of a larger project sponsored by the Ohio Farm Bureau Federation, the Ohio Environmental Protection Agency, and the Ohio Agricultural Research and Development Center.

REFERENCES

1. Forster, D. L., T. J. Logan, R. H. Miller, P. R. Thomas, and R. K. White. "Is Municipal Sewage Sludge a Resource for Agriculture," Department of Agricultural Economics and Rural Sociology, ESS 528, The Ohio State University (1976).
2. U.S. Congress. Public Law 92–500, October 18, 1972.
3. Carroll, T. E., et al. Review of Landspreading of Municipal Sewage Sludge. Environmental Technology Series, EPA–670/2–75–049, U.S. Environmental Protection Agency (1973).
4. Ohio Guide for Land Application of Sewage Sludge. Bulletin 598, Cooperative Extension Service. The Ohio State University, Columbus, Ohio (1975).
5. Ewing, B. B. and R. I. Dick. Disposal of Sludge on Land, Water Quality Improvement by Physical and Chemical Processes, University of Texas Press, Austin (1970).
6. Burd, R. S. A Study of Sludge Handling and Disposal. Federal Water Pollution Control Administration, Publication WP–20–4 (1960).

32

ENVIRONMENT CANADA'S RESEARCH AND DEVELOPMENT ACTIVITIES IN LAND APPLICATION OF SLUDGES

V. K. Chawla, D. B. Cohen and D. N. Bryant
Wastewater Technology Centre, Environmental Protection
Service, Environment Canada, Burlington, Ontario, Canada

INTRODUCTION

In response to the Canada/U.S. Agreement (1) on Great Lakes Water Quality, the governments of Canada and Ontario through the Canada/Ontario Agreement (2), have provided joint funding to a total of six million dollars over a five-year period for research programs. Sludge treatment and disposal is one of the high priority research areas. Field trials with different crops, soils and sludges are being conducted by the University of Guelph. The regional municipality of Niagara is investigating mechanical application methods. The Canada Department of Agriculture is conducting laboratory investigations of soil–sludge complexes. To complement other studies, Environmental Canada at the Wastewater Technology Centre (WTC) has initiated several long-term studies.

The WTC of the Environmental Protection Service, Environment Canada, has the national responsibility to develop technology and provide solutions for treatment and disposal of municipal and industrial wastewaters and residues. A number of projects are being conducted by the staff of the WTC to study the long-term environmental effects of sludge applications on land. A variety of sludges including digested and chemically precipitated phosphorus sludges using alum, iron or lime have been applied to both agricultural and nonagricultural soils.

RECYCLING OF LIQUID DIGESTED SEWAGE SLUDGE ON DREDGED RIVER SAND

A recently completed study (3) assessed the impact of primary digested sludge application on dredged river sand from Vancouver, B.C. The study objectives were:

1. To assess the response of the sand/grass system to different sludge loadings.
2. To monitor the decomposition of sludge constituents in soil and the accumulation of nutrients and metals in soil and plants.
3. To measure the extent of nutrients, metals, minerals and indicator bacteria leaching through the sludged sand.

This study, conducted at the Iona Island Sewage Treatment Plant, Vancouver, B.C. consisted of a four-plot site (0.4 hectares) with each plot polyethylene-lined, covered by 1.2 m of dredged sand (Figure 1). In October 1972, primary digested liquid sludge (1.17 percent TS and 1085 mg TKN/1) was applied to the experimental plots. Plots 2, 3 and 4 were loaded with liquid sludge at 1435, 2635 and 5705 m³/ha, respectively. The corresponding equivalent amounts of TKN were 1560, 2860 and 6190 kg/ha, respectively. Plot 1 was used as a control.

In March 1973, a special grass mixture (15 percent rye grass, 15 percent clover, 30 percent blue grass and 40 percent creeping red fescue) was seeded at a rate of 112 kg/ha. Supplementary irrigation with tap water was applied at the rate of 3.8 cm/week from May to October, 1973.

Forage Composition and Response

As expected, nitrogen-fixing clover was the only successful crop on the control plot. Fescue grass was the dominant variety that became established on all of the sludged plots. The data on total dry matter yield and mean chemical composition of forage are summarized in Table 1. The highest sludge application produced 2170 kg dry matter/ha compared to medium texture agricultural soil in the study area with yields averaging 4000 kg/ha.

Concentrations of N and K in the plant tissue were found to be below normal and these deficiencies may have affected the dry matter yields. Concentrations of all heavy metals analyzed were within the normal concentration range.

Leachate Composition

The experimental period was divided into two stages, the non-growing season (October 1972 to April 1973), and the crop-growing season (May to December 1973). The leachate composi-

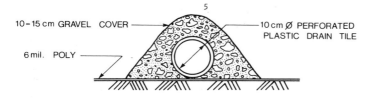

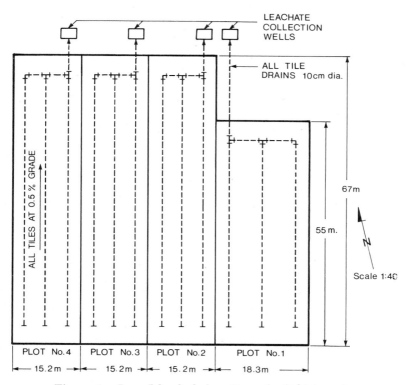

Figure 1. Iona Island sludge disposal—field layout.

tion during both stages is summarized in Table 2. Mean leachate concentrations of TKN, NH_4-N and NO_3-N at the highest sludge application rates were many times higher than the leachate from the control plot.

NH_4-N leachate concentration decreased from 70 mg NH_4-N/l one month after initial application, to less than 1 mg NH_4-N/l nine months after the initial application. At concentrations of 70 mg/l, the undiluted leachate would be acutely toxic to rainbow trout and coarse fish.

Table 1. Forage chemical composition and dry matter yield (April to December 1973).

Constituent		Plot 1 (Control)	Plot 2	Plot 3	Plot 4	Normal Conc.	Ref.
N	(%)	2.57	1.70	2.62	2.82	3.4–3.8	(4)
P	(%)	0.12	0.21	0.32	0.36	0.3–0.4	(4)
K	(%)	2.00	1.61	1.95	1.93	3.0–3.5	(5)
Ca	(%)	1.84	0.49	0.55	0.58	0.9–1.6	(6)
Mg	(%)	0.47	0.17	0.23	0.22	0.3–0.8	(6)
Fe	(μg/g)	508	321	243	219	50–250	(7)
Mn	(μg/g)	295	450	489	559	20–500	(7)
Cu	(μg/g)	9	6	10	11	5–20	(7)
Zn	(μg/g)	37	41	62	69	25–150	(7)
Pb	(μg/g)	6	4	7	8	0.1–10	(8)
Ni	(μg/g)	14	21	25	22	1–50	(8)
Cd	(μg/g)	0.6	0.4	0.4	0.5	0.2–0.8	(8)
Dry Matter (kg/ha)		67	1175	1125	2170		

Table 2. Mean composition of leachate from experimental plots during the nongrowing and growing seasons.

Constituent [a]	Nongrowing Season (Oct 72 to Apr 73)				Growing Season (May to Dec 73)			
	Plot 1	Plot 2	Plot 3	Plot 4	Plot 1	Plot 2	Plot 3	Plot 4
TKN	0.2	0.2	18	86	0.8	0.6	2.5	16
NO_3-N	1	22	26	7	4	48	126	91
NH_4-N	1	2	18	47	0	0.1	21	68
Total P	0.22	0.22	0.53	2.41	0.12	0.06	0.06	0.12
Soluble P	0.03	0.06	0.18	0.60	0.06	0.03	0.04	0.06
Ca	20	71	66	63	18	45	58	45
Mg	9	40	83	67	8	20	38	30
Na	20	26	59	52	6	15	21	16
K	6	12	30	43	6	10	35	41
Fe	10	3	9	18	3	0.8	0.7	0.5
Mn	0.2	0.2	2.4	3.4	0.3	0.2	3.2	5.4
Cu	0.02	0.04	0.06	0.19	0.03	0.05	0.10	0.13
Zn	0.03	0.02	0.04	0.03	0.03	0.04	0.06	0.08
Pb	0.03	0.02	0.03	0.02	0.03	0.03	0.04	0.04
Ni	0.08	0.08	0.20	0.40	0.06	0.06	0.17	0.20
Cd	0.03	0.03	0.03	0.04	0.04	0.04	0.04	0.04
pH	7.0	7.3	7.2	7.3	6.8	6.9	6.2	6.2
Total Coliform (MPN/100 ml)	4.3×10^2	6.7×10^4	1.1×10^5	1.2×10^5	–	1.5×10^4	1.8×10^4	1.8×10^4
Fecal Coliform (MPN/100 ml)	10	20	3.3×10^2	3.7×10^2	2	2	2	2

[a] Mean values in mg/l except pH and coliform bacteria.

While NO_3-N leachate levels from the control plot generally did not exceed 1 mg/l, concentrations up to 240 mg/l were measured from the sludged plots (Figure 2). These concentrations are high compared to nitrified sewage effluent and exceeded the maximum permissible limit for drinking water of 10 mg NO_3-N/l (9).

At the highest sludge application rate, leachate mean metal concentrations were 0.16 mg/l copper, 0.05 mg/l zinc, 0.03 mg/l lead, 0.32 mg/l nickel and 0.03 mg/l cadmium. These concentrations were within the maximum permissible levels for drinking water.

Approximately six months from the time of initial sludge application were required for fecal coliform bacteria counts to decline to the standard acceptable for irrigation water of less than 100 MPN/100 ml (10).

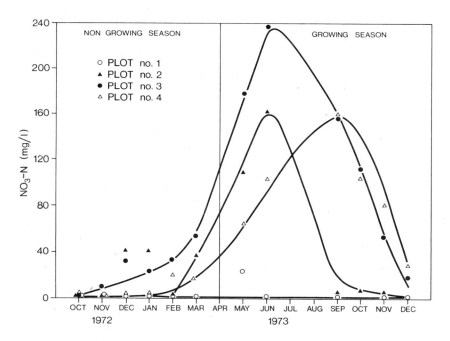

Figure 2.   Concentration of nitrate–N in leachate.

Leachate Nutrient and Metal Losses

Losses of leachate from the sludged plots ranged between 112 to 273 kg N/ha (Table 3) compared to 14 kg N/ha from a fertilized corn plot in Southwestern Ontario (11). The high losses of K, Ca and Mg are significant from a soil fertility point of view.

The losses of all heavy metals analyzed in the leachate were negligible ($\leqslant$0.2 kg/ha) compared to the amounts added. Con-

Table 3.   Leachate nutrient and metal losses (May to December 1973).

Element (kg/ha)	Plot 2	Plot 3	Plot 4
Nitrogen	112	273	154
Phosphorus	0.1	0.1	0.2
Potassium	21	78	55
Calcium	97	117	59
Magnesium	43	72	40
Iron	2.5	0.6	0.2
Manganese	0.3	7.0	5.8
Copper	0.10	0.20	0.20
Zinc	0.07	0.13	0.10
Lead	0.05	0.08	0.05
Cadmium	0.08	0.08	0.05

tinuous accumulation of these metals in sludged sand could pose a soil/plant toxicity problem.

Metal Enrichment of Sludge Sand

The number of years required to reach the normal soil concentrations (8) and maximum recommended concentrations (12) of heavy metals at the highest application rate are summarized in Table 4. The normal concentrations of Cd and Pb could be reached in less than one year, while it would take 16 years before Cd and 225 years before Cu concentrations would reach the maximum recommended concentrations if applied annually at the rate of 68 tdws/ha yr. Based on the above-mentioned criteria, Cd is the most limiting heavy metal.

Table 4.   Comparison of heavy metals in Iona Island sludged sand with normal and maximum recommended concentrations.

Heavy Metals	Plot 4 ($\mu g/g$)	Normal Soil[a] Conc. ($\mu g/g$)	Maximum Recommended[b] Conc. ($\mu g/g$)	Time Required to Reach	
				Normal Conc. (years)	Max. Conc. (years)
Cu	3.6	20	816	5.5	225
Zn	30.8	50	1632	1.6	53
Pb	24.9	10	1632	< 1.0	65
Cd	0.5	0.06	8	< 1.0	16
Mn	8.2	850	1632	104	199

[a] Allaway (8).

[b] Schwing, et al., (12).

Sludged Sand Fertility

The highest sludge application rate increased organic matter content to 1.3 percent, compared to a control level of 0.5 percent (April 1974). A similar increase was reported by Hinesly and Sosewitz (13) in the A horizon of an Illinois sand dune where sludge was applied at the rate of 210 tdws/ha yr.

If the organic matter continued to accumulate in the dredged sand at this rate, the normal soil organic matter level ($\sim$4 percent) would be reached in five to six years of continuous sludge application at the rate of 68 tdws/ha yr.

Loss of available Ca, Mg, and K in the sludged sand was proportional to the amount of liquid sludge applied. In order to prevent depletion of these constituents, supplemental amounts must be added to maintain the long-term fertility of the reclaimed sand.

In summary, the organic matter content of the dredged sand increased from an initial 0.5 percent to 1.3 percent, 18 months after sludge addition at the highest loading rate of 6190 kg TKN/ha (68 tons dry solids/ha). Grass yields increased with increasing rates of sludge application up to 2170 kg total dry matter/ha. No plant toxicity attributable to heavy metals was observed. By the 18th month, heavy metals, nutrients and bacterial concentrations in the leachate met Ontario irrigation water standards.

RECLAMATION OF ACID MINE TAILINGS WITH LIME TREATED SEWAGE SLUDGES

The objectives of this study were:

1. To determine the optimum sludge application rate required to establish vegetation on the tailings.
2. To monitor the changes in the tailings, vegetation and leachate.

Additions of both liquid and dry sludges to the tailings (Figure 3) were based on the lime ($CaCO_3$) required for acid neutralization. In April 1973, sludges were applied to tailings at the rate of 0.3, 0.6 and 1.0 lime requirements (L.R.). Because these rates were not effective, sludges were again applied during September to October 1973 to a cumulative loading rate of 3.5, 7.0 and 10.5 times the lime requirement. These rates were further increased in April 1974 equivalent to a cumulative 5, 10 and 15 times L. R. (Table 5).

Oats seeded on January 23, 1974, showed poor germination and growth at the lower sludge application rates. Even at the highest application rate, dry plant material was insufficient to provide samples for chemical analysis. A second crop (alsike clover) was seeded on May 23, 1974, but again showed very poor germination and growth. A third crop (reed canarygrass) was seeded on June

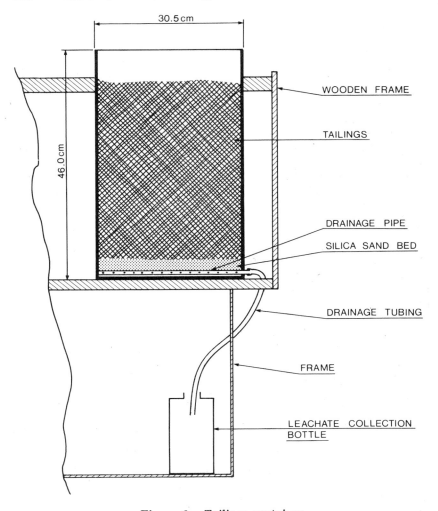

30.5 cm

46.0 cm

WOODEN FRAME

TAILINGS

DRAINAGE PIPE

SILICA SAND BED

DRAINAGE TUBING

FRAME

LEACHATE COLLECTION
BOTTLE

Figure 3. Tailings container.

28, 1974 and harvested on October 23, 1974. Only the higher sludge application rates could sustain plant growth.

Composition and Yield of Reed Canarygrass

The highest dry matter yield (2440 k/ha) was obtained from the highest rate of sludge liquid application (Table 6) and was comparable to yields from poor soils.

Potassium in plant tissue was very deficient (0.1 to 0.33 per-

Table 5. Cumulative sludge application rates.[a]

Liquid Sludge		Dry Sludge	
Lime Requirements	m^3/ha	Lime Requirements	t/ha
5.0	2550	4.25	200
10.0	5100	8.50	400
15.0	7650	12.75	600

[a] Based on lime requirement of tailings (7.4 kg $CaCO_3$/ton of tailings).

cent), Fe and Zn concentrations were high, while Cu and Pb concentrations were very high and possibly toxic.

Surface pH of Tailings

In May 1973, seven weeks after the initial sludge applications, the pH of the tailings were ≥ 4 from the highest liquid and dry sludge application rates. In December 1973, eight weeks after the second sludge application, the pH was ≥ 6 from all sludge applications. By November 1974, only the middle and the high rates of liquid sludge, and the high rate of dry sludge application were able to maintain the pH ≥ 6 (Figure 4).

In summary, a series of crops including oats, clover and reed canarygrass were seeded to establish vegetative cover on acid mine tailings sludged with lime primary digested sludge. The only

Table 6. Composition and yield of reed canarygrass at highest sludge application rates (July to October 1974).

Constituents		Liquid Sludge	Dry Sludge
N	(%)	2.97	3.33
P	(%)	0.11	0.31
Ca	(%)	1.07	2.23
Mg	(%)	0.40	0.39
K	(%)	0.33	0.10
Fe	($\mu g/g$)	1650	1720
Zn	($\mu g/g$)	133	142
Cu	($\mu g/g$)	28	59
Ni	($\mu g/g$)	3	2
Pb	($\mu g/g$)	37	106
Dry matter	(kg/ha)	2440	440
Application rate t/ha		460	600
Application rate m^3/ha		7650	—

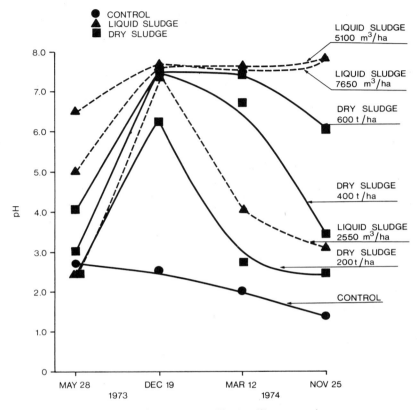

Figure 4. Surface pH of tailings vs. time.

treatments found capable of supporting reed canarygrass were the highest rates of dry and liquid lime sludges. The pH of the sludge surface tailings mixture was $\geqslant 6$ at the highest loading rates.

LIQUID CHEMICAL SEWAGE SLUDGE DISPOSAL ON LAND

A lysimeter study was initiated in 1972 using liquid digested alum, iron or lime sludges. The objectives of this study were:

1. To investigate crop response to sludge application rates.
2. To investigate the effects of sludge on the concentrations of nutrients and metals in plant, soil, and leachate.
3. To monitor the effect of sludge application on microbial quality of soil and leachate.

Material balances were calculated for all the constituents analyzed. Liquid digested alum,, iron or lime sludges were applied
to orchard grass grown on loamy sand and silt loam soils in 66
lysimeters (Figure 5). Three different sludge rates up to a maximum of 2100 kg TKN/ha were applied over two years.

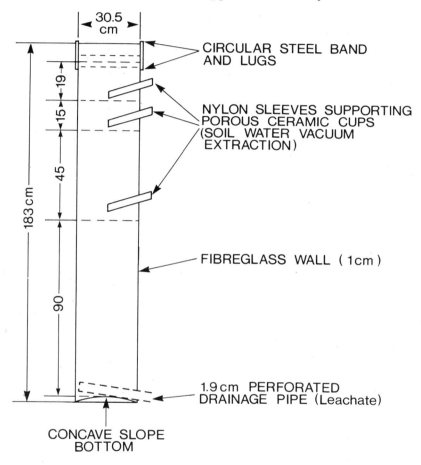

Figure 5. Typical lysimeter.

Alum, ferric chloride and lime treated sewage sludges were
applied at an annual rate of 300, 600 and 900 kg TKN/ha in 1973,
and 400, 800 and 1200 kg TKN/ha in 1974. The NPK annual rate
of application was 336-112-112 kg/ha in 1973, and 400-133-133
kg/ha in 1974.

Dry Matter Yields

Total dry matter yields obtained in 1973/74 are summarized in Table 7. Increasing sludge application rates significantly increased yields on both soils for all sludge types (Figure 6). Yields from all sludge treatments were higher during 1974 than 1973. Highest rate of iron sludge in both years on both soils out-yielded all other treatments. Low 1973 yields from alum sludge treatment may be related to high petroleum hydrocarbon concentrations in this sludge. This was rectified in 1974 by adding a different alum sludge (Tillsonberg).

Concentrations of Chemical Elements in Plant Tissue

The range of mean concentrations of N, P, K, Ca, Mg, Na, Fe, Mn, Al, Cu, Zn, Ni, Cr, Pb and Cd in the orchard grass from all

Table 7.   Orchard grass yields 1973/74 (kg dry matter/ha).

Treatments	Rate (kg/ha) TKN Equiv.		Loamy Sand		Silt Loam	
	1973	1974	1973	1974	1973	1974
Control	0		4262	2516	8378	5256
NPK	336-112-112	400-133-133	10924	12667	13240	15180
Alum Sludge[a]	900	1200	7080	14125	10202	18669
Iron Sludge[a]	900	1200	11635	15362	15175	20540
Lime Sludge[a]	900	1200	10593	14211	14427	17390
S.E.[b]	±		388	521	388	521

[a] Highest rate of sludge application.

[b] S.E. = Standard Error; differences significant at P – 0.01.

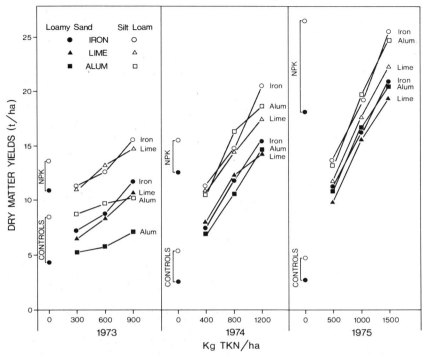

Figure 6. Orchard grass yields vs. sludge application rates 1973–75.

sludge treatments are compared with nonsludge controls as well as the reported normal range in plants (Table 8). The chemical element concentrations in orchard grass were within the normal range except for K, Fe, Al and Pb. No toxic effects were observed for any constituent. K concentrations were lower than normal in 1973 and deficient in 1974 particularly from loamy sand. Supplementary potassium was applied during 1974/75 to improve the N/K ratio thus reducing the K deficiency.

Increasing rates of sludge application increased the plant tissue concentrations of N, Na, Cu and Zn while K and Mn decreased. Concentrations of P, Ca, Mg, Al, Fe, Ni, Cr, Pb and Cd were not affected by sludge application rates. The highest concentrations of Al, Fe and Ca, respectively, were directly related to the chemicals used for P removal.

Leachate Chemical Concentrations

The range of seasonal mean chemical concentrations of NH_4-N, NO_3-N, Total-P, TOC, SO_4, Ca, Mg, Fe, Mn, Cu, Zn, Ni, Cr, Pb

Table 8.   Orchard grass chemical element concentrations (1973/74) compared with normal range in plants.

Constituents	Units	Nonsludge Controls (16)	Sludge Treatments(16)	Normal Range	Ref.
N	%	1.98–2.76	1.93–2.90	3.0	(17)
P	%	0.17–0.30	0.19–0.38	0.23–0.28	(5)
K	%	1.84–3.10	0.60–3.07	2.5	(18)
Ca	%	0.48–0.70	0.51–0.71	0.21–1.00	(19)
Mg	%	0.29–0.38	0.31–0.40	0.31–0.51	(19)
Na	%	0.03–0.49	0.03–0.80	0.46–1.95	(20)
Fe	μg/g	219–317	181–412	50–250	(7)
Mn	μg/g	114–245	57–201	20–500	(7)
Al	μg/g	63–185	57–881	300	(21)
Cu	μg/g	5.2–8.0	5.3–14.8	5–20	(7)
Zn	μg/g	36–53	31–112	25–150	(7)
Ni	μg/g	1.0–1.4	0.7–1.6	1–50	(8)
Cr	μg/g	0.5–1.9	0.5–1.8	0.74–2.07	(22)
Pb	μg/g	5.7–9.9	5.7–16.1	0.1–10.0	(8)
Cd	μg/g	0.16–0.74	0.17–0.63	0.2–0.8	(8)

and Cd in the leachate from all sludge treatments are compared with the nonsludge controls as well as the maximum permissible limits (MPL) for potable water, irrigation water or sewage effluent (Table 9). Concentrations of SO_4, Mg, Fe, Cu, Zn, Cr and Cd were within the Canadian drinking water MPL (9). Mn, Pb and Ni were within the MPL for Ontario irrigation waters (10). No potable or irrigation water standards exist for TOC or Total-P. TOC concentrations in the leachate were therefore compared with the MPL in treated wastewater effluent of 15 mg/l BOD_5 (14). Total phosphorus concentrations were compared with the wastewater effluent MPL of 1.0 mg P/l (1).

Although NO_3-N, Ca and TOC (BOD_5) in leachate from sludge treatments were higher than their respective MPL, they were in the same concentration range as the nonsludge controls. By the second growing season (1974), NO_3-N concentrations did not exceed the MPL for drinking water (10 mg NO_3-N/l). Concentrations of NH_4-N were above the MPL of 0.5 mg/l for drinking water during the 1973 growing season, but dropped and remained below the MPL during 1974. Of all the leachate constituents monitored, TOC is the only one of ongoing concern after the second growing season.

Leachate Bacterial Densities

All species of bacteria in the leachate were of soil origin. The

Table 9.   Leachate chemical concentrations (1973/74) compared with water quality standards.

Constituents	Units	Nonsludge Controls (16)	Sludge Treatments (16)	Maximum Permissible Limit	Ref.
NH_4-N	mg/l	<0.001–0.17	<0.001–1.43	0.5	(9)
NO_3-N	mg/l	0.1–33.3	0.1–34.6	10.0	(9)
Total P-P	mg/l	0.01–0.73	0.01–0.95	1.0	(1)
TOC	mg/l	5.4–28.3	5.9–25.3	15.0 (BOD_5)	(14)
SO_4	mg/l	12–175	49–217	500	(9)
Ca	mg/l	46–269	54–242	200	(9)
Mg	mg/l	11–35	11–42	150	(9)
Fe	μg/l	50–240	42–198	300	(9)
Mn	μg/l	6–212	8–521	$50;20x10_3$	(9,10)
Cu	μg/l	7–94	8–85	1000	(9)
Zn	μg/l	6–93	6–85	5000	(9)
Ni	μg/l	3–19	2–42	2000	(10)
Cr	μg/l	1–5	1–12	50	(9)
Pb	μg/l	9–58	7–59	$50;20x10^3$	(9,10)
Cd	μg/l	1–6	1–7	10	(9)

heterotrophic bacterial concentrations remained fairly constant for all treatments and ranged between 6.0 x 10^4 to 3 x 10^6/ml for both years.

Bacterial Movement in Soil

Separate sludged soil core samples to a depth of 60 cm were taken in April 1974 to determine the vertical distribution of heterotropic bacteria from the highest sludge application rates of 1973 (Figures 7 and 8). Most of the heterotrophic bacteria (> 95 percent) were retained in the upper 15-cm layer, while bacteria of fecal origin (> 95 percent) were retained in the top 7.5-cm layer.

Sludged Soil Chemical Constituent Material Balances

The common tillage depth (15 to 20 cm) is used to calculate the maximum allowable metal concentrations that can be applied to soils (15). Material balances (mg/kg soil) for N, P, Al, Fe, Zn, Cu, Ni, Cr, Pb and Cd were calculated for A horizon (19 cm).

The number of years required to double the initial content of soil chemical constituents are presented in Table 10. The doubling time criteria for maximum metal content permitted in soil is taken from the OMAF guidelines (15).

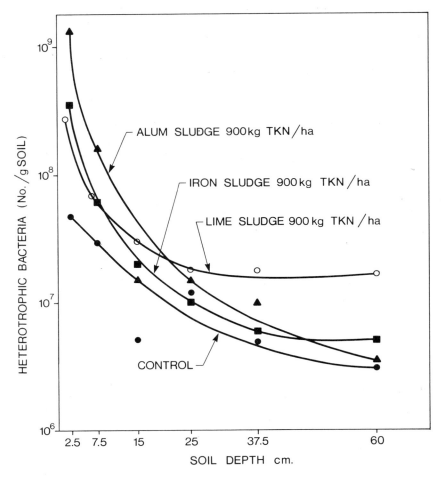

Figure 7. Heterotrophic bacteria in sludged loamy sand—April 1974.

The number of years required to double the initial content of soil is dependent on its initial content and the net annual change [input — (plant uptake + leachate loss)]. As the initial content of chemical constituents in silt loam was higher than loamy sand, the doubling time was usually greater for silt loam. On the basis of sludge constituent concentrations used in this study, the shortest doubling time for Fe, Cu, Zn, Cr, Pb and Cd will occur with iron sludge; N, P and Al with alum sludge and Ni with lime sludge, at the highest application rates.

In summary, liquid digested alum, iron or lime sludges were

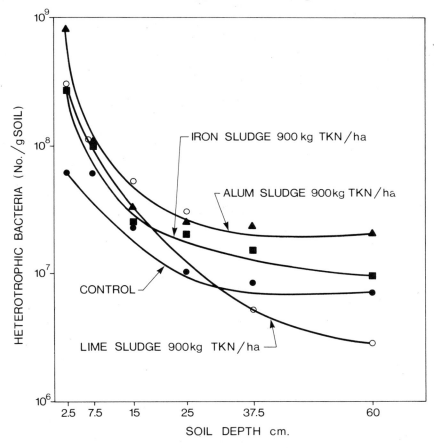

Figure 8. Heterotrophic bacteria in sludged silt loam—April 1974.

applied up to a maximum of 2100 kg TKN/ha over two years to orchard grass grown on loamy sand and silt loam soils. Grass yields were not affected by the chemicals used for P removal, but were directly related to TKN loading rates. Following the first growing season (1973), TOC was the only constituent to increase significantly in the leachate with successive sludge applications. The mean concentrations of other chemical constituents monitored in the leachate were within the maximum permissible Canadian drinking water or Ontario irrigation water limits. Over 95 percent of sludge bacteria applied to soils were retained in the upper 15-cm layer.

Table 10. Number of years required to double the initial metal content of soils.[a]

| Constituent | Initial Content of A Horizon | | Years Required to Double the Content | |
	Loamy Sand (mg/kg) [b]	Silt Loam (mg/kg) [b]	Loamy Sand	Silt Loam
Fe	23000	34000	29	43
Al	45000	48000	202	215
Cu	13	20	1	2
Zn	105	100	4	4
Ni	9	17	31	59
Cr	27	45	5	8
Pb	18	16	1	1
Cd	0.1	0.1	< 1	< 1

[a] At the highest cumulative (1973/74) rate of iron sludge application (2100 kg TKN/ha).
[b] mg/kg x 2.24 = kg/ha (15-cm depth).

UTILIZATION OF AIR-DRIED CHEMICAL SEWAGE SLUDGE ON AGRICULTURAL LANDS

Another long-term study was initiated in 1973 using dry (~5 percent moisture) alum, ferric chloride or lime sludges on sand and clay soils. The major objective of this study was to investigate the effects of heavy metals in chemical sewage sludges on plants, leachate and soil.

Eleven treatments including control, NPK and three air-dried chemical sludges at three different zinc equivalent application rates were used in 44 lysimeters (Figure 9).

In May 1974, air-dried sludges were mixed with the surface 15-cm layer of soils and spring wheat (Glenlea variety) was seeded on May 23, 1974. Sludges were applied in May 1974 prior to spring wheat at rates of 57, 114 and 171 kg zinc equivalent/ha, and similar rates were again applied in September 1974 prior to the seeding of fall wheat. The NPK application rate was 60-60-60 kg/ha for 1974 spring wheat and 120-60-60 kg/ha for 1974/75 fall wheat.

Grain and Straw Dry Matter Yields

Total (grain and straw) dry matter yields of 1974 spring wheat and 1974/75 fall wheat are shown in Figures 10 and 11.

The highest application of iron sludge to spring wheat on both soils depressed the total dry matter (grain and straw) yield as compared to alum and lime sludges. The iron sludge had the lowest TKN amount applied compared to the other sludges at similar

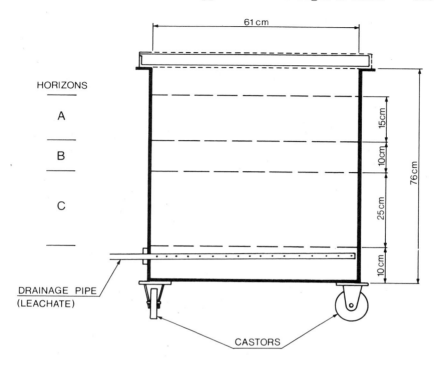

Figure 9.   Experimental lysimeter.

zinc equivalent application rates. Highest application of all three sludges to fall wheat depressed the total yield on clay soil, but significantly increased the yield on sand. The higher fall wheat yields from sandy soil may be related to increased nitrogen availability. The heavy metal concentrations in grain and straw were well below toxic levels.

Leachate Chemical Quality

Concentrations of heavy metals in the leachate were low during 1974/75 and were within irrigation water MPL (10). Concentrations of NO_3-N in leachate at the highest rate of sludge application (cumulative zinc equivalent 342 kg/ha) are shown in Figure 12. During the dormant period (October 1974 to February 1975), the NO_3-N concentrations exceeded 300 mg/l from alum and lime sludges and exceeded 150 mg/l for iron sludge on both soils. Even during the growing period (April to July 1975), the NO_3-N concentrations from both soils for all three sludges exceeded 50 mg/l. These high NO_3-N concentrations may be the major limiting fac-

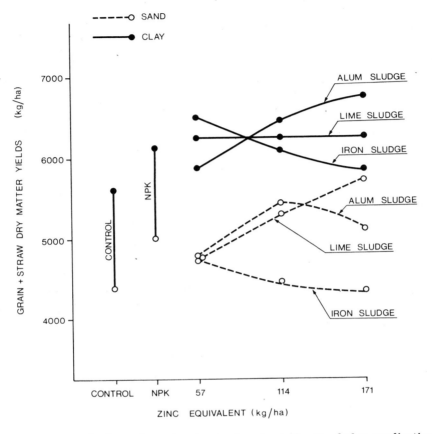

Figure 10. Spring wheat (grain + straw) yields vs. sludge application rates 1974.

tor rather than heavy metals from dry sludge application on agricultural land.

In summary, air-dried digested alum, iron or lime sludges were applied to sand and clay soils up to a maximum of 342 kg zinc equivalent/ha. Spring wheat total dry matter yield increases were proportional to alum sludge application on clay soil and to lime sludge application on sandy soil. The chemical composition of wheat grain and straw was within the normal range in plants. Leachate nitrate-nitrogen concentrations increased with sludge loadings compared to control and NPK treatment. All other chemical constituents monitored in the leachate were within the maximum permissible Canadian drinking water or Ontario irrigation water limits.

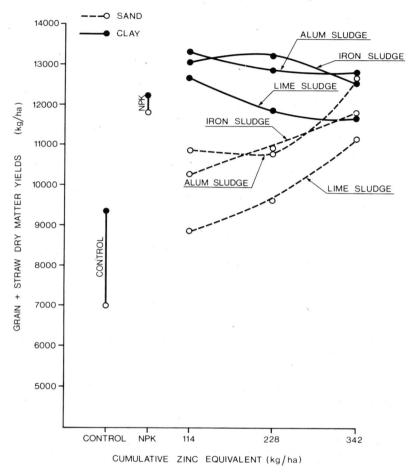

Figure 11. Fall wheat (grain + straw) yields vs. sludge application rates 1974.

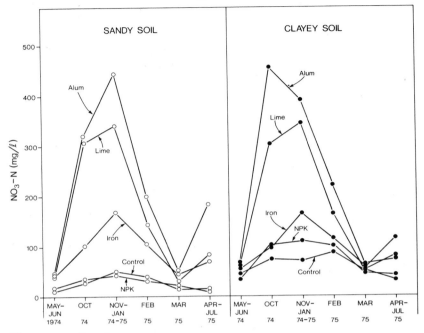

Figure 12. Leachate NO$_3$-N concentrations vs. time at highest sludge loading (342 kg Zn equivalent/ha).

REFERENCES

1. Canada/United States of America. Great Lakes Water Quality Agreement between the United States of America and Canada, U.S. Government Printing Office, Washington, D.C. (1972).
2. Canada/Ontario Agreement (1971).
3. Chawla, V. K., J. Yip and D. B. Cohen. "Recycling of liquid digested sewage sludge on dredged river sand," Technology Development Report EPS 4–WP–76–3 Ottawa (1976).
4. Hallock, D. L., R. H. Brown and R. E. Blaser. "Response of coastal and midland Bermuda grass and Kentucky 31 fescue to nitrogen in southeastern Virginia," Virginia Agr. Exp. St. Res. Rep. 112 (1966).
5. Reid, R. L., G. A. Jung and C. M. Kinsey. "Nutritive value of nitrogen fertilization of orchard grass pasture at different periods of the year," *Agron. J.* 59:519–525 (1967).
6. Jones, J. B., Jr. "Distribution of 15 elements in corn leaves," *Comm. Soil Sci. Plant Anal.* 1:27–34 (1970).
7. Jones, J. B., Jr. "Plant tissue analysis for micronutrients," In: *Micronutrients in Agriculture,* pp. 319–341 SSSA, Madison (1972).
8. Allaway, W. H. "Agronomic controls over environmental cycling of trace elements," *Advances in Agronomy* 20:235–274, Academic Press Inc., New York (1968).

9. National Health and Welfare (NH&W). *Canadian Drinking Water Standards and Objectives,* Queen's Printer, Ottawa (1969).

10. Ontario Ministry of the Environment. (OMOE). 1974 Guidelines and Criteria for Water Quality Management in Ontario, Toronto (1974).

11. Bolton, E. F., J. W. Aylesworth and F. R. Hore. "Nutrient losses through tile lines under three cropping systems and two fertility levels on a Brookston clay soil," *Can. J. Soil Sci.* 50:275–279 (1970).

12. Schwing, J. E. and J. L. Puntenney. "Recycle sludge to feed farms," *Water and Waste Engineering* 11:24–78 (1974).

13. Hinesly, T. D. and B. Sosewitz. "Digested sludge disposal on crop land," *J. Water Poll. Cont. Fed.* 41:822–830 (1969).

14. Ontario Water Resources Commission (OWRC). Industrial Pollution Control in Municipalities, Toronto (1970).

15. Ontario Ministry of Agriculture and Food (OMAF). Provisional Guidelines for Sludge Utilization on Agricultural Lands, Toronto (1975).

16. Chawla, V. K., D. N. Bryant and D. Liu. "Chemical sewage sludge disposal on land, 1, Canada/Ontario Agreement Research Report. (in preparation, 1976).

17. Kresge, C. B. and S. E. Younts. "Response of orchard grass to potassium and nitrogen fertilization of a wickham silt loam," *Agron. J.* 55:161–164 (1963).

18. Robinson, R. R., C. L. Rhykerd and C. F. Gross. "Potassium uptake by orchard grass as affected by time, frequency and rate of potassium fertilization," *Agron. J.* 54:351–353 (1962).

19. Neubert, P., W. Wrazidlo, N. P. Vielemeyer, I. Hunt, F. Gullmick and W. Bergman. *Tabellen Zur Pflanzenanalyze-Erste Orientierende Ubersicht,* Institut fur Pflanzenernährung Jena, Berlin (1969).

20. Bower, C. A. and C. H. Wadleigh. "Growth and cationic accumulation by four species of plants as influenced by various levels of exchangeable sodium," *Soil Sci. Amer. Proc.* 13:218–223 (1949).

21. Jones, L. H. "Aluminum uptake and toxicity in plants," *Plant and Soil* 13:297–310 (1961).

22. Prince, A. L. "Influence of soil types on the mineral composition of corn tissues as determined spectrographically," *Soil Sci.* 83:399–405 (1957).

33

CONTINUOUS SUBSURFACE INJECTION
OF LIQUID ORGANIC WASTES

J. L. Smith and D. B. McWhorter
Department of Agricultural Engineering, Colorado State
University, Fort Collins, Colorado.

INTRODUCTION

Land application of wastewater residuals is gaining renewed interest among waste treatment facility planners and engineers. Concern over environmental pollution and energy costs has made many of the conventional disposal methods difficult to accomplish. Disposal or recycle of increasing large amounts of agricultural wastes also poses a problem. In many instances, the lack of comparable controls and technology has made disposal of agricultural wastes more of a pollution hazard than municipal and industrial wastes.

This chapter discusses the use of continuous subsurface injection to apply wastewater residuals on land. The system can apply municipal or industrial wastewater residuals or liquid manure up to 8 percent solids at rates ranging up to 748,000 liters per hectare per pass. The effects of applying municipal sewage sludge and liquid dairy manure on soils, crops and ground water are also discussed. A mass balance approach was used to predict changes in the levels of potential contaminants in the soil and ground water.

SUBSURFACE INJECTION SYSTEM

The subsurface injection system, manufactured by Briscoe, Maphis, Inc., Boulder, Colorado, is shown in Figure 1. This

Figure 1.   Continuous subsurface injection system.

seven-sweep unit is capable of injecting up to 8 percent solids material at rates up to 3,000 liters per minute. The material is pumped through a 200-m hose to the machine and injected into cavities created by wide high-life sweeps. It is thoroughly mixed with soil as the sweeps move. Injection depths of 10 to 20 cm are typical at rates from 187,000 to 748,000 liters per hectare per pass. Application can be repeated at intervals ranging from one day to one week. General specifications for the unit are given in Table 1. Injectors used at other sites are similar to the one shown in Figure 1, but have different capacities. Additional information on the injectors and their development has been presented by Downs *et al.* (1).

Wastewater residuals are delivered from a treatment or storage facility to the application site through underground pipe. Tanks are used for storage in Boulder, Colorado, and for the liquid dairy manure. In Williamsburg, Virginia, a lagoon with a floating cover is used for storage. Other delivery systems can be used depending upon the particular site and application. Approximately 7.5 hectares can be injected from each hose connection.

APPLICATION SITE

The Boulder site consists of 7.6 hectares immediately east of the 4.3-MGD East Pearl Street Treatment Plant. After four years of operation, only about half of the site has been used. Soils at the site are sandy, silty clays and clayey sands overlaying silty, sandy loam. Soil pH ranges from 7 to 7.2. Ground water is located 1.5 m below the ground surface and flow is from east to west at a rate estimated to be 1 m/day. Anaerobically digested sludge is applied to this site at rates of approximately 65,000 dry kg/ha per year.

The Williamsburg site is located southeast of the city of Williamsburg on 20 hectares of land owned by Anheuser-Busch, Inc. The site serves a 10-MGD treatment plant which treats domestic sewage and wastewater from the Anheuser-Busch Brewery in Williamsburg. Soils at the site consist of sandy clays overlaying clayey loams. Soil pH ranges from 5.5 to 6.5. Ground water is located 8 m below the ground surface and flow is southward toward the James River. Aerobically digested sludge is applied to the site at rates of approximately 100,000 dry kg/ha per year.

Table 1. Subsurface injector specifications.

Operating speed	0.8 to 2.4 km/h
Capacity per sweep	0.23 to 0.45 m^3/min
Operating depth	8 to 20 cm
Field efficiency	95%
Tractor power required	
Crawler type	31 kW
Wheel type	75 kW
Field capacity	190 to 750 m^3/hectare
Maximum area cover per hose attachment	7.5 hectares
Pressure required at hose attachment	480 kPa
Maximum sludge solid content	8%

The manure application site is located north of the town of Timnath, Colorado. Wastes include washwater and manure from holding pens and a milking parlor on a 125-cow dairy. The site covers one hectare. Surface soils are loam and sandy loam with a pH of 7, overlaying sandy gravels. Ground water flows south toward the Poudre River. Application rates at this site are approximately 20,000 dry kg/ha per year.

ENVIRONMENTAL MONITORING PROCEDURES

The effect of applying wastewater residuals was evaluated using soil, crop, leachate and ground water samples. Soil samples were collected with a manual soil probe or Giddings Rig Core Sampler. Seven to 10 cores were collected from each application area. The cores were divided into depths of 0–30 cm, 30–60 cm and greater than 60 cm, and each set of depth samples were composited.

Wells were located on each site to determine the general direction of flow of ground water and to provide samples. At the Boulder site, wells were located on a 150-ft grid as required by local health officials. The manure application site had nine wells —three upstream, three downstream and three on the site. Three wells, situated in a triangle, were located at the edges of the Williamsburg site. Wells were sampled with a portable pump.

Leachate or soil water moving to the ground water table, was sampled using vacuum extractors of the type originally described by Duke and Haise (2). The extractor, shown in Figure 2, consists of two rows of hollow porous ceramic tubes placed near the bottom of a stainless steel pan. The ceramic tubes have a bubbling pressure (vacuum at which desaturation begins) of one atmosphere. When a vacuum, larger than the negative pressure of the soil water but less than the bubbling pressure, is applied to the candles, water will pass through the tube walls without allowing air to pass. The trough is designed so that the vacuum applied to the candles should not interfere with the normal soil water flow (3). Thus, as water moves downward through the soil, it is intercepted by the pan, drawn through the candles and collected in the bottle. The sample volume collected per unit area per unit time is equal to the rate of downward movement of leachate at the depth of the pan and the sample can be analyzed for dissolved constituents. Additional details of the extractor design, installation and use, along with a discussion of potential errors, are given by Trout et al. (4).

RESULTS OF ENVIRONMENTAL MONITORING

Soils

Soils analysis was performed by the Colorado State University Soils Laboratory. Procedures have been summarized by Trout et al. (4).

Control of soil salinity was a minor problem at the Boulder sludge site and a major problem at the manure application site. Richards (5) indicated that when the electroconductivity of a

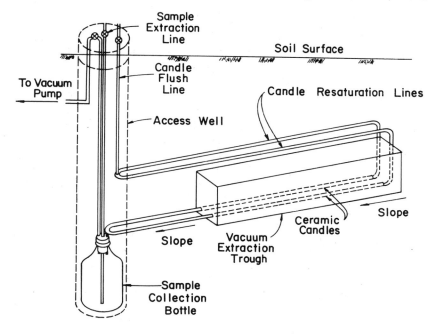

Figure 2. Vacuum extractor system.

saturated soil extract (an indirect measure of soluble salts) exceeded 4.0 mmhos/cm, some plants could be harmed. A high electroconductivity can be created in semi-arid climates where most of the applied water evaporates leaving an accumulation of soluble salts in the surface soils.

A white "salt" crust appears at the Boulder site during most prolonged dry periods. Electroconductivities up to 4.6 mmhos/cm were measured in the early spring of 1974. However, precipitation reduced the electroconductivity to the normal range between 2 and 3 mmhos/cm.

Electroconductivity at the manure site frequently exceeds 7 mmhos/cm during dry periods. In the spring of 1975 it was necessary to delay planting until sufficient irrigation and precipitation could be applied to the site to reduce salt levels. A major problem with this procedure is that contaminants, such as nitrates, may also be transported with the applied waters to the ground water.

It was impossible to correlate the application of wastewater residuals with a measured buildup of heavy metals in the soil at any of the sites. This was attributed to difficulties in obtaining representative soil samples, but may also be due to chemical analysis procedures. That is, the elements may be held within the sludge soil system in forms that are not extractable using standard procedures of the Colorado State University Soils Laboratory.

A mass balance approach was also used to estimate the increase in heavy metals in the soil at the Boulder site. Using this approach, the quantity of each element transported from the site of the ground water is subtracted from the total quantity applied. Results of this analysis are given in Table 2. It should be noted, however, that the fact that certain quantities of metals were retained in the soil does not mean that these materials were in forms available to plants. In fact, it appears that they were not in such forms.

Leachate Analysis

Leachate samples, collected using the vacuum extractors described previously, were analyzed by the Colorado State University Analytical Chemistry Facility. Procedures are summarized by Trout et al. (4).

Average levels of heavy metals in the leachate are shown in Tables 2 and 3 for Boulder and Williamsburg. The percentage of each metal that was leached to the ground water at the Boulder site is also shown in Table 2. Note that, with regard to heavy metals, the leachate generally meets drinking water standards. Analysis of leachate at the manure site is shown in Table 4.

As stated previously, the vacuum extractors used to sample leachate also measure rate of leachate movement. Thus, by measuring the precipitation and knowing the rate of application of sludge, manure and irrigation water, it is possible to develop an estimated water balance for a site or area.

A cumulative water balance is shown in Figure 3 for the Boulder site. Approximately 25 percent of the total water applied before irrigation became deep percolation, and 75 percent of the irrigation water became deep percolation. In contrast, only about 10 percent of the total water applied to the Williamsburg site became deep percolation. The average annual rainfall in Williamsburg is approximately three times (110 cm vs. 35 cm) that in Boulder.

The variation of nitrate concentration in the leachate and total nitrogen added with the sludge are shown in Figure 4 for a plot at the Boulder site. Although there is considerable variability in the data, a definite correlation exists between the nitrogen leachate concentration and the application of sludge.

Table 2. Heavy metal mass balance for Boulder site.[a]

Heavy Metal	Sludge		Soil			Ground Water	
	Average Concentration (kg/1000 dry kg)	Quantity Added (kg/ha)	Mass Remaining in Soil (kg/ha)	Concentration Added to Soil (ppm)	Total Leached (%)	Initial Average Concentration (mg/l)	Resulting Average Concentration (mg/l)
Ag	0.06	3.94	3.94	0.79	<0.025	<0.0001	<0.0001
Cd	0.0087	0.57	0.55	0.11	4.2	0.001	0.0016
Cr	0.639	42.0	42.0	8.4	0.021	0.0025	0.0025
Cu	0.820	53.9	53.6	10.7	0.60	0.0011	0.009
Fe	8.37	573	573	114.6	0.03	<0.05	<0.051
Hg	0.0032	0.21	0.21	1.05	<0.48	<0.0002	<0.0002
Mn	0.162	10.7	10.5	2.12	1.4	0.53	0.57
Ni	0.021	1.38	1.28	0.26	8.7	0.03	0.03
Pb	0.768	50.4	50.4	10.1	0.05	0.0031	0.0035
Zn	1.11	72.9	71.4	14.3	2.0	0.007	0.041

[a] The plot received sludge at the rate of 66,000 dry kg/ha per year over a five-month period.

Table 3.   Analysis of sludge, leachate and ground water for Boulder and Williamsburg.

| | Boulder | | | | Williamsburg | | |
| | Sludge (ppm^a, 4.3% solids) | Ground Water | | Leachate (mg/l) | Sludge (ppm^a, 1.2% solids) | Ground Water (mg/l) | Leachate (mg/l) |
Element		Initial (mg/l)	Affected (mg/l)				
Ag	60	0.0001	0.0001	0.0005	–	–	–
Cd	8.6	0.001	0.0016	0.007	42	0.002	
Cr	639	0.0025	0.0025	0.004	117	–	
Cu	820	0.0011	0.009	0.10	317	0.05	
Fe	0.87%	0.05	0.051	0.07	–	–	–
Hg	3.2	0.0002	0.0005	0.0005	0.58	0.00001	0.0001
Mn	162	0.53	0.57	0.04	–	–	–
Ni	21	0.03	0.03	0.05	42		
Pb	768	0.0031	0.0035	0.008	117	0.05	
Zn	1,110	0.007	0.041	0.54	917	0.05	
B	73	0.48	0.25				
Ca	2.3%	100	103		0.3%	85	
Cl	729	133	102			28	25
K	0.12%					2.0	–
Total N	3.3%	4	4		6.4%	0.56	23
NH$_4$-N	1.3%	0.42	0.36		1.2%		
Total P	4.7%					0.17	–
PO$_4$-P		2	4		0.4%	0.13	0.10

^{a} Unlesss otherwise indicated.

Table 4.   Average analysis of manure, leachate and ground water at manure application site (values are mg/l unless indicated).

| | Manure (1.1% Solids) | Leachate | | Ground Water | |
Element		Treated	Untreated	Upstream	Downstream
TDS	6,400	6,800	4,800	4,800	3,900
K	820	56	34	4	4
Cl	150	190	175	30	30
Total–P		–	–		
PO$_4$-P	140	1.5	0.69	0.1	0.1
Total–N	730	29	24	8	11
NH$_4$–N	580	0.21	0.32	0.1	0.1
Cd	0.08				
Cu	33				
Ni	4.2				
Zn	167				

The decrease in leachate nitrogen during June and July corresponds with the growth of corn on the plot. On an adjacent plot seeded to grass and also irrigated, the leachate nitrogen remained at approximately 250 mg/l for three additional months. It then dropped to a range between 3 and 70 mg/l and remained there for several months. This is the range of concentrations observed on untreated plots.

Leachate nitrogen levels in Williamsburg and at the manure site have generally ranged from 3 to 30 mg/l. The value tends to vary with precipitation at Williamsburg and is higher after precipitation, particularly if no precipitation has occurred for several days.

Ground Water Monitoring and Analysis

Results of analysis of ground water are shown in Tables 2, 3 and 4 for the three sites. No statistical correlations could be established between application of any of the materials and ground water quality. At Boulder and the manure site, only a few weeks were required for leachate from the surface to reach the ground water. However, at Williamsburg, several additional years may be required.

Regulatory agencies often require that ground water quality in the vicinity of an application site be monitored to detect any contamination that may occur. However, reliable interpretation of ground water quality is difficult for several reasons, including dilution, sampling procedures, variation in time and space

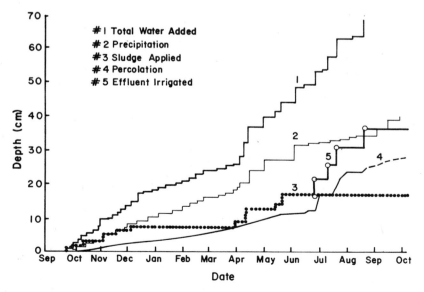

Figure 3.   Water balance for Boulder sludge plot.

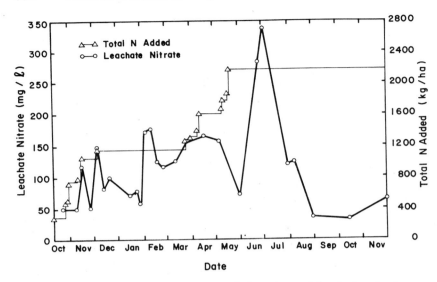

Figure 4. Nitrogen added to Boulder sludge plot and leachate nitrate concentration.

of background levels, variation in ground water flow rates, distribution of contaminants within the aquifer, and contamination from sources external to the site.

Wang and Cheng (6) used a numerical solution of the convection-dispersion equation to study the time and space distribution of contaminants in one-dimensional ground water flow. The source of contaminant in their study was an infinite strip oriented perpendicular to the direction of flow. It was assumed that the contaminant concentration on the strip was constant. Figure 5 shows the concentration distribution resulting from contaminant inflow on a strip with dimensionless width equal to 2. Pe is the Pechlet number which is directly proportional to the seepage velocity of the ground water. Wang and Cheng's calculations indicate that wells to be used for monitoring background concentrations must be placed some distance upstream from the application site to insure that they remain unaffected by the operation. For very low ground water flow rates, the wells should be 4 to 5 half-widths of the application site upstream from the site. For larger ground water flow rates, the wells may be closer to the edge of the site.

Figure 6 shows that there are large concentration differences with depth below the disposal area. This means that sample wells must be very carefully constructed to insure that the desired

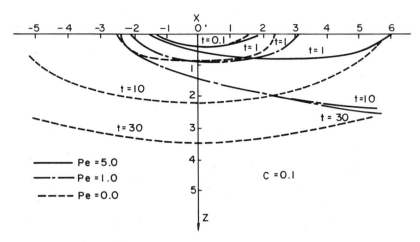

Figure 5.   Isoconcentration distribution at various times, without rainfall [from Wang and Cheng (6)].

information is obtained. Water in a well that extends far down into the aquifer and is open to inflow all along its length, will exhibit a concentration that reflects a composite of the different concentrations adjacent to the well. Rational interpretation of data from the well would be extremely difficult. A well that penetrates only a small distance into the aquifer should have a higher probability of contaminant detection, but the data will have little meaning insofar as indicating the extent and magnitude of water quality degradation in the ground water body as a whole.

The curves in Figure 5 indicate that within a distance of about 5 half-widths downstream, mixing is nearly complete and there is little variation of concentration with depth. It is at this point that the concentration in sample wells will have the highest probability of representing the actual magnitude of any degradation that has occurred. On the other hand, the volume of ground water that has diluted the contaminant may be so large that statistically significant increases in concentration may be difficult to detect.

The problem analyzed by Wang and Cheng (6) was for an aquifer of infinite depth and, therefore, the curves in Figure 5 and 6 tend toward zero concentration at points sufficiently far downstream. A more usual aquifer geometry is that shown in Figure 7. In this case, the concentration at complete mixing will not approach zero. In Figure 7, V_o is the ground water volume required for complete mixing, C_g is the concentration

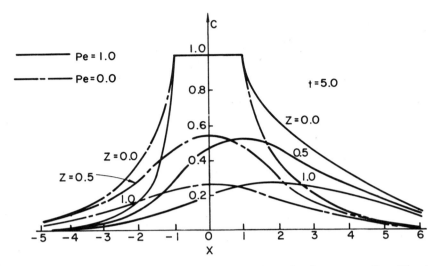

Figure 6.   Concentration distribution at various depths at $t = 5$, without rainfall [from Wang and Cheng (6)].

after mixing, C_o is the background concentration, C_L is the leachate concentration, and Q represents discharge.

The differential equation describing the well-mixed concentration relative to the background concentration is:

$$\frac{d\bar{C}}{dt} + \frac{Q_i}{V_o} \bar{C} = \frac{Q_L \bar{C}_L}{V_o} \tag{1}$$

in which $\bar{C} = C_g - C_o$, $\bar{C}_L = C_L - C_o$, and it has been assumed that $Q_L \ll Q_i$. The solution for constant $Q_L \bar{C}_L$ is:

$$\hat{C} = \frac{Q_L \bar{C}_L}{Q_i} \left[1 - e^{-\frac{Q_i t}{V_o}} \right] \tag{2}$$

The significance of Equation 2 is that it shows the eventual well-mixed ground water concentration will approach

$$C_g = \frac{Q_L}{Q_i} (C_L - C_o) + C_o \tag{3}$$

Using Equation 3, it is possible to determine whether or not C_g will differ sufficiently from C_o to be detected with confidence. If not, there is little to be gained by placing the wells in the downstream locations. Note that the mixing volume V_o affects only the time at which the concentration will approach the value in Equation 3 and not the concentration itself.

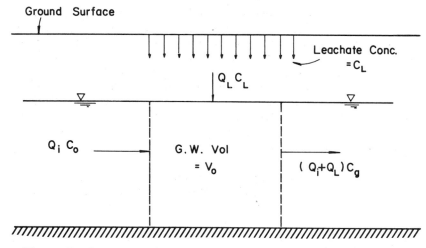

Figure 7. Open ground water system with finite lower boundary.

Using the above analysis, it was predicted that the concentration of nitrate in the ground water flowing under the Boulder site would be increased by 6 mg/l during a period when sludge was applied at regular intervals. This value compared favorably with the increase measured in selected wells. Increases in concentrations of other contaminants were not detectable.

Plant Tissue Analysis

The analysis of sweet corn grain and tissue samples grown on a plot treated with 65,000 dry kg/ha sludge are shown in Table 5. All values are within the range found in the literature

Table 5. Metal content of corn leaf and grain tissues (concentrations in ppm).

| Element | Boulder – Sweet Corn | | From Literature[a] | |
	Leaf	Grain	Leaf	Grain
Cd	0.5	0.3	0.1–5	0.1–1.0
Cu	13	13	2–30	4–17
Fe	190	6	20–250	50–200
Mn	45	24	20–150	1–10
Ni	1	1	0.1–5	0.1–5
Pb	4	1	0.1–12.4	0.03–5
Zn	90	70	5–400	20–100

[a] From Kirkham (8) for field corn.

for field corn grown on untreated soils. The data shown are for 1974. Essentially the same values were observed in 1975. This indicates that the effect of heavy metals in soils may not be additive as is often assumed. Lue-Hing (7) previously reported similar results from more extensive studies.

Silage corn was grown on half of the manure application site during 1975. Yield was identical to that in adjacent fields in spite of a delay in planting.

ACKNOWLEDGMENTS

This research was supported in part by the National Science Foundation, RANN Program, The Environmental Protection Agency, and Anheuser-Busch, Inc.

REFERENCES

1. Downs, H. W., J. L. Smith and D. B. McWhorter. "Continuous subsurface injection of municipal sewage sludge," ASAE Paper No. 75-2530, presented at American Society of Agricultural Engineers Meeting, Chicago, Illinois, December 15-18 (1975).
2. Duke, H. R., and H. R. Haise. "Vacuum extractors to assess deep percolation losses and chemical constituents of soil water," *Soil Sci. Soc. of Am. Proc.* 37 (6):963-964 (1973).
3. Corey, P. R. "Soil water monitoring," unpublished report to Department of Agricultural Engineering, Colorado State University, Fort Collins, Colorado (1974).
4. Trout, T. J., J. L. Smith and D. B. McWhorter. "Environmental effects of land application of digested municipal sewage sludge," report submitted to City of Boulder, Colorado State University, Fort Collins, Colorado (1975).
5. Richards, L. A., Ed. "Diagnosis and improvement of saline and alkali soils," Handbook No. 60, USDA, U.S. Salinity Laboratory, U.S. Government Printing Office, Washington, D.C. (1969).
6. Wang, M., and R. T. Cheng. "A study of convective-dispersion equation by isoparametric finite elements," *J. Hydrology* 24:45-56 (1975).
7. Lue-Hing, C. "Digested sludge utilization in agriculture as a soil amendment," paper presented at Rocky Mountain WPCF, Jackson Lake Lodge, Grand Teton National Park, Wyoming (September 1975).
8. Kirkham, M. B. "Trace elements in corn grown on long-term sludge disposal site," *Environ. Sci. Technol.* 9(8):765-768 (1975).

34

ENGINEERING DESIGN CRITERIA
FOR SLURRY INJECTORS

R. J. Godwin
On leave from NCAE, Silsoe, Bedford, England
E. McKyes, S. Negi, G. V. Eades, J. R. Ogilvie
Department of Agricultural Engineering, Macdonald Campus
of McGill University, Ste. Anne de Bellevue, Quebec, Canada
C. Lovegrove
Ontario Ministry of Agriculture and Food, Embrun, Ontario

INTRODUCTION

The principal objectives of injecting waste slurry below the soil surface are to control odors and prevent nitrogen depletion by ammonia volatilization. Several simple tine injectors have been used by previous workers (1, 2); however, most of these devices open the soil in a section shown in Figure 1a and often do not satisfy the above objectives. The reason for the poor performance of commercially available tillage tines, used as injectors, is that the volume of slurry that has to be injected has not been considered, and neither has the volume of soil which must be disturbed in order for the slurry to be effectively incorporated. The work reported here attempts a more rational approach to the design of injectors, considering: the agronomic and environmental requirements; the volumes of soil that must be disturbed; and the soil forces acting upon the injector.

AGRONOMIC AND ENVIRONMENTAL REQUIREMENT

In general, waste slurry should be placed as near as is practical to the soil surface in order to spread the aerobic stabilization of the organic material, Figure 1c. This principle has been used by the designers of certain injecting devices (3, 4). Near

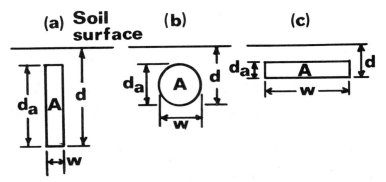

Figure 1.   Some possible shapes of injector tool openings.

the surface, the slurry nutrients are also more readily available to the crop roots. The slurry should be well mixed with the soil and not deposited in a concentrated band. Therefore, the liquid should be distributed as widely as possible, but sufficiently deep to ensure adequate incorporation and cover. The width of a single injector is likely to be limited in the case of row crop application, in order that root damage not be excessive.

Sufficient slurry can be injected to meet the nutrient requirements of growing crops, which, using the example of nitrogen, is in the order of 100–200 lb/acre. The application rate should be less than 300 lb/acre of N lest pollution problems arise (5, 6). The total volume of slurry which can be applied per unit area of land is given by:

$$V = \frac{N}{C}$$

(1)

where V = volume of slurry to be applied per unit area,
N = nutrient requirement weight per area,
C = concentration of nitrogen in the slurry, weight per unit volume.

If expressed as a height of liquid over an area of land:

$$V = 3.68 \times 10^{-5} \frac{N}{C} \text{ in.}$$

(2)

where N is expressed in lb/acre and C is expressed in lb/U.S. gal.

The cross-sectional area of slurry to be injected is given by:

$$A = ZV$$

(3)

where A is the area of slurry on a cross-section, and Z is the distance between injector passes or between multiple injectors. For row crop application, Z is usually equal to the width of rows.

In order that the slurry leave the injector at low pressure, the injector must produce a cavity in the soil at least equal in area to A, in which the slurry can be contained at the instant that it leaves the device. The number of injectors to be used is governed by the delivery pump capacity, the forward speed of the injectors and the desired slurry cross-sectional area to be applied, as follows:

$$n = \frac{P}{AS} \tag{4}$$

where   n = number of parallel injectors,
P = pump capacity in volume per time, and
S = forward speed.

To accommodate U.S. units:

$$n = \frac{0.219P}{AS} \tag{5}$$

when P is expressed in U.S. gal/min, A in in.^{2} and S in mph.

DISTURBED SOIL VOLUME

In addition to providing a cavity of sufficient volume to contain the applied slurry as soon as the liquid is delivered, the injector must also cause soil to be disturbed in order that new voids be created to incorporate the slurry. This is especially true in soils of low permeability and high water content where the slurry would be displaced and driven to the surface by the covering soil, if internal voids were not made available. The magnitude of the necessary cross-sectional area of soil to be disturbed depends on the desired application area A and the change in soil density caused by the injector, as developed below:

$$A = A_f - A_i \tag{6}$$

for complete incorporation in new voids, where A_f is the final area of soil across the injector path, and A_i is the initial area.

Considering unit length, Equation 6 can be written as:

$$A = \frac{W_s}{\gamma_f} - \frac{W_s}{\gamma_i} = \frac{W_s(\gamma_i - \gamma_f)}{\gamma_f \gamma_i} = \frac{A_i(\gamma_i - \gamma_f)}{\gamma_f} \tag{7}$$

where W_s = weight of soil disturbed,
γ_i = initial soil density, and
γ_f = final density of the disturbed soil.

During the cutting and lifting of common topsoils, the change in density divided by the final density, $i.e.$ $(\gamma_i - \gamma_f)/\gamma_f$, is in the order of 0.1 to 0.2. Therefore, the area of soil to be disturbed, A_i, must be 5 to 10 times A, the area of slurry to be injected.

Figure 2a shows the shape of the area of soil disturbed by a

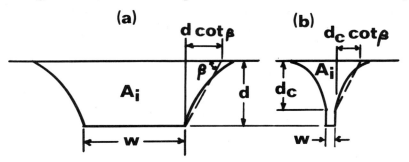

Figure 2.   Areas of soil disturbed by cutting blades of different widths.

wide cutting blade (7, 8), while Figure 2b shows that disturbed by a narrow or deep tool. There is a depth d_c at which soil ceases to be loosened toward the soil surface, but instead is compacted laterally around the tool (9–11).

By approximating the boundaries of disturbed soil by straight lines at angle β to the surface, the areas of soil disturbed by a blade of width W can be given as follows:

$$A_i = d^2\cot\beta + dW \qquad \text{for } d \leqslant d_c \qquad (8)$$

$$A_i = d_c^2\cot\beta + dW \qquad \text{for } d > d_c \qquad (9)$$

From previous research on subsoilers, typical values of d_c and $\cot\beta$ for many soil types have been found to be (8, 10):

$$d_c = 6W$$

$$\cot\beta = 0.59$$

Using these figures, the disturbed area for a range of injector widths has been calculated and presented in Figure 3 as a function of working depth. It is not necessary for an injector to have the same width over the complete depth, as information gained on subsoilers has shown (12) that attachments to the sides of the base of a tool have almost the same effect on increasing the disturbed soil volume as widening the entire length of the tool.

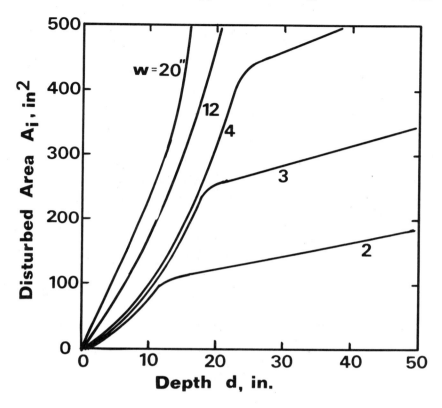

Figure 3.   Areas of soil disturbed by cutting blades of various widths and depths.

SOIL FORCES ACTING ON THE INJECTOR

The area of soil to be disturbed, A_i, can be obtained by a combination of injector width and working depth, as shown in Figure 3. The geometry of an injector should be selected to provide the desired disturbed area, while having the minimum soil reaction forces for operation.

Rigorous mathematical solutions for predicting the soil forces on complicated injector shapes are not available. However, a reasonable approximation can be made by reducing the geometry to simple blade shapes and predicting forces by a semirigorous passive earth pressure theory.

It has been shown (13) that the passive soil force on a semi-infinitely wide, or two-dimensional, cutting blade is given by:

$$\frac{P_p}{W}(\gamma d^2 N\gamma + cdN_c + qdN_q + c_a dN_{c_a}) \qquad (10)$$

where $p_p =$ passive force on tool
$\gamma =$ soil unit weight
$d =$ depth of blade
$c =$ cohesion
$q =$ surcharge pressure
$c_a =$ adhesion to the blade
$W =$ width of blade

and $N\gamma$, N_c, N_q, N_{c_a} are dimensionless factors dependent only upon the rake angle of the blade and internal friction angle of the soil.

For narrow tools or tines, operating at less than the critical depth shown in Figure 2, a crescent-shaped failure zone forms ahead of the tool; and it has been shown (8) that the passive force can be calculated by:

$$P_p = (\gamma d^2 N\gamma + cdN_c + qdN_q + c_a dN_{c_a})r \qquad (11)$$

where $r =$ radius of the crescent failure boundary on the soil surface.

In the case of blades having intermediate width effects, which are of comparable magnitude to the crescent failure edge effects, the combination of Equations 10 and 11 yields:

$$p_p = (\gamma d^2 N\gamma + cdN_c)(r + W) \qquad (12)$$

The surcharge and adhesion terms have been neglected because surcharge is zero for nonbulldozing blades, and the adhesive term is generally very small compared with the weight and cohesive components.

For example, a blade with a rake angle α to the forward horizontal of 45° has a radius of the crescent failure of approximately:

$$r = 1.8d \qquad d \leqslant dc \qquad (13)$$

In that case, comparisons can be made for the forces on blades acting at different depths in (a) a cohesionless soil, and (b) a frictionless soil in which weight is negligible by the equations:

$$\frac{P_p}{\gamma N\gamma} = 1.8d^3 + Wd^2 \qquad (14a)$$

$$\frac{P_p}{cN_c} = 1.8d^2 + Wd \tag{14b}$$

These force ratios have been calculated and plotted in Figures 4 and 5 for a range of injector depths and widths.

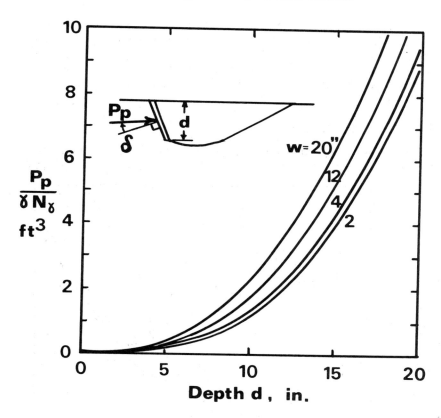

Figure 4. Ratio of force on inclined blades in a frictional soil.

The factors $N\gamma$ and N_c, as well as N_q and N_{c_a}, have been found experimentally and theoretically for blades at different rake angles and in soils of different friction angles (14, 15), and are available in reference soil mechanics literature. These factors are dependent on the internal friction angle of the soil, the rake angle of the tool and the angle of friction between the tool and the soil. Figure 6 shows factor $N\gamma$ as a function of the tool rake angle from the horizontal and at different angles

δ of soil-metal friction, for an example soil internal friction angle ϕ of 30°. It can be seen that there is an optimum rake angle to minimize the required force for movement, depending on the angles of friction involved. This phenomenon is also reported by Kawamura (16) and Soehne (17).

From mechanics theory and experiments (18), the horizontal sweep angle from the direction of travel has very little effect on the forward force required for soil cutting. Therefore the horizontal sweep angle of a straight or vee-shaped blade has a very small influence on the draft force, compared to the vertical rake angle.

DESIGN PROCEDURE

The following procedure shows how the design criteria may be used. An example injector is to place 200 lb of N per acre

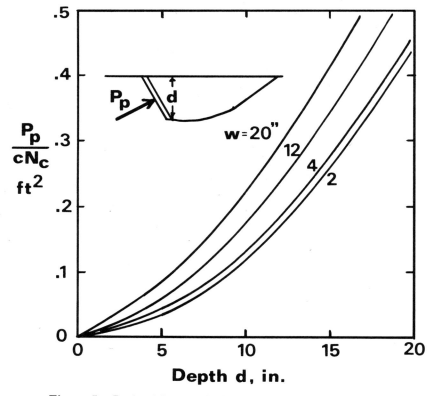

Figure 5.   Ratio of force on inclined blades in a cohesional soil.

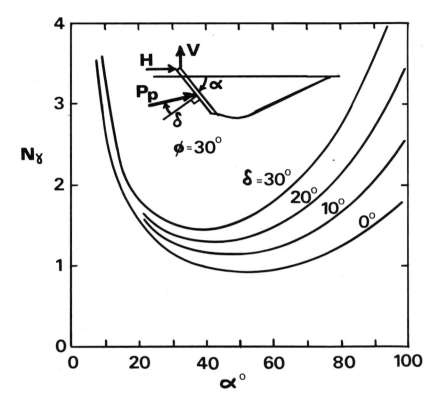

Figure 6.   Force constant *vs.* tool rake angle at different angles of soil–metal friction.

with a slurry having a concentration of 0.014 lb N/U.S. gal. The injector is to work in corn rows 30 in. apart. The soil densities observed before and after soil cutting are 75 and 65 pcf, respectively.

From Equation 2, the depth of application:

$$V = 3.64 \times 10^{-5} \frac{N}{C} = 3.64 \times 10^{-5} \times \frac{200}{0.014} = 0.52 \text{ in.}$$

From Equation 3, the required area of the injector aperature and soil cavity is:

$$A = ZV = 30 \text{ in.} \times 0.52 \text{ in.} = 15.6 \text{ in.}^2$$

From Equation 7, the soil area to be disturbed is:

$$A_i = 15.6 \text{ in.}^2 \text{ x } \frac{65 \text{ pcf}}{(75 - 65) \text{ pcf}} = 101.4 \text{ in.}^2$$

In Figure 3 it is seen that the required area of soil to be disturbed can be achieved by a variety of combinations of tool width, W, and operating depth, d. These different dimensions will, in turn, alter the draft requirements of the injector, as shown in Figures 4 and 5. Tables 1 and 2 list some combinations of width and depth of injector which will disturb the required volume of soil to incorporate the slurry volume. For each of these dimension combinations, the necessary aperture height and horizontal draft force, H, have been calculated and given in the table, using the formulas:

aperture height $d_a = (A/W), r = 1.88d$
upward vertical force $V = P_p \cos(\alpha + \delta)$
horizontal draft force $H = P_p \sin(\alpha + \delta)$
$$= (\gamma d^2 N \gamma + cdN_c)$$
$$\sin(\alpha + \delta)(W + 1.88d)$$

and example soil and injector parameters:

Table 1. Assumes the following parameters:
$c = 0, \phi = 30°, \delta = 20°, \alpha = 30°$; from Figure 6; $N_\gamma = 1.41$.

Design	Width, W (in.)	Working Depth, d (in.)	Aperture Depth, d^a (in.)	Draft H (lb)	Vertical Force, V (lb)
1	20	4.5	0.78	27.0	22.7
2	16	5.6	0.98	39.0	32.7
3	12	6.7	1.30	51.8	43.5
4	8	8.3	1.95	76.2	63.9
5	4	10.2	3.90	113.0	94.8
6	2	11.8	7.80	157.9	132.5

Table 2. Assumes the following parameters:
$c = 2$ psi, $\phi = 20°, \delta = 15°, \alpha = 30°$; from (14); $N_\gamma = 1.26$; $N_c = 1.10$.

Design	Width, W (in.)	Working Depth, d (in.)	Aperture Depth, d^a (in.)	Draft H (lb)	Vertical Force, V (lb)
1	20	4.5	0.78	222	222
2	16	5.6	0.98	263	263
3	12	6.7	1.30	299	299
4	8	8.3	1.95	368	368
5	4	10.2	3.90	461	461
6	2	11.8	7.80	574	574

From both examples in Tables 1 and 2, it is apparent that the principle for obtaining a given disturbed soil volume with minimum draft force is to use a wide injector working at a shallow depth. Fortunately, these engineering results are in agreement with the agronomic benefits of aerobic stabilization, availability to crop roots and wide distribution.

In practice, there may be limits to the width of an injector, which are (a) the damage to roots in grassland and row crops; and (b) the distribution and flow of slurry from an aperture with a large width, W, and small height, d_a.

Considering these constraints, it would be desirable to limit the example injector design to have a width of 10 or 12 in., leaving 9 or 10 in. clear between the injector edge and the crop rows. This will reduce the direct contact with plant roots and allow a greater margin of steering or row alignment error in operation. In addition, the aperture height of 1.3 to 1.56 in. will probably not cause a troublesome restriction in the flow of the slurry.

Then the injector design would have the geometry shown in Figure 7. It would have the capability of injecting 200 gal/min at a forward operating speed of 2.74 mph and depth of 6.7 in. A probable maximum forward speed range for accurate placement in row crops is 4 to 5 mph.

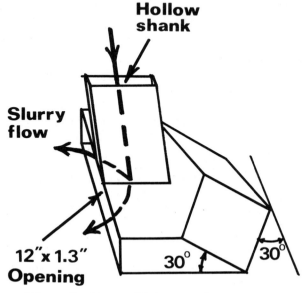

Figure 7. Proposed injector tool geometry.

PRELIMINARY MODEL

A prototype injector based upon the above design criteria was constructed as shown in Figure 8 with a rake angle from the

Figure 8. Injector mounted on the hitch of a 35-hp tractor.

horizontal of 30° and a side sweep angle of 60° from the axis of travel. This device was used in preliminary trials, injecting 200 gal/min of swine manure slurry, having approximately 10 percent solids, into several different soil types and sod, including ground frozen to a few inches depth. Good incorporation of the slurry was observed, as shown in Figures 9 and 10, except in the frozen soil where solid blocks of ground did not loosen as they would when unfrozen. Proper soil cover and odor absorption was also observed, thanks to the flat-top design of the injector and the narrow shank attached. Experienced environmental observers witnessing the trial runs were quite satisfied with the odor control of the very pungent slurry. The draft force was not measured in the first trials, but it was small enough to have a negligible effect on tractor speed, even of a 35-hp tractor at idle throttle setting with the injector working at 7 to 8 in. depth. Furthermore, operating depth control was

Figure 9.   Path of injected swine manure slurry in a clayey field. Slurry was injected at a rate of approximately 1 gal/ft.

Figure 10.   Injection of 200 gal/min in sod.

stable owing to the selection of the 30° rake angle, which results in a moderate downward force from the soil as well as a low draft force as shown in Figure 6.

In the coming year, an exhaustive series of field tests is planned, including the injection of slurries of different consistencies into many different soil and crop conditions. Draft force, incorporation of slurry and soil cover will be measured, and improvements or limitations to the design will be found.

CONCLUSION

The fundamental considerations of this report have shown the criteria for the design of a slurry injector, which will distribute and cover liquid with a minimum of energy input, subject to crop location constraints. The future of such design considerations appears most promising, and hopefully some useful tools for land waste applications have been made available.

ACKNOWLEDGMENTS

The authors wish to thank the Agriculture Canada Engineering Research Service for approving the contract under which the work reported has been a preliminary part, and the Department of Supply and Services for supporting the work described herein under contract number OSW5–0287. Special mention must be made of the assistance and encouragement provided by Mr. Mottie Feldman of Agriculture Canada's ERS during the establishment and execution of this project.

REFERENCES

1. Badger Northland Inc. Kaukauna, Wisconsin 54130 (1975).
2. Pearson Bros. Co. P.O. Box 192. Galva, Illinois 61434 (1975).
3. Bartlett, H. D., and L. F. Marriot. "Subsurface disposal of liquid manure," *Proc. Int. Symposium on Livestock Wastes*, ASAE, St. Joseph, Michigan 49085 (1971).
4. Gold, R. C., J. L. Smith and R. D. Hall. "Development of an organic waste slurry injector," ASAE Paper No. 73–4529. ASAE, St. Joseph, Michigan 49085 (1973).
5. Webber, L. R., and T. H. Lane. "The nitrogen problem in the land disposal of liquid manure," *Proc. Cornell Univ. Conf. Agricultural Waste Management*, Syracuse, New York (1969).
6. United States Army, Corps of Engineers. "Wastewater management by disposal on the land," Special Report 171, CRREL, Hanover, New Hampshire (1972).
7. O'Callaghan, J. R., and K. M. Farrelly. "Cleavage of soil by tined implements," *J. Agric. Eng. Res.* 9(3):259–270 (1964).

8. Godwin, R. J. "An investigation into the mechanics of narrow tines in frictional soils," Unpublished Ph.D. Thesis, University of Reading (1974).
9. Zelenin, A. N. "Basic physics of the theory of soil cutting," *Akademia Nawk USSR*, Moscow-Leningrad (1950).
10. Kostritsyn, A. K. "Cutting of a cohesive soil medium with knives and cones," Translation 58, N.I.A.E., Silsoe, Bedford, England (1956).
11. Payne, P. C. J. "The relationship between the mechanical properties of soil and the performance of simple cultivation implements," *J. Agric. Eng. Res.* 1(1):23–50 (1956).
12. Spoor, G., and R. J. Godwin. Unpublished data, N. C. A. E., Silsoe, Bedford, England (1975).
13. Reece, A. R. "The fundamental equation of earth-moving mechanics," *Symposium on Earthmoving Equipment*, Institution of Mechanical Engineers, Auto. Div., Vol. 179, Part 3F (1975).
14. Hettiaratchi, D. R. P., and A. R. Reece. "Symmetrical three-dimensional soil failure," *J. Terramechanics* 4(3):45–67 (1967).
15. Hettiaratchi, D. R. P. "The calculation of passive earth pressure," Unpublished Ph.D. Thesis, University of Newcastle-upon-Tyne (1969).
16. Kawamura, N. "Study of the plow shape (4), study of soil cutting and pulverization (2)," *Soc. Agr. Mach. J. (Japan)* 15(3):90–94 (1953).
17. Soehne, W. "Some principles of soil mechanics as applied to agricultural engineering," Translation 53, N. I. A. E., Silsoe, Bedford, England (1956).
18. Balaton, J. "Study of the draft of subsoilers," Translation TT74–53143, ARS, USDA and NSF, Washington, D.C. (1971).

35

MUNICIPAL SLUDGE MANAGEMENT:
EPA CONSTRUCTION GRANTS PROGRAM

Robert K. Bastian
U.S. Environmental Protection Agency, Office of Water Program
Operations, Washington, D.C.

INTRODUCTION

Under the Federal Water Pollution Control Act Amendments
of 1972 (PL 92–500), Congress authorized EPA $18 billion in
contract authority to help municipalities construct publicly
owned wastewater treatment works through fiscal year 1977.
The funding authority of PL 92–500 is expected to be continued
for several more years. Over $9 billion has been obligated to date.

An integral part of almost any wastewater treatment plant
is the sludge management system. Residual solids are produced
in nearly every unit process of conventional wastewater treatment, and a significant proportion of both capital outlay and
operations and maintenance (O&M) costs of conventional sewage treatment is associated with sludge production, conditioning
and disposal facilities and operations. As manager of the EPA
Construction Grants Program, the Office of Water Program Operations (OWPO) is deeply involved with municipal sewage
sludge management activities and concerned with the problems
communities are facing with sludge disposal.

This chapter is an attempt to provide a summary of available
information on municipal sewage sludge production and alternatives for the disposal/utilization of municipal sewage sludge
and OWPO activities in this area.

BACKGROUND

A wide variety of publications—journal articles, conference proceedings, agency documents, and even newsletters—are available that deal with various aspects of municipal sewage sludge production, processing and management. However, national recognition of sludge management problems has only recently occurred and the available data base for evaluating the environmental acceptance and cost-effectiveness of various sludge management alternatives at this time is quite limited. Data summarized here represent OWPO's current reference based on nationwide municipal sewage sludge management activities and were derived for the most part from the 1968 Inventory of Municipal Waste Facilities in the U.S., STORET, a 1972 survey of five EPA Regions concerning only land application of liquid sludges, the "Needs" Survey and Construction Grants files. Efforts are currently underway to improve this available information.

Quantities of Sludge

At this time (1976) well over 5 million dry tons of municipal sludge are being produced and disposed of in one manner or another each year. By the time secondary treatment is reached by facilities across the country this volume may reach 9 million dry tons per year. Going to secondary treatment would represent an increase of 80 percent in terms of dry tons of solids produced. With the extensive use of biological secondary treatment processes (such as activated sludge) the increase in wet tons of sludge to be handled would actually increase well over 100 percent, due to the difficulty in dewatering biological sludges.

Although this increase in sludge volume will be faced by treatment plant operators across the country, the major problem areas are in the major cities, especially those facing a phase-out of their current ocean disposal activities. The larger cities have large volumes of sludge to manage and in many cases have run into problems obtaining sites and local approval for implementing any of the acceptable sludge management alternatives. This is not meant to disregard or underplay the problems being faced by the numerous smaller communities across the country.

Size Distribution of Municipal Plants

More than 22,000 municipal treatment plants exist in this country; over 5,000 are pond systems that generally have few if any sludge disposal problems. Of the remaining 17,000 plants, fewer than 350 are larger than 10 MGD. About 65 percent or more of the nation's treatment plants are actually less than 1 MGD in design flow. Well over 65 percent of the known non-pond systems are located in only three EPA Regions (Philadelphia, Atlanta and Chicago).

Current Disposition of Sludge

The current general national breakdown of municipal sewage sludge disposal is estimated as:

Method	% Total Municipal Sludge
Ocean Disposal	15
Incineration	35
Landfill	25
Land Application	25
(croplands)	(20)
(others)	(5)

This estimate does not include the surprising number of operations that simply store sludges in lagoons with no identified future disposal method. The anticipated phase-out of ocean disposal, increasing production of sludge with secondary treatment and increasing fuel costs over the next few years could change this picture dramatically.

Statewide and regionwide sludge disposal practices vary widely as do state regulations (Appendix A). Of course, only those areas located near the oceans have had the ocean option. Most of the communities in the State of Illinois and many other parts of the Midwest are applying sludges to the land, while incineration and landfill are most common in the major inland cities. Chicago and Denver are major inland cities that use land application alternatives.

Costs

Both capital and O&M costs for various sludge management alternatives vary greatly. They are dependent upon numerous variables including energy, transportation, land, and manpower costs as well as monitoring requirements and other criteria established by local, state and federal regulatory agencies. Where sites are available and liquid sludges are accepted, landfilling is often an economical alternative if haul distances are minimal. Where land is not available, incineration is often utilized when air quality criteria allow and fuel allocations are available. However, when land is available and the state and local regulators approve (or at least do not formally disapprove), the land application alternative is being implemented or experimented with by many communities based on cost-effectiveness.

From 30 percent to 50 percent of a conventional treatment plant's capital costs are for the sludge management system. This will involve a major portion of the $12 billion estimated for capital costs of building facilities to meet the 1977 goal (PL 92–500) of secondary treatment. More than $400 million per year is estimated to be required for current O&M costs.

ALTERNATIVES

Currently the sludge management alternatives available to a particular city include various versions of incineration, landfill, land application and possible innovative technologies (e.g., pyrolysis, chemical fixation, strip-mine reclamation, bagged and bulk sales). Ocean dumping has been effectively ruled out for municipalities not currently dumping and to our knowledge no current dumper's sludges meet the ocean dumping requirements (40 CFR 220). Annually renewed interim permits are being granted until these cities develop land-based alternatives. Except for land application and certain innovative technologies, the available options provide little means for the beneficial reuse of municipal sludges.

Ocean Disposal

Although 15 percent of the current sludge volume produced in municipal treatment plants is now disposed of into the oceans, this practice is used by less than 160 cities and towns. There are 17 municipal sludge ocean dumping permits (for dumping at two approved sites—Philadelphia and the New York Bight) at latest count, but additional cities discharge to the ocean through diffusion pipes. Where used, it has represented a least-cost alternative and until recent years has been met by good public acceptance. Both the Federal Water Pollution Control Act Amendments of 1972 (PL 92–500) and the Marine Protection, Research and Sanctuaries Act of 1972 (PL 92–532) require the development of federal standards on materials entering the ocean. Efforts are currently underway to phase-out most of the ocean dumpers and pipe dischargers by 1981.

The concerns over ocean disposal of sewage sludges center on potential impacts upon marine life. Various contaminants often associated with sewage sludges including heavy metals, persistent organic pesticides, PCB's and others, are known to bioaccumulate to where they interfere with reproduction or cause toxic effects to certain marine organisms. The contaminants of immediate concern are mercury and cadmium and TICH (Total Indicated Chlorinated Hydrocarbon compounds).

Incineration

Incineration provides a disposal option where land availability is a problem, but produces an ash (about 10 percent of original volume) to be disposed of. A substantial energy input, either to dewater sludge or in the form of an auxiliary fuel, is required for sludge incineration. The liquid phase removed by dewatering and scrubber waters from stack gas cleaning efforts must also be treated.

Standards exist for particulates and mercury levels in emissions from incinerators and questions remain concerning such chemical contaminants as PCB's. Major public resistance to new incinerators has occurred in several areas in recent years. Co-incineration with solid waste is being considered in many areas for the near future.

Landfill

An option that provides a means of sludge disposal in areas where suitable sites are available is landfilling. This practice avoids the direct public health issues of cropland application of sludge by burying the sludge in conventional sanitary landfills, but can affect ground water quality where substantial leachates are produced.

The EPA Office of Solid Waste Management (OSWMP) has issued guidelines on the design and operation of federally owned solid waste landfills, but not for sludge landfills. As noted (Appendix A), state and local regulatory agencies differ in their requirements for acceptance of municipal sewage sludges. In more than one state, only dewatered sludges are acceptable for landfilling; in certain other areas, little control is imposed on landfilling of sludges. Combining sludge with solid waste in landfills is a common practice in areas where this is acceptable. Many cities have been placing sludge in lagoons (*i.e.*, open landfills) and may eventually cover the lagoons after many years.

Major problems confronting landfill of sludges involve gaining public acceptance of potential landfill sites and preventing ground water pollution from landfill leachates. Metropolitan areas are having difficulty in identifying available and publicly acceptable sites, while medium and smaller sized communities may find this aspect less limiting. Heavy metals, persistent organics and other compounds covered by drinking water standards are of concern in leachates from landfills. Recent data indicate that ground water contamination problems due to leachates from landfills receiving sewage sludges may be more widespread than originally envisioned. Contaminants of most immediate concern are lead and mercury.

Land Application of Sludge

The utilization of sludge by application to croplands, forests and other sites provides a means of sludge disposal and beneficial reuse or utilization at the same time. Although low in nutrient content (approximately 3 percent nitrogen, 2.5 percent phosphorus and 0.3 percent potassium by dry weight), sludge can serve as a valuable soil conditioner (Table 1). Currently,

Table 1. Characteristics of sludge.

	Digested Sludge	
	Range	Typical
Volatile Solids (%)	30–60	40
Nitrogen (N, % TS)	1.6–6.0	3.0
Phosphorus (P_2O_5, % TS)	1.5–4.0	2.5
Potash (K_2O, % TS)	0.0–3.0	0.5
Btu/lb	1700–6800	4000

less than 0.3 percent of the nation's croplands received sewage sludge and if the entire municipal sludge production were to be applied to the land at crop nitrogen requirement rates, less than 1 percent of the agricultural land would be involved.

Land application has been preferred by communities with adequate available land, generally inland cities and smaller rural communities. Both high application rate disposal operations and systems designed to provide supplemental fertilization to agricultural crops have been used. In some cases "dedicated" or publicly owned and controlled sites have been used, but in many cases applications have been made to privately owned and managed farmland. This practice has been commonly used in 30 percent of the smaller communities, some for over 40 years. Over 400 towns in Illinois and 250 in Ohio currently apply sludges to the land. Many more communities simply stockpile dried sludges and allow the public to haul it away for unspecified use.

Larger communities are becoming more interested in land application options due to recent regulatory decisions, increased energy costs, and/or public opposition to other alternatives. There is also a growing interest in the potential for future cash crop returns from agricultural uses to help lower O&M costs. Numerous major municipalities (Table 2) are currently using

Table 2.   Current sludge status in largest U.S. cities.

City	Quantity of Sludge (Dry T/Day)	Present Disposition
New York	230 (600)	Ocean/landfill; pyrolysis prop.
Chicago (MSD), IL	800	Land application/give-away/ lagoon /bulk sales
Los Angeles (& Co.), CA	500	Ocean; compost and bagged
Philadelphia, PA	140 (190)	Ocean; "10-pt plan" prop.
Detroit, MI	160	Incineration
Houston, TX	160	Land application/dried and bulk sales
Baltimore, MD	140	Landfill/lagoon/land application
Dallas, TX	120	(Lagooning); land application prop.
Washington, DC	400	Land application/compost/ injection/bulk sales
Cleveland, OH	200	Landfill/incineration

or experimenting with land application schemes. Several have been conditioning (mainly by heat drying, lagooning or composting) and selling sewage for use as soil conditioners for many years.

The major technical problems facing land application proposals center on the "potential" human health risks involved in growing crops that enter the human food chain on sludge-amended soils. Sewage sludge contains human pathogens and varying amounts of heavy metals (Table 3), persistent organic compounds and a variety of other potentially "toxic" or "hazardous" materials due to the nature of input sources (industry, homes, stormwater) into domestic sewage. Conjecture as to the "potential" health effects of these materials when applied to the land has been extensive while available data and risk interpretations have been limited.

The possible immediate and long-term effects of such materials in sewage sludge when applied to agricultural soils and their translocation into crops that may eventually enter the human food chain are currently being investigated and debated.

Table 3.   Metals in sludges (mg/kg dry sludge).

	Range	Mean	Median
Ag, Silver	nd[a]–960	225	90
As, Arsenic	10–50	9	8
B, Boron	200–1430	430	350
Ba, Barium	nd–3000	1460	1300
Be, Beryllium	nd	nd	nd
Cd, Cadmium	nd–1100	87	20
Co, Cobalt	nd–800	350	100
Cr, Chromium	22–30,000	1800	600
Cu, Copper	45–16,030	1250	700
Hg, Mercury	0.1–89	7	4
Mn, Manganese	100–8800	1190	400
Ni, Nickel	nd–2800	410	100
Pb, Lead	80–26,000	1940	600
Sr, Strontium	nd–2230	440	150
Se, Selenium	10–180	26	20
V, Vanadium	nd–2100	510	400
Zn, Zinc	51–28,360	3483	1800

Other possible contaminants: persistant organics, pathogens, radioactive substances.

[a] nd = not detected.

The fact that the same "materials of concern" currently enter the human food chain through many sources other than agricultural uses of sewage sludge, including conventional agricultural practices (they are present in variable amounts in conventional inorganic fertilizers, animal manures, soil, etc., and lead to highly variable background levels), complicates the issue even further. The problems of establishing current body burdens and acceptable safety factors are involved in current deliberations by the U.S. Food and Drug Administration (FDA).

The fact remains that land application of sewage sludges has been an accepted and unregulated activity for many years—without known significant negative health impacts. Possible "unnoticed problems" with these practices are being investigated and questioned. Well managed systems can be expected to continue their operation into the future without problems—unless regulations are developed that eliminate this alternative.

Prime concerns center on the levels of certain materials that may lead to future plant toxicity (*e.g.*, zinc, copper, nickel and herbicides) or that potentially could lead to public health problems due to bioaccumulation and/or toxicity (*e.g.*, cadmium, lead, PCB's). Agricultural practices that allow for the safe utilization of municipal sewage sludges in agriculture (including nitrate leaching controls) are being recommended and frequently updated by the U.S. Department of Agriculture (USDA). Source controls, pretreatment requirements and monitoring activities are being suggested by EPA and others, but are implemented only at the local level. No EPA, USDA or FDA standards exist (although guidelines are being developed) for most of the areas of current concern. No cost-effective technology exists to remove such contaminants as heavy metals from sludges prior to application to the land—other than source control. Not all states have current regulations controlling this practice.

Additional major limiting factors are public acceptance and the fact that large cities have to transport sludge to rural areas for agricultural use. Odors from poorly managed sludge management systems and perceived odors from anything that has to do with sewage may be the largest single problem. Reluctance of rural areas to receive urban wastes (until adequate economic incentives are offered) is also a significant factor.

Innovative Technologies

Additional municipal sewage sludge management alternatives exist and are being more fully developed under what has been called "innovative technologies." These alternatives include such practices as: strip-mine reclamation projects, pyrolysis, metals recovery, chemical fixation, overseas shipment, and bagged or bulk sales and "give-away" programs. Almost anything that does not fall under the conventional definitions of ocean disposal, incineration, landfill and land application could be considered under this heading.

Municipal sludges are being and have been used as a soil conditioner/stabilizer in reclaiming mine spoils, strip-mined lands and other drastically disturbed areas. Although limited to areas where transport to such sites is cost-effective and locally acceptable, this alternative offers an opportunity to use the basis of one problem to help solve another problem. The experimental use of liquid, dried or composted sludges has been undertaken in such areas as Illinois, Ohio, Pennsylvania, Maryland, Virginia and Florida. The key to successful proposals of this type lies in gaining the support and cooperation of groups such as local, state and federal politicians and regulatory agency personnel, land owners, coal companies, railroads, and public works officials.

Speculation on the potentials of pyrolysis to solve the sludge management problems of major metropolitan areas has occurred for several years. Only recently, however, have funds been made available to move bench-scale research systems for pyrolysis of sludge into pilot-scale demonstration facilities such as the JPL/ ACTS facility involving carbon recovery now in operation at Orange County, California. Other pilot demonstration plants involving sludge pyrolysis or co-pyrolysis with solid wastes have been planned for future construction and testing. Several comprehensive regional sludge management studies have identified pyrolysis as possibly the most cost-effective alternative for handling future sludge volumes. These conclusions have been drawn before scaled-up operational experience has been gained. New York falls into this category while the recommendation for co-pyrolysis in the Minneapolis/St. Paul area has recently been dropped in favor of co-incineration based upon a new look at the economics of by-product values.

Efforts have been and are being undertaken to evaluate the potentials for metals recovery from municipal sewage sludges. Both acid treatment and heat treatment processes are being considered. These processes require considerable chemical and/or energy inputs and result in the loss of organics for uses such as soil conditioners. The problems to date generally relate to cost-effectiveness and individual unit processes problems involved in resource recovery efforts.

Several proprietary chemical treatment processes are available that provide good performance in chemically fixing sewage sludge to allow safe landfilling and possible use in construction of highways. While cost is a major limiting factor for the use of these processes, chemical fixation should be considered as a possible alternative for communities without adequate land availability for land application schemes, or with high levels of contaminants such as heavy metals and PCB's in their sludges.

The overseas shipment of municipal sewage sludges has been extensively discussed. Proposals have been made involving shipment by ore boats, oil tankers, or sludge ships to such areas as the Bahamas and other caribbean areas, Africa's Gold Coast, Egypt and the Middle East for use as soil conditioners. To date no large-scale projects involving overseas shipment of sewage sludge have been implemented.

Conditioning of sludges followed by bagged product of bulk sales has been practiced in several major areas (Los Angeles/ Southern California "Kelloggs"; Milwaukee "Milorganite"; Houston; Chicago) for many years and is being initiated or planned for other areas (Denver and Washington, D.C.). "Give-away" programs have been operated by numerous small communities as a common practice over the years, and new efforts have recently been initiated or planned for several major metropolitan areas (Philadelphia's "Philorganic" and Chicago's "Nu-Earth"). Efforts in areas such as Winston-Salem involve fortification of sludge with nitrogen to allow sale of the final product as a nutrient-rich fertilizer/soil conditioner ("Organiform-SS").

There probably are many other potential "innovative technologies" that exist and should be evaluated as potential sludge management alternatives. Maybe the future will bring a "black box" solution that no one has yet considered which will provide a cost-effective, environmentally acceptable option to current available municipal sewage sludge management alternatives.

OWPO MUNICIPAL SLUDGE MANAGEMENT ACTIVITIES

OWPO prefers to view sludge as a by-product of sewage treatment and, therefore, is interested in identifying and refining available technology as well as encouraging the development of innovative technologies in the sludge management field. These program activities deal with cost-effectiveness and environmental acceptability as well as operational capability of various municipal sludge management alternatives. As a result of helping fund such activities, OWPO also is interested in related activities dealing with economics, social and institutional constraints and environmental impact studies.

At present, the only current legislative mandates involving the regulation of municipal sewage sludge disposal/utilization activities fall under PL 92–500, Sect. 208 (areawide waste management), Sect. 405 (sludges entering navigable waters), and PL 92–532 and its amendments relative to establishing ocean dumping criteria. OWPO does, however, help control sludge disposal activities through control of system design to qualify for federal construction grant funds. An overall philosophy of such activities is to encourage the beneficial utilization of municipal sewage sludges rather than outright disposal) wherever and whenever possible, if shown to be both cost-effective and environmentally acceptable.

PROJECTS AND OUTPUTS

Current and planned OWPO projects and outputs in the area of municipal sewage sludge management are:

1. The technical bulletin, "Municipal Sludge Management: Environmental Factors," has been prepared to assist EPA Regional Administrators in evaluating grant applications for construction of publicly owned treatment works under Section 203(a) of the

Federal Water Pollution Control Act as amended. It also provides designers and municipal engineers with information for selecting a sludge management option. The proposed document for public comment will appear in the *Federal Register* early in June 1976.

2. Program guidance in sludge management was recently released concerning the grant eligibility of land acquisition costs for the ultimate disposal of residues from wastewater treatment processes (Program Guidance Memorandum No. 67). Additional program guidance is being developed concerning grant eligibility for costs of easements and lease options on land for the ultimate disposal of these residues. The use of federal lands by municipalities for sludge disposal activities is also being investigated.

3. At this time approximately $11 million in Step I (or equivalent) construction grants are being used for planning and pilot-scale municipal sludge management projects in major metropolitan areas. Such funding efforts are anticipated to continue and the number of projects to increase.

4. A grant with the Association of Metropolitan Sewerage Agencies (AMSA) will provide detailed information on sludge management practices by the major metropolitan areas. A proposed follow-on effort will provide a survey and evaluation of current municipal sewage sludge management alternatives across the entire country. It is hoped that these and other efforts will provide an improved data base from which future sludge management policy decisions can be better made.

5. A grant to public interest groups is hoped to provide new approaches and insights into the public acceptance of beneficial uses of sewage sludge.

6. Efforts to encourage beneficial utilization alternatives (*e.g.*, strip-mine reclamation) of sludge management are planned which will include sessions at several upcoming sludge management conferences.

7. Current assessment publications covering heavy metals research and human health hazard aspects of cropland application of municipal sewage sludges are planned based on workgroup sessions with leading heavy metals researchers and health officials. These assessment documents will be used to provide current evaluations concerning these two areas of major concern for the land application alternative of sludge management.

8. A technical report on successful examples of sludge utilization on land, complete with detailed technical and institutional information, is planned.

9. Efforts are underway to obtain translations and evaluation

of overseas sludge management activities. In addition, reports of research activities by various major sewerage authorities may be made available for wider distribution.

10. Efforts are underway to develop a documentary film and TV mini-documentaries of available sludge management alternatives. The purpose will be to expose public works officials and the public to the alternatives being used across the country and innovative technologies currently under detailed evaluation. We also hope to alert the engineering community to what alternatives should be considered in development of future municipal sludge management planning efforts.

11. Costs of various sludge utilization or disposal alternatives have been addressed in available generalized documents; reports outlining detailed cost comparisons are planned.

12. A model design report for land application of sludge is planned.

13. Future activities to better define and improve approaches to institutional constraints and public acceptance problems are being planned.

14. Efforts are underway to improve EPA headquarters, regional and state coordination activities in sludge management matters.

15. Efforts are also being made to coordinate activities of the Science Advisory Board, National Academy of Sciences, Government Accounting Office, and other sludge management studies and evaluations that are underway or have been completed.

WORK NEEDED

Major OWPO engineering needs to support the Construction Grants Program involvement in municipal sewage sludge management activities actually boil down to developing the best design criteria and cost information for the available technology and the development of innovative technologies for future implementation. With the current phase-out attitude toward ocean dumping, we are dealing with providing guidance to the regions on the best available land-based technologies for sludge management rather than developing regulatory programs.

The work most urgently needed in the municipal sewage sludge management field includes: resolution of health effects issues; breakthroughs in gaining public acceptance; continued emphasis on innovative technologies leading to beneficial use and resource recovery; information dissemination to design engineers/operators, federal/state/local government personnel and local elected officials.

SUMMARY

A general result of increasing sewage treatment is the more than proportional increase in resulting sludge volumes to be disposed of or utilized. This problem has led to the involvement of almost every segment of the Office of Water and Hazardous Materials, and almost every other office in EPA, in the sludge management arena. These involvements have been most evident during the development of the "Sludge Technical Bulletin." Recently, activities have greatly increased in the areas of regulating ocean disposal, federal pretreatment guidelines development, 208 planning activities, hazardous materials investigations and their concern in sewage treatment residuals (*i.e.,* sewage sludge) and proper management schemes.

The OWPO activities described in this paper can be expected to continue and to increase as guidance is developed on the best available land-based technologies for sludge management.

Further information concerning the items described in this paper and the activities of OWPO can be obtained from: Office of Water Program Operations, Municipal Technology Branch (WH–547), U.S. Environmental Protection Agency, Washington, D.C. 20460.

APPENDIX A

Following is a summary table compiled by the National Commission on Water Quality of the State criteria for land application and other uses of wastewater and sludge. The information was compiled from individual responses to their utilization of section 201(a)–(e) of PL 92–500, from a table on land application prepared by Metcalf and Eddy, from specific state regulations in the *Environment Reporter,* and from telephone contacts.

The summary table lists the restrictions to land application and whether specific regulations are used. It is helpful to state the limitations of this table. It was assumed that a small percentage of states do not have regulations if they did not appear in the state responses to our inquiry or the *Environment Reporter.* Some may indeed have formal health department regulations not indicated in this table. A policy regarding land application of wastewater was not obtained from 5 states, and from 19 states regarding sludges.

For land application of wastewater, from a total of 54 states and territories, 22 states or 41 percent have formal regulations. The State of Washington has draft regulations. Thirty-eight states or 70 percent require a minimum of secondary treatment or more stringent. Application is prohibited in the District of Columbia, discouraged in Rhode Island, and not practiced in Nebraska, Ohio and Iowa.

For land disposal of sludge, 21 states or 39 percent have some form of formal regulations. Of the 35 states with policies regarding sludge disposal, 18 states or 51 percent allow or regulate disposal in landfills, 20 states or 57 percent evaluate disposal on an *ad hoc* or case-by-case basis, 5 states or 14 percent require dewatering, Mississippi and Indiana require some form of stabilization, Wisconsin requires digestion, Idaho requires heat treatment, and Pennsylvania requires digestion for landfill disposal.

LAND APPLICATION OF WASTEWATER AND SLUDGE
SUMMARY TABLE

| State | Wastewater | | Sludge | |
	Regula-tions	Treatment and Other Restrictions	Regula-tions	Restrictions
Alabama	No	Secondary Treatment	No	–
Alaska	Yes	Secondary or Advanced	No	Permit Required
Arizona	Yes	" " "	No	–
Arkansas	No	Secondary	No	–
California	Yes	Primary to Advanced	No	Ad Hoc Basis
Colorado	No	Secondary & Disinfection	No	Allowed in Landfills
Connecticut	No	Ad Hoc Basis	No	–
Delaware	Yes	Secondary & Disinfection	No	Landfill – Ad Hoc
D.C.	Yes	Prohibited	No	–
Florida	Yes	Secondary to Advanced	No	Landfill – Ad Hoc
Georgia	Yes	Secondary & Disinfection	No	–
Guam	Yes	Ad Hoc Basis	–	–
Hawaii	Yes	Secondary & Disinfection	No	–
Idaho	Yes	Secondary or Sec. & Disinfection	Yes	Heat treatment
Illinois	Yes	Secondary	Yes	Ad Hoc Basis
Indiana	No	–	No	Stabilized
Iowa	No	Not Generally Practiced	No	Landfill if Dewatered
Kansas	No	Secondary	Yes	Ad Hoc Basis – Landfill
Kentucky	No	Secondary & Disinfection	No	Ad Hoc Basis
Louisiana	No	Secondary	No	Ad Hoc Basis
Maine	Yes	Secondary & Disinfection	–	–
Maryland	No	Secondary	–	–
Massachusetts	No	–	Yes	Landfill – Ad Hoc
Michigan	No	Prohibited; or Secondary & Disinfection	Yes	Ad Hoc Basis
Minnesota	Yes	Secondary & Disinfection	No	Guidelines in Preparation
Mississippi	No	" " "	No	Stabilized – Ad Hoc
Missouri	Yes	Secondary & Disinfection	Yes	Landfill if Dewatered
Montana	No	–	Yes	Landfill if Dewatered
Nebraska	No	Not Currectly Practiced	No	–
Nevada	No	Ad Hoc by Permits	No	–
New Hampshire	Yes	Secondary & Disinfection	Yes	Permit – Ad Hoc Basis
New Jersey	Yes	Secondary – Toxics Prohib.	Yes	Landfill – Mixed with Refuse
New Mexico	No	–	No	–
New York	No	Secondary – Disinfection	No	Ad Hoc Basis
North Carolina	No	Advanced & Disinfection	No	Ad Hoc Basis
North Dakota	No	Secondary	Yes	Landfill
Ohio	No	If Not Cost Effective – No Disposal	Yes	Landfill – Ad Hoc
Okalahoma	No	Secondary	Yes	Landfill – Ad Hoc
Oregon	No	Secondary & Disinfection or More Stringent	Yes	Landfill – Permit

Summary Table (Continued)

State	Wastewater		Sludge	
	Regula-tions	Treatment and Other Restrictions	Regula-tions	Restrictions
Pennsylvania	Yes	Secondary & Disinfection	Yes	Landfill – Permit, if Digested & Dewatered
Puerto Rico	No	–	No	–
Rhode Island	No	Discouraged	Yes	Landfill – Ad Hoc
South Carolina	No	Ad Hoc Basis	Yes	" " "
South Dakota	No	Secondary & Disinfection	No	Ad Hoc Basis
Tennessee	No	Secondary & Disinfection	Yes	Dewatering
Texas	Yes	Secondary or Secon. & Disin.	Yes	Landfill – Ad Hoc
Trust Territories	No	Ad Hoc Basis	–	–
Utah	Yes	Secondary	Yes	Digested or More Stringent
Vermont	Yes	Secondary & Disinfection to Prohibited	Yes	Landfill – Ad Hoc
Virginia	Yes	Secondary & Disinfection	No	Stabilized or More Stringent
Washington	Draft	Secondary & Disinfection	–	–
West Virginia	No	–	–	–
Wisconsin	Yes	Secondary & Disinfection	Yes	Digestion as a Minimum
Wyoming	No	Secondary	–	–

Source: Compiled by the National Commission on Water Quality.

SECTION VI

INDUSTRIAL
AND
AGRICULTURAL
WASTES

36

OVERLAND RECYCLING
OF ANIMAL WASTE

H. E. Grier, W. Burton, and S. C. Tiwari
Alcorn State University, Lorman, Mississippi

INTRODUCTION

General

A three-year study to evaluate the technical and economical aspects of an alternate disposal and utilization system of swine waste management which provides overland spray irrigation for crop production and effluent improvement was initiated in 1972.

This chapter is presented for the purpose of delineating the status of this study pertaining to changes in the contents of aerobic effluent sprayed overland as reflected in soil water analyses of constituents from such applications. Results of this study, to date, also serve the useful purpose of improving techniques of exploring the ecological effects and concepts of land recycling of animal waste and presenting such ideas more meaningfully. Such knowledge is imperative if we are to develop section or nation-wide guidelines making livestock waste disposal methods compatible with the public interest.

Perspective

Fifteen states produced over one million head of hogs per state in 1969 accounting for a total of 80 million head (1). Thirteen and one-half million of these hogs were grown in confinement. Confinement growing of hogs has been increasing rapidly in the southeastern states. The number of hogs produced in each of these states has climbed above one million. The number of hogs produced in Mississippi in 1969 was 0.49 million. The Mississippi Department of Agriculture and Commerce reported 632,000 head of hogs in the state in 1971. By 1972, the number of hogs over 100 pounds produced in Mississippi had grown to 867,000 head yielding approximately 142,000 tons of waste per year (2). Hog production systems in Mississippi vary from one to five hogs—grown in pens, running loose or in pastures—to more efficient confinement systems with one or two cell lagoons.

The terrain is hilly and wooded in certain sections of Mississippi which gives limited space for pastures or the production of crops for finishing hogs. Therefore, a system of confinement feeding and pasture grazing is promising for economical production (3). An accompanying problem of waste disposal is apparent. Overland application of animal waste offers promise as a method of disposal wherever land is readily available in the U.S. Several studies confirm use of this system as a means of removing concentrated nutrients from wastewater (4).

EXPERIMENTAL PROCEDURE

The site selected for use in this study is 1450 ft from the swine-feeding facility and lagoon area. Three 0.3-acre demonstration fields have been provided along with a 117.6 x 83 ft research area for four replications of three treatments spraying 0.2 in. water, 0.1 in. water with 0.1 in. effluent, and 0.2 in. effluent per application at various weekly frequencies.

A grass work area is between each 20 x 20 ft plot. The prepared irrigation area is on a grass and brush covered ridge that meanders through miles of woodland and varies in width from one to several hundred feet with valley depths of 20 to 80 ft on either side. As is most soil in this area, the soil in the research area is said to have been blown in and deposited during the dust bowl years of the nineteenth century. It is easily eroded and has the capacity to "take in" a large amount of water.

The overland irrigation site (Figure 1) has been prepared to have three fields with different levels. Field one is on one level and fields two and three are on another level. The research plots are located on the highest level. The fields have average slopes of 3 in. and the research area has an average slope of 0.5 in. Effluent is pumped from a 190 x 80 x 5 ft aerobic lagoon with a 10-hp, 80-psi centrifugal pump and improvised aeration is accomplished via a 3-in. outlet on the header assembly.

There are two lagoons—one anaerobic and the other aerobic. Monthly analyses of these two lagoons are conducted for chloride, pH, nitrate, nitrite and ammonia-N, conductance, TS, SS, BOD, DO and coliform. Temperature of the lagoons is also recorded

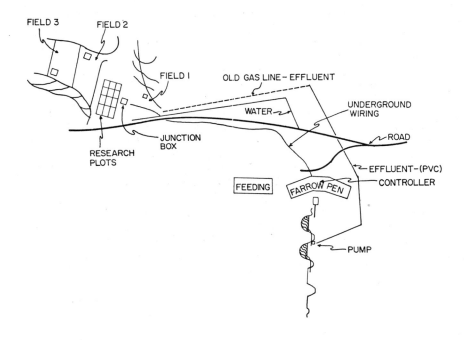

Figure 1.   Feeding and overland treatment area.

monthly. Effluent is transported to the irrigation sites via 1450 ft
of 3-in. PVC and a 3-in. abandoned gas line. Three demonstration fields have grass and legume mixtures and a 20 x 20 ft
plot of edible plants in each field with facilities for catching
runoff and sampling devices for soil water collection. Aerobic
effluent is sprayed on these fields at average rates of 10, 20 and
30 min per application for fields 1, 2 and 3, respectively.

Effluent to demonstration fields is transported through a 4-in.
aluminum aboveground irrigation line. Underground lines continue with the transport of effluent to rainbird pop-up sprinklers on the research plots.

Automation is controller-valve regulated so that 3 day/week applications, 6 day/week (a.m. or p.m.) and 6 day/week (a.m. and p.m.) applications for 1st, 2nd and 3rd periods, respectively, are effected.

Soil water sampling devices have been installed at depths of 6, 12 and 18 in. in each plot. The research plots and fields areas from which soil water and soil samples were obtained are shown in Figure 1.

DISCUSSION

Results of pre-treatment analysis of soil conducted at depths of 6, 12 and 18 in. are shown in Table 1. The results show uniform pH throughout the 12 plots at all depths. The chloride content of the soil overall plots was considerably uniform before irrigation (Table 2). Preliminary analysis of the soil for chloride content showed results which indicated a capacity to reduce the chloride content of aerobic effluent.

Post-treatment of plots with aerobic effluent indicated a trend of soil moisture pH to decrease with penetration downward through the soil accounting for large negative pH values at lower depths. A significant difference in change of pH by overland spray (P < .01) was noted when aerobic effluent was sprayed overland at all application rates (Table 3).

Table 1.   pH of soil at various depths before irrigation.

Replication	Depth (in.)	Rate of Effluent Spray[a]		
		0.2 in. Water	0.1 in. Water 0.1 in. Effluent	0.2 in. Effluent
1	6	5.40	5.30	5.45
	12	5.45	5.65	5.45
	18	5.80	5.45	5.70
2	6	5.40	5.80	5.65
	12	5.65	5.80	5.65
	18	5.90	5.90	6.05
3	6	5.80	5.70	5.45
	12	5.45	5.80	5.45
	18	5.90	5.45	5.70
4	6	5.40	5.40	5.30
	12	5.20	5.45	5.30
	18	5.30	5.70	5.40
	6	5.50	5.55	5.46
Mean Values	12	5.41	5.67	5.41
	18	5.75	5.62	5.71

[a] No significant difference due to depth or rate of effluent spray.

Table 2.   Chloride content of soil at various depths before irrigation.

Replication	Depth (in.)	Rate of Effluent Spray (mg/l)		
		0.2 in. Water	0.1 in. Water 0.1 in. Effluent	0.2 in. Effluent
1	6	6.50	7.50	10.00
	12	7.50	7.50	8.50
	18	7.00	10.00	7.25
2	6	7.50	7.50	6.50
	12	7.50	7.50	6.50
	18	7.00	7.00	6.50
3	6	6.50	7.50	7.50
	12	6.00	7.50	7.50
	18	6.00	7.50	7.50
4	6	6.00	6.50	7.50
	12	6.50	7.50	7.50
	18	6.50	7.50	7.50
	6	6.625	7.50	7.88
Mean Values	12	6.875	7.50	7.50
	18	6.625	8.00	7.18

Table 3.   Change in pH in various volumes of aerobic effluent sprayed overland as measured in soil water at different depths.

Replication	Depth (in.)	Rate of Effluent Spray			
		0.2 in. Water	0.1 in. Water 0.1 in. Effluent	0.2 in. Effluent	
		− − − − − − − pH Change* − − − − − −			
1	6	0.64	1.29	1.01	
	12	0.35	0.99	1.90	
	18	0.77	1.54	1.64	
2	6	0.90	0.10	1.19	
	12	0.60	1.44	1.94	
	18	0.85	0.99	2.04	
3	6	0.07	0.24	1.74	
	12	0.30	0.89	1.67	
	18	+ 0.09	1.64	1.39	
4	6	0.35	1.09	1.64	
	12	0.50	1.12	1.54	
	18	1.00	0.85	1.44	
	6	0.49	0.68	0.85	0.67^a
Mean Values	12	0.44	1.11	1.10	$0.88^{a,b}$
	18	0.63	1.25	1.17	1.02^b
		0.52^a	1.01^b	1.59^c	

*All values negative except where indicated by +.

a, b, cMean values not followed by a common letter differ as determined by DNMRT, (P < .01).

Differences between post-treatment soil water values for chloride concentration and aerobic effluent concentration of chloride are shown in Table 4. The first column of Table 4 is the difference in chloride concentration of soil water post-treatment values and chloride concentration in tap water. There is an increase in negative value from 6 to 12 in. which means less chloride is in the 12-in. depth. More chloride is in the 18-in. depth than the 12-in. depth. Chloride ions and clay particles have the same negative charge; thus the chloride ions should move downward in the soil along with water movement. Water sprayed overland gives least change in chloride content of soil water after treatment. There were significantly greater changes ($P < .05$) in chloride content of soil water after treatment with 0.1 and 0.2 in. of aerobic effluent per application. Significant ($P < .05$) differences in change of chloride content due to depth were noted.

There were no significant differences ($P < .05$) in change of nitrate concentration when aerobic effluent was sprayed overland due to depth of penetration through the soil (Table 5). Nitrate-N should act as chloride ions, $i.e.$, show some apparent increase as the soil depth increases from which samples of soil water are taken. The slight difference in the amount of nitrate-N

Table 4. Change in chloride content in various volumes of aerobic effluent sprayed overland as measured in soil water at various depths.

Replication	Depth (in.)	Rate of Effluent Spray (mg/l)			
		0.2 in. Water	0.1 in. Effluent 0.1 in. Water	0.2 in. Effluent	
1	6	+ 10	−10	+ 5	
	12	− 5	−10	− 25	
	18	0	−25	+40	
2	6	+ 5	− 5	− 25	
	12	0	−15	− 15	
	18	0	−15	− 5	
3	6	+ 20	−15	− 5	
	12	+ 5	−10	− 20	
	18	+ 5	−10	− 15	
4	6	0	−15	+10	
	12	0	−15	− 25	
	18	0	−10	− 25	
Mean Values	6	+ 9	−11	− 4	− 2[a]
	12	0	−12	− 21	−11[b]
	18	+ 1	−15	− 1	− 5[a]
		+3.3[a]	−13[b]	− 9[b]	

[a], [b]Mean values not followed by a common letter differ as determined by DNMRT, ($P < 0.5$).

in the top 6 in. of soil as compared with lower depths is not only due to the addition of effluent but also accounted for by the nitrification of the organic matter which is constantly taking place. Nitrate-N ion concentration increased significantly (P < .05) when both levels of aerobic effluent containing 12.00 mg/l was sprayed overland (Table 5).

The application of aerobic effluent on three demonstration fields was initiated to simulate the effect of high rates of application up to 0.75 in. hour so that minimal tensiometer reading for good growth for the lowest rate of application is maintained.

Adherence to the time 10, 20 and 30 min per field per application is not absolutely necessary so long as the fields receive effluent in these proportions. An increase in the change of all parameters measured, with the exception of NH_3-N, may be observed in Table 6.

Pre-treatment and post-treatment data were compared to determine change in concentration of chloride in soil water after overland application (Table 7). Since a lapse of about 60 days occurred between pre- and post-treatment sampling and analysis, it is quite likely that frequent rainfall affected results and

Table 5. Change in nitrate nitrogen concentration in various volumes of aerobic effluent sprayed overland as measured in soil water at different depths.

Replication	Depth (in.)	Rate of Effluent Spray (mg/l) *		
		0.2 in. Water	0.1 in. Water 0.1 in. Effluent	0.2 in. Effluent
1	6	−4.40	22.00	13.40
	12	−	0.50	11.20
	18	1.76	0.90	20.00
2	6	−4.40	9.00	18.24
	12	−2.20	15.60	46.40
	18	−1.76	9.44	29.20
3	6	4.40	9.00	36.28
	12	−1.76	−	11.20
	18	−4.40	15.60	24.40
4	6	4.40	12.52	9.00
	12	−6.40	11.20	20.00
	18	−4.40	16.48	20.00
	6	0.00	13.13	19.23 10.78[a]
Mean Values	12	−3.45	9.10	22.20 9.28[a]
	18	−2.20	10.61	23.40 12.07[a]
		1.88[a]	10.94[b]	21.61[c]

*All values + except where indicated by −.

[a], [b], [c] Mean values not followed by a common letter differ as determined by DNMRT, (P < 0.5).

Table 6. Change in soil contents (at various depths) after swine waste spray from aerobic lagoon.[a]

Field	Depth (in.)	Time Sprayed (hr)	Cl	NO$_3$	NO$_2$ (mg/l)	NH$_3$	pH	Specific Conductance (μmhos/cm)
					Parameters			
1	36	1.08	+10	+ 2.20	-.033	-.49	+ .55	-90
2	36	2.16	+35	+ 6.60	-.029	-.06	-. .10	+15
3	36	3.25	+25	+11.20	+.182	-.37	+1.50	+111

[a] Aerobic effluent – 25 mg/l chloride, 13.2 mg/l nitrate, 0.0 mg/l nitrite, 25.93 mg/l ammonia, 2 mg/l oxygen, pH 7.6, and 600 μmhos/cm conductance.

the change in mean aerobic effluent concentration after overland application as given in Table 4 may be more meanful.

A study of Table 8 indicates the effectiveness of the nitrification process in converting ammonium salts to other components when aerobic effluent is sprayed overland. The negative values in Table 8 mean that the mg/l of NH$_3$-N shown are that much less in soil water after various treatments than in aerobic effluent.

Table 7. Change in chloride content of soil water at various depths after overland application of aerobic effluent.

Replication	Depth (in.)	0.2 in. Water	0.1 in. Water 0.1 in. Effluent	0.2 in. Effluent	
			Volume of Effluent (mg/l)		
1	6	0.00	–10.00	–8.00	
	12	0.00	–10.00	–2.00	
	18	0.00	0.00	6.00	
2	6	5.00	5.00	–5.00	
	12	0.00	0.00	10.00	
	18	–10.00	0.00	–5.00	
3	6	–5.00	–10.00	0.00	
	12	5.00	–5.00	–5.00	
	18	5.00	–10.00	–10.00	
4	6	–5.00	–	–	
	12	0.00	–10.00	–15.00	
	18	0.00	5.00	–10.00	
	6	–1.25	–5.00	–6.50	–4.17
Mean Values	12	1.25	–6.25	–3.00	–2.67
	18	–1.25	–1.25	–4.75	–2.42
		–0.42[a]	–4.17[b]	–4.75[b]	

a, b Mean values not followed by a common letter differ as determined by DNMRT, (P $<$ 0.05).

The mean analyses for content of treatment material used in this study are shown in Table 9.

Analyses represent parameter averages of treatment media at the time of application. Type of sample collection device for soil water precludes the possibility of obtaining coliform values because of filtering action. Other bacteria are also held from entering the semiporous sample collecting and devices; consequently, BOD analyses of soil water are not practical. BOD values of 200–1600 mg/l for aerobic effluent indicate the demand for oxygen that wastewater from lagoon cells will exert on the land disposal site. Bacteria will be available for action, in the soil, thus lowering BOD, but analyses of soil water for coliform in this study will show 0/ml because of the filtering process that sampled soil water has undergone.

Table 8. Change in ammonia nitrogen concentration in various volumes of aerobic effluent sprayed overland as measured in soil water at various depths.

Replication	Depth (in.)	Rate of Effluent Spray (mg/l) *			
		0.2 in. Water	0.1 in. Water 0.1 in. Effluent	0.2 in. Effluent	
1	6	+0.07	38.61	28.37	
	12	0.00	37.44	38.74	
	18	+0.09	38.74	38.42	
2	6	0.00	38.74	38.41	
	12	0.00	38.74	38.22	
	18	0.00	38.74	38.34	
3	6	+0.24	37.58	38.24	
	12	0.00	38.74	38.51	
	18	0.00	38.00	38.58	
4	6	0.00	38.74	38.62	
	12	0.00	38.74	38.62	
	18	0.00	38.74	38.08	
	6	0.08	38.42	38.41	25.64[a]
Mean Values	12	0.00	38.42	38.52	25.66[a]
	18	0.02	38.55	38.36	25.64[a]
		0.03[a]	38.46[b]	38.43[b]	

* All values negative except where indicated by +. Aerobic effluent contained 38.74 mg/l of NH_3-N.

[a,b] Mean values not followed by a common letter differ as determined by DNMRT, ($p < 0.05$).

Table 9.   Mean content of aerobic effluent and tap water used as irrigation spray on plots and fields.

Parameters	Tap Water	Aerobic Effluent
Chloride (mg/l)	15.00	40.00
NO_3-N (mg/l)	17.60	12.00
NO_2-N (mg/l)	0.00	0.59
NH_3-N (mg/l)	0.00	38.74
BOD (mg/l)	5.00	40.00
Specific Conductance (μmhos, mg/l)	300.00	728.00
pH	7.42	7.54
Coliform (colonies/100 ml)	0.00	200–1600

ACKNOWLEDGMENTS

This project has received support from the U.S. Environmental Protection Agency Research Grant 802336.

REFERENCES

1. Van Arsdall, R. N. "Economic impact of controlling surface water run-off from point sources in U.S. hog production," *Proc. 1974 Cornell Agricultural Waste Management Conference.* Cornell University, Ithaca, New York, Rochester, New York, pp. 97–107 (March 1974).
2. ·Fox, R. "Animal waste management in Mississippi, An Overview of the Problem," Animal Waste Task Group, p. 21 (September 27, 1972).
3. Grier, H. E., B. C. Diggs, and D. C. Carter. "Alternative system of farrowing for feeder pig production," Research Report, Vol. 1, Number 10, Mississippi Agricultural and Forestry Experimental Station (May 1975).
4. Thomas, R. E. "Feasibility of overland-flow treatment of feedlot run-off," National Environment Res. Ctr., U.S. EPA, Corvallis, Oregon (December 1974).

37

SOIL AND CROP RESPONSE
TO APPLIED ANIMAL WASTE

M. L. Horton, R. R. Schnabel, and J. L. Wiersma
Water Resources Institute, South Dakota State University,
Brookings, South Dakota

INTRODUCTION

Present and proposed statutes regarding effluent discharge into waterways, combined with the current state-of-the-art of agricultural waste utilization technology, virtually assure that animal wastes will be applied to the soil in the disposal process.

Guidelines for application of animal wastes to croplands frequently have been based upon the plant availability of the waste nitrogen. The Agronomy Guide (1) for application of animal manures in Ohio uses a decay series concept based on nitrogen mineralization to determine application rates. The North Carolina Agricultural Extension Service (2) determines land application rates for animal waste by calculating the quantity of waste needed to satisfy a soil nitrogen test based on the analyzed nitrogen content of the waste. Powers *et al.* (3) in Kansas used a nitrogen decay series concept to develop application guidelines for beef feedlot waste. In addition, the Kansas guidelines discuss the importance of soluble salts in the soil and suggest ways to account for the salt content of the waste. Under low leaching conditions, the accumulation and movement of salts within the profile of certain naturally saline soils may be as important a parameter in determining application rates as is nutrient transport.

The infiltration and movement of water through the soil is important to plant growth. Some studies have shown increased water infiltration with applied animal waste, while other studies have shown a decrease in infiltration rate after waste application. Manges *et al.* (4) found water infiltration rates increased with beef waste application rates ranging from 93 to 269 MT/ha per year; however, infiltration rates decreased with higher rates of manure applied.

Animal wastes contain nutrients and organic matter which are important to crop yields. Yet, depressed crop yields have been measured where large amounts of animal waste were applied. Increased soil salinity generally has been thought to be responsible. Reddell *et al.* (5) reported yields of corn (*Zea mays*, L.) and grain sorghum (*Sorghum bicolor*, L.) to be good on Pullman clay loam soil up to waste application rates of 672 MT/ha (wet weight), but a 50 percent reduction in crop yields occurred on the 1345 and 2017 MT/ha wet weight plots. Sodium concentration and salinity increased to hazardous levels under the higher rates of applied wastes.

The study reported here was initiated to investigate the effects of various rates of applied wastes produced by animals on different ratios upon soil properties and crop production under subhumid conditions where minimum amounts of leaching water are available.

METHODS

Research initiated in August 1973 has been described by Horton *et al.* (6). The study included feeding trials, field disposal plots and laboratory analyses. The feeding trial consisted of eight pens of beef steers (11 head/pen) in confinement and eight pens (11 head/pen) in the open with no shelter. A common basic ration was used with four levels of added salt (NaCl) which were 0.00, 0.25, 0.50 and 0.75 percent of the ration on a dry-weight basis.

All wastes were collected, held in storage until time of application, and applied to field disposal plots at four rates (44.8, 89.6, 139.4 and 179.2 MT/ha). The wastes applied to field plots consisted of two salinity levels—low (combined wastes from pens receiving 0.00 and 0.25 percent added salt) and high (combined wastes from pens receiving 0.50 and 0.75 percent added salt). Each treatment was replicated four times in field plots 6.1 m by 36.6 m in size.

Laboratory analyses were performed on waste samples collected at the time of field application, on soil samples collected at planting and harvest, and on plant samples collected at tasseling stage.

Soil samples were collected in May and September 1975 to a depth of 9 cm using a power probe. The samples were air-dried, ground and passed through a 2-mm screen in preparation for analyses. Water-soluble and extractable Ca, Mg, K and Na contents of the soil samples were determined by atomic absorption using Isaac and Kerber (6) procedures. Electrical conductivities were determined on soil saturation extracts according to U.S. Salinity Laboratory (8) procedures.

After seedbed preparation, the field plots were planted to corn on May 7, 1975, at a population of 43,000 plants per hectare. Plant leaf samples were collected from all plots in early August 1975. The dried and ground leaf samples were analyzed for major ions by spark-emission spectroscopy. A portion of each plot was harvested for silage yield on September 11, 1975, and a portion was harvested for grain yield on October 8, 1975.

An infiltration study was conducted on half of the plots during the period July 7 to July 18, 1975. Water infiltration was determined in two plots per treatment at three sites per plot. Double-ring cylinder infiltrometers with diameters of 25.4 and 40.6 cm were used during the study. The water delivery system maintained a constant head of 7.5 cm of water inside the inner ring. Modified recording rain gauges calibrated in such a fashion as to continuously record quantity of water supplied to the inner infiltration ring were incorporated into the system. This enabled a continuous record of accumulated infiltration with time. Accumulated infiltration data were averaged within treatments. The average data were then run through a polynomial regression subroutine to develop equations generated by these data. Infiltration rates were calculated by taking the first derivatives of the equations at the points of interest.

RESULTS AND DISCUSSION

Chemical analyses for major cations of wastes produced by steers on a ration with different levels of added salt are shown in Table 1. The values shown are averages for four sampling

Table 1. Major cation content of wastes from beef steers fed a ration with two salinity levels.

Salinity Level	Percent			
	Na	Ca	Mg	K
Low	0.36	1.01	0.89	3.15
High	0.71	1.07	0.93	3.30

dates. The sodium content of the waste varies directly according to the amount of added salt.

Electrical conductivities (EC) of saturation extracts of the soil for several sampling dates are given in Table 2. The EC

Table 2. Electrical conductivities of surface 30 cm in waste management plots.

Proposed Rate (mt/ha)	Salt Level	Fall 1973 (μmhos/cm)	Fall 1974 (μmhos/cm)	Fall 1975 (μmhos/cm)
0		630	1116	1188
44.8	Low	682	2418	2765
44.8	High	460	1928	3347
89.6	Low	724	4873	4502
89.6	High	713	3956	5091
134.4	Low	683	5886	6754
134.4	High	745	5456	6198
179.2	Low	923	4903	6969
179.2	High	830	6016	7520

values for 1973 represent the salinity of the soil prior to application of any waste. The soil where the plots were established are saline (EC exceeds 4000 μmhos/cm) at depths greater than 90 cm. The first year's application of waste increased the EC of the surface 30 cm of soil under all treatments. At rates of 89.6 MT/ha or greater, sufficient salts were added to increase the EC to values in excess of 4000 μmhos/cm for the surface 30 cm, leaving only an approximately 30-cm zone of nonsaline soil. After the second year of waste application, the entire root zone of plots receiving the high wastes were saline.

The exchangeable sodium (Na) of the soil increased with an increase in the amount of salt added; however, the increase in exchangeable sodium for equal amounts applied is not as great for the second year of application. As the soil solution becomes more concentrated, monovalent cations such as Na become better competitors for sites on the exchange complex. The smaller increase in exchangeable Na during the second year may indicate an approaching equilibrium between Na in the soil solution and Na on the soil exchange complex.

The ear corn and silage yield data for 1974 and 1975 are presented in Table 3. There was no significant difference in 1974 silage yields between salt levels nor among waste rates. The 1975 silage yield data showed no significant difference between salt levels; however, among waste rates the yields were significantly different at the 0.05 level.

The 1974 ear corn yield data were significantly different

Table 3. Yield of silage and ear corn from plots receiving beef feedlot wastes.

Salt Level	Proposed Rate (MT/ha) Year	0	44.8	89.6	134.4	179.2
Silage Yield (Dry Weight) (MT/ha)						
Check	1974	5.67				
	1975	7.95				
Low	1974		8.47	7.54	5.98	7.68
	1975		8.00	6.98	5.26	5.65
High	1974		6.72	7.86	6.68	6.37
	1975		8.08	5.11	3.60	3.86
Ear Corn Yield (15.5% Moisture) (hl/ha)						
Check	1974	39.22				
	1975	19.41				
Low	1974		55.19	62.32	52.77	62.19
	1975		31.36	37.18	21.74	27.34
High	1974		35.27	54.50	61.00	53.89
	1975		31.62	33.91	23.44	10.99

among waste rates at the 0.05 level. In 1975 there was no significant difference either between salt levels or among waste rates. The 1974 data results from one year's application with indicated treatments. The EC values of plots receiving higher rates are sufficiently high to adversely affect crop growth, yet there was no significant difference in silage yield. Early season differences in corn height and vigor were visually quite apparent. A reasonable explanation is that corn roots penetrated into a less saline portion of the profile where they were better able to extract moisture and subsequently approached equal dry matter production with plots receiving lower rates of waste. After two years' application, a greater portion of the profile was salinized; therefore, corn roots were unable to penetrate to a region from which to extract moisture. Consequently, there was a significant difference in dry matter production. Lack of difference in ear corn yield in 1974 may also be attributed to generally low rainfall. No differences in 1975 ear corn yields resulted from rainfall distribution and a corn smut problem.

Leaf analysis data are given in Table 4 for both the 1974 and 1975 seasons. The element Mg shows an inverse relationship with amount of waste applied. There was no significant difference in Mg content either with waste rate or with salt level.

Table 4. Leaf analyses from corn plants grown on plots receiving beef feedlot wastes.

Proposed Rate (MT/ha)	Salt Level	K 1974	1975	(%)	Mg 1974	1975
0		2.61	2.30		0.54	0.53
44.8	Low	2.61	2.12		0.39	0.39
44.8	High	2.59	2.18		0.44	0.41
89.6	Low	2.58	2.34		0.36	0.34
89.6	High	2.68	2.24		0.34	0.36
134.4	Low	2.61	2.36		0.32	0.28
134.4	High	2.65	2.20		0.30	0.28
179.2	Low	2.69	2.31		0.32	0.28
179.2	High	2.52	2.23		0.26	0.28

The trend in reduced Mg content with increased rate was not aggravated by the second year application.

Infiltration rates of plots receiving wastes with low and high salinity levels appear in Figures 1 and 2, respectively. The infiltration rates of plots receiving approximately 45 MT/ha at both salinity levels and approximately 89 MT/ha with a low salinity approach or exceed the infiltration rates of the control plots. The infiltration rates of all other treatments are below those of the control plots, with those receiving the highest waste rates going to zero in 3 to 10 hr after initiation of infiltration. Accumulated infiltration at 1 hr was divided by accumulated infiltration at 12 hr to assess the stability of soil to water transport (WSR). The WSR values appear in Table 5 and follow the same trend as the infiltration rates.

Adsorption of excessive amounts of monovalent cations (Na^+, K^+, NH_4^+) on the exchange complex (9, 10) and clogging of pores by organic matter (11) are factors most frequently cited as leading to decreased water transport through soils used as a disposal site for agricultural and municipal wastes.

It is difficult to establish a cause-and-effect relationship regarding the decrease in water transport experienced in the present study. Due to the composition of the wastes, an increase in any of the above factors necessarily involved an increase in the remaining factors. Of these factors only exchangeable Na varies with the salt level within a given application rate as do the WSR values. The conclusion could be reached that the water movement problems encountered may be attributed to exchangeable Na saturation. Sodium saturation values in Table 6 reveal that no treatment exceeded 2.5 percent saturation. While the critical values of Na saturation representing a sodium hazard are by

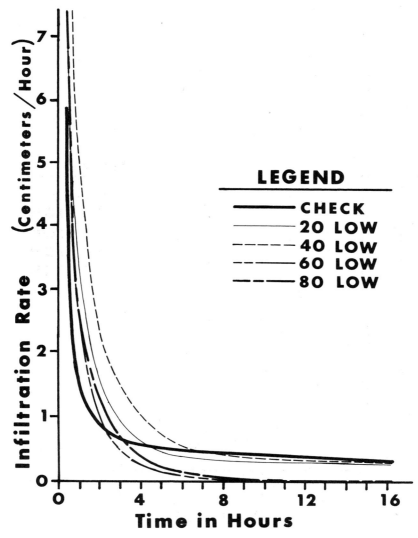

Figure 1.   Water infiltration rate.

no means agreed upon, 2.5 percent sodium saturation cannot be held responsible for soil dispersion, certainly not at the EC values present. At the time the infiltration tests were conducted, a gelatinous substance was observed at a depth of 8–15 cm which was most pronounced in plots receiving high rates of waste. Further research is required to assess the cause(s) of decreased

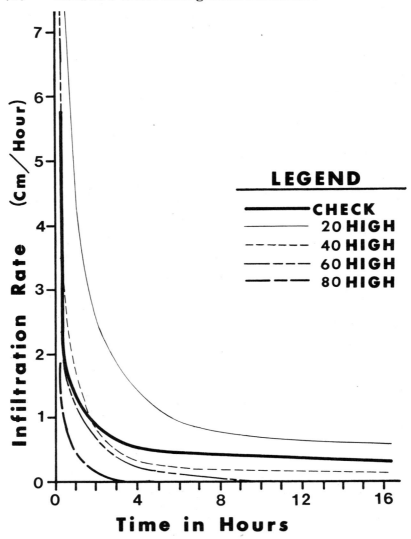

Figure 2. Water infiltration rate.

water infiltration. Preliminary observations suggest that interaction of soluble sodium and the organic matter present in the
soil may be responsible for the decreased ability of the soil to
infiltrate water.

The results confirm that the application of animal waste to
certain soils under conditions of minimal leaching can increase

Table 5.   Water stability ratio.

$$\left(\frac{\text{Accumulated infiltration at 1 hr}}{\text{Accumulated infiltration at 12 hr}} \right)$$

Proposed Rates (MT/ha)	Salt Level	WSR
0		0.33
44.8	Low	0.40
44.8	High	0.37
89.6	Low	0.35
89.6	High	0.68
134.4	Low	0.61
134.4	High	0.86
179.2	Low	0.79
179.2	High	0.87

Table 6.   Sodium saturation of surface 30 cm in waste management plots.

Proposed Rates (MT/ha)	Salt Level	% Na Sat. (Fall 1975)
0		0.36
44.8	Low	0.60
44.8	High	1.09
89.6	Low	0.79
89.6	High	1.56
134.4	Low	1.27
134.4	High	2.20
179.2	Low	1.24
179.2	High	2.33

the salinity to levels harmful to the growth of many plants of agronomic importance.

Applications of animal waste in excess of 90 MT/ha caused reduced infiltration. The cause of the decreased water infiltration appeared to be increased Na levels and dissolution of organic matter.

ACKNOWLEDGMENTS

This study was supported through EPA Demonstration Grant Number R803662–01–0.

Approved for publication by the Director of the South Dakota Agricultural Experiment Station, Brookings, South Dakota, as Journal Paper No. 1431.

REFERENCES

1. Agronomy Guide. Bull. 472, Cooperative Extension Service, Ohio State University, Columbus, Ohio, 46 pp. (1974).
2. North Carolina Agricultural Extension Service. Circular 571, "Beef cattle waste management alternatives," 29 pp. (1973).
3. Powers, W. L., G. W. Wallingford, L. S. Murphy, D. A. Whitney, H. L. Manges and H. E. Jones. "Guidelines for applying beef feedlot manure to fields," Kansas State University, Cooperative Extension Service Bulletin C-502 (1974).
4. Manges, H. L., D. E. Eisenhauer, R. D. Stritzke and E. H. Goering. "Beef feedlot manure and soil water movement," Paper No. 74-2019 presented at the 1974 summer meetings, ASAE, Stillwater (June 25–26, 1974).
5. Reddell, D. L., R. C. Egg, V. L. Smith. "Chemical changes in soil used for beef manure disposal," Paper No. 74-4060 presented at the 1974 summer meeting, ASAE, Stillwater (June 23-26, 1974).
6. Horton, M. L., J. L. Halbeisen, J. L. Wiersma, A. C. Dittman and R. M. Luther. "Land disposal of beef wastes: climate, rates, salinity, and soil," *Proc. 3rd International Symposium on Livestock Wastes*, Urbana, Illinois, pp. 258-260 (1975).
7. Issac, R. A. and J. D. Kerber. "Atomic absorption and flame photometry: techniques and uses in soil, plant, and water analysis," pp. 17–37. In: *Instrumental Methods for Analysis of Soils and Plant Tissue*, L. M. Walsh, ed. Soil Science Society of America, Madison, Wisconsin (1971).
8. U.S. Salinity Laboratory Staff. "Methods for soil characterization," Chapter 6 in USDA Agricultural Handbook No. 60, L. A. Richards, ed., pp. 83-126 (1954).
9. Hinrichs, D. G., A. P. Mazurak and N. P. Swanson. "Effect of effluent from beef feedlots on the physical and chemical properties of soil," *Soil Sci. Soc. Amer. Proc.* 38:661-663 (1974).
10. Travis, D. O., W. L. Powers, L. S. Murphy and R. I. Lipper. "Effect of feedlot lagoon water on some physical and chemical properties of soils," *Soil Sci. Soc. Amer. Proc.* 35:122-126 (1971).
11. Thomas, R. E., W. A. Schwartz and T. W. Bendixen. "Soil chemical changes and infiltration rate reduction under sewage spreading," *Soil Sci. Soc. Amer. Proc.* 30:641-646 (1966).

SALT ACCUMULATION IN SOIL
AS A FACTOR FOR DETERMINING
APPLICATION RATES OF BEEF-FEEDLOT
MANURE AND LAGOON WATER

G. W. Wallingford
Northwest Experiment Station, Crookston, Minnesota
W. L. Powers and L. S. Murphy
Department of Agronomy
H. L. Manges
Department of Agricultural Engineering
  Kansas State University, Manhattan, Kansas

INTRODUCTION

Several factors have been used as the basis for determining application rates of animal manures. Most recommendations have the objective of protecting the environment and maintaining maximum soil productivity without creating soil conditions that are toxic to plant growth. Several state agencies have used nitrogen as the basis for manure-application guidelines. We believe that nitrogen would be the best choice when a single manure constituent is to be used (1). Rates based on a single constituent such as nitrogen, however, may require large initial applications to satisfy crop-nutrient requirements. Various researchers have shown that applying large amounts of manure can create saline soil conditions toxic to plant growth. For that reason, guidelines should enable the user to detect application rates that might cause toxic salt accumulations in the soil.

In experiments in which large amounts of animal manure have been applied, yield depressions have been measured relative to control-plot yields or relative to plots obtaining maximum yields. In most such studies salt toxicity was credited with being responsible for yield reduction (2–5). In experiments in which large applications of manure did not decrease crop growth, soil salinity measurements showed no build-up of soluble salts (6, 7). Crop yields can be lowered as a result of reduced seed germination and seedling growth, and applying manure at high rates has been thought to contribute to reduced yields through increased saline conditions or through toxicity from high ammonium concentrations in the soil (3, 5, 8).

To determine which cations in manures are most responsible for increased soil salinity, saturation-extract determinations were performed on soil samples taken from plots receiving beef feedlot manure and lagoon water. The published measurements from these same plots indicate that applied manure and lagoon water can—by increasing soil salinity—decrease the growth of corn (4, 5).

Knowledge of what cation most influences salt build-up in the soil after animal manure has been applied could help to improve the accuracy of application-rate recommendations. For example, when manures are from rations high in "table salt" (NaCl) or from high-forage rations (which generally contain high amounts of K), they might require management different from that for manures from rations low in salt or high in grain. Regardless of the ration, if analysis shows that the manure contains an unusually high concentration of a certain cation, then application rates could be adjusted to reflect that cation imbalance.

The results of the soil-salinity measurements indicated the need for manure-disposal guidelines based partly on the salt content of the manure (9,10). These guidelines will be summarized briefly here.

METHODS AND MATERIALS

The experimental design was reported in detail previously (4, 5). To summarize, in separate studies, beef-feedlot manure was applied in the fall and beef feedlot lagoon water was applied by furrow irrigation during the summer, on a silty clay loam soil in south-central Kansas. Surface soil samples (0 to 15 cm) were taken from each plot in spring and fall. In the fall of 1973 soil cores were taken to a depth of 3 m from plots in two of the four replications in both the manure and the lagoon-water studies. (For the past four years, these plots had received manure and lagoon-water applications.) The surface and the soil core samples were analyzed for soluble salts by vacuum extraction of a water-saturated paste. The electrical conductivity (EC) of the extract was measured using a Wheatstone bridge; concentration of Na and K in the extract was measured by flame photometry, that of Ca and Mg by atomic absorption techniques.

The lagoon water was sampled prior to each application and

was analyzed for Na and K by flame photometry, for Ca and Mg by atomic absorption, and for EC using the Wheatstone bridge. Manure samples, taken previous to each application, were analyzed for Na, K, Ca and Mg using flame photometry and atomic absorption after a nitric-perchloric acid digestion.

The amount of manure applied to each plot was measured by weighing what fell on a plastic sheet after the passage of a spreader truck. Inflow and outflow measurements were used to determine lagoon-water applications. In the manure study, variation from the intended rates resulted in a fairly uniform distribution between the lowest and highest tonnages applied: rates ranged from 29 to 688 dry mt/ha-yr. Lagoon-water treatments ranged from 7 to 37 cm/yr. In both the manure and the lagoon-water studies, a control plot was included in each of the four replications; it received no chemical fertilizer or waste.

RESULTS AND CONCLUSIONS

Manure and Lagoon Water Analysis

The chemical analyses of the beef feedlot manure and the lagoon water applied to the research plots are presented in Table 1. Both materials came from the same feedlot and show that a

Table 1.   Analysis of beef feedlot manure and lagoon water applied to research plots.

Electrical Conductivity (mmho/cm)		H_2O	Na	K (%)	Ca	Mg
Manure (dry weight basis)						
High	—	39.9	0.49	1.77	1.36	0.50
Low	—	4.6	0.15	0.30	0.36	0.30
Average	—	20.5	0.26	1.14	0.92	0.41
Lagoon Water (wet weight basis)						
High	7.6	98.19	0.0660	0.184	0.0615	0.0239
Low	1.6	99.81	0.0112	0.0259	0.0083	0.0035
Average	3.1	99.52	0.0295	0.0671	0.0225	0.0087

single operation can produce manures of extremely variable composition. To predict accurately how manures will affect soil properties and plant growth, one must have knowledge of the manure composition, and that can be obtained only through periodic chemical analysis.

Surface Soil Analysis

Table 2 and Figures 1 and 2 give results of the saturation-extract analysis of the surface-soil samples taken from the lagoon-water and manure plots. All samples taken previous to the fall of 1973 were analyzed for electrical conductivity (EC) only. The EC of the fall extracts was measured along with concentrations of the individual cations Na, K, Ca and Mg. So that we might compare the relative contribution of those four cations to total salinity, results were tabulated on an equivalent basis. The equivalents of the four cations were summed and the percentage of each was calculated. Expressing cation concentrations by equivalents rather than by atomic weights provides a more accurate assessment of individual cation contribution to total salinity.

The prediction equations for soil EC, using lagoon-water or manure application rate as the independent variable, showed that EC increased linearly with increasing rates of lagoon water and

Table 2. Values of R^2, the slope, and the intercept of straight-line graphs of lagoon-water and manure application rate versus electrical conductivity (EC) and chemical composition of the saturation extracts of the surface soil (0-15 cm). Percentages are based on equivalents. Lagoon-water application began in the summer of 1970; manure application, in the fall of 1969.

Date and Measurement	R^2	Slope	Intercept
Lagoon Water Treatments (cm/yr)			
1970 fall EC (mmho/cm)	0.878	0.0166	0.495
1971 fall EC (mmho/cm)	0.875	0.0382	0.470
1972 fall EC (mmho/cm)	0.914	0.0600	0.391
1973 fall EC (mmho/cm)	0.786	0.0143	0.564
1973 fall total meq/l	0.711	0.0693	3.37
1973 fall % Na	0.019	–	–
1973 fall % K	0.542	0.306	14.2
1973 fall % Ca	0.311	-0.215	26.4
1973 fall % Mg	0.062	–	–
Manure Treatments (dry mt/ha-yr)			
1970 fall EC (mmho/cm)	0.766	0.00389	0.558
1971 fall EC (mmho/cm)	0.782	0.00744	0.575
1972 spring EC (mmho/cm)	0.840	0.0236	0.580
1972 fall EC (mmho/cm)	0.804	0.0109	0.554
1973 spring EC (mmho/cm)	0.882	0.0156	0.195
1973 fall EC (mmho/cm)	0.910	0.00638	0.394
1973 fall total mEq/l	0.910	0.0435	2.01
1973 fall % Na	0.022	–	–
1973 fall % K	0.698	0.0496	14.9
1973 fall % Ca	0.696	-0.0422	32.6
1973 fall % Mg	0.000	–	–

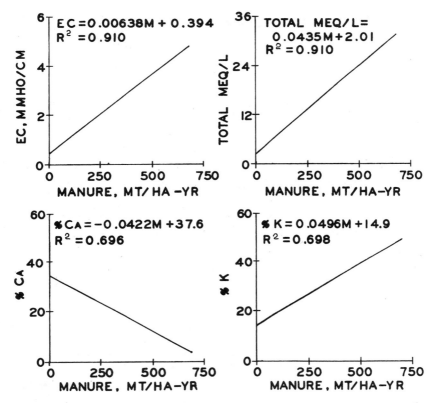

Figure 1. Composition of water-saturated paste extract from surface soil (0 to 15 cm) after four annual applications of beef-feedlot manure. The total meq/l is the sum of Na, K, Ca and Mg, and %K and %Ca are percentages of the total meq/l. EC is electrical conductivity.

manure. The high R^2 values at all sampling dates indicated good correlation between EC and application rate, suggesting that EC measurements could be used as a management tool to monitor the salt status of fields receiving applications of manure and lagoon-water. In data reported earlier (4, 5), lowered yields of corn forage on plots that received high rates of lagoon water and manure were attributed to plant toxicity from elevated salt concentrations in the soil.

The total meq/l of Na, K, Ca and Mg of the 1973 surface-soil saturation extracts increased linearly with increasing rates of lagoon water and manure, and the R^2 values were similar in magnitude to those of the EC prediction equations. That indicated

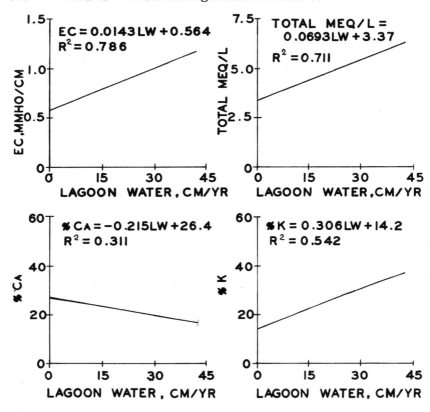

Figure 2. Composition of water-saturated paste extract from surface soil (0 to 15 cm) after four annual applications of beef feedlot lagoon water. The total meq/l is the sum of Na, K, Ca and Mg, and %K and %Ca are percentages of the total meq/l. EC is electrical conductivity.

that the total meq/l increased at the same level of consistency as did the total salinity of the extracts. When the percentage meq/l of the four cations are compared (Table 2, Figures 1 and 2), it is evident that the percentage meq/l of Na and Mg remained constant at all levels of waste applications. In other words, equivalents of Na and Mg in the extracts increased at approximately the same rate as did the total salinity. The meq/l percentage of K, however, increaséd with increasing rates of both lagoon water and manure, while the percentage of Ca decreased with increasing application. At the higher rates, K assumed a larger share of the total salinity of the extracts, while Ca decreased in its share of the total salinity. The relative concentrations of K and Ca applied to the soil were the same, of course, at all application rates, meaning that differences in the soil extracts were due to chemical precipitation and/or cation exchange reactions on soil colloids. Apparently at high application rates, more Ca than K was removed from the soil solution by forming precipitates of low solubility and by attaching to soil colloids.

Soil Core Analysis

Results of the soil core analysis by depth for saturation extract EC and composition are presented in Figures 3 and 4. The EC of saturation extracts was increased (compared with the control) throughout the top two meters by the two highest rates of

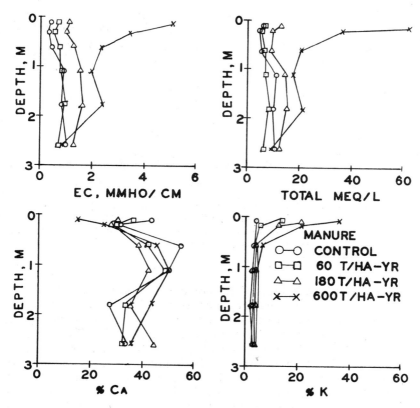

Figure 3.   Composition of water-saturated paste extract from soil cores after four annual applications of beef feedlot manure. The total meq/l is the sum of Na, K, Ca and Mg, and %K and %Ca are percentages of the total meq/l. EC is electrical conductivity.

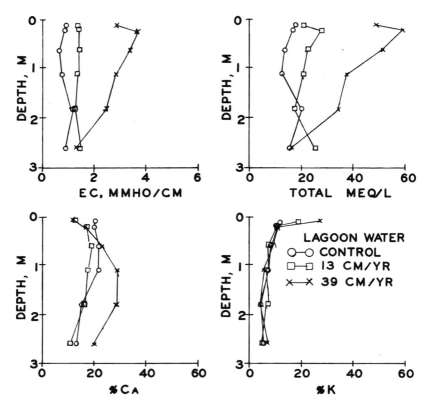

Figure 4. Composition of water-saturated paste extract from soil cores after four annual applications of beef feedlot lagoon water. The total meq/l is the sum of Na, K, Ca and Mg, and %K and %Ca are percentages of the total meq/l. EC is electrical conductivity.

manure and by the highest rate of lagoon water. The lower rates of lagoon water and manure increased EC in the top meter only.

The total meq/l in the soil cores taken from both the manure and the lagoon-water plots had profile distributions very similar to those of EC. The meq/l percentages of K were increased at the surface by increasing rates of both lagoon water and manure. Generally, there were no treatment differences in the meq/l percentage K data below 10 cm in the lagoon water plots or below 20 cm in the manure plots. The meq/l percentage Ca was decreased by increasing application rates at the 10- and 20-cm depths in both lagoon-water and manure plots. Below 20 cm, no trends in the profile distribution of Ca could be attributed to treatment. The meq/l percentage of Na and Mg were not related to treatment, so are not presented here.

The effects on the meq/l percentages of K and Ca in the upper section of the cores paralleled those found in the surface soil samples. Below 20 cm, however, the percentage composition of the saturation extracts was not affected by treatment even though the total salinity of the extracts was increased down to 1.75 meters as indicated by EC and the total meq/l. After four years of lagoon-water and manure applications, the percentage composition of the saturation extracts below 20 cm remained unaffected, indicating that changes in soil cation composition are not likely to happen quickly below 20 cm and that longer-term studies are needed to determine if changes will eventually occur.

APPLICATION—RATE GUIDELINES BASED ON SOIL SALINITY

The increased soil salinity and its detrimental effect on crop growth on the research plots illuminated the fact that some farmers could be lowering the productivity of their soils by applying manures in amounts sufficient to provide salts in excess of what can be removed by natural precipitation or irrigation. That made evident the need for guidelines to show users how to calculate application rates that will not create toxic salt accumulations. Recently two such guidelines were prepared—one for lagoon water, one for manure—based partly on the salt content of the materials (9, 10). For both guidelines, however, salt-balance calculations were based on rainfall and soil conditions found in Kansas; therefore, they should not be used for other areas.

The lagoon-water guidelines are based on maintaining soil salinity in either a low or a medium range on coarse (light), medium or fine (heavy) textured soils. When water is available to dilute the lagoon water before it is applied, users should use dilution factors shown in graphs based on the lagoon water EC and irrigation water EC. When dilution is not possible, they should refer to graphs that can be used to determine the maximum annual irrigation with undiluted lagoon water. Figure 5 is the graph used to determine dilution factors for maintaining a medium salinity on a coarse (light)-textured soil; Figure 6, the graph that allows one to determine maximum annual irrigation of undiluted lagoon water.

The guidelines for applying beef feedlot manure are based primarily on nitrogen content of the manure but recognize that if the nitrogen requirement of the crop is to be satisfied solely by the manure, then large applications may be necessary the first

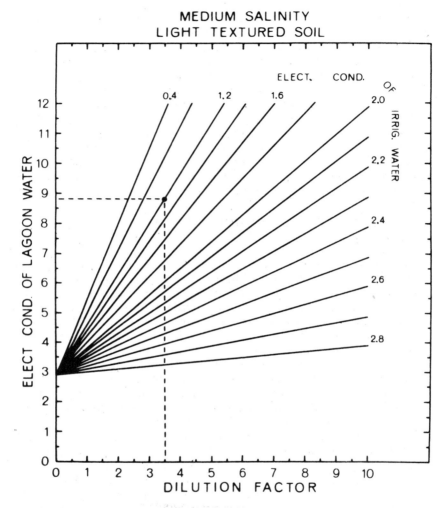

Figure 5.   Lagoon-water dilution factors for a resulting medium salinity
(EC of 6 mmho/cm or less) on a coarse (light)-textured soil.
Dashed lines illustrate an example of a lagoon water EC of 8.8
mmho/cm and an irrigation water EC of 1.2 mmho/cm. A dilution ratio of 1 part lagoon water to 3.5 parts irrigation water
would be used to maintain a medium salinity level.

few years. Because large applications may create toxic salt concentrations in the soil, the guidelines allow for determining maximum allowable application rates on both irrigated and dryland
soils. For irrigated soils, the maximum annual manure application depends on the soil salinity desired (low or medium), soil

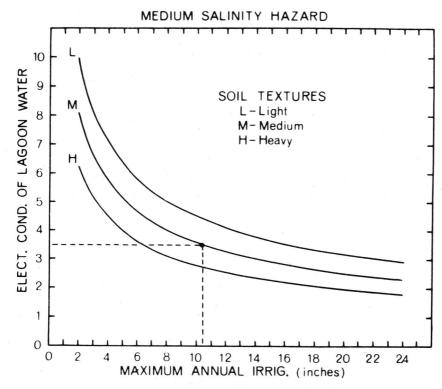

Figure 6. The maximum amount of undiluted lagoon water that can be applied and maintain a medium-soil salinity (EC of 6 mmho/cm or less). Dashed lines illustrate an example where a maximum of 10.5 in./yr of lagoon water with an EC of 3.5 mmho/cm can be applied to a medium-textured soil.

texture, EC of the irrigation water, and percentage of salt in the manure (defined as the total Na, K, Ca and Mg). Maximum annual manure applications for dryland soils are based on the desired salinity level, soil texture, annual precipitation in inches, and the percentage of salt in the manure. Figure 7 is the graph for finding the maximum application for maintaining a low salinity on a medium-textured soil, and Figure 8 the graph for dryland soils.

SUMMARY

Beef feedlot manure and lagoon water were applied, in separate studies, to a silty clay loam soil. Extracts from water-saturated pastes from surface and soil core samples were analyzed for electrical conductivity (EC) and meq/l of Na, K, Ca and Mg. Good correlation between surface soil EC and application rate suggested that EC measurements could be used as a management tool to monitor the salt status of fields receiving manure or lagoon water. In soil cores EC was increased throughout the top two meters by the higher rates of lagoon water and manure. In the surface soil and in the upper sections of the soil cores, the meq/l percentage of K increased with increasing rates

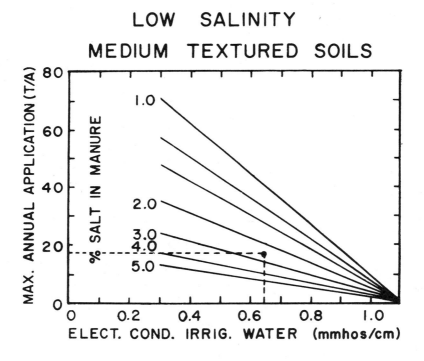

Figure 7. Maximum annual manure applications for a resulting low salinity (EC of 4 mmho/cm or less) on an irrigated, medium-textured soil. Dashed lines illustrate an example where a maximum of 18 dry tons/acre of manure with 25% total salts can be applied when the EC of the irrigation water is 0.65 mmho/cm.

of both lagoon water and manure; the meq/l percentage of Ca decreased with increasing rates. At high application rates, more Ca than K was removed from the soil solution (probably by forming precipitates of low solubility or by attaching to soil colloids). In both the lagoon-water and the manure studies, saturation extracts below 20 cm did not change in percentage composition.

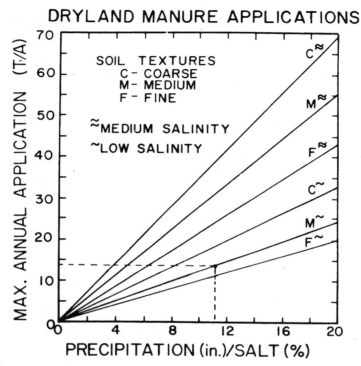

Figure 8. Maximum annual manure applications on nonirrigated soil. Dashed lines illustrate an example where a maximum of 13 dry tons/acre of manure can be applied to a medium-textured soil to maintain a low salinity (EC of 4 mmho/cm or less) when the annual precipitation in inches divided by the % salt in the manure equals 11.

Reduced plant growth can result from toxic salt accumulation in soils after large applications of animal manures. Guidelines for recommending manure application rates, based on salt balance calculations, require the knowledge of the salt composition of the waste, the tolerable salinity in the soil, the salt content of any irrigation water added, the soil texture, and the amount of salt leaching expected from natural precipitation.

ACKNOWLEDGMENTS

Journal paper No. 9515, Department of Soil Science, University of Minnesota, St. Paul, 55108; contribution No. 1583-a, Department of Agronomy; and contribution No. 224-a, Department of Agricultural Engineering, Kansas Agricultural Experiment

Station, Manhattan 66506. The work here was partially supported by the Environmental Protection Agency, Water Quality
Office, under grant No. 13040 DAT and 5800923.

REFERENCES

1. Powers, W. L., G. W. Wallingford and L. S. Murphy. "Research status
   on effects of land application of animal wastes," EPA–660/2–75–010,
   96 p. (1975).
2. Mathers, A. C., and B. A. Stewart. "Corn silage yield and soil chemical
   properties as affected by cattle feedlot manure," *J. Environ. Quality*
   3:143–147 (1974).
3. Shortall, J. G., and W. C. Liebhardt. "Yield and growth of corn as affected by poultry manure," *J. Environ. Quality* 4:186–191 (1975).
4. Wallingford, G. W., L. S. Murphy, W. L. Powers and H. L. Manges.
   "Effect of beef-feedlot-lagoon water on soil chemical properties and
   growth and composition of corn forage, *J. Environ. Quality* 3:74–78
   (1974).
5. Wallingford, G. W., L. S. Murphy, W. L. Powers and H. L. Manges.
   "Disposal of beef-feedlot manure: effects of residual and yearly applications on corn and soil chemical properties," *J. Environ. Quality* 4:
   526–531 (1975).
6. Linderman, C. L., and N. P. Swanson. "Disposal of beef feedlot runoff
   on corn," *Agron. Abstracts*, p. 182 (1972).
7. Swanson, N. P., C. L. Linderman and J. R. Ellis. "Irrigation of perennial forage crops with feedlot runoff," ASAE Paper No. 73–241, presented at the 1973 Annual Meeting ASAE, Lexington, June 17–20, 12
   pp. (1973).
8. Adriano, D. C., A. C. Chang, P. F. Pratt and R. Sharpless. "Effect of
   soil application of dairy manure on germination and emergence of some
   selected crops," *J. Environ. Quality* 2:396–399 (1973).
9. Powers, W. L., R. L. Herpich, L. S. Murphy, D. A. Whitney, H. L.
   Manges and G. W. Wallingford. "Guidelines for land disposal of feedlot
   lagoon water," Kansas State University, Cooperative Extension Service, Manhattan, Kansas, 7 p. (1973).
10. Powers, W. L., G. W. Wallingford, L. S. Murphy, D. A. Whitney, H. L.
    Manges and H. E. Jones. "Guidelines for applying beef feedlot manure
    to fields," Cooperative Extension Service, Kansas State University,
    Manhattan, Kansas, 11 p.m. (1974).

39

ECONOMIC POTENTIAL AND MANAGEMENT CONSIDERATIONS IN LAND APPLICATION OF BEEF FEEDLOT WASTES

Daniel D. Badger
Department of Agricultural Economics, Oklahoma State
University, Stillwater, Oklahoma

INTRODUCTION

Time waits for no man; change is eternal; what works today to solve a problem will not necessarily be an acceptable solution tomorrow. Are these statements value-laden concepts from a university philosophic value theory course? Wherever the statements may emanate, they are representative of an interconnected system of feedlot operator-beef animal-manure-fertilizer price-cropland-EPA regulations very prevalent in our agricultural industry today.

Time indeed waits for no man. If the feedlot operator cannot find immediately acceptable methods of handling his solid and liquid animal wastes, as well as odor problems, he may be out of business tomorrow. Cattle prices change, so cattle placed in feedlots may decline, and less manure is produced. Fertilizer prices increase and fertilizer shortages occur; farmers rediscover both nutrient values and other intrinsic values of manure. Stockpiles of manure disappear and no longer blight the landscape in the southwestern beef-feeding states. More recently, fertilizer prices again declined and fertilizer (particularly nitrogen and phosphorus) became more abundant. What will be the impact on utilization of manure?

The Environmental Protection Agency continues to promulgate regulations to control both point sources and nonpoint sources of agricultural pollution. Currently, EPA wastewater guidelines apply only to feedlot operations of 1,000 animals or more. If proposed new regulations are adopted, as many as 94,500 additional beef, dairy and hog production operations may have to file for a National Pollutant Discharge Elimination System (NPDES) permit (1, p. 12). New guidelines currently are being implemented for Section 208 of PL 92–500, the *Federal Water Pollution Control Act Amendments of 1972,* as related to reducing water pollution from nonpoint sources. The way these Section 208 regulations are implemented will have a significant impact on future application of both solid and liquid beef feedlot wastes to crop and pastureland.

There is an old Chinese proverb:

"Give man a fish, and he can eat a meal.
Teach a man to fish, and he can eat many meals."

This proverb has its modern day counterpart:

"Show an EPA man a pile of manure, and you have given him many pollution problems on which to write laws and regulations.

"Teach an EPA man how society can efficiently use that pile of manure, and you have solved many problems for EPA, for the feedlot operator, and for the farmer."

The emphasis of this book is the exploration of alternative land application solutions to environmentally related problems for both municipal and animal wastes. The objective of this chapter is to explore the economic potential of beef feedlot wastes, primarily solids (manure) and not liquid runoff. Also explored are some of the management problems and special considerations in application of these wastes to the land. The primary area of the research has been the beef feeding operations in the semiarid southwestern feeding region, namely the Oklahoma and Texas Panhandle areas, and to a lesser extent, southwestern Kansas (Garden City area). Thus, the remarks are related primarily to this large-sized feedlot area, of low rainfall, very few navigable streams, and no major runoff problems with liquid waste.

TRENDS IN USE OF FERTILIZER AND BEEF MANURE

Over the past 20–25 years, chemical fertilizer became abundant and prices were relatively inexpensive. Farmers found it more convenient and more pleasant to spread purchased fertilizer rather than to pitchfork and spread animal wastes. Land was abundant in the 1950s and 1960s; much of our cropland was in various reserve programs. Consequently many farmers forgot the nonnutrient values of manure, and failed to pass on to their sons and others the need to continually rebuild the topsoil.

It took an Arab oil embargo in 1973 that drove up oil and petrochemical prices to provide a valuable learning experience for many of our new farmers of the 1970s. They learned that beef feedlot wastes as well as other kinds of manure can provide valuable nutrients. Fertilizer prices that doubled in 1974 also destroyed an axiom or myth that it was not profitable to transport beef wastes more than 10 miles from the feedlot. Larger, tandem trucks with a 15-ton or greater capacity are now transporting manure longer distances from the feedlots and spreading it on pastureland as well as cropland.

FED CATTLE TRENDS AS RELATED TO MANURE PRODUCTION

Seven Key States

The number of cattle on feed as of January 1, was 9.9 million in 1973, 9.4 million in 1974, and only 6.4 million in 1975. Marketing of cattle on feed declined from 18.9 million head in 1973 to 17.4 million in 1974 and 15.0 million in 1975 (2, p. 33). Assuming that one ton of beef feedlot waste (manure) is generated by an animal during its feeding period, then 3.9 million fewer tons of manure were generated in 1975 than in 1973.

The drought that lasted through the winter months of 1976 in several states—notably Texas, parts of Oklahoma, Kansas, and California—forced early movement of stocker and feeder cattle into feedlots. Consequently, in February 1976, placement of cattle on feed in seven key beef feeding states was up 74 percent over a year earlier (3, p. 14). Fed cattle are estimated to have accounted for about 60 percent of commercial cattle slaughter in February, up from 46 percent in February 1975. This significant change reflects a shift from cow slaughter and pasture (nonfed) beef back to fed beef. Fed beef marketings will increase at least through the summer. The obvious implication of this shift is that more manure is being accumulated in beef feedlots than has been the case since the third quarter of 1974.

Numbers of cattle on feed in the seven major cattle feeding states as of March 1 for 1974–1976 are shown in Table 1. Feedlots are again running at full or near-full capacity, at least those feedlots that managed to weather the disastrous market conditions of 1973 and 1974.

Table 1. Cattle on feed in seven major feeding states, as of March 1, 1973–1976.[a]

State	1974 1,000 Head	1975 1,000 Head	1976 1,000 Head	1976/1975 Percent Change
Arizona	597	304	472	+ 55
California	1,064	476	854	+ 79
Colorado	939	685	875	+ 28
Iowa	1,740	1,180	1,550	+ 31
Kansas	1,130	800	1,250	+ 56
Nebraska	1,460	960	1,310	+ 36
Texas	2,318	1,076	1,810	+ 68
Seven States	9,248	5,481	8,121	+ 48

[a] Reference 10, p. 7.

Twenty-three Feeding States

As indicated in Table 2, marketings of beef animals from feedlots declined from 25,281,000 head in 1971 to an estimated 20,-607,000 head in 1975. This is a decline of almost 4.7 million animals during that period. It also indicates that 4.7 million fewer tons of manure were produced in these feedlots in 1975 than in 1971. While this means fewer environmentally related problems for feedlot operators, there was a loss of valuable plant nutrients to cropland adjacent to these feedlots. It is interesting that this reduction in manure tonnage came at a time of higher commercial fertilizer prices, and, particularly in 1975 for many parts of the U.S., temporary shortages of manufactured nitrogen fertilizer.

COMPOSITION OF BEEF FEEDLOT MANURE AND APPLICATION RATES

Nutrient analysis of feedlot waste is almost a daily variable, depending on temperature, dryness condition of the individual pens in the feedlot, slope of the land for each pen, soil moisture conditions, and obviously, rainfall conditions. Composition of the ration generally depends on input price conditions for grain (*e.g.,* grain sorghum versus corn) and supplements. Whether or not the ration has salt added has a significant impact on the amount of sodium in the manure. Current recommendations for the southwestern beef feeding areas is that additional salt is not needed; the components of the ration already contain sufficient salts. The amount of salt in the ration, in turn, will be a determining factor

Table 2. Cattle on feed, placements, and marketings[a] in 23 major beef feeding states.

Year and Quarter	On Feed 1,000 Head	Change Previous Year (%)	Placements[b] 1,000 Head	Change Previous Year (%)	Marketings 1,000 Head	Change Previous Year (%)
1971						
I	12,209	− 0.3	5,734	+12.0	6,231	+ 1.4
II	11,712	+ 0.8	5,455	+ 4.1	6,278	+ 0.9
III	10,889	+ 2.4	6,371	+ 3.7	6,594	+ 4.6
IV	10,666	+ 2.3	8,842	+10.6	6,178	− 0.6
1972						
I	13,330	+ 9.2	5,933	+ 3.5	6,443	+ 3.4
II	12,820	+ 9.5	6,364	+16.7	6,727	+ 7.2
III	12,457	+14.4	6,224	− 2.3	6,907	+ 4.7
IV	11,774	+10.4	8,862	+ 0.2	6,775	+ 9.7
1973						
I	13,861	+ 4.0	6,040	+ 1.8	6,585	+ 2.2
II	13,316	+ 3.9	5,696	−10.5	6,283	− 6.6
III	12,829	+ 2.2	5,283	−15.1	5,958	−13.7
IV	12,054	+ 2.4	7,491	−15.5	6,478	− 4.4
1974						
I	13,067	− 5.7	5,242	−13.2	5,999	− 8.9
II	12,310	− 7.6	4,008	−29.6	6,271	− 0.2
III	10,047	−21.1	4,627	−12.4	5,522	− 7.3
IV	9,152	−24.1	6,005	−19.8	5,538	−14.5
1975						
I	9,619	−26.4	4,376	−16.5	5,522	− 8.0
II	8,473	−31.2	5,082	+26.8	5,013	−20.1
III	8,542	−15.0	5,777	+24.9	5,018	− 9.1
IV	9,301	+ 1.6			5,054[a]	− 8.7

[a] Expected marketings. Reference 11, p. 2.

[b] Beginning 1974 other disappearance subtracted.

in the accumulation of salts in the top layer of the soil after several years of continuous application of beef feedlot waste to a particular field.

Fresh manure loaded directly from the feedlot pens and spread on surrounding cropland may contain up to 3.5 percent nitrogen. The longer manure is stockpiled, the lower the available nitrogen for plant use, due to leaching and decomposition (volatilization). Stockpiled manure may contain less than 1.0 percent nitrogen when it reaches the field (4, p. 3).

The storage conditions for the beef feedlot waste, while stored in the pen or stockpiled outside the pens, as well as the method of spreading and the method of incorporation into the soil also affect the amount of nutrients remaining in a form readily available for use by plants. Nitrogen is the element most easily lost, due to volatilization of the ammonia form as well as through leaching of both ammonia and nitrate forms through the soil. Vitosh *et al.* found that 10 tons of manure per acre per year was the most favorable rate for corn production on Matea sandy loam soil. "There is little doubt that land disposal ... [of beef feedlot wastes] ... is the least expensive and best method of recycling the nutrients in these wastes. The ultimate goal is to recycle the nutrients so contamination of ground water and surface water, odor and spread of disease are minimized" (5, p. 296).

Obviously, high application rates of manure with subsequent intensive rainfall or high irrigation levels may result in increased nitrate in streams and lakes due to surface runoff as well as nitrates in ground water supplies due to leaching. The application of 10 tons per acre of beef feedlot waste each year for several years can be expected to have both beneficial and possibly adverse effects on the soil. Increased capability of the soil to absorb and to hold water in the upper layers is certainly a plus factor. The tilth or soil texture also generally improves. Good soil texture improves aeration on the soil as well as allowing better root penetration.

Very little time-series or cumulative-effects data are available on the buildup of the minor or micronutrients in the soil. Depending on plant uptake, as well as rates of manure applied and levels of these elements in the manure, it may take 20 to 25 years before toxicity levels for arsenic, boron, copper, zinc, heavy metals and soils are reached. More intensive studies of the cumulative effects of these micronutrients in beef feedlot wastes are needed.

It is almost impossible to derive "average" nutrient values for nitrogen (N), phosphate (P_2O_5), and potash (K_2O) for beef feedlot wastes in the Texas-Oklahoma area. According to Tucker, Burton and Baker (6), Oklahoma researchers for many years used the following "rule of thumb" nutrient values for a pound of beef cattle wastes: 0.5 percent N, .25 percent P_2O_5, and 0.5 percent K_2O. The quantities in pounds per ton of waste are 10 pounds of N, 5 pounds of P_2O_5 and 10 pounds of K_2O. However, dramatic changes in beef ration formulations for confined feeding operations occurred in the early and mid-1960s. These changes have resulted in increased nutrient values in beef manure. Typical nutrient values per pound of beef solid wastes in the Texas-Oklahoma area now appear to be: 1.5 percent for N, 1.0 percent for P_2O_5, and 1.5 for K_2O. Other analyses for nearby states are presented in Table 3.

In addition to the three nutrients above, beef feedlot wastes also contain other valuable nutrients essential for crop and pas-

Table 3. Some recent analyses on percent compositions of beef feedlot wastes.

Element	Missouri[a]		Texas[b]		Oklahoma[c]		Kansas[d]	
	%	lb/ton	%	lb/ton	%	lb/ton	%	lb/ton
Nitrogen (N)	0.70	14.0	1.34	26.8	1.50	30.0	1.04	20.8
Phosphorus (P_2O_5)	0.45	9.0	1.22	24.4	1.00	20.0	0.41	8.2
Potassium (K_2O)	0.55	11.0	1.80	36.0	1.50	30.0	1.09	21.8
Calcium (Ca)	0.80	5.6	1.30	26.0	–	–	0.78	15.6
Magnesium (Mg)	0.11	2.2	0.50	10.0	–	–	0.40	8.0
Iron (Fe)	–	0.08	0.21	4.2	0.005	0.1	–	–
Zinc (Zn)	–	0.03	0.009	0.18	0.005	0.1	–	–
Sodium (Na)	–	–	0.74	14.8	–	–	0.23	4.6
Sulfur (S)	0.05	1.0	–	–	0.250	5.0	–	–
Boron (B)	–	0.03	–	–	0.002	0.04	–	–
Copper (Cu)	–	0.01	–	–	0.004	0.08	–	–
Manganese (Mn)	–	0.02	–	–	0.020	0.4	–	–
Molybdenum (Mo)	–	0.002	–	–	–	–	–	–

[a] Reference 12. [c] Reference 13.

[b] Reference 7. [d] Reference 4.

ture production such as sulfur, boron, zinc, copper, manganese and iron. Based on 1976 prices of purchased chemical inputs, the equivalent economic value of these other nutrients is $0.90 in a ton of manure.

ECONOMIC VALUE OF MANURE FROM TEXAS-OKLAHOMA FEEDLOTS

As indicated above, there is no average nutrient analysis for beef feedlot wastes. As rations have improved in feeding value, the nutrient content of the manure for N, P_2O_5, and K have also increased. For purposes of this discussion, the average of several recent sample analyses for the Texas-Oklahoma feedlot area has been used. These averages are presented in Tables 3 and 5. Manure samples should be analyzed frequently, and in particular the stockpile should be sampled just before application and incorporation of the wastes on cropland.

Prices paid by farmers for primary nutrients are also fairly difficult to pinpoint. As indicated in Table 4, both anhydrous ammonia and ammonium nitrate prices declined in 1975, and continue to decline in 1976. Oklahoma prices quoted for ammonium nitrate as of April 15, 1976, range from $115 to $125 per ton,

Table 4.   Average prices per ton and per pound of nitrogen paid by farmers for fertilizer materials for period 1970–75 (prices on September 15).[a]

| | Oklahoma | | | | Texas | | | |
| | Ammonium Nitrate [b] | | Anhydrous Ammonia [c] | | Ammonium Nitrate [b] | | Anhydrous Ammonia [c] | |
Year	Ton	lb N	Ton	lb N	Ton	lb N	Ton	lb N
1970	$ 55	$0.08	$ 72	$0.04	$ 65	$0.10	$ 76	$0.05
1971	57	0.09	70	0.04	68	0.10	77	0.05
1972	59	0.09	70	0.04	67	0.10	77	0.05
1973	75	0.11	89	0.05	78	0.12	84	0.05
1974	170	0.25	220	0.13	175	0.26	235	0.14
1975 [d]	150	0.22	200	0.12	155	0.23	215	0.13

[a] *Agricultural Prices,* Crop Reporting Board, SRS, USDA, Washington, D. C. (September 15 issue for each year cited except for 1975 date changed to October 15.)

[b] Ammonium nitrate is 33.5% N.

[c] Anhydrous ammonia is 82% N.

[d] 1975 prices are for October 15.

or an average price of $120 per ton, plus $1.50 per acre to apply, which averages to $0.18 per pound of nitrogen applied. Since commercial prices may continue down slightly, for comparison purposes, $0.15 per pound of N applied was used as the value equivalent for beef feedlot wastes. The other two major nutrients, P_2O_5 and K_2O continue at about $0.20 and $0.09 per pound, respectively. These latter two prices are based on $108 per ton for 60 percent murate of potash(K_2O) and $180 per ton for 48 percent superphosphate (P_2O_5).

Based on these prices and a representative sample analysis for nutrients in the waste, the equivalent value of a ton of beef feedlot wastes is $12.10 (Table 5). This assumes all the nutrients are available in the first year after application and includes $0.90 as the value of the secondary and micronutrients available in the beef manure.

Only about 50 percent of the three major nutrients (N, P_2O_5, K_2O) is available in the first year of application, with one-half of the remainder available in subsequent years. Thus $6.50 ($5.60 + $.90) is a closer estimate of the immediate economic value of a ton of beef feedlot wastes. After adjusting for the loss of one-half of the remaining primary nutrients, the economic value of that same ton of beef feedlot wastes in succeeding years is an additional $2.80. Thus, total economic value of the wastes is $6.50 + $2.80 = $9.30 per ton. Agronomy specialists at Texas A & M

Table 5.   Economic value[a] of nutrients per ton of beef feedlot wastes for a selected sample, Oklahoma-Texas area, Spring 1976.

Nutrient	Percent Composition	lb/ton	Value of Nutrient (per lb)	Total Value of Nutrient in Ton
Nitrogen	1.500	30	$0.15	$ 4.50
Phosphate	1.000	20	0.20	4.00
Potash	1.500	30	0.09	2.70
Zinc	0.005	0.1	0.50	0.05
Manganese	0.020	0.4	0.50	0.20
Iron	0.005	0.1	0.50	0.05
Sulfur	0.250	5	0.10	0.50
Boron	0.002	0.04	0.50	0.02
Copper	0.004	0.08	1.00	0.08
Total nutrient value/ton				$12.10

[a] This implies all nutrients become available for plant growth immediately upon application. This point is explained in the text.

University found that less than one-half of the nitrogen is lost in subsequent years; they indicate about 77 percent of the nitrogen is available for plant use in the first three crop seasons (7, p. 1).

The consensus of Oklahoma-Texas feedlot operators and commercial manure handlers in a recent survey is that a beef animal will generate one ton of solid wastes during a typical feeding period. As of April 1, 1976, 1,673,000 head of cattle were on feed in Texas and 284,000 head in Oklahoma, a total of 1,957,000 head. Assuming an annual turnover rate of only two times, about 4.0 million head will be fed in these two states in 1976. Thus, approximately 4.0 million tons of beef feedlot wastes will need to be moved from these feedlots and applied on land. Based on the generally recommended rate of 10 tons of manure applied per acre, 400,000 acres of cropland and pasture land near these feedlots could be fertilized. The total economic value of this quantity of manure, based on current market prices for both primary and secondary nutrients is $4,000,000 \times 9.30 = \$37,200,000$ for the Oklahoma-Texas area.

The above survey indicated that the prevailing cost for beef feedlot wastes was $3.00 per ton applied on the land within a 10-mile radius of the feedlot. For lands farther away than 10 miles, the rate varied from $0.05 to $0.07 per ton mile. Thus, for those crop and pasture lands as far away as 35 miles, less than one-half the economic value of the nutrients in a ton of beef feedlot wastes

(Table 5) would justify the $4.25 per ton cost ($3.00 + 25 additional miles $\times$ $.05 = $3.00 + 1.25).

An interesting aspect of beef feedlot waste handling has become more prevalent in the last three or four years. This is the emergence of at least 20 commercial feedlot waste handlers in the confined beef feeding area of Texas-Oklahoma. Three haulers were interviewed to get their concept of waste handling costs and manure value. One method involves the feedlot operator scraping the pens and stockpiling the manure. The commercial hauler loads his own trucks from the stockpile. In this situation, the feedlot operator is selling the manure at $1.00–$1.25 per ton to the hauler. Twelve of the 24 feedlots surveyed (ranging from 10,000 to 100,000 head capacity) sold all of their manure in this manner.

A second method has the feedlot operator contract with a commercial manure handler to rip and load the manure using the hauler's equipment completely. All types of pricing arrangements exist with this method; three feedlot operators charged nothing and paid nothing to have the manure moved, whereas two feedlot operators received $0.50–$0.75 per ton from the commercial hauler to have their pens cleaned in this manner. This is a distinct change from 1971 when some feedlot operators paid $0.40–$0.50 per ton to have their pens cleaned and the manure hauled away.

ADDITIONAL BENEFITS OF BEEF FEEDLOT WASTES

Most of the 24 feedlot operators and 2 farmers interviewed in a recent study commented on advantages of using manure at the rate of 10 tons per acre on their fields in addition to its substitution role for commercial fertilizer. About one-half of the feedlot operators used some of the wastes on their own fields, and their consensus comments were that yields were increased, micronutrients or trace elements in manure improved the appearance of crop and pastures (look healthier, greener), the moisture intake and water-holding capacity of sandy and loam type soils improved, the texture or tilth of soil after a short period of use improved, and the appearance or "richness" of light-colored tight soil typical in parts of Panhandle areas improved.

Most of the farmers commented that with the relative inexpensive commercial fertilizer prices of the late 1960s and early 1970s, they had taken the easy way out and gone to the use of purchased chemicals. Many had tended to forget the side benefits of manure and did not realize the relative ease of application and incorporation into the soil with new, larger equipment. In response to the question, "Will you continue to use manure no matter what happens to commercial fertilizer prices?" 8 of 10 farmers and all 12 of the feedlot operators with land adjacent to their feedlots responded "yes."

MANAGEMENT PROBLEMS
THAT MUST BE CONSIDERED

General

Some of the farmers surveyed indicated problems in using manure. However, none of them indicated the problems as major ones, or of a type that would prevent them from continuing to use manure. Problems listed were: compaction of soil due to large, heavy trucks that deliver and spread manure; slower rate of application compared to commercial fertilizer application; the necessity of working the field one to two times more to incorporate manure into the soil; unevenness of application or "skip" problems, and some imbalance in ratio of nutrients applied relative to soil needs.

Beef feedlot wastes are bulky, low grade sources of nutrients. The nutrient composition changes frequently, even with the same ration being fed. The total plant nutrient content varies from 10 to 20 percent of the content of most purchased (chemical) fertilizer. Thus, it is difficult to determine optimum rates of application. Rules of thumb have been established, a general one being that for several consecutive years of application, 10 tons per acre per year is a safe application rate.

Transportation Costs

Transportation costs have increased in recent years due to many of the same factors that have increased commercial fertilizer prices—oil and labor costs have increased significantly. It is difficult to determine exactly just how far manure can be transported. Based on economic value of the nutrients only, the distance may not be great. However, externalities also must be considered, such as long-run benefits to soil structure and water-holding capacity, as well as on improvements in water quality due to reduction of nutrients in streams, and a more pleasing environment for citizens in the vicinity of feedlots. Thus, from society's standpoint, it may be wise to assist in supplementing these transport costs to get the manure back on the land.

Certainly, the feedlot operator has been forced by both federal and state water, air and other related environmental quality regulations to properly dispose of these wastes and to prevent accumulation in such a manner as to cause actual pollution and/or nuisance problems. More and more regulations are being promulgated to enforce the various sections of Public Law 92-500, *Federal Water Pollution Control Act Amendments of 1972,* as related to both agricultural point and nonpoint sources of pollution.

Removal of beef feedlot wastes from feedlots and transport over longer distances may remain a least-cost option for feedlot operators even though they may have to help subsidize the transportation and spreading costs for land owners who know the value of the manure, and are willing to accept it and to incorporate it into their soil.

The key issue here that needs additional research is to determine the monetary values of some of the nonnutrient benefits of the manure as well as to compare additional transport costs, if paid by the feedlot operation, with other methods of disposal. There is much discussion of the value of beef feedlot wastes for generating methane gas. However, for many feedlots in isolated locations this possibility may not be a real alternative.

Introduction of Unwanted Weeds

Another management problem that varies from area to area is the possibility of introducing different types of weeds on cropland when the beef feedlot manure is applied to these fields for the first time. Farmers generally knew how to control most of the weeds in their row crops, grains, and even in their pasture lands. Due to the movement of hay and grain into the beef feeding areas, the likelihood increases of "exotic" or "strange" weeds being introduced. Some farmers using this manure on their land complained that they have to cope with "unwanted" weeds. Some of the irrigation farmers in western Kansas and in the Texas Panhandle have indicated this manure-related weed problem has become a serious problem. Thus far, this complaint has not been voiced by Oklahoma farmers using the beef feedlot wastes.

Much of this variation would be caused by the differences in rations being fed. Oklahoma Panhandle operators typically use corn and silage. The corn is cooked before feeding, and the silage generates high temperatures in the trench silos; thus any weed seeds in the ration tend to undergo very high temperatures, destroying the germination. The other feeds or supplements normally are pelleted, which also involves a high temperature process.

Conversely, in the Texas Panhandle feedlots, more of the rations utilize chopped hay and milo or grain sorghum. Weed seeds thus can survive the ingestion and digestion process through the beef animal and be deposited in the manure.

Several beef feedlot operators in Oklahoma also indicated that the stockpiled manure generates tremendous heat, which would tend to destroy the viability of any weed seeds. There have been several instances of spontaneous combustion of manure stockpiles; with that kind of heat, any seed would be hard pressed to survive.

Salt Buildup

Salts accumulate in the fields, particularly from inorganic salts of sodium, potassium, calcium, and magnesium. Quantities of these salts in the manure are dependent on the type of ration being fed. Soluble salts in beef cattle manure can trigger some long term, unhealthy side-effects on crop land. When those salts build up on the upper layers of the soil, they restrict water up-take by plants.

Generally in the southwest, beef cattle feeding occurs in low rainfall areas. These are the areas where cropland irrigation already has caused some accumulation of soluble salts in the upper layers of the soil (8, p. 48).

Feedlot operators have reduced the amount of salts being fed in the ration in recent years, with only a very small amount of salt added to the concentrate. Some feedlot operators feel that manure is being used as the "whipping boy" for basic salinity problems being caused by irrigation. These operators do not feel that the small amounts of salt per ton of manure that result from current rations will ever cause toxicity effects on the cropland where manure is being applied. Unfortunately no long term research studies are available to document this point of view. Studies of fields where salinity problems have occurred or have been compounded from manure usage were predicated on fed cattle rations of the 1950s and early 1960s when much higher salt diets were the vogue.

The above discussion does not deny that salts in manure can cause problems both to the soil as well as to the ground water by leaching of these salts. The danger inherent in the accumulation of these salts adds emphasis to the need for moderation in the application of beef feedlot wastes; recent studies show little or no harmful salt effects from the application of 10 tons of beef feedlot wastes per year, over a several year period.

Transporting and Applying Feedlot Wastes To Minimize Nutrient Loss

Research is needed to develop and analyze techniques for more effectively storing animal wastes and applying them to soil so that fewer of the nutrients are lost (8, p. 61). Manure from beef feedlots generally contains 50 percent or more solids; thus it is usually handled as a solid in movement, storage (stockpiling) and transport. Little thought has been given on how to retain as many nutrients as possible in the beef feedlot waste, or to how it may possibly be managed to enhance other values of the manure for the soil.

It has been assumed that the only way to transport the beef feedlot waste is by truck, or if the distance is very short, by tractor and manure spreader. For areas of the southwest, specially designed pipelines and/or open, concrete-lined canals could be used to transport the manure in a slurry form (solids mixed with liquid feedlot runoff and/or irrigation water) from the feedlots to fields 30 or 40 miles away. Particularly for the Texas Highplains and Panhandle areas and the Oklahoma Panhandle areas, such distribution systems could work to a great extent by gravity flow from higher to lower elevations. Both engineering design and economic feasibility studies are needed to determine more efficient methods of storing, transporting and applying beef feedlot wastes, to retain a higher percentage of the nutrients for cropland use than is now the case.

Implementation of the guidelines to enforce Section 208 of PL 92-500 may give a big boost to conservation tillage and minimum tillage practices that are in vogue now. The U.S. Soil Conservation Service estimates that as of June 1974, minimum tillage was being practiced on more than 32.6 million cropland acres in the United States (8, p. 32). Manure needs to be incorporated into the soil quickly once it is spread on the land to avoid additional losses of nitrogen and phosphorus in particular. Incorporation into the soil also helps avoid overland runoff problems of the nutrients during periods of intensive rainfall.

Terracing and contouring of all fields where manure is applied may be a partial solution to abating nonpoint sources of pollution. Minimum tillage also helps in that direction. But neither solves the problem of valuable nutrient loss from applied manure laying on top of the soil. Thus, there appears to be a real need for some research on how to measure the trade-offs involved as well as to find mutually acceptable solutions for cropland utilization of the manure and minimizing runoff to comply with Section 208 of PL 92-500.

SUMMARY

Fertilizer prices increased significantly in late 1973, 1974 and early 1975. Consequently, farmers turned increasingly to beef feedlot wastes as a valuable source of plant nutrients. Beef feedlot operators had a difficult time disposing of these wastes in the late 1960s and early 1970s. Some even reported giving away the manure just to reduce the environmental and esthetic problems. Most were losing money hauling it.

Now farmers in the survey are convinced that the nutrient value of beef feedlot wastes is sufficiently high to pay the current rate of $3.00 to $3.50 per ton of beef feedlot wastes applied to their land. The 1976 economic value of beef feedlot wastes produced in the Texas-Oklahoma feeding area is likely to exceed $37 million. Farmers also have "rediscovered" other related soil benefits of using beef feedlot wastes.

Increasing quantities of beef feedlot wastes in dryer climate zones, such as the Southwest, likely will be applied to both irrigated and dryland crops and pastures. The price structure of chemical fertilizers and transportation costs will determine if these wastes will be transported farther than the current 15–20 mile radius of beef feeding feedlots.

Hopefully in the future, beef feedlot operators will have fewer problems with federal and state environmental agencies concerning odor, insect and health related issues, and fewer nuisance doctrine lawsuits from nearby neighbors. State and federal agencies have not slackened their efforts to implement new federal environmental guidelines that affect feeding operations of 1,000 head capacity or larger (9). Rather significant capital outlays have been expended by beef feedlots in recent years to control runoff. Section 208 of PL 92-500 may result in similar expectations on fields where beef feedlot wastes are applied.

REFERENCES

1. United States Department of Agriculture. *Agricultural Outlook*, ERS, AO–8, Washington, D.C. (March, 1976).
2. United States Department of Agriculture. *Agricultural Outlook*, ERS AO–7, Washington, D.C. (January–February, 1976).
3. United States Department of Agriculture. *Agricultural Outlook*, ERS, AO–9, Washington, D.C. (April, 1976).
4. Powers, L. *et al. Guidelines For Applying Beef Feedlot Manure to Fields*, C–502, Kansas State University, Manhattan, Kansas (May 1974).
5. Vitosh, M. L., J. F. Davis and B. D. Knezek. "Long term effects of manure, fertilizer, and plow depth on chemical properties of soils and nutrient movement in a monoculture corn system," *J. Environ. Quality* 2 (2) :296–298 (1973).
6. Tucker, B. B., C. H. Burton and J. M. Baker. *The Use and Value of Animal Waste as Fertilizer for Crop Production*, Extension Circular E–815, Oklahoma State University, Stillwater (1972).
7. Valentine, J. H., J. R. Supak and F. C. Petr. *Fertilization of Crops with Feedlot Wastes on the Texas High Plains*, Fact Sheet L–1220,

Texas Agricultural Extension Service, College Station, Texas (February, 1974).

8. Committee on Agriculture and Forestry. *Conservation of the Land, and The Use of Waste Materials for Man's Benefits,* Committee Print, United States Congress, 94th Congress, 1st Session, U.S. Government Printing Office, Washington, D.C. (March 24, 1975).

9. Environmental Protection Agency. "Feedlots point source category: effluent guidelines and standards," *Federal Register* **39** (32), Part II, Washington, D.C. (February 14, 1974).

10. United States Department of Agriculture. *Livestock and Meat Situation,* ERS, LMS–208, Washington, D.C. (April, 1976).

11. United States Department of Agriculture. *Livestock and Meat Situation,* ERS, LMS–207, Washington, D.C. (February, 1976).

12. Agronomy Extension Specialists. *Animal Manure for Crop Production,* Science and Technology Guide, Columbia Extension Division, University of Missouri, Columbia, Missouri (July, 1975).

13. Badger, D. D. "Economics of substitution and the demand for beef feedlot wastes: one alternative for solving environmental quality problems," in *Managing Livestock Wastes,* Proceedings of 3rd International Symposium on Livestock Wastes, published by American Society of Agricultural Engineers, St. Joseph, Michigan (1975).

40

LAND DISPOSAL OF FOOD PROCESSING WASTEWATERS IN NEW YORK STATE*

Anthony F. Adamczyk
New York State Department of Environmental Conservation,
Albany, New York

INTRODUCTION

The use of land for the disposal of liquid wastes in New York State in the often referred to context of "the living filter" consists entirely of industrial wastewater spray irrigation systems. These systems serve thirteen fruit and vegetable processors and two dairy product processors. In New York State there are approximately 400 industrial wastewater treatment systems processing a total estimated combined flow of 800 MGD of which the spray irrigators dispose of 6.3 MGD to the land, approximately 1 percent of the total industrial process wastewater flow.

Spray irrigation as a viable means of industrial wastewater treatment evolved in New York primarily because of unique circumstances surrounding these food processors. Their processing schedules primarily occur during late summer and into the fall, coinciding with periods of low stream flows. Additionally, many of these food processors are located in the farming area of central New York away from main watercourses and usually on the

---

*This report has been reviewed by the staff of the Department of Environmental Conservation and approved for publication. Approval does not signify that the comments necessarily reflect the views and policies of the Department, nor does mention of trade names or commercial products constitute endorsement or recommendation for their use.

headwaters of minor tributaries which periodically become intermittent or completely dry in flow. These uncontrollable natural
conditions demand high levels of treatment for short periods of
time in order to protect the surface water quality. Spray irrigation therefore becomes cost-effective under these circumstances.

The design of these spray systems, though apparently simple
and primitive, represents a unique challenge and requires a good
deal of innovativeness on the part of both the designer and regulatory agency. Such an approach requires a design which coordi-
nates the environmental knowledge of engineers, hydrologists,
agronomists and geologists. A multidiscipline approach is necessary to insure proper continuity of the spray system with the
natural cycle into which it is fitted.

It is important to realize that once a spray irrigation system
is built, the responsibility of protecting the water quality of New
York State does not cease. The Department of Environmental
Conservation's (DEC) regulatory monitoring and surveillance
role becomes important to insure proper operation of these systems. Although surveillance and enforcement of environmental
laws and regulations is an important tool, the long-term continued success of these treatment schemes is dependent upon
proper initial design and continued awareness by the system operators and management. It has been our experience that, in
addition to proper design, one of the most important factors directly responsible for the successful operation of a spray system
has been the employment of a conscientious operator, responsibly
assigned.

Our regulatory experience comes from the existing spray irrigation systems which are summarized in Table 1. This summary
reflects the system design parameters and not necessarily continuous operating field conditions. The actual field conditions
vary through the spray season, and for the most part, the fields
will not receive the specified design flow at all times.

The regulatory design and operating requirements for spray
irrigation systems in New York essentially follow two interre-
lated objectives. One is the protection and maintenance of our
ground water classifications and standards; another is prevention
of public nuisance problems. Rationale developed to implement
these requirements focuses on maintaining a healthy growing
crop cover and adequate soil percolation in addition to compliance with our specific groundwater quality standards.

New York ground water standards, Section 703 of Title 6 of
the Official Compilation of Codes of Rules and Regulations of the
State of New York, essentially include a parametric list of drinking water standards which are applied at the point of contact

with the wastewaters and ground surface. This interpretation has limited and will continue to limit spray systems. Without belaboring the point, the limitation is the use of broad quality parameters such as Total Dissolved Solids (TDS) and Carbon Chloroform Extract Residue (CCE) and nonrecognition of potential crop and soil assimilative capacity for other parameters. In general, food processing operations which have not added to their wastewaters pathogenic organisms or chemicals which will affect the potability of ground water should be given such consideration as well as the water renovation potential of crop and topsoil cover.

DESIGN GUIDELINES

As noted, DEC design guidelines for spray irrigation have evolved from an objective of maintaining surface and ground water standards and prevention of nuisance conditions. The specific wastewater characteristics of a proposed spray irrigation system are first checked against the ground water standards. A review is then conducted with the following guidelines to promote maintenance of these ground water standards and enhance the operation, inspection and control of these systems. The specific guidelines which are generally specified in a system discharge permit issued by our department are:

1. Spray irrigation should be practiced only during the period May 1 to November 30.
2. Spray irrigation should be practiced only during daylight hours.
3. Spray irrigation should not be practiced during periods of measurable natural precipitation.
4. No area of the spray irrigation fields should be irrigated on two consecutive days.
5. Surface runoff of irrigated wastewater from the spray field shall not be permitted, nor shall surface runoff be permitted to enter the spray fields.
6. A viable cover crop shall be maintained on all spray irrigation fields.
7. Spray irrigation fields should be designed and operated to prevent surface accumulation of wastewaters.

The decision prohibiting winter and night spray irrigation is designed particularly to enhance the operation, inspection and control of these systems. The decision to prohibit winter spray irrigation is based on the objective of maintaining a good system with a healthy cover crop and adequate soil percolation capacity. A growing cover crop is essential for removal of wastewater nutrients and maintaining aerobic conditions in the soil. A cover crop also will provide evapotranspiration and increase soil surface infiltration.

Table 1. Industrial wastewater treatment facilities utilizing spray irrigation.

Company	Design Flow (MGD)	Overall Appl Rate (in./wk)	Spray Field Scheduling				No. of Spray Fields	Crop Cover	Soil	Pre-Treatment	Product
			Appl. Rate (in./hr)	(Use/ Rest) Ratio	Spray Time (hr)	Total Spray Area (acres)					
Deltown Foods Delhi (T) Delaware Co	0.2	2.7	0.25	1:3	0.33	23	4	reed canarygrass	coarse and fine sediments	activated sludge	dairy-cream butter, cheese powdered milk
Comstock Foods Waterloo (V) Seneca Co	0.189	0.68	0.037	1:8	5.25	85.6	6	reed canarygrass	low plastic clay	lagoons	beets cabbage
C–B Foods Leicester (T) Livingston Co	1.8	5.5	0.25	1:5	20	61	5	forested with heavy undergrowth	Genesee fine sandy loam	screening and pH adjustment	beets corn peas
C–B Foods Bergen (V) Genesee Co	1.0	3.15	0.30	1:7	12	80	4	reed canarygrass and clover	Ontario loam and Cazenovia silt loam	aerated lagoons	corn, beans peas, carrots potatoes
C–B Foods Alton Wayne Co	0.64	3.53	0.21	1:28	6	73.5	7	reed canarygrass	Williamson and Wallington loam	pH adjustment and aerated lagoons	beans, cherries beets, apples
Libby McNeill and Libby Geneva (T) Ontario Co	0.8	2.7	0.15	1:8	8	80	9	reed canarygrass	Schoharie silty loam	lagoons	sauerkraut beans
Marion Foods Marion (T) Wayne Co	0.26	3.36	0.2	1:7	12	20	4	reed canarygrass	gravelly loam	screening aerated lagoons	beans apples

Company										
Lohmann Foods Gorham (T) Ontario Co	0.058	2.0	0.12	1:9	12	7.5	5	reed canarygrass	sandy loam	pH adjustment aerated lagoon screening	beets cabbage
HC Hemingway Clyde (V) Wayne Co	0.124	1.7	0.1	1:9	12	35	5	reed canarygrass	Ontario loam Palmyra-fine sandy loam	screening pH adjustment aerated lagoon	peas, beans pumpkin applesauce
Cuba Cheese Cuba (T) Allegany Co	0.025	2.6	0.11	1:6	18	9	5	reed canarygrass	Chenango gravel	screening	cheddar cheese
Gro-Pac Eden (T) Erie Co	0.17	2.4	0.13		12	4		rye grass	gravel and silt	lagoons	beans peas
Sodus Fruit Farm Sodus (T) Wayne Co	0.1	1.0	0.1	1:17	12	27	9	Japanese millet	fine, sandy loam and silt loam	lagoons	cherries. prunes apples
Marion Foods Williamson (T) Wayne Co	0.396	2.4	0.14	1:9	12	47	5	reed canarygrass	sandy loam	screening pH adjustment aerated lagoon	peas, beans corn, apples
C–B Foods Oakfield (T) Genesee Co	0.46	3.36	0.2	1:9	12	35	5	rye grass reed canarygrass	Ontario loam	pH adjustment lagoons screening	beans, corn carrots, sauerkraut, carbonated beverage
Indian Summer Sodus (T) Wayne Co	0.047	2.02	0.12	1:9	12	6	6	reed canarygrass	sandy silt	screening pH adjustment lagoon	cider vinegar

Adequate soil percolation capacity is obviously necessary to provide disposal of the wastewaters to the hydrological cycle via the ground waters. The path to the ground water through the soil mantle is necessary for reduction of wastewater nutrients.

Winter climatic conditions for upper New York State greatly reduce or eliminate the above two components of a good spray irrigation system. The absence of a growing crop cover would greatly reduce the removal of nutrients from the wastewater, eliminate evapotranspiration and reduce the soil surface infiltration. Frozen soil conditions, especially of saturated soils, would greatly reduce the soil infiltration and percolation capacity. The result would be ice storage and potential surface runoff. Additionally, we have experienced after spring snow melt a choking organic slime layer deposited on the fields which has prevented adequate spray field recovery (1) (Figures 1 and 2).

Figure 1.   Winter spraying. During winter months an ice buildup occurs to depths which may reach 6 ft. The ice sheet itself is generally coarse and honeycomb in structure. Melting generally occurs continuously through the winter from the bottom up with no noticeable runoff beyond the ice shield. During the spring melt, considerable runoff and ponding occurs with a buildup of sludge on the crop cover.

A growing cover crop and adequate soil percolation are extremely important for wastewaters in question. The wastewaters of many of these systems are untreated and some contain high nutrient levels, well in excess of those in domestic sewage. Inadequate nutrient reduction in the soil mantle or short circuiting of the ground water portions of the hydrological cycle, i.e., surface runoff, may result in uncontrollable pollution of the waters of the state.

It is recognized that year-round spray irrigation is permitted by some states and also that some states prohibit winter spraying. Additionally some literature recommends winter storage (2) in colder climates. However, winter spraying is usually practiced in less severe winter climates or with wastewaters that have been treated for a high degree of nutrient removal or both.

The growing cover crop season and the unfrozen ground water period will vary somewhat with year-to-year climatic variation. For this reason, we are flexible regarding consideration of the spray irrigation season on a year-to-year basis. However, spray irrigation systems are required to be designed for the potential prohibition of spray irrigation from November 1 to May 1.

While the current policy of the DEC is to not permit winter spray irrigation, we remain receptive to evaluations of winter spray irrigation that would solve the potential problem described above.

The decision regarding prohibition of night spraying is based on the opinion that visual inspection of spray irrigation systems is essential to their proper operation. State permit violations or problems that occur during darkness would likely go undetected until daylight or, if they terminate before daylight, might remain unattributable to the source. Because of this, it is felt necessary to restrict spraying to daylight hours when the operation can be carefully observed. Additionally, an evaporation benefit results from daylight spraying.

The cost of the night prohibition to the discharger is nonexist-

Figure 2.   Damaged crop cover. Suffocation of crop cover by dairy wastes has caused considerable deterioration of spray fields. Excessive ponding and runoff can be noted in background. The crop cover under the spray has nearly been completely destroyed.

ent in most cases. Spray irrigation hydraulic application rates are based primarily on overall long-term (weeks) irrigation rates rather than instantaneous rates. The nighttime prohibition does not require the discharger to increase the spray irrigation field capacity because New York State will permit instantaneous spraying at rates that exceed acceptable overall rates by a factor greater than the ratio of total hours to daylight hours.

Natural rainfall may at times exceed the infiltration capacity of a field and produce conditions that could result in permit violations if irrigation was practiced at that time. Facilities are required to store the wastewater during these wet periods. The required storage time is usually in the order of weeks and greatly exceeds the half-day storage requirement resulting from the nighttime prohibition. The additional storage resulting from the nighttime prohibition is insignificant and has usually been neglected by design engineers.

The policy regarding no winter spraying and no spraying during periods of measurable precipitation does require design consideration for storage. Though the spray season extends through the month of November, storage is required for the period of November 1 through May 1. Additional storage in the neighborhood of approximately two weeks, depending on specific site characteristics, also is required because of rainfall possibly occurring just prior to and through the prohibited winter period. The rationale for this storage requirement has been developed by Mr. Zambrano of the DEC staff, and is shown in Figure 3 (3).

Because of local odor problems positive aeration is required in storage facilities that may be adjacent to local residents. The system design if possible should attempt to deliver fresh, nonseptic wastewaters to fields in order to prevent the possible development of odor problems and minimize requirements for mechanical aeration. This consideration must, however, be weighed against possible additional treatment requirements such as BOD removal, depending on waste characteristics.

Choice and development of an adequate crop cover should also be given consideration. The general recommended and successfully used crop cover for spray fields in New York State is reed canary grass (*Phalaris arundinacea*). This grass develops an extensive deep root system, has a relatively large leaf area, and is tolerant to adverse conditions, making it ideal for spray irrigation. It is, however, sensitive to adverse storage conditions and requires a fairly long (40–50 days) germination period and therefore requires some care in initial plantings. If nutrient removals are also a consideration, then routine harvesting and crop management is recommended.

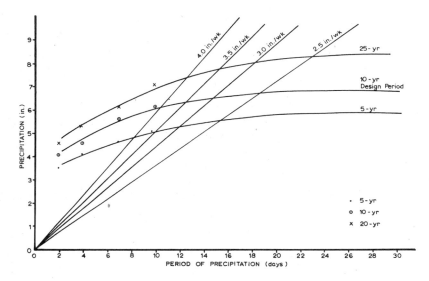

Figure 3. Storage for precipitation, Rochester, N.Y.

Up to now, guidelines have been discussed which were developed to promote an acceptable spray irrigation system through the maintenance of a healthy viable crop cover and aerobic soil conditions. Other specific but related considerations must be made for hydraulic application rates, organic application rates, sodium adsorption ratios, nutrient loadings and pH.

Hydraulic application rates generally cannot be directly related to soil percolation. Such a comparison may provide misleading results. A site with high soil percolation generally has surface infiltration as the limiting factor when one is concerned with the prevention of surface runoff and maintenance of aerobic soil conditions. In order to achieve objectives of aerobic soil conditions and renovative capabilities, it has been the policy to limit the overall hydraulic application rate to 3 in. per week. The basis of this limitation is somewhat subjective but generally reflects conditions of successfully operating systems and is recommended by the literature (4) and other states such as Pennsylvania (5). If a spray field under consideration becomes limiting with respect to soil percolation, the technique for determining hydraulic disposal capacity becomes critical and should include the services of a soil geologist and plant scientist. Design considerations must also include water mounding and effects of increased ground water elevation on adjacent areas. There is one system, because of extremely low soil porosity, designed for evapotranspiration utilizing the Thornthwaite method (6). This system has been successful but requires backup runoff collection and storage facilities. In designing a specific spray application scheme, it is desirable to have relatively short spray application periods followed by rest periods in ratio of approximately 1:4 or greater (*e.g.*, 0.5 hr spray: 2-hr rest).

The hydraulic application rate must be considered in relation to the quality of water being sprayed. A sodium adsorption ratio (SAR) must be determined because of the possibility of soil swelling and becoming less permeable. Additionally, consideration should be made for sodium, TDS and excessive bicarbonate.

Acceptable organic application rates have been much more difficult to rationally define. The ground water classifications and standards of New York require that no discharge shall render them unsuitable for a potable water supply. High organic levels in ground water would obviously violate this general provision. Much literature indicates that the crop cover and upper soil surface does have renovating capabilities. Systems in New York State spraying at relatively low organic application rates up to 500 lb BOD_5 per acre per day apparently have not caused problems. Some literature also suggests this as an upper limit. We have had problems with systems at higher loadings (2000–5000 lb BOD_5/acre/day (1, 7). The higher loaded system did achieve an overall BOD_5 removal of 85 percent, but still caused ground water problems because of the remaining dissolved organics. The rationale for determining organic assimilative capacity must be further refined so as to respond more adequately to the concern for potable ground water supplies and yet provide a cost-effective alternative for wastewater treatment. This specific issue is now being evaluated by researchers at Cornell University (8).

Consideration of crop and soil assimilative capacity regarding nitrates and pH is restricted to the specific parameter limitations in the ground water classifications and standards applied at the point of contact with the wastewater and ground. This aspect of the standards should be given reconsideration in the context of system renovative capabilities.

MONITORING REQUIREMENTS

As part of an approved wastewater treatment facility, DEC requires monitoring of spray systems. The requirements for such monitoring include raw and sprayed wastewater characteristics, continuous flow recording, spray system water balance (*i.e.*, sprayed flow, rainfall, runoff), and ground water quality monitoring.

Ground water monitoring is accomplished by sampling wells so as to determine water flow and characteristics into and out of a spray site. The purpose is to provide assurance for the continued maintenance of the ground water quality standards. The sampling wells must be constructed so as to provide a representative sample of the ground water at the ground water surface.

DESIGN AND REVIEW GUIDELINES

Wastewater treatment systems in New York State must be reviewed formally by DEC to reasonably assure compliance with the water quality requirements. As part of this review, guidelines for the design of spray irrigation have been compiled. The recommendations are summarized as follows.

Application Rates

1. *Hydraulic*—design to prevent surface runoff, ponding, soil clogging and maintain aerobic soil conditions. Consider the following: infiltration, percolation, soil classifications and site geology, crop cover, evapotranspiration, local climatic conditions, maximum recommended application 3 in./wk.
2. *Organic*—design to prevent anaerobic conditions and good BOD removal. Maximum recommended organic loading 500 lb BOD/acre/day.
3. *Inorganic*—design to meet ground water quality standards, prevent crop damage and soil clogging due to excessive soluble salts. Check sodium adsorption ratio (SAR). Maximum recommended salinity 1000 mg/l.

Ground Water

1. *Quality*—Class GA ground water standards apply to effluent at point of contact with ground surface.
2. *Ground water table*—estimate mounding and provide flow net projections.

Spray Scheduling

1. Spray duration—recommend 8 hr or less.
2. Spray use/rest ratio—1:4 or greater (*e.g.*, 3-hr spray: 32-hr rest).
3. Daylight spray only.
4. No winter operation (November 30–May 1).
5. No spraying during rainfall.

Spray Field Layout

1. No overlapping of wetted area.
2. Minimum separation to classified surface waters, dwelling and public roadways—200 ft.
3. Provide for runoff collection and recirculation.

Crop Cover Maintenance

1. Crop cover—select cover having tolerance to adverse conditions, relatively large leaf area; and extensive deep root system. Recommended crop—reed canarygrass.
2. Crop harvesting—cut before going to seed or at least twice per season.
3. Subsoiling—recommended seasonally.

Pretreatment

1. Grit, stone and debris protection.
2. Equalization and pH control if required.
3. Storage—required for wastes generated outside of spray season and for periods of continuous rainfall.
4. Secondary treatment and disinfection for sanitary waste.

Monitoring

1. Flow measurement and water characteristics: raw process effluent (flow and characteristics); spray effluent (log of flow scheduling and wastewater characteristics); rainfall (log during spray season); runoff (estimate of accumulation); and ground water (elevation and quality).

CONCLUSIONS

Spray irrigation has been used in New York State as a cost-effective wastewater disposal scheme. Continued use of such systems must be accompanied by more rigorous and reliable design criteria for the protection of critical ground water quality parameters. These criteria include hydraulic application rates, effects of sodium, determination of organic assimilative capacity, pH buffering capacity, nitrogen removal capacities and the interaction of these elements.

A closer surveillance of the operating systems should be conducted to prevent possible long-term pollution of our ground waters and also to provide crop and soil credit for renovative potential where justified. The solutions lie in an enlightened restructuring of the ground water standards coupled with a scientifically rigorous procedure to evaluate the assimilative potential of these systems regarding the critical water quality parameters.

All these activities would be meaningless if the existing systems are not managed as good neighbors. At this time much of the DEC contact with the public regarding the existing systems has been due to nuisance conditions such as runoff, ponding and odors. The public notice and participation procedures of New York laws are such that should we be faced with a spray irrigation proposal we can expect objections from potential adjacent property owners due to nuisances and bad experiences from other systems. In many cases it is these issues which determine the fate of proposed spray systems and not protection of our ground water resources, or potential renovative capabilities of crop and soil cover.

ACKNOWLEDGMENTS

The author wishes to extend credit to Mr. John Zambrano and Mr. William Schaff, Department of Environmental Conservation staff members, who have contributed material used in this presentation.

REFERENCES

1. Bogedain, F. O., *et al.* "Land disposal of wastewater in New York State," New York State Department of Environmental Conservation (March, 1974).
2. Malhotra, S. K. and E. A. Myers. "Design operation and monitoring of municipal irrigation systems," *J. Water Pollution Control Fed.* 47:2627 (1975).
3. U.S. Department of Commerce. "Two to ten-day precipitation for return periods of 2 to 100 years in the contiguous United States," Technical Paper No. 49, Superintendent of Documents, U.S. Government Printing Office (1964).
4. Chapman, S. W., *et al.* "Nitrogen mass balance determination for stimulated wastewater land-spreading operations," Water Resources Center, Texas Tech University, Lubbock, Texas 79400 (1975).
5. Pennsylvania Department of Environmental Resources. "Spray irrigation manual," Bureau of Water Quality Management Publication No. 31 (1972).
6. Thornthwaite, C. W. and J. R. Mather. *Instructions and Tables for Computing Potential Evapotranspiration and the Water Balance*, Vol. X(3), 5th Printing, Laboratory of Climatology, Centerton, N.J. (1957).
7. MacNeill, J. S. "Spray irrigation system for disposal of industrial waste," Engineering Report, Deltwon Foods, Inc., Fraser, N.Y. (December 1971).
8. Jewell, W. J. "Interim report—limitations of land disposal of wastes in the food processing and canning industries," Cornell University, Ithaca, N.Y. 14853 (January 1976).

41

LAND DISPOSAL OF OILY WASTEWATER
BY MEANS OF SPRAY IRRIGATION

D. M. Neal
Kaiser Aluminum and Chemical Corporation, Ravenswood,
    West Virginia
R. L. Glover and P. G. Moe
West Virginia University, Morgantown, West Virginia

INTRODUCTION

The biodegradation of oil was first reported 70 years ago (1), but little practical application of this knowledge has been made to date. One of the first investigations into the utilization of microorganisms for the decomposition of hydrocarbons was made by Sohngen in 1913 (2). He reported that gasoline, kerosene, paraffin oil and paraffin wax could be oxidized to carbon dioxide, water and organic acids by commonly occurring microorganisms. Stone *et al.* (3) found that oils containing a high percentage of paraffinic hydrocarbons were more readily decomposed than oils containing aromatic compounds and that naphthenic fractions exhibited an intermediate susceptibility to microbial degradation.

Other researchers have indicated that most crude oils can be degraded biologically in the presence of mixed cultures of bacteria. Mixed cultures are most effective because there appears to be some selectivity in the ability of certain microbial species to attack various components of the oil. More than 100 species of bacteria, yeasts and fungi have been found capable of oxidizing one or more hydrocarbons. These microorganisms are widely distributed in nature, particularly in areas subjected to frequent pollution by petroleum or refinery products.

A method for the disposal of oily wastes by mixing the oily wastes with soil and then allowing the soil microorganisms to decompose the oil has been demonstrated at the Shell Oil Company's Houston Refinery. They reported an oil decomposition rate of approximately one pound per cubic foot of soil per month when commercial fertilizers were used to enrich the soil with nitrogen and phosphorus (4). Their test indicated that the oil and fertilizer compounds did not infiltrate vertically through the soil at the test location under prevailing climatic conditions. Differences in decomposition rate and microbial species for the hydrocarbon types tested were minimal.

In December 1972, Kaiser Aluminum and Chemical Corporation decided to try land application as a means of disposal of the oily wastes generated at their Ravenswood Works. The main constituent of the oily wastes generated in this plant is an oil emulsion used to cool and lubricate the large rolling mills. This emulsion initially consists of 6 percent "rolling oil" suspended in deionized water. The rolling oil used is a complex mixture and varies from one rolling mill to another, but in general it consists of 85 percent mineral oil of naphthenic base stock, 12 percent fatty acids used as emulsifiers, and 3 percent alcohol, glycerine and antimicrobial agents. Approximately 70,000 gallons of this oil emulsion are used daily in the Ravenswood Works. Other oily wastes in the plan include hydraulic fluids, waste-lubricating oils, and various solvents.

In the past, chemical treatments had been used to break oil emulsion. The oil phase had been applied to road surfaces and the water phase had been dumped into the nearby Ohio River. The oil and water separation was never complete, however, and objectionably large amounts of oil were being dumped into the river with the water phase. In recent years, due to a greater emphasis being placed on environment protection, alternative means of waste disposal were sought.

PROJECT DETAILS

A permit was obtained from the West Virginia Department of Natural Resources to establish on a trial basis a land disposal system for these oily wastes. A disposal site was selected in a 138-acre field of Huntington silt loam soil located on the flood plain of the river. The layout of the disposal system is diagrammed in Figure 1.

All oily wastes from throughout the plant are first collected in a centrally located 550,000-gallon capacity steel storage tank. The oil is then pumped to three holding ponds for reclamation and

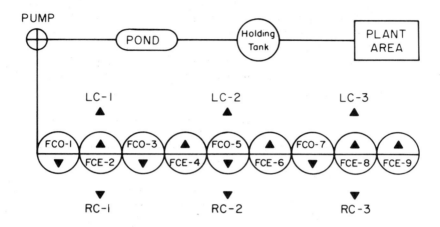

Figure 1. Physical arrangement of the oily waste disposal system in operation at the Ravenswood Works.

disposal. The three ponds have a combined storage capacity of over three million gallons. The first pond is the primary storage pond where the emulsion breaks down spontaneously upon standing and natural separation of the oily phase occurs. The oil layer is skimmed off the surface of this pond and pumped to the second pond which is a polishing pond for collecting salvaged oil. Further separation of the oil and water phases occurs in this pond upon prolonged standing. The usable oil is then pumped through a filter to a storage tank, where it is held until it is used as a supplementary fuel in the plant boilers. The water phase is periodically returned to the primary holding pond. The third pond is only used to provide additional storage capacity during periods of weather unsuitable for spray irrigation.

The oily water phase from the bottom of the primary holding pond is pumped out to the disposal site by means of a gasoline-powered irrigation pump capable of moving 400 gallons per minute at a pressure of 90 pounds per square inch. The oily waste-water is pumped through an aluminum irrigation line decreasing from an 8-in. diameter at the pump to a 4-in. diameter out in the field. A series of nine hydrants are permanently mounted at 300-ft intervals along this line in the field. Three mobile sprinklers, each capable of irrigating a 300-ft-diameter circular area, are rotated systematically among these nine hydrants. The total area of the field actually irrigated is 14.6 acres. The irrigation system is normally three hours per day for three days per week, weather permitting. Over the past three years, the application rate has averaged just under one-half an acre-inch per week for the entire disposal area. The field is mowed and disced as needed to control the growth of weeds.

The oily wastewater that was actually applied to the field was sampled and analyzed periodically. The results of these analyses are summarized in Table 1. The total organic matter content of

Table 1.   Analyses of waste oil emulsion (mean values of four individual analyses).

pH	6.85
Five day BOD	298 ppm
Aerobic plate count	29.5×10^5 org/ml
Inorganic carbon	42 ppm
Organic carbon	2058 ppm

Mineral Ion Analyses (ppm)

Na	22.50	Al	2.27
Ca	45.70	Mn	0.29
Mg	17.40	Zn	0.29
Fe	4.20	Cu	0.16
K	3.50	Ni	0.10

the waste material at this point varies from about 0.2 to 0.5 percent and consists of a complex mixture of residual components of the original oils, fatty acids, soaps, and intermediate breakdown products of microbial degradation. A gas chromatograph of the oily waste material (Figure 2), indicates that most of the hydrocarbons remaining have carbon chains ranging in length from 12 to 20 carbon atoms. The relatively high aerobic plate count, 3 million cells per milliliter, and the high biological oxygen demand (BOD) indicate a readily available nutrient source and a general lack of toxic factors. The concentrations of sodium, calcium and magnesium are relatively high, but the concentrations of the heavy metals are relatively low. It is thought that some of the metal ions, particularly iron and aluminum, probably remain with the oil phase during the physical separation that occurs in the holding tank and ponds.

Soil samples were collected periodically from sampling sites in and around the disposal area as indicated by the triangular symbols in Figure 1. These samples were analyzed at West Virginia University for pH, bacterial population, total carbon, total nitrogen, and exchangeable cations. The results of these analyses are presented in Tables 2 and 3. The only significant effects associ-

GAS CHROMATOGRAPHIC ANALYSIS
OF OIL EMULSION WASTE

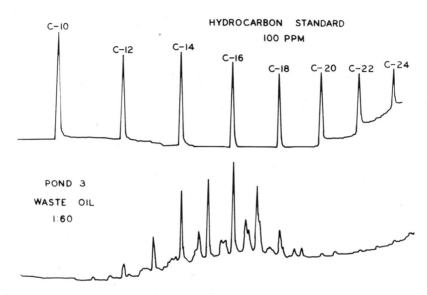

Figure 2. Comparison of gas chromatographs obtained with the oily waste
material and straight-chain hydrocarbon standards having chain
lengths varying from 10 to 24 carbon atoms.

Table 2. Summary of field experiment soil analyses (mean values for four
separate sampling dates).

Depth (in.)	Area	Plate Count x 10^5/g	pH	Carbon (%)	Nitrogen (%)
0–18	Control	40.2	6.21	2.52	0.092
	Treated	37.2	6.19	2.53	0.112
18–36	Control	19.8	6.48	1.85	0.064
	Treated	23.1	6.41	2.03	0.083

ated with waste application treatment at either soil depth sampled
were increases in the sodium, manganese and potassium concentrations. The values are not unusually high, however, and indicate
no permanent injury to the soil environment.

Ground water samples were also collected routinely from four
special wells sited in and around the waste disposal area. Here

Table 3.   Summary of field experiment soil analyses (mean values for four separate sampling dates).

Depth (in.)	Area	Mineral Ions in Milliequivalents/100 g					
		Ca	Na	Mn	K	Mg	Al
0–18	Control	9.80	0.44	0.13	0.36	2.55	0.04
	Treated	9.24	0.80	0.32	0.52	2.42	0.02
18–36	Control	9.67	0.41	0.10	0.22	2.40	0.05
	Treated	10.13	0.65	0.23	0.28	2.85	0.02

again, there was never any indication of any effects of treatment on the environment. All analyses to date indicate that all of the oily wastewater has been absorbed and decomposed within the waste disposal area without any contamination of the surrounding environment.

LABORATORY EXPERIMENTS

Since the waste disposal area did not demonstrate any harmful effects of the waste application treatments, it was decided to conduct a laboratory experiment using higher rates of application, under carefully controlled conditions, to determine at what application rate detrimental effects on the environment might occur.

Twelve columns of polyvinyl chloride pipe, 5 ft high and 1 ft in diameter, were constructed and filled with soil taken from the waste disposal area. Oily wastewater was leached through these soil columns for a period of three months, collecting and analyzing all of the leachate recovered. Application rates varied from 0.5 in./week, which is the rate currently being used in the waste disposal area, up to 8.0 in./week. A control treatment received only deionized water at a rate equivalent to normal precipitation for the area. Total organic carbon added in the oily wastewater and recovered in the leachate was determined on a Beckman 915 Total Carbon Analyzer.

The total organic carbon recovered in the leachate, as shown in Table 4, was less than 1 percent of that added in all columns except those receiving the highest application rates. The soil was obviously very effective in absorbing the organic compounds from the oily wastewater. These results indicate that a higher rate of application could probably be utilized in the waste disposal area with little or no detrimental effect to ground water quality.

After three months, the soil columns were cut open and the soil was divided up into 6-in. segments and analyzed separately. For the sake of simplicity, the analytical results of the control samples are compared only with columns receiving the lowest and the

Table 4.   Total organic carbon recovery obtained in the leachate collected from the soil columns in the leaching experiment (mean values of two replicates).

Application Rate (in./week)	Total Organic Carbon Added (mg C/column)	Total Organic Carbon Recovered (mg C/column)	Percent Organic Carbon Recovered (%)
0.5	14,750	98	0.66
1.0	29,500	162	0.54
2.0	59,000	283	0.47
4.0	118,000	1,118	0.94
8.0	236,000	4,820	2.03

highest rates of oily wastewater application in Figures 3, 4 and 5. Intermediate rates of application tended to give intermediate results.

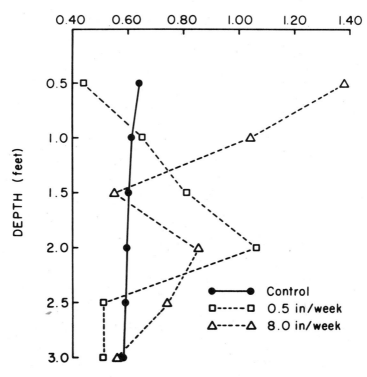

Figure 3.   Comparison of distribution of organic carbon in soil columns receiving different application rates of oily wastewater.

The control column showed a constant total organic carbon content throughout its depth (Figure 3). The 0.5-in. week application rate resulted in a lower carbon content in the upper soil segments and a higher carbon content in the lower soil segments, in contrast to the 8.0-in./week application rate which showed a very high carbon content in the upper segments, but gradual decreases with depth. The lower carbon content in the upper segments of the column receiving the lower application rate is believed to be due to increased biological activity occurring in this portion of the soil column. The accumulation of organic carbon with the high application rate is believed to be due in part to the natural filtering action of the soil accompanied perhaps by some inhibition of biological activity.

Figure 4 shows that the size of the bacterial population is

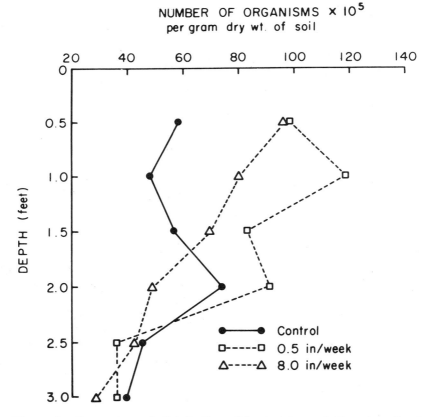

Figure 4. Comparison of distribution of bacterial populations in soil columns receiving different application rates of oily wastewater.

greater in the upper segments of the treated columns than in the control, but decreases with depth to become quite similar at the foot of the columns. The bacterial population is greater in the columns receiving the lower application rate than in the columns receiving the highest application rate, again indicating a toxic condition developing in the soil in these columns. This could be simply decreased aeration due to the heavier hydraulic loading of these columns.

The soil pH, as shown in Figure 5, was consistently lower in

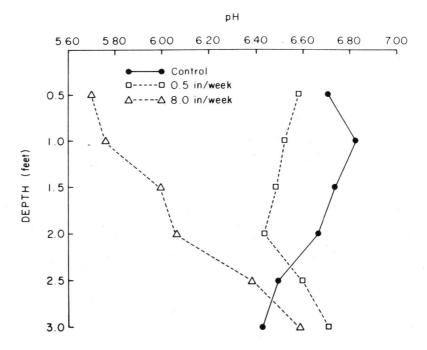

Figure 5. Comparison of soil pH values in soil columns receiving different application rates of oily wastewater.

the treated columns. This is undoubtedly due to the higher biological activity and the production of organic acids and alcohols as intermediate metabolic products. The soil pH correlates well with the application rates and bacterial population distribution in the soil columns.

Table 5 lists the dominant bacterial genera isolated from the oily wastewater and from the treated soil in the waste disposal area. Most of the organisms are common inhabitants of the soil.

Table 5.   Organisms isolated from waste oil emulsion and soil.

	Source	
Genus	Waste Oil Emulsion	Soil
Achromobacter	Yes	Yes
Arthrobacter	Yes	No
Bacillus	Yes	Yes
Flavobacterium	Yes	No
Micrococcus	Yes	No
Pseudomonas	Yes	Yes

The genera *Achromobacter, Bacillus* and *Pseudomonas* were dominant in both the oily wastewater and in the treated soil, while *Arthrobacter, Flavobacterium* and *Micrococcus* were dominant only in the oily wastewater.

SUMMARY

In general, the Kaiser Aluminum and Chemical Corporation is pleased with the success of their land application system of oily waste disposal. Since December 1972, over 36 million gallons of oily wastewater have been disposed of successfully through this system and approximately 600,000 gallons of oil have been reclaimed and used successfully as fuel in the plant boilers. The value of the salvaged oil has more than offset the total costs of the entire project to date. Plans for the future include the installation of a permanent underground piping system and automated pumping system. Further studies are now underway to determine the compatibility of this method of waste disposal with normal cropping use of the soil.

It is not meant to imply, however, that this system would work equally well in other locations with other types of oily waste materials. Soil is a very complex environment and all of the factors affecting the success or failure of a waste disposal system such as this are not yet well enough understood to predict the outcome of any specific project. Each site must be experimentally tested to determine the suitability for land disposal of waste materials, and from the experimental results a decision can be made as to the merits of this type of system.

ACKNOWLEDGMENTS

Published with the approval of the Director of the West Virginia University Agricultural Experiment Station as Scientific Paper No. 1429.

REFERENCES

1. Sohngen, N. L. "Bacteria which use methane as a carbonaceous energy source," *Zentr. Bakt. Parasitenk., Abt.* II:15:513–517 (1906).
2. Sohngen, N. L. "Gasoline, kerosine, paraffin oil and paraffin wax as carbon energy sources for microbes," *Zentr. Bakt. Parasitenk., Abt.* II:37:595–609 (1913).
3. Stone, R. W., A. G. White and M. R. Fenske. "Microorganisms attacking petroleum and petroleum fractions," *J. Bact.* 39:91–92 (1940).
4. Kincannon, C. B. "Oily waste disposal by soil cultivation process," Environmental Protection Tech. Series, Report No. EPA–R2–72–110 (1972).

42

TREATMENT OF POTATO PROCESSING WASTEWATER ON AGRICULTURAL LAND: WATER AND ORGANIC LOADING, AND THE FATE OF APPLIED PLANT NUTRIENTS

J. H. Smith, C. W. Robbins, J. A. Bondurant, and C. W. Hayden
USDA-ARS-WR, Snake River Conservation Center,
Kimberly, Idaho

INTRODUCTION

Potato processors discharge large volumes of wastewater that contain large amounts of organic matter, suspended solids, and various inorganic constituents including nitrogen, phosphorus, and potassium (1, 2, 3). Until recently, the wastewater was discharged into rivers and streams, but governmental regulations now prohibit this practice. The potato processors must either treat their wastewater to meet established quality standards before discharging it into streams or find an alternative disposal method. Irrigating agricultural cropland is the disposal and treatment method that many potato processors have chosen.

Little was known about the chemical composition of potato processing wastewater until recently. Loehr (4) cited data on water requirements for processing and waste loading per ton of processed potatoes. More recently Smith and associates published the nutrient content of potato processing wastewater (5), water loading, organic loading, reduction of COD and nitrates in soil (6), denitrification in potato processing waste treatment fields (7) and a guideline for irrigation with potato processing wastewater (8). De Haan and Zwerman (9) reported research results in the Netherlands on land disposal of potato starch wastewater. They concluded that the systems worked well and that oxygen demand and other constituents except potassium were satisfactorily removed in sandy soil if the applications did not exceed 100 to 500 mm per dose.

Sprinkler irrigation with food processing wastes was first tried in 1947, and since that time its use has greatly increased in the U.S. (10, 11, 12). Several potato processors formerly using other systems, such as secondary treatment, recently have converted to land disposal. Many newer potato processing plants are using some form of land disposal for their wastewater.

The processing season begins with freshly harvested potatoes and continues throughout the winter months and part of the next summer using potatoes from storage. Irrigation with the wastewater has been as successful with flooding of graded fields as with sprinkling when using equipment designed to operate at temperatures below freezing.

The objectives of this chapter are to summarize data for sprinkler and flood irrigation with potato processing wastewater, loading with nutrients and organic matter, water cleanup through filtration and microbiological activity, some aspects of nutrient utilization, and to discuss the feasibility of continued irrigation with these wastewaters.

METHODS AND MATERIALS

This study was conducted at five potato processing plants in southern Idaho where the wastewater is used to irrigate cropped fields. Three fields are irrigated by flooding bordered, nearly level land and two are irrigated by sprinkling. Orchard grass, tall fescue, reed canary grass, or mixtures of these species are grown on the fields and harvested for hay or grazed by livestock. Wastewater was sampled at each potato processing plant at monthly intervals during most of three processing seasons. An automatic sampler, activated at 20-minute intervals for 24 hours, delivered water into a freezer where it was frozen in a plastic container for storage until analyzed in the laboratory (13).

Soil water was sampled monthly using 3.8-cm diameter polyvinylchloride sampling tubes with porous ceramic cups cemented to one end. The sampling tubes were inserted vertically into the soil to depths of 15, 30, 60, 90, 120, and 150 cm at each sampling site. When taking samples, approximately 0.7 bar suction was applied to the tubes for about 48 hours. The extracted water was pumped into a suction flask, transferred to a plastic bottle, and taken to the laboratory for refrigerated storage until analyzed. Not every tube yielded a water sample at every sampling.

The water samples were analyzed for COD according to *Standard Methods for the Examination of Water and Wastewater* (14).

Nitrate-N was determined with a nitrate-specific ion electrode, and total nitrogen was determined by a Kjeldahl procedure, modified by substitution of copper for the mercury catalyst (14). Total phosphorus was determined using persulfate oxidation (15) and potassium by flame photometry. Water applications to the fields were measured by the treatment field operators using meters or other devices. Processing plant waste effluents, water samples extracted with extraction cups, and saturated soil extracts were also analyzed for sodium, calcium, magnesium, chloride, bicarbonate, sulfate, pH, and electrical conductivity. Soils sampled annually were analyzed for the above constituents and the first sampling for cation exchange capacity. Particle-size analyses of the soils were made from each sampling depth (Table 1). One site, where a shallow water table developed in the sum-

Table 1. Soil mechanical analyses of wastewater treatment fields.

Treatment	Fields	Sampling Depth (cm)	Clay	Sand	Silt	Textural Class
				(Percent)		
1-F	Site 1	0–150	18	43	39	Loam
	Site 2	0–60	17	50	33	Loam
		60–150	3	89	8	Sand
2-F	Site 1	0–30	12	65	23	Sandy loam
		30–150	2	97	1	Sand
	Site 2	0–150	10	62	28	Sandy loam
3-S	Site 1	0–150	13	63	24	Sandy loam
	Site 2	0–150	11	62	27	Sandy loam
4-S	Site 1	0–60	19	46	35	Loam
		60–150	6	81	13	Loamy sand
5-F	Site 1	0–150	15	56	29	Sandy loam
	Site 2	0–150	19	37	44	Loam

mer because of irrigating, the surrounding agricultural area was instrumented for redox potential measurements in the soil (7).

The processing plants with the flood irrigated fields are referred to as 1-F, 2-F, and 5-F and the sprinkler fields as 3-S and 4-S. Plants 2-F, 4-S, and 5-F use steam peeling and produce dehydrated potato products. Plant 1-F uses dry lye peeling and produces frozen french fried potatoes and other products. Plant 3-S used wet lye peeling the first season of the study, and then converted to dry lye peeling. It produces dehydrated potato products and starch.

RESULTS AND DISCUSSION

Waste Effluent Analyses and Application

The nitrogen, phosphorus, and potassium concentrations in the wastewater are reported in Table 2 as averages by years of all the

Table 2. Annual wastewater applications; mean nitrogen, phosphorus and potassium concentrations, and annual applications in wastewater from five potato processing plants.[a]

Treatment Fields	Year	Water Applied (cm)	Nitrogen (mg/l)	Nitrogen (kg/ha)	Phosphorus (mg/l)	Phosphorus (kg/ha)	Potassium (mg/l)	Potassium (kg/ha)
1-F	1973	546	52	2550	10	630	114	5750
	1974	460	47	2130	13	630	162	7730
	1975	260	50	1500	12	300	130	3180
2-F	1973	115	32	400	6	80	75	930
	1974	209	33	610	6	110	94	1880
	1975	174	35	640	6	120	88	1840
3-S	1974	119	91	1500	21	150	180	2670
	1975	161	133	1720	16	220	250	3540
4-S	1973	246	52	760	9	120	132	2490
	1974	78	52	670	8	110	111	1910
	1975	27	43	350	8	70	77	680
5-F	1973	266	59	980	9	160	150	2540
	1974	201	44	950	8	170	158	2670
	1975	278	51	1420	10	280	104	2830

[a]Monthly applications and concentrations were used for calculating annual values.

samples obtained from each processing plant during 1973, 1974, and 1975. The nitrogen is primarily organic, with mean concentrations less than 2 mg nitrate-N/l. Phosphorus in the wastewater averaged 32 percent ortho, 22 percent acid hydrolyzable, and 46 percent organic. Nitrogen in the wastewater ranged from 32 to 133 mg/l, phosphorus from 6 to 21 mg/l and potassium from 75 to 158 mg/l for all plants.

The amount of wastewater applied annually ranged from 27 to 546 cm (Table 2). The wastewater at most of the potato processing plants is screened for removal of rocks and potato pieces, then passed through a clarifier, and the settled solids are removed by a vacuum filter. The filter cake containing 10–15 percent solid material was ensiled and later used for livestock feed.

Nitrogen applied in the wastewater ranged from 350 kg/ha

(310 lb/acre) to 2550 kg/ha (2280 lb/acre) annually. The lowest nitrogen application can probably be utilized by a good grass crop in this climatic area but the highest rate exceeds crop requirements.

Phosphorus applied in the wastewater ranged from 70 kg/ha (60 lb/acre) to 630 kg/ha (565 lb/acre). These applications exceed phosphorus requirements for most crops, and phosphorus will increase in the soil as a result of irrigation with potato processing wastewater. During three years of irrigation with these wastes, the bicarbonate extractable phosphorus increased about 40 ppm in the top 30 cm of soil, with much smaller increases below that depth (Table 3).

Table 3. Bicarbonate extractable soil phosphorus in two potato processing waste treatment fields for three years.

Depth in Soil (cm)	Bicarbonate Extractable (P, ppm)					
	Treatment Field 5-F			Treatment Field 3-S		
	1972	1973	1974	1973	1974	1975
0–30	11.0	25.7	51.3	16.6	54.4	53.0
30–60	4.1	17.6	15.7	10.0	21.9	19.2
60–90	—	18.2	17.5	3.8	10.7	6.3
90–120	3.6	16.9	15.4	2.9	12.6	5.6
120–150	3.5	15.1	14.7	1.6	8.8	4.5

Potassium applied in the wastewater exceeded the potassium requirements of grass. Potassium concentration in the soil solution is expected to increase until it reaches an equilibrium with the soil, after which the excess applied will be leached.

Infiltration occurred at each flood irrigation in the winter with 15°C water even when the air temperature was below minus 40°C. With sprinkler irrigation in cold weather, ice accumulated in mounds around the sprinklers, remaining until air temperatures were above freezing. Ice accumulation occurs because water leaving the sprinkler nozzle approaches dewpoint temperature before the droplets reach the ground regardless of the water temperature in the sprinkler nozzle (16). Melting was usually slow enough to allow infiltration, and no major runoff problems were observed. Results with both irrigation methods were similar, although higher nitrates were found in the soils under sprinkling than under flood irrigation. This probably did not result from different irrigation methods. Usually less water was applied on the sprinkler fields than when irrigating by flooding.

COD Applications

Mean concentrations of COD in the wastewater ranged from 765 to 3080 mg/l (Table 4). These differences result from dif-

Table 4.   COD in potato processing wastewater used for irrigation.

Processing Plant	Year	COD (mg/l)			COD Applied (tons/ha)
		max.	min.	mean	
1-F	1973	1990	950	1480	58.6
	1974	2680	800	1570	85.1
	1975	2500	740	1440	29.9
2-F	1973	1220	730	940	9.5
	1974	1040	530	760	15.6
	1975	1530	450	880	15.6
3-S	1974	7400	1280	3080	35.2
	1975	2940	1070	2110	34.8
4-S	1973	3510	400	2540	20.2
	1974	3830	490	2540	15.4
	1975	2970	1410	2020	12.1
5-F	1973	2040	1080	1540	40.9
	1974	1660	850	1370	27.0
	1975	1640	410	1250	35.9

ferent peeling and potato handling processes in the potato processing plants. The high COD concentrations in the wastewater at plant 3-S resulted from not using vacuum filtration. Maximum COD concentrations usually occurred when poorer quality potatoes were processed late in the winter. Also late in the season, processing plants often operated at less than maximum capacity. Under these conditions they used more water than normal per ton of potatoes and the wastewater was less concentrated in COD and other constituents than during normal operation.

Annual COD applied to the treatment field varied from approximately 10 to 85 tons per hectare. At plant 1-F the high application rates in 1973 and 1974 decreased as more land was used for wastewater irrigation. The lower amounts were applied at Plant 2-F because the disposal system was designed and constructed to utilize the total plant effluent. When construction was complete and a grass cover was established, irrigation with wastewater was initiated, and the system has worked exceptionally well from the beginning.

Plant 5-F had the first potato processing wastewater irrigation system in Idaho. Settling of recently leveled land caused ponding in the low areas, which killed the grass. Long stretches of open ditch carrying wastewater in the field became anaerobic and created noxious odors. This was corrected by installing underground pipe for the field water distribution system and eliminating the open ditches. Low spots in the field were filled with soil and reseeded with grass.

COD in Extracted Water Samples

Obtaining soil water samples from the sprinkled fields was difficult. In the winter the soil and sampling sites were covered much of the time with ice. Even when the sites were not covered with ice, generally little or no water could be extracted. In the summer when the processing plants were not running, the fields frequently were under-irrigated for a normal grass crop and the soil was too dry to obtain water samples. Under these conditions, ground water pollution could not be a problem because very little water was passing through the soil. In the surface-irrigated fields water samples could be extracted consistently from the soil during the entire year.

At most of the sampling sites a large number of extracted samples were obtained during two or three years sampling and for this report maximum, minimum, and mean COD are reported at each of six depths from 15 to 150 cm deep in soil (Table 5). At locations 1-F through 5-F the means of COD removal ranged from 95 to 98 percent at the 150 cm depth. The maximum COD remaining at the 150 cm depth ranged from 80 to 98 percent COD removal based on maximum COD concentrations applied to the sites. The minimum COD concentrations reported in Table 5 probably represent normal background, although no data are available for determining normal background COD concentrations.

Nitrogen in Extracted Water Samples

Nitrate-nitrogen in the wastewater was low with the average at all locations and sampling being 1.2 mg/l (Table 6). A few samples were as high as 4.9 to 9 mg N/l. Nitrates in the soil water would correlate more closely with the total nitrogen in the wastewater than with the nitrates in the wastewater because most of the organic nitrogen eventually becomes nitrate through microbiological breakdown of the organic wastes. At two plants, 1-F and 5-F, denitrification probably was an important factor in the

Table 5. Chemical Oxygen Demand (COD) and Kjeldahl-N in water samples extracted from potato processing wastewater treatment fields at five locations for three years.

Treatment Field	Depth in Soil (cm)	Chemical Oxygen Demand (mg/l)			Kjeldahl-N (mg/l)		
		max.	min.	mean	max.	min.	mean
1-F	15	400	35	160	10.9	0	4.0
	30	780	25	230	26.5	0.7	10.5
	60	820	20	165	21.8	0.4	4.7
	90	730	25	190	37.1	0	10.1
	120	460	15	155	45.0	0.6	11.2
	150	170	5	70	4.5	0.2	2.0
2-F	15	575	7	74	10.6	0	2.4
	30	485	8	85	15.8	0	4.3
	60	470	5	85	14.0	0	3.7
	90	555	2	120	20.0	0	4.0
	120	425	2	60	2.4	0	0.7
	150	140	1	35	2.0	0	0.6
3-S	15	900	17	124	33.4	2.0	8.5
	30	405	1	120	8.4	0.7	2.4
	60	1180	14	130	4.6	0.7	1.8
	90	1300	12	145	19.2	0.1	4.0
	120	705	10	70	4.3	0	1.3
	150	90	12	40	1.0	0.4	0.7
4-S	15	335	15	70	2.9	0	2.6
	30	235	5	60	3.7	0.1	1.1
	60	100	20	45	1.6	0.7	1.1
	90	110	20	40	2.8	0.1	1.1
	120	70	25	50	--	--	--
	150	145	20	45	1.0	0.3	0.6
5-F	15	200	10	65	4.4	0.4	1.9
	30	585	10	80	4.3	0	1.3
	60	595	10	65	3.0	0	0.8
	90	280	5	50	12.1	0	2.4
	120	170	5	55	19.0	0	1.4
	150	180	4	35	2.6	0	0.7

low nitrate concentrations observed because the water table ranged from 1 to 3 meters below the surface.

At field 5-F, a site was instrumented with platinum electrodes from the surface to 150 cm, and redox potentials were measured. The potentials confirmed that conditions were favorable for denitrification and that denitrification was a major factor in accounting for the low nitrates (7).

At processing plant 2-F the water table is very deep. Redox measurements are now being made and there is some indication that conditions suitable for denitrification occur following some

Table 6. Nitrate-N in potato processing effluent water and water samples extracted from wastewater treatment fields at five locations for three years.

Processing Plant	Depth in Soil (cm)	Nitrate-N (mg/l)		
		maximum	minimum	mean
1-F	0[a]	3.6	0	0.6
	15	26.0	0.1	4.6
	30	35.0	0.1	6.4
	60	29.8	0.1	9.0
	90	38.0	0	6.5
	120	12.5	0	2.3
	150	9.0	0	2.0
2-F	0	3.0	0	0.7
	15	33.5	0	4.0
	30	22.5	0	3.9
	60	53.0	0	4.7
	90	19.5	0	3.2
	120	30.0	0	3.1
	150	22.4	0	3.3
3-S	0	4.8	0.1	1.2
	15	110.0	0.2	14.6
	30	59.4	0	17.6
	60	140.0	0	16.9
	90	73.0	0.1	15.6
	120	36.0	0	10.1
	150	17.0	0	7.1
4-S	0	3.6	0	1.0
	15	48.0	0.1	12.8
	30	50.0	0.1	15.8
	60	125.0	3.2	32.7
	90	95.0	1.0	26.0
	120	67.0	18.5	41.4
	150	100.0	6.0	35.3
5-F	0	9.0	0	1.1
	15	25.0	0	2.5
	30	35.0	0	3.3
	60	4.9	0	0.6
	90	2.6	0	1.1
	120	3.4	0	0.8
	150	3.6	0	0.6

[a] The 0 depth water samples are processing wastewater.

irrigations with wastewater during warm weather. The potential for nitrate leaching and for ground water pollution are greater at locations with a deep water table than at locations with shallow water tables because denitrification is greater with a shallow water table. Two water samples at the 120- and 150-cm soil depth had nitrate-nitrogen concentrations of 30 and 22 mg/l. The concentrations of nitrate in the soil water samples were highest in fields 3-S and 4-S. Equipment maintenance shutdown during the early summer at both locations prevented irrigations for maximum growth of grass, and consequently nitrates accumulated to greater concentrations than they might have if the grass had grown all summer.

Kjeldahl nitrogen determined in the water samples extracted from the treatment field soils did not include nitrate-nitrogen. Most Kjeldahl nitrogen concentrations in the soil water were 2 mg N/l or less at the 150-cm depth (Table 5). Kjeldahl nitrogen concentrations in the soil water samples followed the same trends as the COD. Most water samples had COD:N rations of 20:25. These ratios indicate that nitrogen was not a limited factor in decomposition and that the organic wastes should decompose rapidly when the temperature is favorable. The grass crops grown on these fields when irrigated properly in the summer were excellent, with yields of five to ten tons of hay per acre. This hay was high in protein with some samples containing as much as 18 to 20 percent crude protein (total N $\times$ 6.25). Nitrates were generally within acceptable concentrations with most samples below 1800 ppm NO_3N.

Inorganic Constituents

Electrical conductivity, sodium, calcium, and magnesium, were measured in the wastewater, the soil water, and in soil samples. Sample measurements of electrical conductivity in the wastewater and soil water are reported in Table 7. When potatoes are peeled with steam, no sodium hydroxide is used and sodium problems are unlikely from irrigating with the wastewater. Plants 2-F, 4-S, and 5-F used steam peel, and the fields had no previous salinity problems. Plant 3-S used wet lye peeling in 1974 and then was converted to dry lye peeling. When wet lye peeling was used, sodium and salinity in the fields increased. Changing to dry lye peeling should prevent further sodium buildup.

At processing plant 1-F, dry lye peeling was used, with enough of the lye escaping to increase the sodium absorption ratio of the wastewater to values ranging from 3 to 12. The soils that were irrigated with wastewater had a history of salinity problems. Irrigation with wastewater decreased the electrical conductivity markedly (Table 7), and after three years the electrical conductivity of the soil water samples was about the same as the electrical conductivity of the wastewater being applied to the fields. Electrical conductivity in the fields of processing plant 2-F was low at the beginning of wastewater irrigation and increased slightly to equilibrium with the wastewater used for irrigation. These electrical conductivity values are satisfactory for growing most crops and should pose no future problems.

Table 7. Sample electrical conductivity measurements in wastewater and in water extracted from potato processing wastewater disposal fields.

Processing Plant	Depth in Soil (cm)	EC (μmhos/cm)			
		12/21/72	10/4/73	9/6/74	9/15/75
1-F	0	1215	1040	1260	880
	15	9250	2900	3000	–
	30	8200	2350	1900	1370
	60	7250	4550	–	–
	90	5380	3500	1800	1360
	120	1640	3900	2600	2100
	150	825	1420	2530	–
		4/27/73	4/5/74	12/12/74	12/12/75
2-F	0	1235	1000	1160	1000
	15	790	1000	470	900
	30	970	930	750	935
	60	590	1000	810	890
	90	545	1100	820	1080
	120	510	1060	680	1080
	150	695	1150	710	1000

SUMMARY

Nitrogen, phosphorus, and potassium concentrations in the potato processing wastewater vary widely but, with the amounts of water being applied in these wastewater irrigation systems, provide large amounts of nutrients for crops grown on these fields. In some cases the applications are excessive, and much more efficient use could be made of the nutrients by irrigating larger land areas.

The amount of wastewater applied ranged from approximately 25 to 550 cm per year. With proper land preparation to avoid ponding and with drying periods between irrigations to avoid development of anaerobic conditions in the fields, water for these irrigations infiltrated the fields without waterlogging problems developing.

The organic loading of 10 to 85 tons COD/ha were assimilated by the soils without anaerobiosis developing near the surface, and therefore organic loading is not a limiting factor in operating wastewater treatment and disposal fields like those studied.

With wastewater treatment and disposal fields where a water table lies within 1 to 3 meters of the surface, nitrates are not likely to be a problem even when 2 to 3 tons of N/ha are applied annually. In fields with a deep water table, nitrogen application may have to be limited to prevent excessive nitrate leaching and possible ground water pollution.

Wastewater from wet lye peel potato processing is not suitable for long term irrigation of agricultural land for growing crops. Wastewater from dry lye peeling systems, which keeps the sodium hydroxide separated from the wastewater effluent, and from steam peel potato processing plants can be used successfully for irrigating cropped agricultural land. Irrigating with wastewater utilizes the water and some of the nutrients that were wasted when the water was discharged into streams and rivers. The research and observations made during several years indicate that wastewater irrigation of cropped agricultural land can be used successfully for a long time to come.

REFERENCES

1. Knapp, C. E. "Agriculture poses waste problems," *Environ. Sci. Technol.* 4:1098–1099 (1970).
2. Pearson, G. A., W. G. J. Knibbe and H. L. Worley. "Composition and variation of wastewater from food processing plants," *USDA–ARS* 41–186 (1972).
3. Smith, J. H. "Decomposition in soil of waste cooking oils used in potato processing," *J. Environ. Quality* 3(3):279–281 (1974).
4. Loehr, Raymond C. *Agricultural Waste Management: Problems, Processes, Approaches*, Academic Press, Inc., New York, 576 pp. (1974).
5. Smith, J. H., C. W. Robbins, and C. W. Hayden. "Plant Nutrients in potato processing wastewater used for irrigation," *Proc. 26th Ann. Pacific Northwest Fertilizer Conference*, Salt Lake City, Utah, July 15–17 (1975) pp. 159–165.
6. Smith, J. H. "Treatment of potato processing wastewater on agricultural land," *J. Environ. Quality* 5(1):113–116 (1976).
7. Smith, J. H., R. G. Gilbert and J. B. Miller. "Redox potentials and denitrification in a cropped potato processing wastewater disposal field," *J. Environ. Quality* (1976), accepted for publication.
8. Smith, J. H. "Sprinkler irrigation of potato processing waste for treatment and disposal of land," *In: Waste Water Resource Manual*, Edward Norum, Ed., Sprinkler Irrigation Assoc., Silver Spring, Maryland, 2C2/21–29 (1975).
9. de Haan, F. A. M., and P. J. Zwerman. "Land disposal of potato starch processing wastewater in the Netherlands," *In: Food Processing Waste Management.* Proc. of the 1973 Cornell Agric. Waste Management Conf., pp. 222–228 (1973).
10. Anderson, D. J., and A. T. Wallace. "Innovations in industrial disposal of steam peel potato wastewater," *Pacific Northwest Ind. Waste Mgt. Conf. Proc.*, University of Idaho, Moscow, pp. 45–61 (1971).
11. Bell, J. W. "Spray irrigation for poultry and canning waste," *Public Works* 86(9):111–112 (1955).
12. Bolton, P. "Cannery waste disposal by field irrigation," *Food Packer* 38:42–43, 46 (1947).

13. Fisher, H. D., and J. H. Smith. "An automatic system for sampling processing wastewater," *Soil Sci. Soc. Amer. Proc.* 39(2):382–384 (1975).
14. American Public Health Association, Inc. *Standard Methods for the Examination of Water and Wastewater,* 13th edition, New York, 874 pp. (1971).
15. U.S. Environmental Protection Agency. *Methods for Chemical Analysis of Water and Wastes,* 298 pp. (1974).
16. Pair, C. H., J. L. Wright and M. E. Jensen. "Sprinkler irrigation spray temperatures," *ASAE Trans.* 12(3):314–315 (1969).

43

CONTINUOUS WATERSHED MODELING OF WASTEWATER STORAGE AND LAND APPLICATION TO IMPROVE DESIGN PARAMETERS

J. A. Anschutz, J. K. Koelliker
Department of Agricultural Engineering
J. J. Zovne, T. A. Bean
Department of Civil Engineering
W. H. Neibling
Department of Agricultural Engineering
  Kansas State University, Manhattan, Kansas

INTRODUCTION

Management of wastewaters to meet standards of performance developed for point source categories by the U.S. Environmental Protection Agency (EPA) has required a radical reconsideration of how to properly design certain kinds of control facilities. In particular, those systems which interact with the uncontrollable variable of the local climate at some point in the generation, storage, and final treatment of wastewater must also adequately account for the effects of this variable. One such industry is the feedlot category. Systems which use the soil as the ultimate treatment system and open ponds or lagoons for storage also fall into this area.

Due to the unpredictable nature of both the weather-related runoff phenomena from open feedlots and the water quality of the resulting runoff, prior to 1973 several states chose to adopt design standards which required a certain size of runoff containment basin as their regulatory tool. Since it appeared economically infeasible to design a retention basin capable of containing all feedlot runoff under all climatological conditions, most designs were based on a 24-hr storm event that had a specified probability of occurrence. This resulted in a system capable of operating under a specified set of conditions. However, soon after such basins were installed, engineers realized the need for subsequent land disposal from such basins. Therefore, complete designs included management of the runoff from the time it started to cross the lot surface until it was ultimately disposed onto cropland (1).

In 1973, the EPA developed effluent guidelines for feedlots (2). In doing so, the design criteria of 24-hr storms were translated into part of the performance criteria for the effluent guidelines. As a result, instead of designing to control the design storm, the standard of performance appears to allow discharges from feedlots only when a storm in excess of the 10-yr, 24-hr storm occurred. This radical change in approach left designers and regulatory agencies with no information on the performance of facilities properly designed and managed under programs in existence prior to 1973. Koelliker *et al.* (3) first developed a simplistic continuous watershed model to evaluate the long-term performance of feedlot runoff control facilities in 1973. Development and testing of similar models have been reported, *i.e.*, Larsen *et al.* (4), and Wensink and Miner (5). Modeling was the only tool available to adequately assess the performance of such facilities.

Continuous watershed modeling provides an inexpensive tool to fit together the various sizes of components of a waste management scheme to assess the ability of the overall system to control discharges. The effects of various management schemes, climatic variables from a continuous series of weather data, and other important site-specific parameters, such as soil type, when added to the model provide additional constraints on the operation of the system.

Much of the design of wastewater control facilities, heretofore, has relied on a combination of specific conditions to determine design parameters. An example would be an evaporation pond sizing based upon some probability of maximum precipitation coupled with an equal probability of minimum evaporation over some time period. Whether such discrete events do in fact occur simultaneously is unknown. It appears that a system such as feedlot runoff control facilities which are required to operate over the continuum can best be designed if the design considers operation over the continuum. For these reasons a continuous watershed model is being developed. As such, the model contains no new approaches to watershed modeling. The best and most physically based processes and input parameters from various disciplines are being drawn together into a model specifically adapted to the feedlot simulation. This chapter is a description of this physically based model for open cattle feedlot systems.

THE MODEL

Components of the model include the feedlot surface (wastewater generator), wastewater storage facility, disposal system, and disposal/treatment area. The movement of wastewater into and out of, as well as the amount stored in each component, is estimated each day. The necessary daily weather data and input parameters are estimated each day. The daily weather data and input parameters needed for the model are listed in Table 1.

Table 1.   Input parameters for model.

Daily average air temperature (F°)
Daily precipitation (in.)
Mean monthly relative humidity (%)
Mean monthly percent sunshine (%)
Mean monthly wind speed (MPH)
Mid-monthly intensity of solar radiation (mm of water evaporated per day)
Feedlot area (acres)
Disposal area (acres)
Disposal area soil type [irrigation design group from Kansas Irrigation Guide (23)]
Disposal rate (in./day)
Crop type
Percentage of soil moisture depletion for disposal (% field capacity)
Pond size:
    length of base (ft)
    width of base (ft)
    side slope (run-to-rise)
    maximum pond depth (ft)
Allowable disposal dates

Continuous weather data tapes are available from the National Oceanic and Atmospheric Administration, Asheville, North Carolina. These tapes contain the same data published in Climatological Data but on a continuous basis. Length of records available are usually at least 25 years, with many stations available with considerably longer records.

Calculation of Feedlot Runoff

Feedlot runoff is determined by the Soil Conservation Service Equation (6). Runoff curve numbers (RCN) are based on 3-day rather than 5-day antecedent moisture conditions (AMC). During the growing season (April–October), a RCN of 91 was used when the 3-day antecedent moisture (AM) is less than 0.5 in.

A RCN of 97 is used when AM is 0.75 in. or greater. When a rainfall event exceeds 1.25 in., the SCS equation is applied to the first 1.25 in. The total runoff is that calculated by the SCS equation plus the amount of rainfall exceeding 1.25 in. Runoff from rainfall events less than 1.25 in. is calculated directly from the equation. During the dormant season (January–March, November and December) if the AM is less than 0.5 in., a RCN of 91 is used. Other times 97 is used. Runoff is calculated on the first inch of rainfall with the remainder of rainfall added as runoff to that calculated. Runoff is assumed to be zero for rainfall events less than 0.5 in. and 3-day AMC less than those specified for the growing or dormant season.

Since the precipitation data on the tapes do not reliably differentiate rainfall from snow, precipitation occurring when the ground is frozen is assumed to be snow. Snow is accumulated on the feedlot until the ground thaws at which time the runoff is calculated by the SCS equation using a RCN of 97. The total snow accumulated is input as one precipitation event. The ground is judged to be frozen when the present and previous days' air temperature averages less than 32°F. It is assumed to thaw when the present and previous two days' air temperatures average 38°F or greater (3).

The above-mentioned modifications of the SCS method were determined by correlating actual runoff data from feedlots at Gretna, Nebraska (7), Bushland, Texas (8), and Pratt, Kansas (9). Based on these correlations, further refinement was made by subtracting 0.06 in. from calculated feedlot runoffs exceeding that amount. This resulted in our best correlation coefficient (R^2) of 0.85 for the regression equation of actual versus calculated runoff.

Wastewater Storage Facility

A storage facility is sized to store the feedlot runoff until it can be disposed of, either by irrigation or evaporation from the pond. The model is capable of handling many different shapes and sizes of ponds and will determine the surface area of the stored wastewater subject to evaporation. The evaporation from the wastewater surface is calculated as lake evaporation. Lake evaporation is determined by a modification of the net radiation calculated by the Penman equation (10). This modification is necessary since the reflection coefficient used in the Penman equation is not the same for a water surface as it is for bare soil. Seepage of the storage facility is assumed to be zero.

Disposal

The decision of whether to dispose is based on the following parameters:

1. Volume of wastewater in storage pond
2. Time of year
3. Disposal area frozen or thawed
4. Ambient air temperature
5. Soil moisture conditions of the disposal area

Any of these parameters can be varied to simulate a wide variation of management practices used for wastewater disposal.

Soil Moisture Accounting of the Disposal Area

Recharged by precipitation and irrigation waters, the disposal area can be conceived as a soil-water reservoir which is depleted by evapotranspiration. Most annual row crops and grasses develop their root zone in the upper 5 ft of soil. Since soil-water is extracted through the plant roots, most transpiration occurs from above this level. Root extraction pattern studies by Russell (11) and a consumptive use study by Manges [by personal communication] conducted in England and Kansas, respectively, have shown that approximately 70 percent is extracted from the upper 1 ft of soil with nearly all of the remaining 30 percent being taken from the next 4 ft. On this basis, the disposal area soil-moisture model simulates soil-water interactions above the 5-ft level, unless the soil profile is shallower, in the manner described with two zones being referred to as upper and lower. Evapotranspiration is proportioned accordingly and the surface is handled separately. Water which moves vertically through both zones is accounted as percolation out of the root zone.

Transfer of water from the soil back into the atmosphere is handled by two processes, direct soil evaporation and evapotranspiration. When soil lies fallow or annual row crops are in the early growth stage providing little vegetative cover, the transfer is accomplished by means of soil moisture evaporation. As plant leaf surface increases through the growth stage, direct soil moisture evaporation decreases while plant transpiration increasingly gives rise to evapotranspiration.

Soil Moisture Evaporation

Studies by Ritchie (12) and field work by Kanemasu (13) indicate that soil moisture evaporation occurs in two stages. First-stage evaporation occurs when the soil is sufficiently wet to readily transport water to the surface. Termed as the constant rate stage, evaporation proceeds at the potential rate calculated for bare soil. The calculation of the potential evapotranspiration will be discussed later. As the soil moisture is depleted, the hydraulic properties of the soil start to limit the constant evaporation rate. As approached by Richie (12); second-stage evaporation is calculated by:

$$E_s = 0.0039(Ct^{1/2} - C(t-1)^{1/2})$$

where E_s = second-stage evaporation (inches of water/day)
       C = hydraulic coefficient of soil
       t = time after first-stage evaporation in days

Kanemasu's field work has provided accurate values for C and the point at which first-stage evaporation enters into second-stage evaporation.

Potential Evapotranspiration (PET)

The Penman combination method is used to calculate PET. Penman's equation is the most theoretical approach for predicting PET (10). Very good correlation has been obtained between computed and measured amounts of consumptive use. Inputs to Penman's equation include daily average ambient air temperature and relative humidity, wind travel, net radiation, and percent sunshine. The principal limitation of the Penman approach is the lack of sufficient weather measurements in most localities. Only a few climatological stations record the needed data on a daily basis.

Using monthly mean values for daily average relative humidity, percent sunshine, wind travel (14) and net radiation (10) as well as daily average temperature the Penman equation can be used to compute PET for any station recording daily temperatures. For localities not listing monthly means, interpolations can be made from two or three nearby stations having the required data.

Potential evapotranspiration was calculated by two different ways for Topeka, Kansas for the 10-yr period of 1954–1963. One way used daily inputs to the Penman equation while the second used monthly mean values of relative humidity, percent sunshine, wind travel, and net radiation. Statistical analysis by the differential t-test (15) of the two sets of PET data on a day-to-day basis showed there was significant variation between daily values in 42 percent of the months at the 95 percent level. However, when comparisons were made on total PET of 5-day blocks calculated both ways, the differential t-test showed that on the average only 17 percent of the 5-day blocks were judged to be significantly different at the 95 percent confidence interval.

The first analysis made on a day-to-day basis was determined from the difference of each day's PET calculated by the two different methods. Monthly totals of these differences were then calculated and the differential t-test was determined for each month. In the second analysis the data was grouped into 5-day blocks. Totals of these blocks were determined and a difference was found between corresponding groups. The differential t-test was also computed for these differences.

Since the model we are using is a continuous model, we feel that since the PET predictions within a 5-day period are usually not significantly different, this method using monthly averages gives the model sufficient accuracy to predict PET. Since the data requirements are considerably less, this method has been adopted.

Evapotranspiration

The actual rate of evapotranspiration (AET) is dependent upon atmospheric, plant and soil factors (12). Atmospheric factors are involved in the computation of PET by the Penman method. Application of the Blaney-Criddle method (16) is widely accepted in account for plant factors. This method is used to modify the PET rate by a plant consumptive factor, k, which has been determined from experimental data for each crop. Results of application to dry-land agricultural watersheds (17) and irrigated lands (18) display the versatility of the Blaney-Criddle method.

Under wet soil conditions evapotranspiration will proceed at the maximum rate. As approached by Kanemasu (13), this relation continues until the soil-water deficit reaches a level which is equivalent to 0.3 of the maximum available moisture (θ_{max}). Available moisture or capillary water, as defined by Israelsen (10), is the difference in moisture content of the soil between field capacity and permanent wilting point. When the soil-water deficit falls below 0.3 θ_{max}, Kanemasu modifies the potential rate by

$$K_s = \theta_a/0.3\,\theta_{max}$$

where θ_a is the actual available soil moisture content. With application of K_s to the PET rate, the soil-water deficit allows AET to be at the maximum rate above 0.3 θ_{max} and limits it to zero at the permanent wilting point.

Disposal Area Runoff, Interception and Infiltration

The Soil Conservation Service method (6) is used to predict runoff from the disposal area. For user convenience, the model uses input runoff curve numbers which are specified by input soil and crop types. The curve numbers are based on condition II antecedent moisture, between 0.5 and 0.8 of field capacity, for the specified soil. When soil moisture is less than 0.5 of field capacity the curve number (N) is modified to condition I antecedent moisture by

$$N = N \times 0.39 \, e^{(-0.009 \times N)}$$

and, when the soil moisture is greater than 0.8 field capacity, by

$$N = N \times 1.95 \, e^{(-0.00663 \times N)}$$

The use of these equations makes the SCS method efficient for computer simulation.

After calculating surface runoff, interception losses must be separated from the remainder of the initial abstraction. A simple approach for computer simulations (19, 20) requires fixing an interception storage capacity which must first be satisfied before water from any event is available for infiltration. For purposes of this model, the interception-storage is fixed at 0.1 in. and depleted at the potential rate whenever evaporative energy is available. According to Ward (20), this storage should account for 10–20 percent of annual precipitation. Of the previously tested Kansas and Oregon stations, approximately 16–18 percent of the annual precipitation was accounted as interception-storage losses which evaporated back into the atmosphere.

After runoff and interception-storage losses are calculated, the remainder of the water will infiltrate into the soil. Infiltration rates and redistribution of water within the soil are dependent upon hydraulic properties of the soil (10). Actual rates are difficult to simulate because of the variability of influencing factors. Saxton (19) has modeled infiltration by means of physical redistribution. Using this scheme, a soil layer is allowed to take on an equivalent amount of water to raise the soil moisture level of the upper zone to 0.9 saturation with any excess amount being cascaded to the next successive layer. This continues until all available water for infiltration is stored. Gravitational water is then computed as the difference between 0.9 saturation and field capacity. This water percolates into the succeeding lower zone the next day.

Snow on the Disposal Area

At many locations in the U.S., a sizeable snow pack accumulates during winter months. For this reason, a rational method of accumulating and dissipating snow in the disposal area must be incorporated to closer approximate actual conditions on the disposal area. Of the numerous snowmelt methods existing, the degree-day approach requires the least amount of input. This method requires only dry bulb temperature which makes it quite simple to model. In spite of its simplicity and limited input, the degree-day approach has obtained good results (21).

Operation of the Disposal Area

Various management techniques are incorporated in the model for scheduling the disposal of wastewater. Having calculated the soil moisture of the disposal area, irrigation scheduling can be accomplished on a soil moisture depletion basis. Optimum crop growth will vary somewhat and aeration, waterholding capacity of the soil and the type of crop grown but is normally at optimum conditions when the soil moisture is slightly less than field capacity (10). By specifying a soil moisture level of 90 percent of field capacity before irrigation is allowed, we can maximize disposal and at the same time maintain soil moisture conditions conducive to optimum crop growth, provided there is sufficient wastewater in the storage facility during the growing season to meet crop needs. This management procedure also prevents irrigating when the disposal area is saturated, thus eliminating runoff of excess wastewater.

The disposal rate can be varied to simulate different irrigation practices. The model has been tested using disposal rates of one-tenth the amount of the 25-yr, 24-hr rainfall for the specific site. These rates would be typical of sprinkler irrigation. Any time the storage pond contains less than one day's irrigation, disposal does not occur. This would reduce labor requirements for actual operational systems.

Dates allowable for disposal can be specified to simulate different irrigation management practices. Some irrigators may not irrigate during the winter months, especially in regions with cold seasons. With this management scheme the storage facility would need to be sized large enough to hold feedlot runoff during the winter months to prevent possible discharge. In regions with milder winter seasons, winter irrigation would increase disposal capabilities, therefore eliminating the extra needed storage capacity to retain feedlot runoff during the winter months.

The model presently simulates irrigation uniformly over the entire disposal area. Further development of the model is proposed to accommodate disposal on different segments of the disposal area. This approach would increase disposal days and would permit the use of smaller pumping plants. Gravity irrigation systems could be simulated easily by this method. The flexibility of imputing various irrigation management practices allows the model to predict effects these practices have when using land as a disposal area for feedlot runoff wastewater. Design criteria for disposal systems can be improved with these predictions.

RESULTS AND DISCUSSION

For this presentation the model was run for three stations: Topeka, Kansas (25 years), Garden City, Kansas (50 years), and Corvallis, Oregon (50 years). A printout of each year's annual summary estimates the effectiveness of the management practices specified in controlling feedlot runoff at each specific site. Table 2 shows a sample of the computer output of the annual summary for Topeka, Kansas for the year 1968. This particular year was chosen because it illustrates a year having discharge of wastewater. The dates and amounts of discharge are listed on the top of the summary.

Inflows and outflows of the storage facility are listed by months and have units of inches over the disposal area. For example the precipitation listed under the water account for the storage facility is the amount of the precipitation collected in the storage facility expressed in inches of depth had it been spread uniformly over the disposal area.

The water balance in the disposal area lists the inputs, outputs, and change in soil moisture in inches in the disposal area. AET under this heading is the actual evapotranspiration. The percent of feedlot runoff controlled is listed indicating the effectiveness of the specified management practice in controlling feedlot runoff for the particular year. Potential disposal days were all days disposal would have occurred had there been ample wastewater in the storage facility.

Table 3 shows the results for the three stations. Three different pond sizes were specified at each location. The pond sizes are designated 1X, 2X and 3X, where X is the volume of the 25-yr, 24-hr rainfall event applied over the feedlot area. Comparisons of the three locations indicate that different design criteria are required to control discharges of feedlot runoff.

Garden City, which had the least average annual runoff, required the smallest sized pond to effectively control all runoff. This location also had the largest number of disposal days since it had the least average annual precipitation of the three stations run. Even with a 1X facility size, discharge occurred in only 8 out of 50 years.

Topeka and Corvallis had about twice the amount of average

Table 2.   Sample computer output for Topeka, Kansas.

ANNUAL SUMMARY

8/ 2/68 –	Discharge of	57.61 Acre-in.
8/ 9/68 –	Discharge of	10.13 Acre-in.
8/10/68 –	Discharge of	37.69 Acre-in.

Water Account for Storage Facility
(In Inches Over Disposal Area) – 1968

| Month | Inflows | | | Outflows | | | |
	Precipi-tation	Feedlot Runoff	No. Dispo-sal Days	Disposal Volume	Surface Evap.	Discharge	Change in Vol.
Jan	0.03	0.21	0	0.0	0.02	0.0	0.23
Feb	0.02	0.0	0	0.0	0.02	0.0	0.00
Mar	0.02	0.09	0	0.0	0.09	0.0	0.02
Apr	0.14	0.52	3	1.83	0.14	0.0	−1.31
May	0.11	0.12	1	0.61	0.14	0.0	−0.52
June	0.11	0.61	1	0.61	0.20	0.0	−0.09
July	0.34	2.96	1	0.61	0.20	0.0	2.50
Aug	0.24	1.98	5	3.05	0.20	1.32	−2.34
Sept	0.08	0.48	1	0.61	0.14	0.0	−0.18
Oct	0.14	1.07	0	0.0	0.10	0.0	1.11
Nov	0.05	0.21	1	0.61	0.04	0.0	−0.39
Dec	0.07	0.0	0	0.0	0.01	0.0	0.06
Tot	1.34	8.26	13	7.93	1.28	1.32	−0.92

Water Balance (Inches) in the Disposal Area – 1968

| Month | Inputs | | Outputs | | | | |
	Precipi-tation	Irrigation	Intercep-tion	Surface Runoff	Percola-tion	AET	Change in SM
Jan	0.89	0.0	0.34	0.13	0.0	0.10	0.65
Feb	0.56	0.0	0.25	0.03	0.0	0.19	0.10
Mar	0.46	0.0	0.26	0.02	0.0	0.10	0.08
Apr	4.20	1.83	1.03	1.16	0.57	2.73	0.54
May	3.37	0.61	0.92	0.39	0.07	2.29	0.31
June	3.18	0.61	0.43	1.55	0.53	1.62	−0.34
July	10.17	0.61	1.08	4.87	0.0	5.58	−0.74
Aug	7.40	3.05	1.31	3.67	0.0	6.22	−0.74
Sept	2.50	0.61	0.62	0.48	0.0	3.54	−1.53
Oct	4.19	0.0	0.58	2.05	0.0	0.94	0.62
Nov	1.56	0.61	0.50	0.38	0.0	1.04	0.25
Dec	2.10	0.0	0.30	0.67	0.0	0.11	0.80
Tot	40.58	7.93	7.62	15.40	1.17	24.44	−0.02

Percent of feedlot runoff controlled = 84.04.

Potential Disposal Days = 60.

Pack on December 31 = 0.22; change in snow storage = −0.09.

Inputs–Outputs–change in snow storage = change in soil moisture.

Percent of maximum pond volume required = 100.00.

Estimated lake evaporation, inches = 47.58.

Table 3. Results for three locations.

Location: Garden City, Kansas   Soil Class: 3   Crop: Grain Sorghum
Years of Records Used: 1921-1970 (50 years)   Feedlot Size: 160 ac
Mode of Operation: Dispose Jan-Dec (90% Field Capacity)   Disposal Rate: 0.47 in./day
Pond Size: 1X = 4.7 in.   Ratio Disposal Area/Feedlot: 1/1

	Precipitation, in.	Runoff, in.	Discharge, in.			Disposal Days		
			1X	2X	3X	1X	2X	3X
Totals:	903.77	263.14	13.16	0	0	448	461	460
Avg:	18.07	5.26	–	–	–	8.96	9.22	9.20
Range:	5.68 – 36.19	0.38 – 15.70				0-24	0-31	0-30
No. Years with Discharge:			8	0	0			
Avg. No. Discharge/Yr with Discharge:			4.5	0	0			
Range:			1-8	0	0			

Maximum Pond Size Required for No Discharge: 8.33 in.

Location: Topeka, Kansas   Soil Class: 3   Crop: Grain Sorghum
Years of Records Used: 1949-1973 (25 years)   Feedlot Size: 40 ac
Mode of Operation: Dispose Jan-Dec (90% Field Capacity)   Disposal Rate: 0.61 in./day
Pond Size: 1X = 6.0 in.   Ratio Disposal Area/Feedlot: 2/1

	Precipitation, in.	Runoff, in.	Discharge, in.			Disposal Days		
			1X	2X	3X	1X	2X	3X
Totals:	856.82	161.44	19.16	3.67	0	225	237	247
Avg:	34.28	6.46	–	–	–	9.00	9.48	9.88
Range:	19.07 – 60.89	1.98 – 15.71				3-16	3-18	3-19
No. Years with Discharge:			9	3	0			
Avg. No. Discharge/Yr with Discharge:			8.9	5.3	0			
Range:			1-27	1-13	0			

Maximum Pond Size Required for No Discharge: 15.10 in.

Location: Corvallis, Oregon   Soil Class: 5   Crop: Corn
Years of Records Used: 1919-1968 (50 years)   Feedlot Size: 40 ac
Mode of Operation: Dispose Jan-Dec (90% Field Capacity)   Disposal Rate: 0.60 in./day
Pond Size: 1X = 6.0 in.   Ratio Disposal Area/Feedlot: 2/1

	Precipitation, in.	Runoff, in.	Discharge, in.			Disposal Days		
			1X	2X	3X	1X	2X	3X
Totals:	1943.81	326.53	211.39	138.71	40.87	213	381	541
Ave:	38.88	6.53	–	–	–	4.26	7.62	10.82
Range:	22.54–66.03	1.77–14.80				4-7	4-11	4-14
No. Years with Discharge:			48	45	21			
Avg. No. Discharge/Yr with Discharge:			44.7	27.3	17.5			
Range:			17-95	2-50	2-50			

Maximum Pond Size Required for No Discharge: Not Determined.

annual rainfall; therefore, the disposal area was sized twice the
size of the feedlot area. However, with higher average rainfall,
actual disposal days were decreased considerably as compared

to Garden City. Topeka did have more disposal days allowing a smaller facility size to control feedlot runoff. At Topeka the 2X facility discharged 12 percent of the years and none with the 3X size. At Corvallis the 2X size discharged 90 percent of the years and the 3X, 42 percent of the years.

At this point the quantity of wastewater moving through the system is addressed. Quality of that same wastewater is considered only where it places limitations on the amount of wastewater that may be safely applied or discharged over some period of time. We are purposely attempting to develop this model so that the ultimate user can provide as many of the input parameters as possible. The model can then be site-specific. We feel that with only minor modifications this model ultimately would be a usable tool for many wastewater treatment systems that must interact with weather-related phenomena.

ACKNOWLEDGMENTS

Partial support of this research was provided by the Environmental Protection Agency under Grant No. R-803797-01-0.

REFERENCES

1. Missouri Water Pollution Board and Extension Service, University of Missouri-Columbia. "The Missouri approach to animal waste management," Columbia, Missouri (1971).
2. U.S. Environmental Protection Agency. "Effluent limitation guidelines for existing sources and standards of performance and pretreatment standards for new sources for the feedlots category," *Federal Register* 39:32, pp. 5704–5710, Washington, D.C. (1974).
3. Koelliker, J. K., H. L. Manges and R. I. Lipper. "Modeling the performance of feedlot-runoff-control facilities," *Trans. ASAE* 18:1118–1121 (1975).
4. Larsen, C. L., L. G. James, P. R. Goodrich and J. A. Bosch. "Performance of feedlot runoff control systems in Minnesota," ASAE Paper No. 74–4013, St. Joseph, Michigan (1974).
5. Wensink, R. B. and J. R. Miner. "A model to predict the performance of feedlot runoff control facilities at specific Oregon locations," ASAE Paper No. 75–4027, St. Joseph, Michigan (1975).
6. U.S. Soil Conservation Service. "Hydrology," *National Engineering Handbook*, Section 4, Pt. 1, Watershed Planning, Washington, D.C. (1964).
7. Swanson, N. P., L. N. Mielke, J. C. Lorimor, T. M. McCalla and J. R. Ellis. "Transport of pollutants from sloping cattle feedlots as affected by rainfall intensity, duration, and recurrence" (1971).
8. Clark, R. N., A. D. Schneider and B. A. Stewart. Analysis of runoff from southern Great Plains feedlots. Technical Report No. 12 (1972).
9. Fields, W. H. Hydrologic and water quality characteristics of beef feedlot runoff (1970).
10. Israelsen and Hansen. *Irrigation Principles and Practices* (1962).

11. Russell, R. S. and F. B. Ellis. "Estimation of the distribution of plant roots in soil," *Nature*, London (1968).
12. Ritchie, J. T. "Model for predicting evaporation from a row crop with incomplete cover," USDA Soil and Water Conservation Research Division, Blackland Conservation Research Center, Temple, Texas (1972).
13. Kanemasu, E. T. "Application of information on water-soil-plant relations to use and conservation of water," Evapotranspiration Laboratory, Kansas State University (1975).
14. National Oceanic and Atomospheric Administration, U.S. Department of Commerce. *Climates of the States*, Volumes I and II.
15. Snedecor, G. W. and W. G. Cochran. *Statistical Methods*, Sixth Ed. (1973).
16. Schwab, G. O., R. K. Frevert, T. W. Edminister and K. K. Barnes. *Soil and Water Conservation Engineering* (1966).
17. Zovne, J. J. and A. Nawaz. "Predicting evapotranspiration from agricultural watersheds under dry conditions," Kansas Water Resources Research Institute (1975).
18. Jensen, M. E., J. L. Wright and B. J. Pratt. "Estimating soil moisture depletion from climate, crop and soil data," *Trans. ASAE* (1971).
19. Saxton, K. E., H. P. Johnson and R. H. Shaw "Modeling evapotranspiration and soil moisture," *Trans. ASAE* (1974).
20. Ward, R. C. *Principles of Hydrology*, Second Ed. (1975).
21. Linsley, R. K., Jr., M. A. Kohler and J. L. H. Pavlhus. *Hydrology for Engineers*, Second Ed. (1975).

44

WASTE MANAGEMENT PROBLEMS AND THEIR IMPACT ON THE ENVIRONMENT

I. Fratrič and E. Parraková
Czechoslovak Research and Development Centre for
  Environmental Pollution Control, UNO/WHO Programme,
  Bratislava, Czechoslovakia

INTRODUCTION

The age in which we are living is earmarked by intensive scientific and technological progress which enables cosmic flights, utilization of atomic energy and management of production processes through cybernetics. The scientific, technological and social progress, the continuous improvement of the standard of living and intensive industrialization also have their disadvantages on environmental quality, *e.g.*, an increase in the quantity and volume of solid, liquid and gaseous wastes which can endanger further technical and social development. In this context, the side effects caused by herbicides, fertilizers, as well as the adverse effects of road, rail, water and air transport also can be pointed out. Land settlements and exploitation of raw materials, regardless of the natural potential sources, also contribute to environmental deterioration. What was meant to serve as a tool for man in capturing nature is now endangering man himself.

In line with other parts of the world, the production of solid wastes has also increased in Czechoslovakia in its quality and volume. In Czechoslovakia about 2.2 billion m^3 of wastewater is produced per year. Out of this total, 14 million tons of sludge are produced in wastewater treatment plants. However 4.2 million tons of sludge leak into river waters. The production of manure is about 70 million tons per year, from which leakage in river waters is nearly 2 million tons. Production of inorganic industrial wastes composed of slag and ash is 12 million tons per year.

Czechoslovakia has a great quantity of ash production from power plants and is third in this problem after the U.S. and the German Federal Republic. The quantity of municipal solid wastes represents nearly 2 million tons. These quantities of wastes, if the costs for their transport, tipping or discharging are added, result in an enormous financial burden which must be carried by the national economy. It is understood that the economical pressure is closely connected with environmental protection (1).

The conditions in Czechoslovakia enable only the utilization of soil for the eventual disposal of wastes to protect water quality. The air can be used as a recipient of solid wastes only in the case of incineration designed to ensure admissible air pollution concentration.

The occurrence of an acute situation in Czechoslovakia, in connection with health and water management sanitation and technology caused by municipal and industrial solid wastes, has led to studies for the elimination of these problems.

A technical and economical study was elaborated in 1970 by the Water Research Institute in Bratislava and by the Hydroproject in Prague. On the basis of this study the task entitled "Investigation of Utilization and Disposal of Waste from Settlements and Industry" was included in the state plan.

This task is coordinated by the Czechoslovakia Research and Development Centre for Environmental Pollution Control, Programme UNO/WHO, and the research work is undertaken by the two above-mentioned institutes and the Institutes for Plant Production, Prague, Institute of Hygiene, Bratislava and the Research and Development Institute of Municipal Economy, Prague.

Presently only part of the solid wastes are incinerated in Prague. An incineration plant for Bratislava is under construction and will start operation this year. Construction of incineration plants are proposed for greater town agglomerations. From the whole quantity of produced solid wastes only a small part is burned; about 8 percent is used for composting and the rest is dumped on many tipping locations. However, solid waste disposal or utilization brings many technical and hygienic problems. The studies on waste management within the framework of the UNDP/WHO Project covers the following: controlled tipping, composting, recycling, sludge stabilization and animal waste management.

CONTROLLED TIPPING

As already mentioned, the greater quantity of solid wastes is tipped. While doing so, attention was paid to the problem of controlled tipping and its influence on the surrounding soil and ground water contamination.

In the field of soil hygiene, studies on selective and precise chemical and bacteriological methods for the comprehensive evaluation of soil contamination from quantitative and qualitative aspects were performed. For the qualitative assessment of soil contamination, studies were made on the occurrence of a certain group of microorganisms supplemented by various chemical intermediates characterizing the self-purification processes. The dependence of oxidoreduction processes on the activity of microorganisms has led to methodology for the establishment of the redox potential of the contaminated soil.

For the quantitative assessment of soil contamination, proposals have been to divide the indicators into three groups. The values of each indicator were classified into four categories according to the degree of contamination. The grouping of the selected indicators into three groups was done with respect to their hygienic importance, as follows: the first group represents general indicators of contamination; the second group, indicators of recent contamination; the third group, classification of indicators of advanced organic matter decomposition stages.

The results obtained from experimental studies are fully applied for checking the effects of controlled tipping on soil and percolated water. Attention is paid especially to the dynamic changes of microorganisms during the past two years in the surroundings of the experimental tipping location in Northern Bohemia.

According to the ascertained number of microorganisms—the general indicator of contamination—the surrounding soil of the tipping location was in the reduction zone, and almost half of the investigated soil samples had a degree of medium and heavy contamination.

Indicators of recent contamination were classified in the second group. The occurrence of microbial indicators was dependent on the degree of contamination. In view of the origin of the contamination—solid wastes—it was characterized by the occurrence of thermophilic bacteria, associated with *enterococci*, coliforms and the anaerobic group of *Clostridium*. The numerical distribution of thermophilic bacteria in heavily and moderately contaminated soil samples indicated a sign of coincidence with the distribution of coliforms.

The high content of thermophilic bacteria reflects the organic waste contamination, especially at the bottom of the tipping location, caused by continuous soil contamination with wastes. The occurrence of *enterococci* was variable due to their limited survival in the environment. Therefore, their occurrence was higher during the first year of investigation. On the other hand the occurrence of the *E. coli* indicated a heavy degree of contaminated soil during the second year of investigation. Uniform distribution of *Clostridium* in the soil can be explained by the property of these bacteria to indicate the equalization of values reached at various parts of tipping location.

The last group of indicators of advanced stages of organic matter decomposition correspond to the degree of contamination at the surrounding area of the tipping location. It was manifested by partial inhibition of the development of nitrifying bacteria and spore-forming microorganisms which are causally dependent on each other. This is caused by overloading the soil milieu with the intermediates of the decomposition of organic matter.

The influence of the sanitary landfill on the percolating water was tested. Two ditches were dug, one above and one below the tip from which samples were collected. Bacteriological analyses showed that the contamination of soil was higher than that of water. This was caused by low rainfall and the soil barrier between the tipping location and the ditches. The results are presented in Tables 1 and 2.

Table 1. Average values of microorganisms in the soil and percolating water.

Microorganisms	Soil (in 1 g)	Number of Microorganisms (in ml) Percolating Water	
		Upper Ditch	Lower Ditch
E.coli	4,500	103	245
Enterococci	260	–	–
Clostridium perfr.	650	30	48
Mezophilic bacteria	886,000	112	346
Thermophilic bacteria	19,530	110	166

The necessity of waste disposal is an extensive problem as after incineration or composting, the waste is reduced to only about 60–80 percent of its original quantity. Sanitary landfill is often economically and rationally the best system for solid waste disposal. It must be noted that the wastes are usable not only for landscape filling, but they can be used also as construction material for the protection of embankments or for artificial hills built for recreational purposes. These solutions are being utilized in the Capital of the Slovak Socialistic Republik (Bratislava). In this case the old dumping place had to be removed to another suit-

Table 2.   Colititer freques in soil and percolating water.

Colititer	Soil	Percolating Water	
		Upper Ditch	Lower Ditch
10^{-6}	6x	—	—
10^{-5}	7x	—	—
10^{-4}	14x	—	5x
10^{-3}	20x	4x	6x
10^{-2}	16x	5x	5x
10^{-1}	31x	6x	3x
1.0	19x	4x	2x
2.0	29x	3x	1x

able place. In the preparatory stage all environmental requirements were taken into account.

To protect the soil, the dislocated wastes are tipped at the hill.
Using this process the waste becomes a structural element for
positive landscape creation. The place for such controlled tipping
must be chosen carefully and in line with the country design. This
leads to the conclusion that the integration of waste management
studies within the frame of system analysis is of great importance.
This can, however, be possible only by means of extensive inter-
disciplinary planning. This type of controlled tipping is already
in operation, and presently detailed analyses are being carried out
in connection with their effects on the environment.

COMPOSTING

Part of solid wastes and sludge is utilized for the production of
organic fertilizers for agricultural purposes. For example, in
1969 the production of compost in Czechoslovakia was some 912,-
000 tons. The main requirements for compost production is the destruction of pathogens which must be specified by the responsible
hygienist in each case of waste disposal. During our experiments
it was found that aerobic composting is accompanied by microbiological exotherm processes, which result in disinfection of the
initial material.

Pathogenic microorganisms were used as a bacteriological-
hygenic testing system. The survival of pathogenic germs in waste
compost depends on many factors such as humidity, temperature,
fresh solid waste content and inhibitor formation. Decontamination is based on the formation of inhibitors acting as antibiotics
during the process of decomposition. These active materials are
produced by various microbiological species contained in the ini-

tial material under certain conditions. They can be isolated and tested *in vitro* due to the survival of pathogenic germs and their sensibility. Thus different systems of waste composting can be checked. Experiments confirmed that at 50°C, composts produced from fresh solid wastes and sludge destroy *Salmonella* within seven days.

RECYCLING

The third approach for waste management is the industrial circulation of raw material. While population growth is 2 to 2.5 percent per year, the growth of industrial production is more than 6 percent.

Raw material sources are not inexhaustible, and therefore it is necessary to slow down their output. In the future we shall be forced to reclaim the raw material content of wastes by its integration into the industrial circulation of substances. The recycling and reutilization of waste of all kinds is becoming more and more imperative.

It is important to economize raw materials not only from the point of view of their recirculation but also from the point of view of not contaminating the environment with wastes.

It is not possible to destroy wastes; they can only be changed and disposed of in the soil, water or atmosphere. The tipping sites have only limited capacity.

Long before the word "recycling" came into common usage, recirculation, reutilization or reuse of wastes had been practiced. Recycling has a positive perspective in the future, and its importance is continuously growing due to the limited amount of raw materials.

Already almost each second ton of steel or paper is produced from iron scrap or paper waste. From the viewpoint of environmental pollution control it is necessary to regain almost all the wastes. For industrial wastes new approaches are being applied, *e.g.*, the so-called waste exchanges which are ensuring the reuse of wastes.

It is understood that recycling is mainly in connection with industrial wastes, but it would be very useful to consider the feasibility of reusing also the different components of municipal solid wastes, *e.g.*, domestic refuse. If we take into account the fact that in Czechoslovakia the yearly quantity of solid wastes is over 2 million tons including 315,000 tons of paper, 168,000 tons of glass and 168,000 tons of metal waste, it would be advantageous to develop a suitable technology of mixed domestic refuse classification, with the purpose of further reuse of regained raw materials.

It is difficult to understand why the greater part of wastes is tipped without reutilization. A suitable technology system should make it possible to take advantage of nature's existing facilities. Far-reaching microbial processes should be developed. This should be done logically and in accordance with natural conditions in order to suggest new technical recirculation processes.

If we consider the contemporary recycling techniques, attention should be given to the following materials: iron and steel, other metals, nonferrous metals, paper, textiles, chemical products, oils, plastics, glass and domestic refuse.

Steel, with its extraordinary large scale of technological utilization, can hardly be replaced by other materials. Therefore, recycling in this case is well justified. The actual technologies enable the recirculation of almost all iron-containing materials and of most parts of metallurgical dross. Other refuse can be well utilized in road or dam construction. For recycling of iron and metals the supply of materials has to be ensured in accordance with the different classification degrees in order to make their immediate usage possible.

For the paper industry, recycling is quite well known. Paper was discovered more than 2,000 years ago. At that time paper production already had been involved with waste reuse, namely in the utilization of tree barks, fibers and rags. It is clear that even at the early stages of the paper industry, recycling had occurred.

SLUDGE STABILIZATION

Solid waste disposal also covers sludge disposal from waste treatment plants. Great efforts are being made to increase the number and capacity of wastewater treatment plants to ensure the purity of water sources. From the hygienic point of view, sludge disposal from wastewater treatment plants is one of the most important stages of waste management. In this connection great attention was paid to the hygienic efficiency of the aerobic sludge stabilization system by means of pilot-plant wastewater treatment equipment. In addition, several laboratory experiments with *Salmonella* survival, including a one-year investigation of microorganism changes from influent samples, activation tank, and effluent from the stabilization tank and sludge water were performed.

For a more exact evaluation of the influence of aerobic sludge stabilization or *Salmonella* survival, an experiment with artificial contamination was undertaken. The results have shown that the longest survival period of the tested *Salmonella* was 20 days in stabilized sludge and 9 days in the effluent.

Our results differ from those of Takacs and Knack (1968). These scientists exclude the existence of *Salmonella* in the system of aerobic sludge stabilization. Our results correspond more with those of Obrist, Hess and Lott (2, 3) *Salmonella* survival in aerobic sludge stabilization was shorter than in anaerobic digestion. Müller (4) showed that the survival of *Salmonella* in an unheated sludge digester with alkaline anaerobic digestion can be up to 11 months.

The tested *Salmonella* survival in aerobic stabilized sludge was as follows: *S.enteritidis*, 20 days; *S.heidelberg*, 9 days; and *S.cholerae suis*, 6 days. Besides the *Salmonella* survival, changes in the total plate count of *E.coli* and *enterococci* in the influent and effluent during wastewater treatment were followed. The results have shown a reduction of the total plate count from 50–95 percent and of *E.coli* from 50–90 percent. The occurrence of *Enterococcus* was variable.

The last experiment was prepared with the above *Salmonella*. The tested organisms were left in the influent 1 hr, in the activation 3 hr, in the sedimentation tank 2 hr and in the stabilization tank 21 days. The decay of *Salmonella* was obseved. During the 21-day period, the number and titer of tested *Salmonella* showed an intensive reduction. This becomes evident if we follow the inoculum:

	Beginning	End
Salm. enteritidis	80 mil/ml	40/ml
Salm. heidelberg	17 mil/ml	70/ml
Salm. cholerae suis	1.4 mil/ml	140/ml

When comparing the results obtained from the last two experiments, the *Salmonella* disappeared earlier in the previous experiments, completely during 20 days, whereas in the latter experiments they existed after 21 days, although in lower quantities. The aerobic sludge stabilization gave better results than anaerobic digestion. Aerobic stabilized sludge must be decontaminated before using it as fertilizer for vegetables and fruits.

ANIMAL WASTE MANAGEMENT

Incorporation of wastes into the natural cycle is one of the measures as to how to lower the environmental risk. In this connection it is important to mention one branch of waste production which becomes more and more important. This is the production of animals in cattle, pig and poultry feedlots. Large-scale animal breeding results in the concentration of a great number of animals within a limited space with frequent replacement of generations, the operation being maintained by a minimum number of workers with the application of mechanization, automation and with a maximum use of high-value feeds.

The nutrient cycle in the new feedlots produces waste in the form of manure and slurry which are accumulated at the location, contrary to the older types of agricultural enterprise. Areas surrounding feedlots normally do not have sufficient space to accumulate the quantity of feedlot wastes. Consideration was given to treating these wastes as part of sewerage in wastewater treatment plants. This seems to be a questionable process.

Feedlot development is a modern and widening situation. In connection with this question a seminar was held in Bratislava in September 1975, SEMAW'75. The seminar was organized by the Czechoslovak Research and Development Centre for Environmental Pollution Control (joint UNDP/WHO project) and was sponsored by the Czechoslovak Federal Ministry of Agriculture and Food and the Federal Ministry for Technical Development and Investment in collaboration with the WHO Regional Office for Europe. Four main topics were discussed: large-scale animal industry and environmental quality; technologies for processing animal wastes; utilization and disposal of animal wastes; and economic aspects of animal waste management.

In relation to the operation of animal feedlots, great attention has been given to the degree of soil contamination with emphasis on the enteropathogenic *E.coli,* changes in the sensibility of the strains to antibiotics as a consequence of using fortified nutrition, and the effects on the environment and personnel of large-scale pig farms.

The results appear to indicate that the large-scale technology of breeding contaminates the living environment, mainly the soil and the air by increasing the frequency of resistant strains of *E.coli*. Further, the increased number of *Staphylococcus aureus* on the skin and *E.coli haemolytica* in the feces of the personnel was observed. These significantly increased indicators of environmental contamination had not affected the general health of the personnel.

SUMMARY

The economical solution of waste problems in all systems requires reutilization or disposal and the reduction of waste volume —this fact can not be disputed. Waste management is one of the elements of environmental control and creation. While environmental control emphasizes one side only—namely, the passive defense of existing values of human life—the creation of environment presupposes an active approach to solve the problems evoked by generally accepted production priorities.

Therefore, comprehensive environmental control is the purposeful synthesis of different biosphere resources utilized by man to ensure his existence, whereby accepted hygienic principles also have to be fully respected.

REFERENCES

1. Holy, M. "Kaly a odpady v zivotnim prostredi," *Kalova problematika* 73, Usti n. Labem (1973).
2. Obrist, W. "Die Klarschlammbehandlung durch Pasteurisation und Bestrahlung," *ISWA* 6:96–104 (1971).
3. Hess, E. and G. Lott. "Klarschlamm aus der Sicht des Veterinar hygienikers," *Gas–Wasser–Abwasser* 51(2):42–44 (1971).
4. Müller, W. "Die Abwasserbeseitigung aus der Sicht der Seuchenhygiene," *Stadtehygiene* 7–12 (1971).

INDEX

807